NUTRITION AND EPIGENETICS

OXIDATIVE STRESS AND DISEASE

Series Editors

LESTER PACKER, PhD
ENRIQUE CADENAS, MD, PhD

UNIVERSITY OF SOUTHERN CALIFORNIA SCHOOL OF PHARMACY
LOS ANGELES, CALIFORNIA

1. Oxidative Stress in Cancer, AIDS, and Neurodegenerative Diseases, *edited by Luc Montagnier, René Olivier, and Catherine Pasquier*
2. Understanding the Process of Aging: The Roles of Mitochondria, Free Radicals, and Antioxidants, *edited by Enrique Cadenas and Lester Packer*
3. Redox Regulation of Cell Signaling and Its Clinical Application, *edited by Lester Packer and Junji Yodoi*
4. Antioxidants in Diabetes Management, *edited by Lester Packer, Peter Rösen, Hans J. Tritschler, George L. King, and Angelo Azzi*
5. Free Radicals in Brain Pathophysiology, *edited by Giuseppe Poli, Enrique Cadenas, and Lester Packer*
6. Nutraceuticals in Health and Disease Prevention, *edited by Klaus Krämer, Peter-Paul Hoppe, and Lester Packer*
7. Environmental Stressors in Health and Disease, *edited by Jürgen Fuchs and Lester Packer*
8. Handbook of Antioxidants: Second Edition, Revised and Expanded, *edited by Enrique Cadenas and Lester Packer*
9. Flavonoids in Health and Disease: Second Edition, Revised and Expanded, *edited by Catherine A. Rice-Evans and Lester Packer*
10. Redox–Genome Interactions in Health and Disease, *edited by Jürgen Fuchs, Maurizio Podda, and Lester Packer*
11. Thiamine: Catalytic Mechanisms in Normal and Disease States, *edited by Frank Jordan and Mulchand S. Patel*
12. Phytochemicals in Health and Disease, *edited by Yongping Bao and Roger Fenwick*
13. Carotenoids in Health and Disease, *edited by Norman I. Krinsky, Susan T. Mayne, and Helmut Sies*
14. Herbal and Traditional Medicine: Molecular Aspects of Health, *edited by Lester Packer, Choon Nam Ong, and Barry Halliwell*
15. Nutrients and Cell Signaling, *edited by Janos Zempleni and Krishnamurti Dakshinamurti*
16. Mitochondria in Health and Disease, *edited by Carolyn D. Berdanier*
17. Nutrigenomics, *edited by Gerald Rimbach, Jürgen Fuchs, and Lester Packer*
18. Oxidative Stress, Inflammation, and Health, *edited by Young-Joon Surh and Lester Packer*

19. Nitric Oxide, Cell Signaling, and Gene Expression, *edited by Santiago Lamas and Enrique Cadenas*
20. Resveratrol in Health and Disease, *edited by Bharat B. Aggarwal and Shishir Shishodia*
21. Oxidative Stress and Age-Related Neurodegeneration, *edited by Yuan Luo and Lester Packer*
22. Molecular Interventions in Lifestyle-Related Diseases, *edited by Midori Hiramatsu, Toshikazu Yoshikawa, and Lester Packer*
23. Oxidative Stress and Inflammatory Mechanisms in Obesity, Diabetes, and the Metabolic Syndrome, *edited by Lester Packer and Helmut Sies*
24. Lipoic Acid: Energy Production, Antioxidant Activity and Health Effects, *edited by Mulchand S. Patel and Lester Packer*
25. Dietary Modulation of Cell Signaling Pathways, *edited by Young-Joon Surh, Zigang Dong, Enrique Cadenas, and Lester Packer*
26. Micronutrients and Brain Health, *edited by Lester Packer, Helmut Sies, Manfred Eggersdorfer, and Enrique Cadenas*
27. Adipose Tissue and Inflammation, *edited by Atif B. Awad and Peter G. Bradford*
28. Herbal Medicine: Biomolecular and Clinical Aspects, Second Edition, *edited by Iris F. F. Benzie and Sissi Wachtel-Galor*
29. Inflammation, Lifestyle and Chronic Diseases: The Silent Link, *edited by Bharat B. Aggarwal, Sunil Krishnan, and Sushovan Guha*
30. Flavonoids and Related Compounds: Bioavailability and Function, *edited by Jeremy P. E. Spencer and Alan Crozier*
31. Mitochondrial Signaling in Health and Disease, *edited by Sten Orrenius, Lester Packer, and Enrique Cadenas*
32. Vitamin D: Oxidative Stress, Immunity, and Aging, *edited by Adrian F. Gombart*
33. Carotenoids and Vitamin A in Translational Medicine, *edited by Olaf Sommerburg, Werner Siems, and Klaus Kraemer*
34. Hormesis in Health and Disease, *edited by Suresh I. S. Rattan and Éric Le Bourg*
35. Liver Metabolism and Fatty Liver Disease, *edited by Oren Tirosh*

NUTRITION AND EPIGENETICS

EDITED BY

EMILY HO
FREDERICK DOMANN

CRC Press
Taylor & Francis Group
Boca Raton London New York

CRC Press is an imprint of the
Taylor & Francis Group, an **informa** business

CRC Press
Taylor & Francis Group
6000 Broken Sound Parkway NW, Suite 300
Boca Raton, FL 33487-2742

Version Date: 20140825

International Standard Book Number-13: 978-1-4822-0382-0 (eBook - PDF)

Visit the Taylor & Francis Web site at
http://www.taylorandfrancis.com

and the CRC Press Web site at
http://www.crcpress.com

Contents

SECTION III Other Essential Nutrients

SECTION IV Phytochemicals

Series Preface

OXYGEN BIOLOGY AND MEDICINE

Through evolution, oxygen—itself a free radical—was chosen as the terminal electron acceptor for respiration. The two unpaired electrons of oxygen spin in the same direction; thus, oxygen is a biradical. Other oxygen-derived free radicals, such as superoxide anion or hydroxyl radicals, formed during metabolism or by ionizing radiation, are stronger *oxidants*, i.e., endowed with a higher chemical reactivity. Oxygen-derived free radicals are generated during metabolism and energy production in the body and are involved in regulation of signal transduction and gene expression, activation of receptors and nuclear transcription factors, oxidative damage to cell components, antimicrobial and cytotoxic action of immune system cells, as well as aging and age-related degenerative diseases. Conversely, the cell conserves antioxidant mechanisms to counteract the effect of oxidants; these *antioxidants* may remove oxidants either in a highly specific manner as, for example, by superoxide dismutases or in a less specific manner (for example, small molecules such as vitamin E, vitamin C, and glutathione). *Oxidative stress* as classically defined is an *imbalance between oxidants and antioxidants.* Overwhelming evidence indicates that oxidative stress can lead to cell and tissue injury. However, the same free radicals that are generated during oxidative stress are produced during normal metabolism and, as a corollary, are involved in both human health and disease.

Oxygen is a dangerous friend and virtually all diseases thus far examined involve oxygen free radicals and redox reactions. In most cases, free radicals are secondary to the disease process, but in some instances, causality by free radicals themselves is known. It has been established that low levels of oxygen free radicals and reactive oxygen species are essential for some cell signaling pathways Thus, a delicate balance among oxidants, antioxidants, and redox status is essential for healthy aging.

Reactive oxygen species, i.e., nonradical oxidants such as H_2O_2 and $^1O_2$, and other species reflect the rich variety of reactive species in free radical biology and medicine, i.e., encompassing nitrogen-, sulfur-, oxygen-, and carbon-centered radicals. With the discovery of nitric oxide, nitrogen-centered radicals gathered momentum and have matured into an area of enormous importance in biology and medicine. Nitric oxide or nitrogen monoxide (NO), a free radical generated in a variety of cell types by nitric oxide synthases (NOS), is involved in a wide array of physiological and pathophysiological phenomena, such as vasodilation, neuronal signaling, and inflammation. Of great importance is the reaction of nitric oxide with superoxide anion: this is among the most rapid nonenzymatic reactions in biology (well over the diffusion-controlled limits) and yields the potent nonradical oxidant, peroxynitrite, involved in tissue injury through oxidation and nitration reactions.

Gaseous species such as hydrogen sulfide and molecular hydrogen produced at low levels in tissues by enzymes and by intestinal and tissue microbiota (microbiome) are also recognized as important regulators of oxidative stress and in maintaining tissue homeostasis.

Reactive oxygen and nitrogen species and the hydrogen gases are involved in the redox regulation of cell function. Oxidative stress is increasingly viewed as a major upstream component in the signaling cascade involved in inflammatory responses, stimulation of cell adhesion molecules, and chemoattractant production and as an early component in age-related neurodegenerative disorders, such as Alzheimer's, Parkinson's, and Huntington's disease and amyotrophic lateral sclerosis. Hydrogen peroxide is probably the most important redox signaling molecule that, among others, can activate NF-κB, Nrf2, and other universal transcription factors. Increasing steady-state levels of hydrogen peroxide have been linked to the cell's redox status with clear involvement in adaptation, proliferation, differentiation, apoptosis, and necrosis. The identification of oxidants in regulation of redox cell signaling and gene expression was a significant breakthrough in the field of oxidative stress: the classical definition of oxidative stress as an *imbalance between the production of oxidants and the occurrence of cell antioxidant defenses* proposed by Sies in 1985 now seems to provide a limited concept of oxidative stress, but it emphasizes the significance of the cell's redox status, because individual signaling and control events occur through discreet redox pathways rather than through global balances.

Thus, a new definition of oxidative stress was advanced by Dean P. Jones (*Antioxidants & Redox Signaling*, 2006) as a disruption of redox signaling and control that recognizes the occurrence of compartmentalized cellular redox circuits. Measurements of GSH/GSSG or cysteine/cystine, $\text{thioredoxin}_{\text{reduced}}/\text{thioredoxin}_{\text{oxidized}}$ provide a quantitative definition of oxidative stress. The redox status is thus dependent on the degree to which tissue-specific cell components are in the oxidized state. In general, the reducing environment inside cells helps to prevent oxidative damage. In this reducing environment, disulfide bonds (S—S) do not spontaneously form because sulfhydryl groups are maintained in the reduced state (SH), thus preventing protein misfolding or aggregation. The reducing environment is maintained by metabolism and by the enzymes involved in maintenance of thiol/disulfide balance and substances, such as glutathione, thioredoxin, vitamins E and C, and enzymes such as superoxide dismutases, catalase, and the selenium-dependent glutathione reductase and glutathione and thioredoxin-dependent hydroperoxidases (perire-doxins), which serve to remove reactive oxygen species (hydroperoxides). Also of importance is the recognition of the existence of many tissue-specific and cell compartment–specific isoforms of antioxidant enzymes and proteins.

Compelling support for the involvement of free radicals in disease development originates from epidemiological studies that show that an enhanced antioxidant status is associated with reduced risk of several diseases. Of high significance is the role that micronutrients play in modulation of redox cell signaling: this establishes a strong link between diet and health centered on the ability of micronutrients to regulate redox cell signaling and modify gene expression.

These concepts are anticipated notions to serve as a platform for the development of tissue-specific therapeutics tailored to discreet, compartmentalized redox circuits. This, in essence, dictates principles of drug development guided knowledge of mechanisms of oxidative stress. Hence, successful interventions will take advantage of new knowledge of compartmentalized redox control and free radical scavenging.

In this series of books, the importance of oxidative stress and disease associated with organ systems of the body is highlighted by exploring the scientific evidence and clinical applications of this knowledge. These books are intended for researchers in the basic biomedical sciences, health professionals, and clinicians. The potential of such information for healthy aging and disease prevention warrants further knowledge about how oxidants and antioxidants modulate cell and tissue function.

"The Role of Nutrition and Metabolism on Epigenetic Regulation" by Emily Ho and Frederick Domann

Jean-Baptiste Larmar proposed a popular idea in the early nineteenth century, which advanced the idea that characteristics acquired during a lifetime of an organism could be passed on to future generations. The concept was largely dismissed later in the nineteenth century with the origin of the Mendelian inheritance conceived by the work of Gregor Johann Mendel. However, since the twentieth century, environmental effects on genes, epigenetics research, and the role of the epigenome have dramatically changed our modern concepts of inheritance where the human diet plays an important role.

To appreciate this rapidly developing and important area of research requires a book that brings together and highlights the major nutrients and their epigenetic relevance to health and disease. The editors of this volume and the internationally renowned authors who have contributed are to be congratulated for bringing together "the most important dietary factors and specific nutrients that can modulate epigenetic alterations and their susceptibility to disease" as stated by the editors.

Lester Packer
Enrique Cadenas

Preface

It is increasingly realized that gene–environment interactions, particularly gene–nutrient interactions, contribute significantly to the health, aging, and disease. We now understand that not only macronutrients but also micronutrients and phytochemicals impose their effects on gene expression in large part through epigenetic control of gene expression. The term "epigenetics," first coined by Conrad Waddington, is broadly defined as acquisition and maintenance of heritable states of gene expression that occur above (epi) the level of genetics through alterations in chromatin structure and accessibility. These heritable chromatin architectural states are largely governed by posttranslational modifications to histone proteins that make up the scaffold for DNA winding and accessibility and also by enzymatic changes to the methylation state of cytosines in CpG dinucleotides within the DNA itself. Since these modifications are brought about by enzymes whose cofactors are typically intermediary metabolites and often essential nutrients, it is clear to see how nutritional variables can affect the integrity and function of the epigenome to elicit abnormal gene expression portending to disease states.

As the field of epigenetics grows, a paradigm shift has emerged in the way we view the impact of dietary factors on regulation of gene expression and disease susceptibility. The role of epigenetic alterations during development and chronic disease development has gained increasing attention and has resulted in an increase in our understanding of mechanisms leading to disease susceptibility and healthy aging. Importantly, essential nutrients and phytochemicals can impact epigenetic events as a novel mechanism of action. The classes of nutrients that have classically been thought to modulate the epigenome focused on nutrients that modulate one-carbon metabolism, either by affecting methyl donor pools or by affecting enzymes involved in methylation reactions. New research has discovered numerous other nutrients and nonnutrients that also affect the epigenetic machinery in other distinct ways. Maintenance of epigenomic integrity is a critical factor regulating gene expression related to cell proliferation, differentiation, apoptosis (programmed cell death), and necrosis. The nutritional epigenetics is a center of interest for its role in integrating signals from the nutritional environment to gene expression patterns that determine an individuals' susceptibility to unhealthy aging and disease. This work has made it increasingly clear that in addition to genetic changes, epigenetic changes are just as critical to healthy human development and physiology. More importantly, dietary factors and specific nutrients can modulate epigenetic alterations and alter susceptibility to disease. The goal of this book is to highlight the interactions among nutrients, epigenetics, and health.

This book covers new information on the action of diet and nutritional determinants in regulating the epigenetic control of gene expression in health and disease. Each chapter gives a unique perspective on a different nutritional or dietary component or group of components and reveals novel mechanisms by which dietary factors modulate the epigenome and affect development processes, chronic disease, or the aging process. This book will be a resource to basic scientists and clinical researchers interested in nutrition, aging, and metabolic diseases.

Emily Ho
Frederick Domann

Editors

Emily Ho is the endowed director of the Moore Family Center for Whole Grain Foods, Nutrition and Preventive Health. She is a full professor in the College of Public Health and Human Sciences, with an emphasis in nutrition, and is a principal investigator at the Linus Pauling Institute at Oregon State University. She joined the Oregon State University faculty in 2003 after receiving her PhD in human nutrition from the Ohio State University and completing her postdoctoral research at Children's Hospital Oakland Research Institute and UC Berkeley. During the course of her career, Dr. Ho has published 97 peer-reviewed articles and abstracts and four book chapters. She has been invited to give more than 40 presentations and has mentored more than 40 undergraduate and graduate students. She currently serves on the editorial board for *Journal of Nutritional Biochemistry and Frontiers in Epigenomics*. Her research interests are in the area of antioxidants and gene expression and dietary chemoprevention strategies. More specifically, she focuses on the effects of zinc status on DNA damage, DNA repair, and stress–response signal pathways. Another major focus in her laboratory is investigating the genetic and epigenetic mechanisms by which foods such as soy, tea, and cruciferous vegetables may protect against prostate cancer.

Frederick Domann is a professor of radiation oncology in the Free Radical and Radiation Biology Program at the University of Iowa in Iowa City, Iowa. He is a distinguished alumnus from the University of Wisconsin where he earned his PhD in 1991 in human cancer biology under Kelly H. Clifton, studying radiation-induced thyroid and mammary carcinogenesis. He subsequently pursued postdoctoral research at the Arizona Cancer Center with G. Tim Bowden, where he studied the redox biology of gene expression. He joined the faculty at the University of Iowa in 1993 and has since become an internationally recognized expert in free radical biology, cancer metabolism, and epigenetics. His work has led to more than 140 peer-reviewed publications. Dr. Domann codirects the Free Radical Cancer Biology Program in the Holden Comprehensive Cancer Center at the University of Iowa. He serves on the Councils of the Society for Experimental Biology and Medicine and the Society for Free Radical Biology and Medicine. He is the recipient of the mentoring awards from the Radiation Research Society and the Graduate College of the University of Iowa. His research interests include transcriptional regulation, redox metabolism, and epigenetics of human diseases including cancer.

Contributors

Olivia S. Anderson
Department of Environmental Health Sciences
University of Michigan
Ann Arbor, Michigan

Moray J. Campbell
Department of Pharmacology and Therapeutics
Roswell Park Cancer Institute
Buffalo, New York

Lindsay G. Carter
Graduate Center for Nutritional Sciences
University of Kentucky
Lexington, Kentucky

Laura Cartularo
Department of Environmental Medicine
New York University School of Medicine
New York, New York

Hong Chen
Department of Food Science and Human Nutrition
University of Illinois Urbana-Champaign
Urbana, Illinois

Wen-Hsing Cheng
Department of Food Science, Nutrition and Health Promotion
Mississippi State University
Mississippi State, Mississippi

Sang-Woon Choi
Chaum Life Center
CHA University, School of Medicine
Seoul, South Korea

Eric D. Ciappio
DSM Nutritional Products
Parsippany, New Jersey

Kate J. Claycombe
Grand Forks Human Nutrition Research Center
USDA Agricultural Research Service
Grand Forks, North Dakota

Gerald F. Combs Jr.
Grand Forks Human Nutrition Research Center
USDA Agricultural Research Service
Grand Forks, North Dakota

Max Costa
Department of Environmental Medicine
and
Department of Biochemistry and Molecular Pharmacology
New York University
New York, New York

Jimmy W. Crott
Vitamins and Carcinogenesis Laboratory
USDA Human Nutrition Research Center on Aging
Tufts University
Boston, Massachusetts

Dana C. Dolinoy
Department of Environmental Health Sciences
University of Michigan
Ann Arbor, Michigan

Simonetta Friso
Department of Medicine
University of Verona
Verona, Italy

Chad Hancock
Center for Cancer Research
National Cancer Institute
Frederick, Maryland

Meredith Harman
Center for Cancer Research
National Cancer Institute
Frederick, Maryland

Sung-Eun Kim
Department of Nutritional Sciences
University of Toronto
and
Keenan Research Center of the Li Ka Shing Knowledge Institute
St. Michael's Hospital
Toronto, Ontario, Canada

Young-In Kim
Departments of Medicine and Nutritional Sciences
University of Toronto
and
Keenan Research Center for Biomedical Sciences
Division of Gastroenterology
St. Michael's Hospital
Toronto, Ontario, Canada

Ji-Young Lee
Department of Nutritional Sciences
University of Connecticut
Storrs, Connecticut

Wei Liu
Center for Cancer Research
National Cancer Institute
Frederick, Maryland

Elizabeth H. Marchlewicz
Department of Environmental Health Sciences
University of Michigan
Ann Arbor, Michigan

Shannon Masih
Department of Nutritional Sciences
University of Toronto
and
Keenan Research Center of the Li Ka Shing Knowledge Institute
St. Michael's Hospital
Toronto, Ontario, Canada

Meltem Muftuoglu
Department of Molecular Biology and Genetics
Acibadem University
Atasehir, Istanbul, Turkey

Yuan-Xiang Pan
Department of Food Science and Human Nutrition
University of Illinois Urbana-Champaign
Urbana, Illinois

Kevin J. Pearson
Department of Pharmacology and Nutritional Sciences
University of Kentucky
Lexington, Kentucky

Tho X. Pham
Department of Nutritional Sciences
University of Connecticut
Storrs, Connecticut

James M. Phang
Center for Cancer Research
National Cancer Institute
Frederick, Maryland

Donato F. Romagnolo
Department of Nutritional Sciences and Department of Cancer Biology
University of Arizona
Tucson, Arizona

Julia A. Sabet
Vitamins and Carcinogenesis Laboratory
USDA Human Nutrition Research Center on Aging
and
Freidman School of Nutrition Science and Policy
Tufts University
Boston, Massachusetts

Ornella I. Selmin
Department of Nutritional Sciences
University of Arizona
Tucson, Arizona

Prashant K. Singh
Department of Pharmacology and Therapeutics
Roswell Park Cancer Institute
Buffalo, New York

Rita S. Strakovsky
Department of Food Science and Human Nutrition
University of Illinois Urbana-Champaign
Urbana, Illinois

Kellie L. K. Tamashiro
Department of Psychiatry and Behavioral Sciences
Johns Hopkins University School of Medicine
Baltimore, Maryland

Nilay Y. Thakar
Stem Cell Engineering Group
Australian Institute for Bioengineering and Nanotechnology
University of Queensland
St. Lucia, Queensland, Australia

Silvia Udali
Department of Medicine
University of Verona
Verona, Italy

Huan Wang
Department of Food Science and Human Nutrition
University of Illinois Urbana-Champaign
Urbana, Illinois

Ernst J. Wolvetang
Australian Institute for Bioengineering and Nanotechnology
University of Queensland
St. Lucia, Queensland, Australia

Ryan T. Y. Wu
Department of Nutrition and Food Science
University of Maryland
College Park, Maryland

Huawei Zeng
Grand Forks Human Nutrition Research Center
USDA Agricultural Research Service
Grand Forks, North Dakota

Dan Zhou
Department of Food Science and Human Nutrition
University of Illinois Urbana-Champaign
Urbana, Illinois

Section I

Fetal Programming

1 Early-Life Exposures and the Epigenome

Interactions between Nutrients and the Environment

Elizabeth H. Marchlewicz, Olivia S. Anderson, and Dana C. Dolinoy

CONTENTS

1.1 INTRODUCTION

The developmental origins of health and disease (DOHaD) hypothesis postulates that early-life environmental and/or nutritional exposures can influence developmental plasticity resulting in altered susceptibility to chronic disease development in adulthood (Barker 1997, 2004; Bateson et al. 2004). The mechanistic link between what individuals are exposed to in early life and disease formation throughout the life course appears to involve alterations to epigenetic modifications, such as chromatin folding and attachment to the nuclear matrix, packaging of DNA around nucleosomes, covalent modifications of histone tails, DNA methylation, and noncoding RNA. Epigenetics is the study of mitotically heritable yet potentially reversible, molecular modifications to DNA, and chromatin without alteration to the underlying DNA sequence (Reik et al. 2001; Li 2002). Aberrant epigenetic gene regulation has been proposed as a mechanism of action for nongenotoxic carcinogenesis (Silva Lima and Van der Laan 2000), imprinting disorders (Murphy and Jirtle 2003), and complex disorders including Alzheimer disease (Bassett et al. 2006), schizophrenia (Petronis 2004), asthma (Vercelli 2004), and autism (Lopez-Rangel and Lewis 2006). Although DNA sequence is fairly permanent, epigenetic modifications are dynamic throughout the life course and can be heavily influenced by external factors such as the environment and nutrition (Reik et al. 2001). Thus, external effects on the epigenome may alter gene expression, potentially giving rise to phenotypic disparity including disease formation.

Environmental and nutritional exposures are factors that humans and animals encounter from the everyday environment (e.g., water, air, food, workplace) and can pose hazardous side effects after single (acute) or continuous (chronic) contact (Evans and Kantrowitz 2002). Environmental exposures fall into four categories including physical, chemical, nutritional, and biological. Exposures within these categories can be natural or man-made. Physical exposures are factors that can harm the physical safety of an animal, such as radiation, noise, abuse, stress, and temperature (Evans and Kantrowitz 2002). Chemical exposures come in the form of gas, solid, or liquids. Examples of chemical exposures include endocrine disrupting compounds (e.g., bisphenol A [BPA], phthalates), pesticides, persistent organic pollutants, and metals (Vandenberg et al. 2012; Wallace 2012). Nutritional factors include dietary intake of phytochemicals, micronutrients, fats, and proteins, as well as total caloric intake (Choi and Friso 2010). Lastly, biological exposures are organisms and/or their by-products that are hazardous to animals, such as viruses, bacteria, and parasites, and the toxins produced and released by these organisms (Rhodehamel 1995).

To understand the role of environmental and nutritional exposures on developmental plasticity through epigenetic mechanisms, careful consideration of exposure timing must be considered (Figure 1.1). The epigenome undergoes extensive

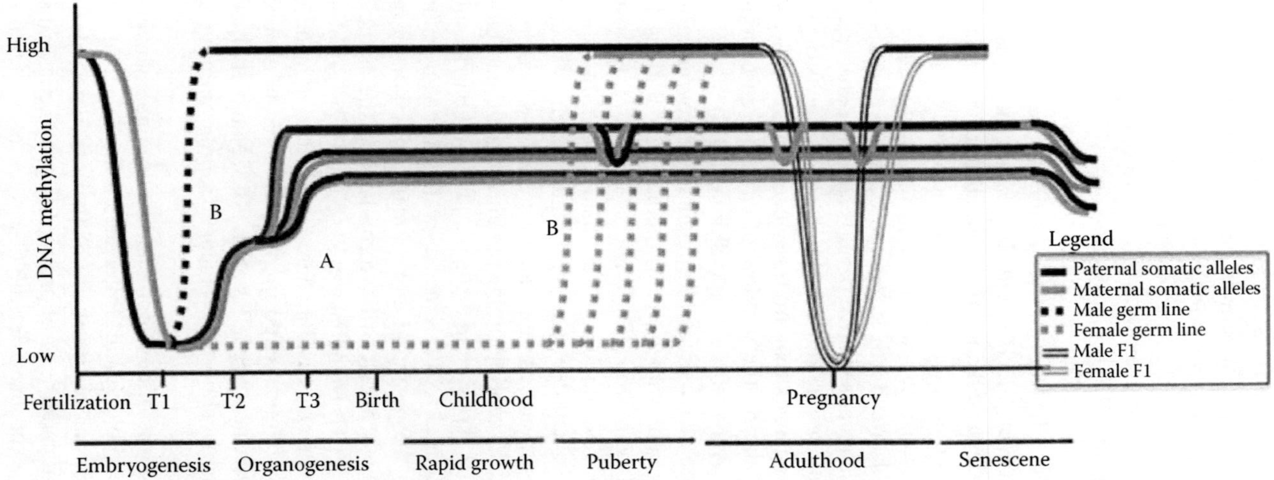

FIGURE 1.1 This DNA methylation timeline begins with fertilization and follows the offspring through the two waves of methylation reprogramming: during early embryogenesis in pre-implantation zygotes and in primordial germ cells (PGCs). Active demethylation occurs in the paternal genome followed by passive demethylation of the maternal genome. (A) Somatic lineage embryonic cells are reprogrammed. (B) The PGC lineage is reprogrammed according to the sex of the offspring, in males prior to birth, and in females, after puberty as each oocyte matures.

reprogramming during two key time-points in early gestation with the purpose of establishing cell- and tissue-specific gene expression. Following fertilization, two waves of methylation reprogramming take place in the developing embryo. The first wave of demethylation begins shortly after fertilization with active demethylation of the paternal genome, followed by slower passive demethylation of the maternal genome and is global with the exception of imprinted genes, which retain their parent-of-origin methylation marks. Some other loci also escape demethylation on both parental genomes, such as transposons, which are only partially demethylated at this time point. Tissue-specific DNA methylation occurs later during fetal organogenesis (Jirtle and Skinner 2007). The second wave of reprogramming is restricted to the primordial germ cells of the fetus and occurs during their migration to the genital ridge, which occurs between GD 11.5 and GD 12.5 in the mouse. In these cells, parental imprinted marks are erased and de novo methylation is established according to the sex of the fetus. In male mouse fetuses de novo methylation occurs around GD 16–18.5, whereas in female mouse offspring, the germ cells are halted after meiosis 1 and remethylated after birth in mature oocytes. Fluctuation in somatic tissue DNA methylation occurs at sensitive periods such as puberty and pregnancy; hypomethylation occurs with aging. Thus, to elucidate the effects of an environmental exposure on epigenetic changes that are transgenerational versus heritable from cell to cell, for example, defining the window of exposure susceptibility within a study design is crucial (Guerrero-Bosagna and Skinner 2012).

Due to these developmental windows of epigenetic reprogramming, women of childbearing age and pregnant women are vulnerable populations for epigenetic reprogramming associated with DOHaD (Wigle et al. 2007). Exposure levels, in addition to the absorption, distribution, metabolism, and elimination of exposed factors, may vary among women due to several factors such as occupation, transportation, dietary practices, genetics, and health status. Additionally, throughout gestation, the fetus is more susceptible to chemical, physical, and biological exposures due to the ability of the exposure to cross the placental barrier into the uterine environment (Jiménez-Díaz et al. 2010; Krueger et al. 2012; Nguyen and Goodman 2012; King et al. 2013). In the postnatal environment, infants may be exposed to environmental insults through breast milk, formula, and the home environment (Koyashiki 2010; Lin et al. 2011; Mendonca et al. 2014; Rojas-Squella et al. 2013). The fetus and pregnant women also exhibit altered metabolizing machinery and biotransformation capacity for chemical exposures as compared to nonpregnant adults (Vandenberg et al. 2012; Martinez-Arguelles et al. 2013; Nahar et al. 2013), thereby, potentially increasing the risk of abnormal epigenetic reprogramming resulting from environmental exposures.

This chapter explores the role of developmental environmental exposures (physical, chemical, nutritional, and biological) influence on the epigenome. We focus on animal and human work that was designed around the critical periods of reprogramming of somatic methylation patterns or of primordial germ cells (transgenerational studies). Studies that begin exposure after the first week of gestation miss the first wave of epigenetic reprogramming that is critical for germ layer and tissue differentiation; therefore, these later life exposures are not examined in depth in this chapter. Exposures in adult human and animal studies may impact epigenetic regulation of gene expression, but they are not involved in fetal programming and thus do not fit within the DOHaD paradigm. Adult models of environment × epigenetics are therefore outside the scope of this chapter also. As nutritional factors often serve as important mediators of the effects of exposures on biological outcomes, we pay close attention to the evidence of nutritional exposures counteracting or enhancing the effects of developmental exposures to the epigenome.

1.2 PHYSICAL EXPOSURES

Increasingly, we are recognizing that environmental influences on the epigenome are diverse and include physical (behavior, stress, temperature, species density, radiation) effects (Table 1.1). The vast majority of studies examining the effects of physical exposure on the epigenome are conducted within in vitro or animal populations, although a growing number of stress and behavioral studies utilize human populations.

1.2.1 Behavior and Stress

Behavior- and stress-induced effects on the developing and adult epigenome are widespread from insects to mammals. For example, the desert locust (*Schistocerca gregaria*) produces more offspring of the gregarious swarming phenotype when bred in crowded conditions (Maeno and Tanaka 2010). Similarly, rats show persistent DNA methylation changes of the glucocorticoid receptor and many other gene loci in the hippocampus in response to high versus low levels of maternal care in the first week of life (Weaver et al. 2004; McGowan et al. 2011). Adult humans abused in early life display increased DNA methylation at the *NR3C1* glucocorticoid receptor promoter in the hippocampus (McGowan et al. 2009). After being near cats, rats exhibit symptoms of posttraumatic stress syndrome concomitant with increased methylation in the *Bdnf* gene in the hippocampus (Roth et al. 2011), and increased methylation of this same gene is seen in human suicide victims (Keller et al. 2010). In mice, stress from maternal separation results in depressive behavior coupled with both increased and decreased DNA methylation in a number of genes (Murgatroyd et al. 2009; Franklin et al. 2010). Interestingly, these mice can transmit the DNA methylation pattern and phenotype transgenerationally through the male line (Franklin et al. 2010). Finally, early life stress in humans is also linked with gene expression changes for a polymorphic form of serotonin receptor (Caspi et al. 2003).

1.2.2 Radiation

Humans are exposed to low-dose ionizing radiation (LDIR) from a number of environmental and medical sources. In addition to inducing genetic mutations, there is emerging concern that LDIR may also alter epigenetic programming. For example, high doses of ionizing radiation result in epigenetic modifications in adult mice (Tawa et al. 1998). They also induce genomic instability and the bystander effect, a phenomenon in which nonirradiated cells exhibit radiation damage even though

TABLE 1.1
Epigenetic Impact of Physical Exposures

Species	Physical Exposure	Timing of Exposure	Exposure Dose	Epigenetic Impact	Nutrient Interaction	Reference
			Animal Models			
Desert locust, *Schistocerca gregaria*	Crowded conditions in colony during reproductive cycle	Sensitive period: 4 days when oocytes are 1.5-4 mm long	Pairing of females with 4 males instead of 1	Larger the increase in egg size after exposure to crowding, smaller the proportion of green hatchlings (more gregarized dark hatchlings)	Not investigated in this study	(Maeno and Tanaka 2010)
C57BL/6J mice	Whole-body irradiation	10 weeks old	4, 7, or 10 Gy of X-rays	All levels of radiation corresponded to decreased methylation within 8 h; no effect in brain or spleen	Not investigated in this study	(Tawa et al. 1998)
Viable yellow agouti (A^{vy}) mice	Low-dose ionizing radiation (LDIR)	Radiation at GD 4.5; AO diet 1 week before mating through lactation	Radiation: 0.4, 0.7, 1.4, 3.0, 7.6 cGy AO diet: Harlan Teklad TD.95092	LDIR increased DNA methylation at A^{vy} locus in male offspring; AO mitigated methylation changes	Maternal dietary AO supplements	(Bernal et al. 2013)
C57BL/6J mice	Early life stress via maternal separation	PND 1–14	Chronic, unpredictable maternal separation; 3 h daily or left undistrubed	Altered DNA methylation at *MeCP2*, *CB1*, *CRFR2* in germ line of separated males; comparable changes in methylation and gene expression are present in brain of offspring	Not investigated in this study	(Franklin et al. 2010)
C57BL/6J mice	Ionizing radiation bystander effects (examine spleen following head exposure)	One time exposure to adult animals	5 cGy	Global hypomethylation in males 6 and 96 h after exposure, seen in females only after 96 h	Not investigated in this study	(Koturbash et al. 2008)

C57BL6/J mice	Early life stress	PND 1–10	3 h per day in cage without mother	Induces persistent hypomethylation of arginine vasopressin (*Avp*) gene enhancer with upregulated *Avp* expression	Not investigated in this study	(Murgatroyd et al. 2009)
Long-Evans hooded rats	Maternal care	Birth to weaning	Low versus high care	Altered pattern of H3K9 acetylation, DNA methylation and higher transcriptional activity in *NR3C1* locus of adult offspring with high maternal care	Not investigated in this study	(McGowan et al. 2011)
Sprague-Dawley rats	Posttraumatic stress disorder	Adulthood	2 acute cat exposures + 31 days of daily social instability	Increased *Bdnf* DNA methylation and decreased expression in dorsal hippocampus	Not investigated in this study	(Roth et al. 2011)
Human Studies						
Suicide completers (n = 44) and nonsuicide controls (n = 33)	Suicide	Adulthood	Successful suicide or nonsuicide patients	Increased DNA methylation at *BDNF* promoter and exon IV with correspondingly reduced mRNA levels	Not investigated in this study	(Keller et al. 2010)
Quebec Suicide Brain Bank history of child abuse (n = 12), no history of child abuse (n = 12)	Child abuse	Childhood	Severe sexual or physical abuse or severe neglect assessed by survey	Decreased glucocorticoid receptor (*NR3C1*) mRNA and increased methylation at promoter in abused suicide victims	Not investigated in this study	(McGowan et al. 2009)

Note: AO, antioxidant; GD, gestational day; PND, postnatal day; TBHQ, tert-butylhydroquinone.

they are not directly exposed (Koturbash et al. 2008). The bystander effect has been shown to be associated with epigenetic signaling, since targeted disruption of DNA methyltransferases, *Dnmt1* and *Dnmt3a*, in cultured cells eliminates the transmission of genomic instability (Rugo et al. 2011). Recent evidence in mice now indicates in vivo LDIR exposure during early development causes both dose- and sex-dependent epigenetically induced adaptive changes in mice that are in part dependent upon a cellular oxidative stress response (Bernal et al. 2013).

1.2.3 Nutrition Modification

The only study to date that examines the impact of nutrient intake on physical exposures is the recent Bernal et al. (2013) publication. In this study, viable yellow agouti dams were split with half receiving a control chow diet and half on an antioxidant (AO) rich diet, containing vitamin C, vitamin E, α-lipoic acid, *N*-acetyl-L-cysteine, *tert*-butylhydroquinone, and seleno-L-methionine, beginning 1 week prior to mating and continuing for the duration of pregnancy and lactation. Dams were then exposed to one of six radiation doses (0, 0.4, 0.7, 1.4, 3.0, and 7.6 cGy) on GD 4.5, thereby providing perinatal exposure to LDIR for the offspring. The LDIR induced a sex-specific increase in DNA methylation at the *Avy* locus in male offspring only. However, those male offspring whose mothers had been on the AO diet had smaller changes in methylation at that gene locus (Bernal et al. 2013). Other gene loci were not explored, but this study suggests that consuming an AO rich diet may protect against the epigenetic-altering affects if LDIR.

PHYSICAL EXPOSURES: SUMMARY

- Exposure to behavior, stress, temperature, species density, and radiation during the fetal and early childhood periods induce epigenetic changes in the offspring that persist across the lifespan.
- Physical exposures act primarily through tissue-specific DNA methylation or altered DNMT activity. Little evidence exists relating physical exposures to histone modifications or other epigenetic mechanisms.
- Only one study has investigated the epigenetic interaction of dietary intake and physical exposures. Maternal supplementation with an antioxidant rich diet prior to and during pregnancy and lactation decreased the impact of concomitant radiation exposure on DNA methylation.

1.3 CHEMICAL EXPOSURES

Chemical exposures can be natural or man-made and include plastic monomers and plasticizers such as BPA and phthalates, pesticides and persistent organic pollutants like organophosphates, dioxins, and polychlorinated biphenyls (PCBs), and metals

(Diamanti-Kandarakis et al. 2009). Dependent upon the exposure scenario, these chemicals may be ingested, inhaled, or dermally absorbed. Once inside the body, they exert their actions through nuclear hormone receptors, membrane-associated receptors, transcription factors, and numerous enzymatic pathways (De Coster and van Larebeke 2012). Evidence has shown that fetal exposure to chemicals results in adverse health effects that can persist into adulthood (Diamanti-Kandarakis et al. 2009; Braw-Tal 2010; Meeker 2012). Developmental chemical exposures have been associated with neurodevelopment, reproductive function, cancer, obesity, and immune function (Muczynski et al. 2012; Weiss 2012; Hao 2013; Rogers et al. 2013; Soto et al. 2013). In addition to disease phenotypes arising from a chemical's ability to interfere with receptor, transcription factor, and enzyme activity, an emerging mechanism of interest is epigenetic programming (Bernal and Jirtle 2010; Fleisch et al. 2012). Here we discuss animal in vivo and human epidemiological studies that have demonstrated the ability of chemicals to alter DNA methylation, histone function, and noncoding RNA expression, summarized in Table 1.2. We pay close attention to studies that were designed to cover key epigenetic reprogramming events (Figure 1.1).

1.3.1 Bisphenol A

BPA is a chemical produced in high volumes and found in consumer products, such as food and water containers, receipt paper, dental composites, and metal can linings (Vandenberg et al. 2010). BPA can leach from consumer products, which presents many opportunities for human exposure. Accumulating work suggests that early BPA exposure increases susceptibility for adverse phenotypic outcomes via epigenetic mechanisms (Jirtle and Skinner 2007; Kundakovic and Champagne 2011).

Several studies have focused on the effects of early life BPA exposure on DNA methylation machinery. Establishment of DNA methylation marks during late fertilization relies on the efficiency of this machinery. Following maternal exposure to 2, 20, and 200 μg BPA/kg body weight (BW)/day in mice, the expression of both DNA methyl transferase 1 (*Dnmt1*), the maintenance methyltransferase enzyme, and DNA methyl transferase 3 (*Dnmt3a*), the enzyme involved in de novo DNA methylation in early development, were significantly altered in the prefrontal cortex and the hypothalamus of offspring (Kundakovic et al. 2013). Additionally, exposed male offspring displayed a decrease in DNA methylation at estrogen receptor 1 (*Esr1*) in the prefrontal cortex while females exposed to 20 μg/kg/day displayed a decrease in methylation at *Esr1* in the hypothalamus. Moreover, maternal exposure to 1.25-mg BPA/kg diet from GD 0 to GD 18.5 led to a decrease in expression of *Dnmt1* in GD 18.5 female brain tissue (Wolstenholme et al. 2011). The decrease in *Dnmt1* expression was associated with an increase in *Sc1a1* expression in the same brain regions. These studies indicate that BPA has an effect on DNA methylation programming by altering the function of DNA methylation machinery.

Imprinted genes are also of interest, since they are functionally haploid and the health consequences of genomic imprinting are potentially disastrous if epigenetic dysregulation were to occur (Murphy and Jirtle 2003). Mice exposed to 40, 80, and 160 μg BPA/kg BW during GD 0.5 to 12.5 resulted in hypomethylation of primordial

TABLE 1.2
Epigenetic Impact of Chemical Exposures

Species	Chemical Agent	Timing of Exposure	Exposure Dose	Epigenetic Impact	Nutrient Interaction	Reference
			Animal Models			
BALB/c mice, brain	BPA	Gestation	2, 20, and 200 μg/kg/day	Expression changes in *Dnmt1* and *Dnmt3a*, and hypomethylation at *Esr1* in two distinct brain regions	Not included in the study	(Kundakovic et al. 2013)
C57BL/6J mice, brain	BPA	ED 0–18.5	1.25 mg BPA/kg diet	Expression changes of *Dnmt1* in brain	Not included in the study	(Wolstenholme et al. 2011)
CD-1 mice, primordial germ cells	BPA	ED 0.5–12.5	40, 80, and 160 μg BPA/kg/day	Hypomethylation at *Igf2r*, *Peg3*, and *H19* in primordial germ cells	Not included in the study	(Zhang et al. 2012)
C57BL/6J mice, placenta, and embryonic tissue	BPA	ED 0–9.5	50 mg BPA/kg diet	Altered DNA methylation of *Snrpn* and *Igf2* in placental and embryonic tissue, respectively, and global hypomethylation in placenta	Not included in the study	(Susiarjo et al. 2013)
A^{vy}/a and *a/a* mice, tail	BPA	Gestation and lactation	50 mg BPA/kg diet	Hypomethylation at A^{vy} locus	Methyl donor and genistein counteracted hypomethylation	(Dolinoy et al. 2007)

A^{vy}/a and *a/a* mice, tail	BPA	Gestation and lactation	50 ng, 50 µg, and 50 mg BPA/kg diet	Hypomethylation at A^{vy} locus	Not included in the study	(Anderson et al. 2012a)
C57BL/6J mice, brain	BPA	Gestation and lactation	10 nM BPA	Decreased binding of H3K9ac to the *Kcc2* promoter	Not included in the study	(Yeo et al. 2013)
CD-1 mice, mammary tissue	BPA	ED 9–26	10 µg/kg/day	Increased EZ2H expression, and increased H3 trimethylation	Not included in the study	(Doherty et al. 2010)
Sheep, fetal ovaries	BPA	ED 30–90	0.5 mg/kg/day	Expression changes in 56 miRNAs	Not included in the study	(Veiga-Lopez et al. 2013)
Mecp2(308/+) mice, brain	PBDE	Gestation and lactation	0.03 mg/kg/day	Decreased global methylation	Not included in the study	(Woods et al. 2012)
Sprague-Dawley rats, sperm	2,3,7,8-Tetrachlorodibenzo-*p*-dioxin	ED 8–14	100 ng/kg/day	50 DMRs	Not included in the study	(Manikkam et al. 2012b)
Sprague-Dawley rats, sperm	Vinclozolin	ED 8–14	100 mg/kg/day	52 DMRs	Not included in the study	(Guerrero-Bosagna et al. 2010)
Sprague-Dawley rats, granulosa cells	Vinclozolin	ED 8–14	100 mg/kg/day	43 DMRs	Not included in the study	(Nilsson et al. 2012)
Sprague-Dawley rats, sperm	Pesticide mixture: permethrin and DEET	ED 8–14	39% for permethrin and 2% of DEET of LD50 dose	363 DMRs	Not included in the study	(Manikkam et al. 2012b)
CD-1 mice, liver	Triadimefon, propiconazole, or myclobutanil	30 day adult treatment	1800 ppm, 2500 ppm, or 2000 ppm, respectively	Expression changes of 19 miRNAs	Not included in the study	(Ross et al. 2010)

(continued)

TABLE 1.2 (Continued)
Epigenetic Impact of Chemical Exposures

Species	Chemical Agent	Timing of Exposure	Exposure Dose	Epigenetic Impact	Nutrient Interaction	Reference
			Animal Models			
Zebrafish,	Fipronil, triazophos, or mixture (fipronil and triazophos)	96-h treatment in adults	30%, 1%, or 31%, respectively	Expression changes of 14 miRNAs	Not included in the study	(Wang 2010)
Sprague-Dawley rats, sperm	BPA, DBP, and DEHP	ED 8–14	50 mg/kg/day, 750 mg/kg/day and, 66 mg/kg/day, respectively	197 DMRs	Not included in the study	(Manikkam et al. 2013)
Long Evans rats, brain	Lead	Gestation and lactation	0, 150, 375, and 750 ppm	Expression changes of Dnmt1 and Dnmt3a	Not included in the study	(Schneider et al. 2013)
Wistar rats, liver	Cadmium	ED 0 to birth	50 ppm	Hypomethylation at glucorticoid receptor and expression changes of Dnmt3a	Not included in the study	(Castillo et al. 2012)
Chicken, embryo	Cadmium		50 μM	Expression changes of Dnmt3a	Not included in the study	(Doi et al. 2011)
			Human Studies			
n/a	n/a	n/a	n/a	n/a	n/a	n/a

germ cells (PGCs) of the differentially methylated regions (DMR) of *Igf2r*, *Peg3*, and *H19* (Zhang et al. 2012). Furthermore, DNA methylation at the imprinting control region (ICR) of *Snrpn* was decreased in GD 9.5 placental tissue and increased at the DMR1 of *Igf2* in GD 9.5 embryonic tissue following exposure to 50-mg BPA/kg diet from conception to GD 9.5 via maternal diet in a mouse model (Susiarjo et al. 2013). In addition to changes in methylation status of the imprinted genes, BPA exposure significantly reduced global DNA methylation levels in the placenta.

Metastable epialleles are loci of identical sequence harboring different epigenetic marks that are established early in development. Thus, metastable epialleles have been used extensively to investigate the impacts of BPA and nutrition on the epigenome and to facilitate the understanding of DOHaD (Dolinoy 2008). The methylation status at the murine *viable yellow agouti* (A^{vy}) metastable epiallele is inversely correlated to coat color. The A^{vy} mouse model has demonstrated that gestational and lactational exposure to 50-ng, 50-μg, and 50-mg BPA/kg diet shifts the distribution of coat color by altering methylation levels at the A^{vy} allele (Dolinoy et al. 2007; Anderson et al. 2012a). However, nutritional supplementation of methyl donors and genistein was able to counteract the shift of the distribution of coat color in offspring exposed to 50-mg BPA/kg diet (Dolinoy et al. 2007). Additionally, using BPA as a representative chemical exposure during development, novel murine metastable epialleles have been discovered that are important for assessing risk to chemical exposures that elicit effects via epigenetic pathways (Weinhouse et al. 2011).

Few studies have explored the effects of developmental BPA exposure on epigenetic marks other than DNA methylation. Exposure to 10 nM of BPA during gestation and lactation resulted in decreased binding of acetylated H3K9 to the *Kcc2* promoter in mouse offspring, leading to diminished *Kcc2* expression, a gene that plays an important role in neurodevelopment (Yeo et al. 2013). Mice offspring displayed an increase in EZ2H, a histone methyltransferase, expression in adult mammary tissue after exposure to 10 μg BPA/kg BW with a concomitant increase in H3 trimethylation (Doherty et al. 2010). MicroRNA (miRNA) expression was measured in fetal ovaries of sheep after in utero exposure to 0.5 mg/kg/day of BPA (Veiga-Lopez et al. 2013). BPA Exposure resulted in 45 and 11 miRNAs that were differentially expressed at GD 65 and 90 fetal tissue, respectively.

1.3.2 Phthalates

Phthalates are high-production esters that are found in a range of consumer products. High-molecular-weight phthalates like diethylhexyl phthalate (DEHP) are added to plastics while low-molecular-weight phthalates like dibutyl phthalate (DBP) are more likely used in adhesives, solvents, and personal care products. Due to phthalates' ubiquitous presence in the environment, urinary concentrations of the metabolites are found among the general population (Silva 2004). Phthalate exposure has been associated with altered reproductive function, allergies, cancer, and neurodevelopment (Braun 2013).

The focus of phthalate exposure and its effects on epigenetics has been mainly limited to in vitro studies (Chan Kang and Mu Lee 2005; Hsieh et al. 2012a, b; Sun et al. 2012). Additionally, two in vivo rodent models have demonstrated alterations to DNA methylation in testis after exposure to DEHP resulting in abnormal testosterone production in adulthood (Martinez-Arguelles et al. 2009; Wu 2010).

1.3.3 Persistent Organic Pollutants

Persistent organic pollutants (POPs) are lipophilic synthetic chemical compounds including PCBs, dioxins, dichlorodiphenyldichloroethylene (DDE), hexachlorobenzene (HCB), and organochlorine pesticides (OCPs) (La Merrill 2013). Owing to their lipophilic nature, POPs have been found to accumulate in biological systems, and additionally have the ability to cross the placenta resulting in fetal exposure (Covaci et al. 2002). Early life exposures to POPs have been associated with adverse health outcomes such as delayed cognitive development, behavioral issues, delayed growth, and immune-related disease (Guo et al. 2004; Stølevik et al. 2013).

The effect of POPs on DNA methylation has been explored in vivo. Female mice that were exposed to 0.03 mg polybrominated diphenyl ether (PBDE) per kg BW 28 days prior to mating through lactation exhibited a decrease in global DNA methylation levels in the brain in adulthood (Woods et al. 2012). The alterations to global DNA methylation correlated with decreased social behavior.

A handful of transgenerational studies have been conducted in rodents looking at DNA methylation and exposure to 2,3,7,8-tetrachlorodibenzo-*p*-dioxin (TCDD). The F3 generation offspring from female rats (F0 generation) exposed to 100 ng/kg BW/day from GD 8 to 14, displayed 50 regions of altered methylation following an epigenome-wide scan of the sperm (Manikkam et al. 2012a, b). These epigenetic modifications in the F3 generation were associated with testicular abnormalities and delayed onset of puberty.

1.3.4 Pesticides

Pesticides are wide class of substances used to kill, repel, or control animals or plants that are unwanted (Mostafalou and Abdollahi 2013). The different classes of pesticides include herbicides, insecticides, fungicides, and disinfectants and are used to control vegetation, insects, molds, and bacteria, respectively. Pesticide exposure has been linked to a variety of disease outcomes such as neurodegenerative diseases like Alzheimer and Parkinson, reproductive function, behavioral disorders, cancer, and diabetes. Current studies are now demonstrating pesticides' ability to interfere with epigenetic mechanisms leading to adverse health effects (Collotta et al. 2013).

Several transgenerational studies have been conducted in rodents looking at maternal pesticide exposure and resulting epigenetic changes in the F3 generation of offspring. Treatment of 100 mg/kg/day vinclozolin, a fungicide, during GD 8 through 14 resulted in 43 and 52 DMRs following an epigenome-wide scan in promoter regions of granulose cells and sperm, respectively (Guerrero-Bosagna et al. 2010; Nilsson et al. 2012). Furthermore, exposure to the pesticide mixture containing permethrin and DEET (39% for permethrin and 2% of DEET of LD50 dose) during GD 8 through 14 resulted 363 DMRs in sperm (Manikkam et al. 2012c). These epigenetic modifications in the F3 generation were correlated to ovarian and testicular abnormalities.

In addition to alteration of DNA methylation following pesticide exposure, the effects of exposure on microRNAs (miRNAs) has been explored. Male mice 30 days of age were treated with 1800 ppm triadimefon, 2500 ppm propiconazole, or 2000 ppm myclobutanil, fungicides used for agricultural applications, via diet for 90 days (Ross et al. 2010). Following fungicide exposure to triadimefon and propiconazole, evaluation of miRNA expression revealed an overlap of 19 miRNAs that were differentially expressed in the liver. The expression changes correlated to tumorigenic outcome at 2 years old. A separate study measured the effect of two insecticides, 30% fipronil, 1% triazophos, or a 31% mixture (fipronil and triazophos), on miRNA expression in zebrafish (Wang 2010). Ninety-six hours after exposure treatment, 14 miRNAs were differentially expressed. These studies indicate that pesticide exposure induced alterations to miRNA are potential biomarkers for chronic disease states.

1.3.5 Mixture Studies

In addition to looking at chemical exposures on an individual basis, there are some studies that have used a mixture of these compounds since an individual is likely to have traces of metabolites in their system of several compounds at any given time (Hauser and Calafat 2005; Calafat et al. 2008). The F3 generation offspring descended from female rats (F0 generation) exposed to a mixture of BPA and two phthalates (DBP and DEHP) from GD 8 to 14 displayed 197 DMRs in the sperm (Manikkam et al. 2012a, 2013). Additionally, of these DMRs, a few of the regions were correlated to a number of disease pathologies found among these rats like testis disease and pubertal abnormalities.

1.3.6 Metals

Exposure to metals or metal-containing compounds has been shown to interact with the epigenome (Cheng et al. 2012). In particular, arsenic, cadmium, chromium, zinc, nickel, and lead have been studied using in vitro, in vivo, and epidemiological approaches. Although there is sufficient evidence that metal exposure results in an altered epigenome, there have been limited studies exploring metal exposure during the critical epigenetic reprogramming period at postfertilization and how it affects epigenetics later in life. For example, although there has been research conducted on the effects of arsenic, chromium, and zinc on epigenetic modifications, the exposure periods of these studies do not cover the critical window of early epigenetic reprogramming in early development or are limited to in vitro models (Chervona et al. 2012; Perng et al. 2012; Tsang et al. 2012). In this section, our main focus will be on the limited studies covering this critical period of development.

The naturally occurring metal, lead (Pb), is found in a variety of products such as paint, ceramics, pipes, batteries, and gasoline as well as soil, dust, and drinking water (Brown 2012). Female rat offspring exposed to 150, 375, and 750 ppm lead had decreased *Dnmt1* expression in the hippocampus with a corresponding decrease in the methyl cytosine-binding protein (Schneider et al. 2013). Males exposed to 150 and 750 ppm lead displayed increased expression of *Dnmt3a*. This study indicates that lead may alter the epigenome by disrupting the methylation machinery needed for DNA methylation programming.

Cadmium is released into the environment most commonly from burning fossil fuels, cigarette smoke, and as a by-product of zinc, lead, or copper ores (Satarug 2010). Fetal exposure (GD 0 to birth) to 50 ppm of cadmium through maternal drinking water resulted in hypomethylation of the hepatic glucocorticoid receptor (GR) in female rat offspring while males displayed hypermethylation at the same locus (Castillo et al. 2012). The changes in DNA methylation are correlated to an increase and decrease of GR mRNA expression, respectively, as well as corresponding changes in *Dnmt3a* expression. Chick embryos treated with 50 μM of cadmium for 4 h had decreased mRNA expression levels of *Dnmt3a* when compared with control embryos (Doi et al. 2011). Also, the immunoreactivity of 5′-methyl cytosine (5′-MeC) was decreased and reflects the associated changes in global methylation.

Nickel can be released into the environment from natural sources like rock weathering or by anthropogenic sources like fossil fuel combustion (Chervona and Costa 2012; Chervona et al. 2012). Nickel has been shown to inhibit histone methylation by interfering with iron absorption resulting in inhibition of iron-dependent histone demethylases (Chervona et al. 2012). Alterations to miRNA expression have been noted in a nickel exposure study. Human bronchial epithelial cells treated with nickel sulfide resulted in downregulation of miRNA 152 expression (Ji et al. 2013). miRNA 152 was also shown to repress *Dnmt1* expression displaying cross-talk between this noncoding RNA and methyltransferase.

1.3.7 Nutrition Modification

To our current knowledge, only one study of chemical exposure includes a nutrition intervention introduced in prenatal period. Two weeks prior to mating, viable yellow agouti (A^{vy}) dams were assigned to one of four diets: (1) corn oil control chow, (2) corn oil chow + 50-mg BPA/kg diet, (3) corn oil chow + 50-mg BPA/kg diet + 250-mg genistein/kg diet, or (4) corn oil chow + 50-mg BPA/kg diet + methyl donor supplement in chow (including folic acid, vitamin B_{12}, betaine, and choline). The two nutritionally supplemented BPA diets did not alter general offspring characteristics (e.g., litter size or survival, sex ratio, wean weight, genotypic ratio), but did shift offspring coat color distribution back to control litter distribution despite also containing BPA. Both genistein and methyl donor diets moderated the DNA hypomethylation in BPA-exposed offspring at the A^{vy} intracisternal A particle (IAP) retrotransposon at the 5′-end of the *Agouti* gene locus (Dolinoy et al. 2007). Genistein and methyl donor levels included in the supplemented diets are comparable to those women would receive if consuming high soy or consuming a whole foods diet high in methyl donors. Thus, maternal dietary intake during pregnancy may be able to counteract some of the deleterious epigenetic programming that occurs in response to chemical exposures. However, this study did not follow offspring past PND 22, the day after weaning, so life course health consequences and epigenetic patterns following BPA exposure and modification by genistein- and methyl donor-supplemented diets could not be assessed. Also, changes in methylation status of non-metastable epiallele loci were not examined. Further work investigating the impact of genistein and methyl donor diets on genes related to disease risk and health outcomes over a longer follow-up period are critical to help elucidate the potential for nutritional modification of metabolic programming during pregnancy.

1.3.8 Alcohol

Alcohol has also been explored as chemical exposure affecting epigenetic gene regulation. Using the A^{vy} murine model, ethanol exposure at GD 0.5–8.5 resulted in hypermethylation at the A^{vy} metastable epiallele in addition to craniofacial dysmorphology resembling fetal alcohol syndrome (Kaminen-Ahola et al. 2010a, b). Further epigenetics studies have been shown to play a role in the spectrum of fetal alcohol and liver disorders following exposure to alcohol by decreasing the NAD+/NADH ratio. The resulting deficit of NAD+ alters the activity of histone deacetylases dependent on NAD+ for enzymatic activity (Imai et al. 2000; Cunningham and Bailey 2001). Alcohol exposure has also been shown to interfere with one-carbon metabolism, the cycle required for DNA methylation, by inhibiting methionine production and degrading DNA methyl transferase activity (Martínez-Chantar 2002; Mukhopadhyay et al. 2013). However, epigenetic studies following alcohol exposure are limited to perinatal periods outside of the window of epigenetic reprogramming and to adulthood or to in vitro studies.

CHEMICAL EXPOSURES: SUMMARY

- Perinatal exposure to plasticizers, pesticides, persistent organic pollutants, and metals induce epigenetic modifications in the form of altered DNA methylation, DNMT activity, histone tail modifications, and miRNA expression.
- Many studies utilize prenatal chemical exposures to assess the impact on adult disease as suggested by the DOHaD paradigm; however, few studies begin the exposure prior to conception and therefore miss the early epigenetic reprogramming events occurring in the first few days of development.
- Only one study has investigated the epigenetic interaction of dietary intake and a chemical exposure. Maternal supplementation with a methyl-donor rich diet or diet supplemented with genistein had a similar effect; both dietary exposures counteracted the BPA-induced DNA hypomethylation and the BPA-associated shift in coat color distribution towards yellow.

1.4 NUTRITIONAL EXPOSURES

Nutrition has long been known to be important in health promotion and disease prevention. In the 1940s, Ancel Keys began to examine dietary habits of middle-aged men to determine if dietary intake patterns were associated with risk of cardiovascular disease (CVD). His Seven Countries Study, begun in 1958, clearly identified the association between total serum cholesterol and rates of heart attacks and strokes at both the individual and population levels across all countries. Men in the seven countries were observed for 10 years; thus, a change in Mediterranean diet and lifestyle toward the Western diet and more sedentary lifestyle occurred and was accompanied by a significant increase in CVD (Keys et al. 1980). Today we know that many of the health benefits and risks from diet result from their biochemical interactions and epigenetic alterations within the body. We are now able to discern the impact of dietary intake on gene expression and subsequent downstream functional differences (Lillycrop 2011; Lillycrop and Burdge 2011; Anderson et al. 2012b; Koletzko et al. 2012).

The potential for nutrition to impact epigenetic reprogramming during early fetal development has gained support over the past decade, putting additional emphasis on the importance of maternal nutrient intake during the perinatal period. In addition to methyl donor compounds exerting these modifying effects, maternal fat intake and protein or total caloric restriction have been shown to alter metabolic programming across the lifespan. Therefore, the fetal nutritional environment in utero is a critical exposure in the DOHaD paradigm, reviewed extensively elsewhere (Lillycrop 2011; Lillycrop and Burdge 2011; Anderson et al. 2012b; Koletzko et al. 2012). Maternal overnutrition and undernutrition both regulate gene expression of enzymes involved in lipid and carbohydrate metabolism. Therefore, it is not surprising that prenatal maternal diet have been associated with long-lasting effects of metabolic dysregulation, such as obesity, insulin resistance, type 2 diabetes, dyslipidemia, hypertension, and CVD (Canani et al. 2011). In some cases, nutritional intake of the offspring after birth can modify the epigenetic programming effects of the fetal nutritional environment. Evidence for developmental programming by fetal nutrition and later modifying effects of diet will be reviewed in this section and are summarized in Table 1.3.

1.4.1 Methyl Donor Nutrients

Generation of *S*-adenosylmethionine (SAM), the cellular molecule that donates a methyl group to DNA, which is covalently bound via DNA methyltransferases (DNMTs) relies upon methyl donor nutrients, which include vitamins and micronutrients involved in one-carbon metabolism, such as vitamins B_2, B_6, and B_{12}, amino acid methionine, and micronutrients folate, choline, and betaine (Selhub 1999; Anderson et al. 2012b). Low maternal folate intake (0.4-mg/kg diet) during pregnancy resulted in global hypomethylation in the proximal small intestine of PND100 mice, and post-weaning folate repletion was not able to rescue the hypomethylation (McKay et al. 2011). This supports the DOHaD paradigm, by illustrating the importance of nutritional exposure during critical periods in embryonic development; once epigenetic reprogramming has occurred at these early stages, it is not always possible to alter epigenetic and gene expression profiles later in life. In viable yellow agouti mice that have an epigenetic propensity for obesity, weight gain is exacerbated in each generation; however, supplementation of F0 dams with a methyl donor–rich diet that is continued through to the F3 generation of offspring prevents the amplification of body weight observed in genetically identical mice following the same breeding scheme (Waterland et al. 2008).

Human birth cohort data suggest that maternal intake of methyl donor nutrients may impact DNA methylation only in initially depleted mothers, with no added benefit of intake in women who already meet daily intake requirements. In Project Viva,

TABLE 1.3
Epigenetic Impact of Nutritional Exposures

Species	Nutritional Exposure	Timing of Exposure	Exposure Dose	Epigenetic Impact	Nutrient Interaction	Reference
			Animal Models			
C57BL/6J mice	Maternal low versus normal folate diets Offspring weaned to low versus normal folate diet	Diets begun during mating, continued through pregnancy and lactation Measured at PND100	Low folate: 0.4 mg FA/kg diet Control folate: 2 mg FA/kg diet	Low folate: proximal small intestine tissue was globally hypomethylated	Postweaning folate repletion did not modify effects from developmental folate deficiency	(McKay et al. 2011)
Viable yellow agouti (A^{vy}) mice	Methyl donor diet (Harlan Teklad TD.01308)	Chow versus methyl donor diet began at PND21 in F0 dams	Diets were maintained across three generations (F0–F3)	Epigenetic propensity for obesity was exacerbated with each generation in A^{vy}, benefit of methyl donor diet was independent of epigenetic changes at A^{vy} locus	Methyl donor diet prevented amplification of body weight	(Waterland et al. 2008)
Sprague-Dawley rats	Methyl donor supplementation	21–28 days prior to mating through pregnancy and lactation	12 g methionine, 15 g choline, 15 g betaine, 1000 mg B_{12}, 15 mg folic acid/kg diet	Increased methylation at *Lep* and reduced leptin secretion in offspring; joint effect impaired postnatal growth in both genders and lower long-term weight gain in males	Protein restriction did not affect *Lep* methylation, but did have higher insulin secretion	(Giudicelli et al. 2013)

(continued)

TABLE 1.3 (Continued)
Epigenetic Impact of Nutritional Exposures

Species	Nutritional Exposure	Timing of Exposure	Exposure Dose	Epigenetic Impact	Nutrient Interaction	Reference
			Animal Models			
Wistar rats	Maternal protein restriction	Diets assigned following mating, continued through pregnancy and lactation	Control: 20% Low: 8% protein	Decreased H3K9 acetylation at *Cyp7a1* accompanied by reduced fetal hepatic expression of *Jmjd2a* (H3K9 demethylase)	Not included in this study	(Sohi et al. 2011)
Wistar rats	Maternal protein restriction	Diets assigned following mating, continued through pregnancy and lactation	Control: 20% Low: 8% protein	H3K9 acetylation decreased at *Lxrα*, with decreased *Lxrα* expression and increased *Hsd11b1* and *G6pc* expression due to decreased binding of LXRα to the putative LXRE on *Hsd11b1* and *G6pc*	Not included in this study	(Vo et al. 2013)
Wistar rats	Protein restriction	Diets assigned following mating	Control: 180 g/kg + 1 mg/kg FA Ptn-restricted: 90 g/kg + 1 mg/kg FA Ptn-restricted and folate-supplemented: 90 g/kg + 5 mg/kg FA	*Ppara* and *Nr3c1* methylation lower, expression higher in protein-restricted	Folate supplementation with protein restriction prevented altered methylation	(Lillycrop et al. 2005)

German Landrace sows	Maternal protein restriction	Diets assigned following mating	Control: 12% Low: 6% High: 30% protein	Global hepatic DNA methylation decreased in LP fetal samples, and decreased in HP finisher pigs; *Dnmt1*, *3a*, *3b* levels were influenced by HP and LP diets	Not included in this study	(Altmann et al. 2012)
Wistar rats	Maternal high fat diet	Following confirmation of mating		PND2 livers: hypomethylated *Cdkn1a* with increase in mRNA expression	Not included in this study	(Dudley et al. 2011)
Crl:CD(SD) rats	Maternal high fat diet	Diets assigned following mating to ED20	Control: 16% HF diet: 45%	Offspring of HF diet dams increased mRNA levels of gluconeogenic genes; decrease in H3 acetylase, H3K9Me3; and increase in H4 acetylase, H3K4Me2	Not included in this study	(Strakovsky et al. 2011)
Wistar rats	High-fat–high-sucrose diet, then randomized to remain on HFHS diet or change to control chow (Research Diets, D12451)	36-week-old rats on each diet for 10 weeks	HFHS diet: 45% fat, 35% carbohydrate, (23.1% sucrose) Chow diet: 10.3% fat, 73.1% carbohydrate (6.5% sucrose)	HFHS diet 20 weeks: hypermethylated *Lep* CpG sites 6.7 and 29.30 and hypomethylated site 15; *Fasn* hypomethylated at CpG site 1, no methylation change at *Ppargc1a*, *Srebf1*, aquaporin 7 promoter regions	HFHS + chow diet: reversed methylation at *Lep* CpGs and hypomethylated CpG sites of *Srebf1*, hypermethylated sites in *Ppargc1a* and *Fasn* promoters	(Uriarte et al. 2013)

(continued)

TABLE 1.3 (Continued)
Epigenetic Impact of Nutritional Exposures

Species	Nutritional Exposure	Timing of Exposure	Exposure Dose	Epigenetic Impact	Nutrient Interaction	Reference
			Animal Models			
Viable yellow agouti (A^{vy}) mice	Maternal excess calories (obesity)	A^{vy} dams were obese at time of mating versus a/a dams who were not obese	Methylation pattern at A^{vy} locus determines risk of dam overweight	Obese dams: downregulated *ATPase* and *Cytb*; widespread DNA methylation changes in male offspring of obese dams	Offspring weaned to control chow or HF diet; HF diet unmasked latent metabolic phenotype from obese dams	(Li et al. 2013)
C57BL/6J mice	Maternal excess calories (obesity) (Research Diets: HF diet D12492)	Began control or HF diet at weaning (6 months prior to mating)	Control: 10% HF diet: 60% fat	Obesity-resistant F2 offspring: repressed *Ppara* and other lipid uptake, storage, utilization genes	Offspring weaned to HF diet for 20 weeks	(Attig et al. 2013)
			Human Studies			
Project Viva, human birth cohort (n = 516)	Gestational intake of folate, vitamin B_{12}, betaine, choline, cadmium, zinc, iron	First and second trimester for mother	Folate: T1 786 µg/day and T3 1244 µg/day assessed via FFQ with supplement questions	Periconceptional betaine and choline (males only) inversely associated with cord blood methylation, in folate-replete population, no association with methyl donors overall	Not included in this study	(Boeke et al. 2012)
Human birth cohort (n = 913)	Dietary and supplement folate intake	First trimester	FFQ with supplement questions; energy adjusted mean 350 µg/day	After week 12 of gestation, supplement use associated with higher *IGF2* methylation and reduced methylation at *PEG3* and LINE-1	Not included in this study	(Haggarty et al. 2013)

NEST, human birth cohort (n = 438)	Dietary and supplement folate intake	First trimester	No FA: 0 µg/day Moderate FA: 600–1200 µg/day High FA: 1000–1400 µg/day + diet via FFQ	*H19* methylation decreased with greater FA intake (males > females); no association between FA intake and *IGF2* methylation	Not included in this study	(Hoyo et al. 2011)
Human birth cohort (n = 120)	Periconceptional folic acid (FA) supplementation	FA supplement: 4 weeks before through 8 weeks after conception Child at testing: 17 months	400 µg/day	Children of mothers taking FA had 4.5% higher methylation at *IGF2* DMR compared with kids of mothers not taking FA; *IGF2* methylation of child associated with maternal SAM blood level and inverse association with birth weight	Not included in this study	(Steegers-Theunissen et al. 2009)
Human birth cohort (n = 99)	Maternal and child serum folate	Serum levels measured at delivery	Cord blood mean: 7.29 ng/mL; Maternal blood mean: 2.29 ng/mL		Not included in this study	(Ba et al. 2011)
Human birth cohort (n = 99)	Maternal and child serum vitamin B_{12}	Serum levels measured at delivery	Cord blood mean: 240 pg/mL; Maternal blood mean: 175 pg/mL	Increased maternal serum B_{12} levels correlated with decreased methylation at *IGF2* promoter P3	Not included in this study	(Ba et al. 2011)

Note: FASN, fatty acid synthase; FFQ, food frequency questionnaire; PND, postnatal day.

a folate-replete Boston birth cohort, there was no association among maternal folate, vitamin B_{12}, cadmium, zinc, or iron, and long interspersed nuclear element 1 (LINE-1) DNA methylation in first- or second-trimester maternal blood or infant cord blood. However, higher levels of periconceptional choline, estimated by first trimester maternal food frequency questionnaire (FFQ) self-report, were associated with lower LINE-1 methylation levels in the cord blood of male infants (Boeke et al. 2012). The Newborn Epigenetics STudy (NEST) at Duke University found total perinatal folic acid intake is associated with hypomethylation at the maternally expressed imprinted gene, *H19*, with a greater effect seen in male compared with female infants. Maternal folic acid intake did not impact methylation at the paternally expressed counterpart, *IGF2* (Hoyo et al. 2011). Another study found no impact on DNA methylation with folate supplementation 400 μg/day in the first 12 weeks of pregnancy (the current health-care recommendation), but did observe altered methylation in women supplementing after their 12th week of pregnancy (Haggarty et al. 2013). Folate supplementation use after 12 weeks of gestation had higher methylation at *IGF2* and reduced methylation at *PEG* and LINE-1 sites, suggesting that methyl donor levels may play a critical role in epigenetic programming throughout pregnancy (Haggarty et al. 2013).

1.4.2 Protein Intake

A recent study investigated the interaction of protein restriction and excess methyl donor nutrients in rats. Female rats began one of four diets 3 to 4 weeks prior to mating and continued throughout pregnancy and lactation on a control chow diet, control + methyl donor diet, protein-restricted diet, or protein-restricted + methyl donor diet. Leptin promoter methylation was increased and secretion reduced in offspring of methyl donor–supplemented dams. Leptin methylation and secretion were not impacted by protein restriction; however, these pups did have depressed circulating insulin levels. Offspring of dams on the protein-restricted diet with methyl donor supplementation had significantly decreased postnatal growth despite no change in food intake; weight gain to 22 weeks was restricted in male offspring (Giudicelli et al. 2013). Gestational protein restriction (90 g/kg versus 180 g/kg casein) resulted in decreased promoter methylation and increased expression of hepatic peroxisomal proliferator–activated receptor (*Ppara*) and glucocorticoid receptor (*GR*) compared with control chow diets in rat offspring post-weaning. However, when folate supplementation (5 mg/kg versus 1 mg/kg) was added to the protein-restricted diet, methylation, and expression changes were prevented (Lillycrop et al. 2005). Both maternal dietary protein restriction and protein excess have been shown to impact offspring DNA methylation levels via alteration in *Dnmt1*, *Dnmt3a*, and *Dnmt3b* expression. For example, global hepatic methylation was decreased in fetal pigs whose mothers ate a low protein (LP; 6%) diet. However, at a later age, the association reversed; at PND188 the pigs whose mothers ate a high protein (HP; 30%) diet displayed a decrease in global hepatic methylation suggesting that protein composition of the maternal gestational diet may have long-term impacts on offspring hepatic methylation patterns and therefore liver function (Altmann et al. 2012).

Gestational protein restriction also impacts chromatin folding by inducing a decrease in fetal hepatic expression of the histone H3K9 demethylase, Jmjd2a, which drives increased H3K9 methylation and decreased H3K9 acetylation at the *Cyp7a1* promoter. Hepatic cholesterol 7α-hydroxylase (Cyp7a1) is responsible for cholesterol catabolism into bile acids. Following the repressive histone modifications, Cyp7a1 levels are also decreased in protein-restricted offspring and an elevation in serum cholesterol is observed (Sohi et al. 2011). Maternal protein restriction during pregnancy also decreased the acetylation of H3K9 at the liver X receptor α (LXRα) transcription start site, leading to lower *LXRα* expression. LXRα regulates glucose homeostasis through inhibition of gluconeogenic genes such as *G6pase*, *11β-Hsd1*, and *Pepck*. Lower *LXRα* transcript levels decrease binding of LXRα to the LXR response element (LXRE) in the promoter regions of gluconeogenic genes; if this epigenetic alteration persists, increased gluconeogenesis and glucose intolerance would develop as the offspring age (Vo et al. 2013).

1.4.3 Fat Intake

Maternal high-fat diet during pregnancy has long been postulated to impact offspring metabolism. Recent research suggests this may at least partially be explained by the epigenetic control of hepatic development. In a rat model of maternal high-fat (HF) diet, hypomethylation of the hepatic cell cycle inhibitor, *Cdkn1a* gene promoter was found along with an increase in *Cdkn1a* mRNA transcripts in PND 2 offspring liver tissue (Dudley et al. 2011). Maternal high-fat diet in rats also alters histone tail modifications at gluconeogenic gene loci (*G6Pase*, *Cebpα*, *Srebp1a*, and *Pgc1a*) resulting in significantly increased levels of these transcripts, thus driving in utero gluconeogenesis (Strakovsky et al. 2011). This study assessed fetal livers at GD 20, so long-term programming impact cannot be assessed, but if the increased expression of gluconeogenic genes continues into adulthood, it would lead to glucose dysregulation and an increased potential for type 2 diabetes development. Mouse dams exposed to an HF diet during pregnancy had increased histone methylation and decreased acetylation at H3K9 at the adiponectin promoter in the adipose tissue of OH offspring at 2, 12, and 24 weeks of age compared with control offspring. At the leptin promoter in adipose tissue, HF diet–exposed offspring also had increased methylation at H4K20 at 2, 12, and 24 weeks compared with control offspring (Masuyama and Hiramatsu 2012). These histone modifications were accompanied by significantly increased in leptin and decreased adiponectin mRNA levels in HF diet–exposed offspring compared with controls at all time points; serum triglycerides and leptin increased while serum adiponectin decreased significantly at 12 and 24 weeks, indicating in utero HF diet exposure could induce metabolic alterations in offspring up to 6 months of age.

Dietary omega 3 (n-3) fatty acids have been associated with epigenetic changes in adult rodent models. For example, following 35 days of flaxseed oil–supplemented chow adult male diabetic Wistar rats peroxisome proliferator–activated receptor α (PPAR-α) was upregulated and sterol regulatory element binding protein 1 (SREBP-1) was downregulated, resulting in improved β-oxidation of free fatty acids and decreased hepatic lipogenesis. In this study, both flaxseed oil and fish oil increased D5 and D6 desaturase activity, downregulated cytokines (TNF-α, IL-6) and adipocyte fatty acid–binding protein, thus ameliorating lipid abnormalities normally observed in adult diabetic Wistar rats (Devarshi et al. 2013). In a human supplementation trial, women in their second trimester of pregnancy were randomized to take 400 mg DHA daily or to placebo pills; there was no difference in DNA methylation of nonsmoking mothers. However, among mothers who smoked, those taking DHA supplements had infants with increased LINE-1, IFN-γ, and IL-13 methylation, suggesting that dietary omega-3 intake may stabilize the infant cord blood methylome following the stress of prenatal smoking (Lee et al. 2013). Since DHA supplementation occurred after the first wave of epigenetic programming in the first week of pregnancy and offspring were not followed past infancy, this study does not demonstrate the full DOHaD paradigm of developmental programming, but is indicative of omega-3 fatty acids' ability to impact DNA methylation.

1.4.4 Total Caloric Intake

Maternal caloric excess is an increasing problem worldwide as rates of obesity increase. Maternal caloric abundance plays a role in metabolic programming of the offspring. Maternal obesity is associated with widespread changes in DNA methylation of male offspring; for example, murine offspring of an obese mother weaned to a high-fat diet displayed exacerbated metabolic dysregulation and increased risk of developing obesity, metabolic syndrome, and type 2 diabetes-like symptoms (Li et al. 2013). However, a recent study in mice suggests that switching an obese dam to a healthy, control diet during pregnancy may confer resistance to a high-fat diet in female offspring. The obesity-resistant phenotype was associated with DNA methylation changes in multiple genes involved in hepatic anti-steatosis pathways; these animals also lacked the dysregulation of lipid storage and utilization pathways observed in obese offspring (Attig et al. 2013). Thus, although maternal gestational obesity increases risk of child obesity development, altering the maternal diet during pregnancy may provide some resistance to this propagation of metabolic dysregulation. If these findings are repeated in human birth cohort studies, this could lead to an improvement in nutrition counseling for overweight pregnant women.

1.4.5 Nutrition Modification

More research exists on the interaction between perinatal nutrition exposures and simultaneous or later life nutrition interventions than exists for other exposure types. However, many studies still begin the nutrition exposure after mating has occurred, which means that the first few days of embryonic development may occur before the exposure; since the initial wave of epigenetic reprogramming occurs in the first few days following oocyte fertilization, these later exposures miss a sensitive period for potential lifelong programming. Up to this point, the majority of nutrition intervention studies have also been performed in animal models. Confirmation of findings in human birth cohort studies will be necessary prior to translating nutrition modifications into clinical recommendations for pregnant women. Nutritional exposures are often considered to be more under the control of the individual exposed, compared with physical, chemical, and biological exposures; thus, making this area of research even more critical due to the potential for preventive guidelines that could be implemented to provide the most health-promoting epigenetic program possible to future children.

NUTRITIONAL EXPOSURES: SUMMARY

- Perinatal exposure to methyl donor nutrients, protein restriction, dietary lipid, and total caloric intake impact DNA methylation, DNMT activity, histone tail modifications, and miRNA expression.
- Many studies that begin dietary exposures during pregnancy do so after mating has occurred, thus missing the sensitive period of initial epigenetic reprogramming in the first few days of embryonic development.
- While some nutrition exposure and later nutrition modification studies have occurred in human birth cohorts, the majority have been in animal models. The few studies examining the interaction in humans focus on methyl donors with little research on other nutrients as modifiers of exposure.

1.5 BIOLOGICAL EXPOSURES

Viral and bacterial infectious agents utilize host epigenetic machinery to regulate their own epigenome and to decrease host immune defense systems thereby increasing chances of survival and propagation. Examples of epigenetic hijacking by infectious agents are reviewed in this section, but biological exposure research designs rarely begin during early development. Therefore, the potential for epigenetic reprogramming during the fetal period to impact disease susceptibility and other health outcomes across the lifespan cannot be addressed yet. The interaction of nutrient intake and the epigenetic modifications induced by infectious agents has not been studied to date. However, recent research interest in the interaction between environmental and nutritional exposures and the host microbiome has begun to elucidate potential mechanisms, which may provide useful insight into a potential interaction between nutritional and pathogenic exposures. For example, the development of inflammatory bowel disease (IBD) is now thought to have a fetal origin, based on DNA methylation changes of intestinal mucosa in response to microbiota. Emerging microbiome research will be discussed at the end of this section.

1.5.1 In Vitro Studies: Epigenetic Mechanisms Elucidated

Despite a lack of research in the developmental origins and nutritional interactions of biologic exposures, in vitro studies investigating epigenetic mechanisms of pathogenic agents abound (Table 1.4). Common epigenetic modifications utilized by viral and bacterial pathogens include increased transcription of DNA methyltransferases (DNMT1, DNMT3a, and DNMT3b) and subsequent hypermethylation of tumor suppressor promoter regions, thereby decreasing the activity of tumor suppressor genes. The altered control of cell cycle regulation increases the risk of unchecked growth and is a key mechanistic factor connecting infection to cancer. Cancers currently identified to have a viral- or bacterial-initiated development and progression include hematologic cancers such as leukemias and lymphomas (EBV, HTLV), gastric

TABLE 1.4
Epigenetic Impact of Biological Exposures

Species	Biological Exposure	Timing of Exposure	Exposure Dose	Epigenetic Impact	Nutrient Interaction	Reference
In Vitro: Cell Culture						
COS-7 and A549 cells	*Acinetobacter baumannii* (transposase protein)	Cells incubated for 24 h	Cells transfected with 1.6 μg bacterial DNA	Increased methylation of E-cadherin promoter and subsequent decreased expression	Not included in the study	(Moon et al. 2012)
293T cells	Adenovirus 5 (E1A protein)	Cells incubated for 48 h	Cells transfected with Ad5 E1A 12S	N-terminus of viral protein E1A associates with *DNMT1* and can stimulate *DNMT1* activity	Not included in the study	(Burgers et al. 2007)
Carcinoma cell line: CCL185	Epstein-Barr virus (EBV)	Selective pressure for EBV for 10 passages	Co-culture infection	Expression changes in >1000 genes resulting from DNA hypermethylation and decreased histone acetylation at specific loci	Not included in the study	(Queen et al. 2013)
Lymphoblastoid cell lines: LCLs	Epstein-Barr virus (EBV)	Cells incubated for 4 days	Transfection with virus to titres of 50–250 raji green units/μL cell culture supernatant	Activation of viral protein EBNA3C represses $p16^{INK4A}$ transcription, by interaction with CtBP (transcription corepressor) and increased H3K27me3 repressive histone modification	Not included in the study	(Skalska et al. 2010)
MDCK, MDA-MB-468, MCF-7 cells	Epstein-Barr virus (EBV)	Cells incubated for 48 h	Transfection with 0.5–1 μg EBV reporter	Latent membrane protein (LMP1) of EBV increases *DNMT1, 3a, 3b* transcription, resulting in hypermethylation of E-cadherin promoter and decreased expression	Not included in the study	(Tsai et al. 2002)

Human gastric cancer cell lines: TMK-1, MKN-74, MKN-7	*Helicobacter pylori*	Cells incubated for 48 to 72 h	*H. pylori*/cell ratios of 0, 1/1, 5/1, 10/1, 50/1, 100/1 added to cell culture wells	E-cadherin promoter methylated, mediated by IL-1β	Not included in the study	(Qian et al. 2008)
Human hepatoma cell line: HepG2	Hepatitis B virus (HBV)	Cells incubated for 48 h	2 μg HBV expressing plasmid transfected into cells	Hypermethylation of $p16^{INK4A}$ promoter with decreased expression, triggers transcriptional activation of Dnmt1via pRB-E2F1 pathway; E-cadherin promoter hypermethylation and expression decreased	Not included in the study	(Jung et al. 2007)
Human hepatoma cell line: HepG2	Hepatitis B virus (HBV)	Cells incubated for 48 h	1 μg HBV expressing plasmid transfected into cells	Hypermethylation of *RARB* promoter by upregulating Dnmt1 and Dnmt3a	Not included in the study	(Jung et al. 2010)
Human hepatoma cell line: HepG2	Hepatitis B virus (HBV)	Cells incubated for 48 h	2 μg HBV expressing plasmid transfected into cells	Promoter hypermethylation and decreased transcription of E-cadherin; *DNMT1* transcription induced	Not included in the study	(Lee et al. 2005)
Human hepatoma cell line: HepG2	Hepatitis C virus (HCV)	Cells incubated for 48 h	2 μg HCV expressing plasmid transfected into cells	E-cadherin transcription decreased via hypermethylation of promoter by Dnmt1 and Dnmt3b	Not included in the study	(Arora et al. 2008)
Colon cancer cell line: HCT-116	Human Cytomegalovirus (HCMV)	Cells incubated for 24 h	At 50% confluency, cells were infected in triplicate at a MOI 1 and 10	*DNMT1* and *DNMT3b* knockouts susceptible to HCMV infection, wild type NOT susceptible	Not included in the study	(Esteki-Zadeh et al. 2012)
Human embryonic lung fibroblast cell line: MRC-5	Human Cytomegalovirus (HCMV)	Cells incubated for 3 days	At 80% confluency, cells were infected in triplicate at a MOI 5 or left uninfected	Global hypomethylation associated with relocation of Dnmt1 and Dnmt3b from nucleus to cytoplasm	Not included in the study	(Esteki-Zadeh et al. 2012)

(*continued*)

TABLE 1.4 (Continued)
Epigenetic Impact of Biological Exposures

Species	Biological Exposure	Timing of Exposure	Exposure Dose	Epigenetic Impact	Nutrient Interaction	Reference
			In Vitro: Cell Culture			
T-cell lymphoma cell line: Hut 78	Human Immunodeficiency virus type 1 (HIV-1)	Cells incubated for 3–5 days	10 to 30 ng of p24 antigen from HIV-1 added to cell culture	Increased expression of *DNMT1*, no change in Dnmt3a or 3b levels; increase in genome-wide methylation, hypermethylation, and decreased $p16^{INK4A}$ expression	Not included in the study	(Fang et al. 2001)
JMO and CS-3 CD4+ cell lines	Human Immunodeficiency virus type 1 (HIV-1)	Cells incubated for 72 h	Cells infected with 100 μL isolate from HIV-1	De novo methylation of *IFNG* promoter, subsequent downregulation	Not included in the study	(Mikovits et al. 1998)
293T cells	Human Papillomavirus (HPV) (E7 protein)	Cells incubated for 48 h	Cells transfected with HPV-16 E7	Zinc-finger domain of viral protein E7 associates with Dnmt1 and can stimulate Dnmt1 activity	Not included in the study	(Burgers et al. 2007)
Human colon carcinoma cell line: HCT116	Human Papillomavirus (HPV) (E6 protein)	Cells incubated for 48 h	Cells transfected with 1μg of plasmid DNA	E-cadherin expression decreased, *DNMT* activity increased but promoter not directly methylated, but general hypermethylation and reduced binging of transactivators Sp1 and AML1	Not included in the study	(D'Costa et al. 2012)
Human keratinocyte progenitor cell line: NIKS	Human Papillomavirus 16 (HPV16) (E7 protein)	Cells incubated for 48 h	Cells transfected with 0.8 μg HPV16 DNA	Reduction in E-cadherin levels and increase in Dnmt1 activity; no change in retinoblastoma protein or AP-2a degradation	Not included in the study	(Laurson et al. 2010)

Human dermal microvascular endothelial cells: TIME	Kaposi's sarcoma-associated herpesvirus (KSHV)	Cells incubated for 5 days	Cells transfected with 2.5 μg LANA	LANA stimulated relocalization of Dnmt3a from nuclear matrix to chromatin, promoted de novo methylation of H-cadherin	Not included in the study	(Shamay et al. 2006)
Balb/c 3T3 cells	Simian virus 40 (SV40)	Cells incubated for 48 h	Cells transfected with 1 μg SV40 DNA	SV40 T antigen induces increase in *DNMT* transcription and an increase in genome-wide methylation	Not included in the study	(Slack 1999)
			Animal Models			
BALB/C mice	*Campylobacter rectus*	Injected E7.5; sacrificed E16.5	Intramuscular injection of 100 μL of 10^9 CFU/mL live *C. rectus*	Downregulation of Igf2 via hypermethylation of the promoter region, in placental tissue of infected mice	Not included in the study	(Bobetsis et al. 2007)
Male Mongolian gerbils: gastric epithelial cells	*Helicobacter pylori*	Inoculated at 5 weeks, measured at 25 and 50 weeks of age	4×10^8 colony-forming units per gerbil	DNA hypermethylation at 10 CpG islands – methylation decreased when *HP* was eradicated	Not included in the study	(Niwa et al. 2010)
Male Mongolian gerbils: gastric epithelial cells	*Helicobacter pylori*	Inoculated at 5 weeks, measured at 25 and 50 weeks of age	4×10^8 colony-forming units per gerbil	DNA hypermethylation at *HE6*, *HG2*, *SB1*, *SB5*, *SF12*, *SH6*, treatment with 5-aza-dC prevented gastric cancer incidence	Not included in the study	(Niwa et al. 2013)

(continued)

TABLE 1.4 (Continued)
Epigenetic Impact of Biological Exposures

Species	Biological Exposure	Timing of Exposure	Exposure Dose	Epigenetic Impact	Nutrient Interaction	Reference
			Animal Models			
BALB/C mice	*Campylobacter rectus*	Injected E7.5; sacrificed E16.5	Intramuscular injection of 100 μL of 10^9 CFU/mL live *C. rectus*	Microarray analysis showed 74 genes differentially expressed during maternal infection, only 9/74 were upregulated, downregulation of growth and development genes	Not included in the study	(Bobetsis et al. 2010)
			Human Studies			
Chinese and South African HBV carriers with and without liver tumors	Hepatitis B virus (HBV)	Adults	Naturally occurring levels were measured in tumor and nontumor tissue samples	Increase in E-cadherin promoter methylation in HBV-XAg-positive patients in both tumor and nontumor tissue, no change in E-cadherin methylation in HBV negative patients	Not included in the study	(Liu et al. 2006)
Liver tumor biopsy tissue: HCC and adjacent nontumor tissue	Hepatitis B virus (HBV)	Adults	Naturally occurring levels were measured	HBx incorporates into HDAC1/DNMT3a protein complex in vivo via pull down and immunoflourescence	Not included in the study	(Zheng et al. 2009)
Liver tumor biopsy tissue	Hepatitis B virus (HBV) and hepatitis C virus (HCV)	Adults	Naturally occurring levels were measured	50% decrease in TIRAP expression in liver tissue of chronic HBV- and HCV-infected people	Not included in the study	(Vivekanandan et al. 2010)
Peripheral blood mononuclear cells (PBMCs): healthy blood donors	Human Cytomegalovirus (HCMV)	Adults	Differentiated macrophages were infected at MOI of 10 or left uninfected for 96 h	Cytoplasmic accumulation of DNMT1 apparent	Not included in the study	(Esteki-Zadeh et al. 2012)

Healthy volunteers, HTLV carriers, ATLL patients: PBMCs or lymph node tissues	Human T-lymphotropic virus type I (HTLV-1)	Adults	Naturally occurring levels measured	Promoter methylation increased with disease progression at *SHP1*, *p15*, *p16*, *p73*, *HCAD*, *DAPK*, *hMLH-1*, *MGMT* with the greatest methylation in ATLL patients compared with healthy volunteers and HTLV carriers	Not included in the study	(Sato et al. 2010)
Leukemia and lymphoma patients: tumors and tissues	Simian virus 40 (SV-40)	Adults	Naturally occurring levels measured in oncology patients; healthy control tissues were infected at MOI of 50–100 plaque-forming units	Increased methylation in SV40+ versus SV40– at seven loci (*CDH1*, *CDH13*, *p16*, *DcR1*, *DcR2*, *CRBP*, *DAP kinase*); no differentially methylated regions in EBV+ versus EBV– cases	Not included in the study	(Shivapurkar et al. 2004)
Cord blood from preterm births in the NEST cohort	Chorioamnionitis and funisitis	Prenatal period	Presence versus absence of infection	DNA methylation levels increased at maternally expressed imprinted gene: *PLAGL1*	Not included in the study	(Liu et al. 2013)
Cord blood from preterm births in the NEST cohort	Maternal periconceptional antibiotic use	3 months before to 3 months after conception	Yes/no taking antibiotics, provided name	Differential DNA methylation at *IGF2*, *H19*, *PLAGL1*, *MEG3*, *PEG3* with antibiotic exposure	Not included in the study	(Vidal et al. 2013)

Note: MOI, multiplicity of infection; 5-aza-dC, 5-aza-2′-deoxycytidine.

cancer (*H. pylori*), hepatocellular carcinomas (HBV and HCV), cervical, and head and neck cancer (HPV).

In addition to alteration of DNMT activity and gene promoter methylation, some viral and bacterial pathogens recruit and bind histone-modifying proteins. The H3K27me3 methyltransferase, enhancer of zeste homolog (EZH2), is induced by HPV's E7 proto-oncogene protein (Holland et al. 2008) and by EBV (Paschos et al. 2009). The KSHV protein, LANA, also initiates transcriptional repression via activation of H3K9 methyltransferase, SUV39H1 (Sakakibara et al. 2004). *Anaplasma phagocytophilum* is associated with increased expression of HDAC1 followed by an increase in H3 methylation (Garcia-Garcia et al. 2009). Other bacterial pathogens express effector proteins, which interact with and modify host histones and chromatin structure, including *Shigella flexneri*, *Listeria monocytogenes*, *Chlamydia trachomatis*, *Toxoplasmosis gondii*, and *Mycobacterium* spp. (Rennoll-Bankert and Dumler 2012).

1.5.2 Animal Models: In Vivo Mechanisms

Murine experiments in which pregnant dams were orally infected with *Campylobacter rectus* showed that translocation of the bacteria from the oral cavity to the placenta, followed by a decrease in placental *Igf2* expression accompanied by increased methylation at the *Igf2* promoter region (Bobetsis et al. 2007, 2010). In humans, periodontal *C. rectus* infection is associated with increased risk of premature delivery and low-birth-weight infants (Buduneli et al. 2005). Experiments in juvenile male Mongolian gerbils with exposure to *H. pylori* beginning at 5 weeks of age found DNA hypermethylation in the promoter region of 10 gene loci, which abated with the eradication of *H. pylori* (Niwa et al. 2010). Treatment of the gerbils with the DNA methylation inhibitor 5-aza-2′-deoxycytidine (5-aza-dC) prevented the incidence of gastric cancer, further supporting the role of hypermethylation of tumor suppressor gene promoter regions by *H. pylori* in the development of gastric cancer (Niwa et al. 2013).

1.5.3 Human Studies: Translation to Clinical Applications

Human tissue extracted via biopsy has been used to study in vivo associations between infections and epigenetic alterations in human clinical populations. Study findings have confirmed mechanisms identified in cell culture experiments; however, few studies have begun to explore the potential role for infection in the prenatal environment and epigenetic reprogramming effects on the developing fetus or child later in life. Two large, epidemiological cohort studies (Newborn Epigenetics Study [NEST] and Epigenetic Impact of Childbirth [EPIIC]) have begun to investigate the developmental impact of perinatal maternal infections.

In South African and Chinese adults, an increase in E-cadherin promoter methylation occurred in patients positive for HBV X-antigen in both tumor and nontumor surrounding tissue, whereas HBV negative adults had no change in E-cadherin methylation (Liu et al. 2006). In liver biopsy tissue from adults, HBV X-antigen was found to incorporate into the HDAC1/DNMT3a regulatory protein complex (Zheng et al. 2009), suggesting that X-antigen directly impacts tumor suppressor gene methylation status in vivo via recruitment of DNA methylation-promoting protein complexes. Expression of toll-interleukin 1 receptor domain containing adaptor (TIRAP) is 50% lower in hepatic tissue of individuals with chronic HBV or HCV infections (Vivekanandan et al. 2010). Acting as an adaptor for toll-like receptors in the innate immune system, TIRAP activation initiates cytokine secretion (e.g., NF-κB, JNK, MAPK) and the inflammatory response. A study comparing promoter methylation of tumor suppressor genes among healthy volunteers, carriers of HTLV, and adult T-cell leukemia/lymphoma (ATLL) patients found an increasing percent methylation with disease progression suggesting methylation status is critical to disease status (Sato et al. 2010).

The NEST at Duke University has investigated the impact of intrauterine infections, chorioamnionitis, and funisitis, on preterm birth and DNA methylation at imprinted gene loci. They found no difference in DNA methylation from infant cord blood between term and preterm infants but did find methylation levels increased at the *PLAGL1* loci for infants with chorioamnionitis and funisitis compared with infants without these infections (Liu et al. 2013). *PLAGL1* encodes a transcription factor involved in growth and development via *IGF2* signaling (Van Dyck et al. 2007). *PLAGL1* methylation status is correlated a growth restriction and hyperglycemia disorder, transient neonatal diabetes mellitus (TNDM), and is associated with tumor development and progression (Temple and Shield 2010). If the epigenetic changes observed at *PLAGL1* associated with in utero infection persist across the life course, this could lead to increased risk of metabolic dysfunction and cancer development. Antibiotic use in the periconceptional period, approximately 3 months before and after conception, was assessed by maternal self-report in the NEST cohort. Use of antibiotics during this epigenetically sensitive period was associated with a significant decrease in infant birth weight (138 g lower) compared with nonexposed infants; the association was stronger in women reporting use of nonpenicillin antibiotics. Differential methylation at five imprinted gene loci, *IGF2*, *H19*, *PLAGL1*, *MEG3*, and *PEG3*, was associated with antibiotic use, suggesting the potential for this antibiotic use to impact gene expression later in life (Vidal et al. 2013). This study did not include a nutritional intervention so an interaction among intrapartum infection, periconceptional antibiotic use, and nutrients cannot be assessed in these human cohort studies. A new international birth cohort, EPIIC is designed to follow mother–child dyads longitudinally to determine the impact of intrapartum environmental conditions on epigenetic programming in both the mother and infant. EPIIC is collecting information on infection during pregnancy and will therefore provide additional insight to the DOHaD perspective on reprogramming potential of fetal biologic exposures (Dahlen et al. 2013).

1.5.4 Challenges and Limitations

No studies currently exist that investigate the potential interaction of a nutritional intervention with epigenetic changes resulting from biological infection. However, the potential for chronic inflammation to initiate aberrant DNA methylation at polycomb group proteins (PcG) suggests that individuals with greater baseline levels of inflammatory markers may be at increased risk of infection pathogenicity (Paschos and Allday 2010). Therefore, it is possible that consuming an anti-inflammatory diet, including n-3 polyunsaturated fatty acids (PUFAs), polyphenols, catechins, sulforaphanes, phytoestrogens, and performing regular physical activity could lower incidence and severity of biological infections.

Few studies examine the ability of biological pathogen exposure during fetal development to impact epigenetic programming across the lifespan. However, it is well documented that the maternal inflammatory response system is downregulated during pregnancy to prevent rejection of the fetus by the placenta (Warning et al. 2011). In both the fetal and amniotic compartments, tumor necrosis factor α (TNF-α), interleukin 1β (IL-1β), and T-helper cell 1 (Th1) inflammatory immune modulators are downregulated, while anti-inflammatory and anti-microbial IL-23, IL-10, IL-6, and Th17 are upregulated (Witkin et al. 2011). Thus, the maternal–fetal environment is more sensitive to combating extracellular biologic pathogens than to inflammatory processes. Once the infant is born, the neonate's own immune system must take over the role of protection from biological exposures. Breastmilk contains secretory immunoglobulin A (IgA), lactoferrin, and IL-7, which are critical in early development of the gastrointestinal immune systemand are associated with decreased incidence of diarrheal and respiratory diseases, neonatal meningitis, septicemia, and infant mortality (Hanson and Korotkova 2002). Breastfeeding, compared with formula feeding, also promotes the growth of commensal microbiota. Breastfed infants had greater intestinal expression of 660 genes involved in immune defense and had greater heterogeneity of microbiota composition compared with formula-fed infants (Schwartz et al. 2012).

BIOLOGICAL EXPOSURES: SUMMARY

- Exposure to infectious biologic agents, including bacteria and viruses can induce epigenetic changes in the host epigenome.
- Bacteria and viruses have been shown to alter DNMT activity, resulting in altered DNA methylation; decreased expression of host tumor suppressor genes is a common mechanisms.
- Viral and bacterial pathogens also promote histone deacetylase and histone methyltransferase activity.
- No studies have investigated the epigenetic interaction of dietary intake and infectious exposures.

1.6 EXPERIMENTAL DESIGN CONSIDERATIONS AND FUTURE DIRECTIONS

To fully succeed in identifying the role of epigenetic mechanisms in DOHaD, scientists must integrate animal model, human clinical, and human population approaches, paying close attention to windows of vulnerability, environmental and nutritional assessment, and cell-specific epigenetic patterns (Dolinoy and Faulk 2012). Animal models, in which exposures are well controlled and characterized, will continue to help inform the evaluation of dose–response effects on the epigenome as well as vulnerable time periods, including gestation and multigenerational or transgenerational effects (Skinner et al. 2011; Manikkam et al. 2012a; Nilsson et al. 2012). Further, animal models and clinical samples are often useful for proteomic and chromatin structure evaluation (Dolinoy et al. 2010), which may be limited in human population approaches due to sample storage and processing limitations. Finally, as the epigenome in contrast to the genome is particularly subjected to tissue- and cell-type specificity, animal model and human clinical studies, in which cell-type specificity is more readily evaluated can serve as important proof-of-principle approaches to evaluate the use of peripheral tissue (e.g., blood, saliva) in human epigenetic epidemiology studies.

The majority of current animal and human studies report associations between environmental exposures and epigenetic alterations, but the mechanism by which the exposures impact epigenetic modifications are not well described. To accurately assess mechanistic links in vivo direct dosing, exposure tracing, and longitudinal tissue collection are required. These invasive techniques are challenging in human studies due to ethical concerns for participant rights and health. In animal models, exposure dose–response experiments can be assessed and removal of a critical mechanistic step via genetic knockout species or pharmaceutical knockdown models is possible. These studies are expensive, time-intensive, and resource-intensive and are therefore conducted after in vitro mechanistic links have been identified. This transition into mechanistic animal model studies of environmental impacts on epigenetic modifications is a crucial next step in the future progress of this field. Additionally, the majority of environmental epigenetic studies have investigated associations with DNA methylation; developmental studies examining histone modifications, miRNA regulation, and the enzymes involved in epigenetic modifications are needed to provide a more complete picture of the potential impact of environmental exposures on epigenetic regulation of gene expression.

Utilizing a multipronged approach with a mouse model, human clinical samples, and ongoing longitudinal epidemiological studies and BPA and lead (Pb) as representative early environmental exposures, we have developed a comprehensive strategy for evaluating effects on the developing epigenome, with a focus on life course metabolic syndrome and neuropathology risks. This approach combines multidose studies in animal models with careful environmental characterization of human clinical samples and extended analysis of epidemiologically characterized human population samples. Integral to this strategy and important for identifying biomarkers for disease risk and progression, environmentally induced changes on the developmental epigenome are correlated with subsequent gene expression as well as phenotypic effects across the life course. First, epigenetic reprogramming plays an important role in development, although the repertoire of epigenetically labile targets has not yet been defined in any model species or in humans. Recently, we developed an approach to identify genes called metastable epialleles in mice whose epigenetic dysregulation early in development predisposes individuals to disease later in life (Weinhouse et al. 2011). This work builds upon our earlier epigenetics work with the mouse as a model of perinatal exposure (Dolinoy et al. 2006, 2007) leading to epigenetic alterations at candidate metastable epialleles that are reflected in variation of coat color, adult-onset obesity, and/or adult-onset diabetes. As a follow-up to the identification of epigenetic labile genic regions, we have also developed a novel approach in mice identifying epigenetically labile repetitive DNA content (Faulk et al. 2013). Parallel efforts are underway to characterize labile regions in the human genome, which is crucial for the development of novel screening and therapeutic targets for human disease prevention. Second, studies claiming epigenetics as a mechanism in DOHaD often fail to follow individuals with well characterized exposures beyond birth outcomes. Using physiologically relevant levels of dietary BPA exposure, in 2012, we first showed that perinatal BPA leads to dose dependent nonmonotonic effects on the fetal epigenome, with higher dietary exposure to BPA leading to hypomethylation at two candidate loci, but lower dietary levels being less effective at hypomethylating, and leading to the opposite—hypermethylating—effects (Anderson et al. 2012a). In a follow-up approach, we next demonstrated that among females, perinatal exposure to BPA is associated with hyperactive and lean phenotypes, with improved hormone and free fatty acid profiles (Anderson et al. 2013). Our observations of hyperactivity and lean body mass following developmental BPA exposure are in contrast to cross-sectional epidemiological studies associating BPA with higher body weight and increased obesity (Lang et al. 2008; Trasande et al. 2012), which may be confounded by food consumption practices and altered BPA metabolism associated with body composition. Ongoing studies are now closing the loop between perinatal exposure and adult offspring phenotypes by evaluating epigenome-wide methylation and chromatin profiles, to understand the mechanism linking early BPA exposure to later-in-life disease risk.

As identified in this chapter, surprisingly few studies have examined the potential for perinatal or postnatal dietary intake to modify epigenetic alterations induced by perinatal physical, chemical, or biological environmental exposures (Table 1.5). Dietary interventions following specific in utero nutritional conditions are currently the most commonly studied interaction, but even there, more work would be helpful in confirming current associations. There is a distinct need for more studies that (1) begin exposure prior to conception to ensure coverage of the initial epigenetic reprogramming events, (2) characterize exposures and health outcomes overtime with corresponding measures of potential epigenetic drift, (3) investigate tissue- and cell-specific differences in epigenetic responses to environmental exposures, (4) examine exposure mixtures that more closely approximate real-world exposures that people face, and (5) explore mechanistic links between environmental exposures and associated epigenetic alterations. In addition to these general experimental design considerations, the realization that many chemical exposures act as endocrine disrupting compounds (EDC) that mimic or antagonize endogenous hormone activity suggests that additional considerations should be included in these exposure studies. Investigation of sexually dimorphic epigenetic alterations and their downstream health effects following EDC exposure are critical since EDC activity often differ depending on the background environment of hormone and hormone receptor levels dictated by sex of the individual. Low-dose and nonmonotonic responses are also characteristic of EDC exposures and may therefore pose additional layers of complexity to the nutrient–toxicant interactions, especially since nutrients themselves often exert differential effects at low doses and may act as signaling molecules in the body (e.g., retinoic acid for RAR and RXR, lipid oxidation products for PPARs).

TABLE 1.5
Nutrition Modification of Environmental Exposures

Nutrient or Functional Food Group	Nutrient Level Causing Effect	Timing of Nutrient Intake	Epigenetic Impact	Gene Loci Impacted	Exposure That Nutrient Modifies	Reference
Antioxidants (Harlan Teklad TD.95092)	408 mg vitamin C, 71.4 mg vitamin E, 57 mg α-lipoic acid, 171.4 mg *N*-acetyl-l-cysteine, 14 mg *tert*-butylhydroquinone, 6 mg seleno-L-methionine/kg diet	1 week prior to mating (concurrent with LDIR exposure)	Prevented increase in DNA methylation seen after LDIR exposure	A^{vy} (in male offspring only)	LDIR	(Bernal et al. 2013)
Methyl donors (Harlan Teklad TD.06310)	4.3 mg folic acid, 0.53 mg vitamin B_{12}, 5 g betaine, 7.97 g choline chloride/kg diet	2 weeks prior to mating (concurrent with BPA exposure)	Prevented DNA hypomethylation after BPA exposure	A^{vy}	BPA	(Dolinoy et al. 2007)
Methyl donors (Harlan Teklad TD.01308)	High methyl donor	PND21 in F0 dams, continued to F3 offspring	Reversed amplification of body weight across generations	A^{vy}	Maternal obesity	(Waterland et al. 2008)
Methyl donor supplementation	12 g methionine, 15 g choline, 15 g betaine, 1000 mg B_{12}, 15 mg folic acid/kg diet	21–28 days prior to mating through pregnancy and lactation	Increased DNA methylation and expression	*Lep*	Protein-restricted diet	(Giudicelli et al. 2013)

(continued)

TABLE 1.5 (Continued)
Nutrition Modification of Environmental Exposures

Nutrient or Functional Food Group	Nutrient Level Causing Effect	Timing of Nutrient Intake	Epigenetic Impact	Gene Loci Impacted	Exposure That Nutrient Modifies	Reference
Folate	0.4- versus 2-mg folic acid/kg diet	Postweaning repletion of folate	Not able to reverse epigenetic programming resulting from prenatal folate deficiency	Global hypomethylation	Prenatal folate deficiency	(McKay et al. 2011)
Folate	Control: 180 g/kg + 1 mg/kg FA Ptn-restricted: 90 g/kg + 1 mg/kg FA Ptn-restricted and folate-supplemented: 90 g/kg + 5 mg/kg FA	After mating through pregnancy and lactation	Hypomethylation and increased mRNA transcript levels	*Ppara* and *Nr3c1*	Protein-restricted diet	(Lillycrop et al. 2005)
Genestein (Harlan Teklad TD.06039)	250-mg/kg diet	2 weeks prior to mating (concurrent with BPA exposure)	Prevented DNA hypomethylation after BPA exposure	A^{vy}	BPA	(Dolinoy et al. 2007)
Healthy diet (standard chow)	10.3% fat, 73.1% carbohydrate (6.5% sucrose)	36-week-old rats on each diet for 10 weeks	Reversed methylation caused by high fat high sucrose diet	*Lep*, *Srebf1*, *Ppargc1a*, *Fasn*	High fat high sucrose diet	(Uriarte et al. 2013)

The U.S. Environmental Protection Agency (EPA) is now undertaking multiple projects to more comprehensively test potentially harmful environmental exposures and to catalog the findings in a publicly available, searchable database. First, Tox21 is a joint initiative between the EPA and the National Institutes of Health (NIH) and Food and Drug Administration (FDA) using robotic technology to test the potential toxicity of over 10,000 chemicals with over 50 high-throughput screening assays scheduled per year over the next several years. The second EPA effort, Distributed Structure-Searchable Toxicity (DSSTox) Database Network, aims to create a centralized, publicly accessible source of data using computational toxicology to improve the foundation of knowledge in structure–activity and predictive toxicology. ToxCast, launched in 2007, is another computational toxicology program created to develop methods to predict the potential toxicity of chemicals and to prioritize thousands of chemicals that require toxicity testing in a cost-effective manner. The integration of these efforts will help to create a foundation upon which investigators may expand examinations of nutrient–toxicant and nutrient–mixture analyses in the future.

ACKNOWLEDGMENTS

This work was supported by NIH grant ES017524 and the University of Michigan NIEHS Core Center P30 ES017885 as well as NIH/EPA P20 grant ES018171/RD 83480001. Support for EHM was provided by NIEHS Institutional Training Grant T32 ES007062. The author's thank Dr. Chris Faulk for his assistance with Figure 1.1.

REFERENCES

Altmann, S., E. Murani et al. (2012). "Maternal dietary protein restriction and excess affects offspring gene expression and methylation of non-SMC subunits of condensin I in liver and skeletal muscle." *Epigenetics* **7**(3): 239–252.

Anderson, O. S., M. S. Nahar et al. (2012a). "Epigenetic responses following maternal dietary exposure to physiologically relevant levels of bisphenol A." *Environmental and Molecular Mutagenesis* **53**(5): 334–342.

Anderson, O. S., K. E. Peterson et al. (2013). "Perinatal bisphenol A exposure promotes hyperactivity, lean body composition, and hormonal responses across the murine life course." *FASEB Journal: Official Publication of the Federation of American Societies for Experimental Biology* **27**(4): 1784–1792.

Anderson, O. S., K. E. Sant et al. (2012b). "Nutrition and epigenetics: An interplay of dietary methyl donors, one-carbon metabolism and DNA methylation." *Journal of Nutritional Biochemistry* **23**(8): 853–859.

Arora, P., E.-O. Kim et al. (2008). "Hepatitis C virus core protein downregulates E-cadherin expression via activation of DNA methyltransferase 1 and 3b." *Cancer Letters* **261**: 244–252.

Attig, L., A. Vige et al. (2013). "Dietary alleviation of maternal obesity and diabetes: Increased resistance to diet-induced obesity transcriptional and epigenetic signatures." *PLoS One* **8**(6): e66816.

Ba, Y., H. Yu et al. (2011). "Relationship of folate, vitamin B12 and methylation of insulin-like growth factor-II in maternal and cord blood." *European Journal of Clinical Nutrition* **65**(4): 480–485.

Barker, D. J. (1997). "Intrauterine programming of coronary heart disease and stroke." *Acta Pædiatrica (Oslo, Norway: 1992). Supplement* **423**: 178–182; discussion 183.

Barker, D. J. (2004). "The developmental origins of adult disease." *Journal of the American College of Nutrition* **23**(S6): 588–595.

Bassett, S. S., D. Avramopoulos et al. (2006). "Further evidence of a maternal parent-of-origin effect on chromosome 10 in late-onset Alzheimer's disease." *American Journal of Medical Genetics Part B: Neuropsychiatric Genetics* **141B**(5): 537–540.

Bateson, P., D. Barker et al. (2004). "Developmental plasticity and human health." *Nature* **430**(6998): 419–421.

Bernal, A. J., D. C. Dolinoy et al. (2013). "Adaptive radiation-induced epigenetic alterations mitigated by antioxidants." *FASEB Journal: Official Publication of the Federation of American Societies for Experimental Biology* **27**(2): 665–671.

Bernal, A. J. and R. L. Jirtle (2010). "Epigenomic disruption: The effects of early developmental exposures." *Birth Defects Research Part A: Clinical and Molecular Teratology* **88**(10): 938–944.

Bobetsis, Y. A., S. P. Barros et al. (2007). "Bacterial infection promotes DNA hypermethylation." *Journal of Dental Research* **86**: 169–174.

Bobetsis, Y. A., S. P. Barros et al. (2010). "Altered gene expression in murine placentas in an infection-induced intrauterine growth restriction model: A microarray analysis." *Journal of Reproductive Immunology* **85**: 140–148.

Boeke, C., A. Baccarelli et al. (2012). "Gestational intake of methyl donors and global LINE-1 DNA methylation in maternal and cord blood." *Epigenetics* **7**(3): 253–260.

Braun, J. M. (2013). "Phthalate exposure and children's health." *Current Opinion in Pediatrics* **25**(2): 247–254.

Braw-Tal, R. (2010). "Endocrine disruptors and timing of human exposure." *Pediatric Endocrinology Reviews: PER* **8**(1): 41–46.

Brown, M. J. (2012). "Lead in drinking water and human blood lead levels in the United States." *MMWR. Surveillance Summaries* **61 Suppl**: 1–9.

Buduneli, N., H. Baylas et al. (2005). "Periodontal infections and pre-term low birth weight: A case-control study." *Journal of Clinical Periodontology* **32**: 174–181.

Burgers, W. A., L. Blanchon et al. (2007). "Viral oncoproteins target the DNA methyltransferases." *Oncogene* **26**: 1650–1655.

Calafat, A., X. Ye et al. (2008). "Exposure of the U.S. population to bisphenol A and 4-tertiary-octylphenol: 2003–2004." *Environmental Health Perspectives* **116**(1): 39–44.

Canani, R. B., M. D. Costanzo et al. (2011). "Epigenetic mechanisms elicited by nutrition in early life." *Nutrition Research Reviews* **24**(2): 198–205.

Caspi, A., K. Sugden et al. (2003). "Influence of life stress on depression: Moderation by a polymorphism in the 5-HTT gene." *Science* **301**(5631): 386–389.

Castillo, P., F. Ibáñez et al. (2012). "Impact of cadmium exposure during pregnancy on hepatic glucocorticoid receptor methylation and expression in rat fetus." *PLoS One* **7**(9): e44139.

Chan Kang, S. and B. Mu Lee (2005). "DNA methylation of estrogen receptor α gene by phthalates." *Journal of Toxicology and Environmental Health, Part A* **68**(23–24): 1995–2003.

Cheng, T.-F., S. Choudhuri et al. (2012). "Epigenetic targets of some toxicologically relevant metals: A review of the literature." *Journal of Applied Toxicology* **32**(9): 643–653.

Chervona, Y., A. Arita et al. (2012). "Carcinogenic metals and the epigenome: Understanding the effect of nickel, arsenic, and chromium." *Metallomics* **4**(7): 619–627.

Chervona, Y. and M. Costa (2012). "The control of histone methylation and gene expression by oxidative stress, hypoxia, and metals." *Free Radical Biology and Medicine* **53**(5): 1041–1047.

Choi, S. W. and S. Friso (2010). "Epigenetics: A new bridge between nutrition and health." *Advances in Nutrition* **1**(1): 8–16.

Collotta, M., P. A. Bertazzi et al. (2013). "Epigenetics and pesticides." *Toxicology* **307**: 35–41.

Covaci, A., P. Jorens et al. (2002). "Distribution of PCBs and organochlorine pesticides in umbilical cord and maternal serum." *Science of the Total Environment* **298**(1–3): 45–53.

Cunningham, C. C. and S. M. Bailey (2001). "Ethanol consuption and liver mitochondria function." *Biological Signals and Receptors* **10**(3–4): 271–282.

D'Costa, Z. J., C. Jolly et al. (2012). "Transcriptional repression of E-cadherin by human papillomavirus type 16 E6." *PLoS One* **7**: e48954.

Dahlen, H. G., H. P. Kennedy et al. (2013). "The EPIIC hypothesis: Intrapartum effects on the neonatal epigenome and consequent health outcomes." *Medical Hypotheses* **80**: 656–662.

De Coster, S. and N. van Larebeke (2012). "Endocrine-disrupting chemicals: Associated disorders and mechanisms of action." *Journal of Environmental and Public Health* **2012**: 52.

Devarshi, P., N. Jangale et al. (2013). "Beneficial effects of flaxseed oil and fish oil diet are through modulation of different hepatic genes involved in lipid metabolism in streptozotocin-nicotinamde induced diabetic rats." *Genes & Nutrition* **8**(3): 329–342.

Diamanti-Kandarakis, E., J.-P. Bourguignon et al. (2009). "Endocrine-disrupting chemicals: An endocrine society scientific statement." *Endocrine Reviews* **30**(4): 293–342.

Doherty, L., J. Bromer et al. (2010). "In utero exposure to diethylstilbestrol (DES) or bisphenol-A (BPA) increases EZH2 expression in the mammary gland: an epigenetic mechanism linking endocrine disruptors to breast cancer." *Hormones and Cancer* **1**(3): 146–155.

Doi, T., P. Puri et al. (2011). "Epigenetic effect of cadmium on global de novo DNA hypomethylation in the cadmium-induced ventral body wall defect (VBWD) in the chick model." *Toxicological Sciences* **120**(2): 475–480.

Dolinoy, D. C. (2008). "The agouti mouse model: An epigenetic biosensor for nutritional and environmental alterations on the fetal epigenome." *Nutrition Reviews* **66**: S7–S11.

Dolinoy, D. C. and C. Faulk (2012). "The use of animal models to advance epigenetic science." *The ILAR Journal* **53**(3–4): 227–231.

Dolinoy, D. C., D. Huang et al. (2007). "Maternal nutrient supplementation counteracts bisphenol A-induced DNA hypomethylation in early development." *Proceedings of the National Academy of Sciences of the United States of America* **104**(32): 13056–13061.

Dolinoy, D. C., C. Weinhouse et al. (2010). "Variable histone modifications at the A^{vy} metastable epiallele." *Epigenetics* **5**(7): 637–644.

Dolinoy, D. C., J. Wiedman et al. (2006). "Maternal genistein alters coat color and protects Avy mouse offspring from obesity by modifying the fetal epigenome." *Environmental Health Perspectives* **114**(4): 567–572.

Dudley, K., D. Sloboda et al. (2011). "Offspring of mothers fed a high fat diet display hepatic cell cycle inhibition and associated changes in gene expression and DNA methylation." *PLoS One* **6**(7): e21662.

Esteki-Zadeh, A., M. Karimi et al. (2012). "Human cytomegalovirus infection is sensitive to the host cell DNA methylation state and alters global DNA methylation capacity." *Epigenetics* **7**: 585–593.

Evans, G. W. and E. Kantrowitz (2002). "Socioeconomic status and health: The potential role of environmental risk exposure." *Annual Review of Public Health* **23**(1): 303–331.

Fang, J.-Y., J. A. Mikovits et al. (2001). "Infection of lymphoid cells by integration-defective human immunodeficiency virus type 1 increases de novo methylation." *Journal of Virology* **75**: 9753–9761.

Faulk, C., A. Barks et al. (2013). "Phylogenetic and DNA methylation analysis reveal novel regions of variable methylation in the mouse IAP class of transposons." *BMC Genomics* **14**: 48.

Fleisch, A. F., R. O. Wright et al. (2012). "Environmental epigenetics: A role in endocrine disease?" *Journal of Molecular Endocrinology* **49**(2): R61–R67.

Franklin, T. B., H. Russig et al. (2010). "Epigenetic transmission of the impact of early stress across generations." *Biological Psychiatry* **68**(5): 408–415.

Garcia-Garcia, J. C., N. C. Barat et al. (2009). "Epigenetic silencing of host cell defense genes enhances intracellular survival of the rickettsial pathogen Anaplasma phagocytophilum." *PLoS Pathogens* **5**: e1000488.

Giudicelli, F., A. L. Brabant et al. (2013). "Excess of methyl donor in the perinatal period reduces postnatal leptin secretion in rat and interacts with the effect of protein content in diet." *PLoS One* **8**(7): e68268.

Guerrero-Bosagna, C., M. Settles et al. (2010). "Epigenetic transgenerational actions of vinclozolin on promoter regions of the sperm epigenome." *PLoS One* **5**(9): e13100.

Guerrero-Bosagna, C. and M. K. Skinner (2012). "Environmentally induced epigenetic transgenerational inheritance of phenotype and disease." *Molecular and Cellular Endocrinology* **354**(1–2): 3–8.

Guo, Y., G. Lambert et al. (2004). "Yucheng: Health effects of prenatal exposure to polychlorinated biphenyls and dibenzofurans." *International Archives of Occupational and Environmental Health* **77**(3): 153–158.

Haggarty, P., G. Hoad et al. (2013). "Folate in pregnancy and imprinted gene and repeat element methylation in the offspring." *The American Journal of Clinical Nutrition* **97**: 94–99.

Hanson, L. Å. and M. Korotkova (2002). "The role of breastfeeding in prevention of neonatal infection." *Seminars in Neonatology* **7**: 275–281.

Hao, C. (2013). "Perinatal exposure to diethyl-hexyl-phthalate induces obesity in mice." *Frontiers in Bioscience (Elite Edition)* **5**: 725–733.

Hauser, R. and A. M. Calafat (2005). "Phthalates and human health." *Occupational and Environmental Medicine* **62**(11): 806–818.

Holland, D., K. Hoppe-Seyler et al. (2008). "Activation of the enhancer of zeste homologue 2 gene by the human papillomavirus E7 oncoprotein." *Cancer Research* **68**: 9964–9972.

Hoyo, C., A. P. Murtha et al. (2011). "Methylation variation at IGF2 differentially methylated regions and maternal folic acid use before and during pregnancy." *Epigenetics* **6**(7): 928–936.

Hsieh, T.-H., C.-F. Tsai et al. (2012a). "Phthalates stimulate the epithelial to mesenchymal transition through an HDAC6-dependent mechanism in human breast epithelial stem cells." *Toxicological Sciences* **128**(2): 365–376.

Hsieh, T.-H., C.-F. Tsai et al. (2012b). "Phthalates induce proliferation and invasiveness of estrogen receptor-negative breast cancer through the AhR/HDAC6/c-Myc signaling pathway." *FASEB Journal: Official Publication of the Federation of American Societies for Experimental Biology* **26**(2): 778–787.

Imai, S., C. Armstrong et al. (2000). "Transcriptional silencing and longevity protein Sir2 is an NAD-dependent histone deacetylase." *Nature* **403**(6771): 795–800.

Ji, W., L. Yang et al. (2013). "MicroRNA-152 targets DNA methyltransferase 1 in NiS-transformed cells via a feedback mechanism." *Carcinogenesis* **34**(2): 446–453.

Jiménez-Díaz, I., A. Zafra-Gómez et al. (2010). "Determination of bisphenol A and its chlorinated derivatives in placental tissue samples by liquid chromatography–tandem mass spectrometry." *Journal of Chromatography B* **878**(32): 3363–3369.

Jirtle, R. L. and M. K. Skinner (2007). "Environmental epigenomics and disease susceptibility." *Nature Reviews. Genetics* **8**(4): 253–262.

Jung, J. K., P. Arora et al. (2007). "Expression of DNA methyltransferase 1 is activated by hepatitis B virus X protein via a regulatory circuit involving the p16INK4a-cyclin D1-CDK 4/6-pRb-E2F1 pathway." *Cancer Research* **67**: 5771–5778.

Jung, J. K., S.-H. Park et al. (2010). "Hepatitis B virus X protein overcomes the growth-inhibitory potential of retinoic acid by downregulating retinoic acid receptor-beta2 expression via DNA methylation." *The Journal of General Virology* **91**: 493–500.

Kaminen-Ahola, N., Ahola, A. et al. (2010a). "Postnatal growth restriction and gene expression changes in a mouse model of fetal alcohol syndrome." *Birth Defects Research. Part A, Clinical and Molecular Teratology* **88**(10): 818–826.

Kaminen-Ahola, N., Ahola, A. et al. (2010b). "Maternal ethanol consumption alters the epigenotype and the phenotype of offspring in a mouse model." *PLoS Genetics* **15**(6): e100811.

Keller, S., M. Sarchiapone et al. (2010). "Increased BDNF promoter methylation in the Wernicke area of suicide subjects." *Archives of General Psychiatry* **67**(3): 258–267.

Keys, A., C. Aravanis et al. (1980). *Seven Countries. A Multivariate Analysis of Death and Coronary Heart Disease*. Harvard University Press, Cambridge, MA.

King, E., G. Shih et al. (2013). "Mercury, lead, and cadmium in umbilical cord blood." *Journal of Environmental Health* **75**(6): 38–43.

Koletzko, B., B. Brands et al. (2012). "Early nutrition programming of long-term health." *Proceedings of the Nutrition Society* **71**(3): 371–378.

Koturbash, I., K. Kutanzi et al. (2008). "Radiation-induced bystander effects in vivo are sex specific." *Mutation Research* **642**(1–2): 28–36.

Koyashiki, G. A. K. (2010). "Lead levels in human milk and children's health risk: A systematic review." *Reviews on Environmental Health* **25**(3): 243–253.

Krueger, C., E. Horesh et al. (2012). "Safe sound exposure in the fetus and preterm infant." *Journal of Obstetric, Gynecologic, & Neonatal Nursing* **41**(2): 166–170.

Kundakovic, M. and F. A. Champagne (2011). "Epigenetic perspective on the developmental effects of bisphenol A." *Brain, Behavior, and Immunity* **25**(6): 1084–1093.

Kundakovic, M., K. Gudsnuk et al. (2013). "Sex-specific epigenetic disruption and behavioral changes following low-dose in utero bisphenol A exposure." *Proceedings of the National Academy of Sciences of the United States of America* **110**(24): 9956–9961.

La Merrill, M. (2013). "Toxicological function of adipose tissue: Focus on persistent organic pollutants." *Environmental Health Perspectives* **121**(2): 162–169.

Lang, I. A., T. S. Galloway et al. (2008). "Association of urinary bisphenol A concentration with medical disorders and laboratory abnormalities in adults." *JAMA: The Journal of the American Medical Association* **300**(11): 1303–1310.

Laurson, J., S. Khan et al. (2010). "Epigenetic repression of E-cadherin by human papillomavirus 16 E7 protein." *Carcinogenesis* **31**: 918–926.

Lee, H.-S., A. Varraza-Vaillarreal et al. (2013). "Modulation of DNA methylation states and infant immune system by dietary supplementation with n-3 PUFA during pregnancy in an intervention study." *The American Journal of Clinical Nutrition* **98**: 480–487.

Lee, J.-O., H. J. Kwun et al. (2005). "Hepatitis B virus X protein represses E-cadherin expression via activation of DNA methyltransferase 1." *Oncogene* **24**: 6617–6625.

Li, C. C. Y., P. E. Young et al. (2013). "Maternal obesity and diabetes induces latent metabolic defects and widespread epigenetic changes in isogenic mice." *Epigenetics* **8**(6): 602–611.

Li, E. (2002). "Chromatin modification and epigenetic reprogramming in mammalian development." *Nature Reviews. Genetics* **3**(9): 662–673.

Lillycrop, K., E. Phillips et al. (2005). "Dietary proteinn restriction of pregnant rats induces and folic acid supplementation prevents epigenetic modification of hepatic gene expression in the offspring." *The Journal of Nutrition* **135**: 1382–1396.

Lillycrop, K. A. (2011). "Effect of maternal diet on the epigenome: Implications for human metabolic disease." *Proceedings of the Nutrition Society* **70**(1): 64–72.

Lillycrop, K. A. and G. C. Burdge (2011). "The effect of nutrition during early life on the epigenetic regulation of transcription and implications for human diseases." *Journal of Nutrigenetics and Nutrigenomics* **4**(5): 248–260.

Lin, S., H.-Y. Ku et al. (2011). "Phthalate exposure in pregnant women and their children in central Taiwan." *Chemosphere* **82**(7): 947–955.

Liu, J., Z. Lian et al. (2006). "Downregulation of E-cadherin by hepatitis B virus X antigen in hepatocellullar carcinoma." *Oncogene* **25**(7): 1008–1017.

Liu, Y., C. Hoyo et al. (2013). "DNA methylation at imprint regulatory regions in preterm birth and infection." *American Journal of Obstetrics and Gynecology* **208**: 395.e391–397.

Lopez-Rangel, E. and M. E. S. Lewis (2006). "Loud and clear evidence for gene silencing by epigenetic mechanisms in autism spectrum and related neurodevelopmental disorders." *Clinical Genetics* **69**(1): 21–22.

Maeno, K. and S. Tanaka (2010). "Epigenetic transmission of phase in the desert locust, Schistocerca gregaria: Determining the stage sensitive to crowding for the maternal determination of progeny characteristics." *Journal of Insect Physiology* **56**(12): 1883–1888.

Manikkam, M., C. Guerrero-Bosagna et al. (2012a). "Transgenerational actions of environmental compounds on reproductive disease and identification of epigenetic biomarkers of ancestral exposures." *PLoS One* **7**(2): e31901.

Manikkam, M., R. Tracey et al. (2012b). "Dioxin (TCDD) induces epigenetic transgenerational inheritance of adult onset disease and sperm epimutations." *PLoS One* **7**(9): e46249.

Manikkam, M., R. Tracey et al. (2012c). "Pesticide and insect repellent mixture (permethrin and DEET) induces epigenetic transgenerational inheritance of disease and sperm epimutations." *Reproductive Toxicology* **34**(4): 708–719.

Manikkam, M., R. Tracey et al. (2013). "Plastics derived endocrine disruptors (BPA, DEHP and DBP) induce epigenetic transgenerational inheritance of obesity, reproductive disease and sperm epimutations." *PLoS One* **8**(1): e55387.

Martinez-Arguelles, D. B., E. Campioli et al. (2013). "Fetal origin of endocrine dysfunction in the adult: The phthalate model." *The Journal of Steroid Biochemistry and Molecular Biology* **137**: 5–17.

Martinez-Arguelles, D. B., M. Culty et al. (2009). "In utero exposure to di-(2-ethylhexyl) phthalate decreases mineralocorticoid receptor expression in the adult testis." *Endocrinology* **150**(12): 5575–5585.

Martínez-Chantar, M. L., F. J. Corrales et al. (2002). "Spontaneous oxidative stress and liver tumors in mice lacking methionine adenosyltransferase 1A." *FASEB Journal: Official Publication of the Federation of American Societies for Experimental Biology* **16**(10): 1292–1294.

Masuyama, H. and Y. Hiramatsu (2012). "Effects of a high-fat diet exposure *in utero* on the metabolic syndrome-like phenomenon in mouse offspring through epigenetic changes in adipocytokine gene expression." *Endocrinology* **153**: 2823–2830.

McGowan, P. O., A. Sasaki et al. (2009). "Epigenetic regulation of the glucocorticoid receptor in human brain associates with childhood abuse." *Nature Neuroscience* **12**(3): 342–348.

McGowan, P. O., M. Suderman et al. (2011). "Broad epigenetic signature of maternal care in the brain of adult rats." *PLoS One* **6**(2): e14739.

McKay, J., K. Waltham et al. (2011). "Folate depletion during pregnancy and lactation reduces genomic DNA methylation in murine adult offspring." *Genes & Nutrition* **6**: 189–196.

Meeker, J. D. (2012). "Exposure to environmental endocrine disruptors and child development." *Archives of Pediatrics & Adolescent Medicine* **166**(10): 952–958.

Mendonca, K., R. Hauser et al. (2014). "Bisphenol A concentrations in maternal breast milk and infant urine." *International Archives of Occupational and Environmental Health* **87**(1): 13–20.

Mikovits, J. A., H. A. Young et al. (1998). "Infection with human immunodeficiency virus type 1 upregulates DNA methyltransferase, resulting in de novo methylation of the gamma interferon (IFN-gamma) promoter and subsequent downregulation of IFN-gamma production." *Molecular and Cellular Biology* **18**: 5166–5177.

Moon, D. C., C. H. Choi et al. (2012). "Nuclear translocation of Acinetobacter baumannii transposase induces DNA methylation of CpG regions in the promoters of E-cadherin gene." *PLoS One* **7**: e38974.

Mostafalou, S. and M. Abdollahi (2013). "Pesticides and human chronic diseases: Evidences, mechanisms, and perspectives." *Toxicology and Applied Pharmacology* **268**(2): 157–177.

Muczynski, V., C. Lecureuil et al. (2012). "Cellular and molecular effect of MEHP involving *LXRα* in human fetal testis and ovary." *PLoS One* **7**(10): e48266.

Mukhopadhyay, P., F. Rezzoug et al. (2013). "Alcohol modulates expression of DNA methyltranferases and methyl CpG-/CpG domain-binding proteins in murine embryonic fibroblasts." *Reproductive Toxicology* **37**: 40–48.

Murgatroyd, C., A. V. Patchev et al. (2009). "Dynamic DNA methylation programs persistent adverse effects of early-life stress." *Nature Neuroscience* **12**(12): 1559–1566.

Murphy, S. K. and R. L. Jirtle (2003). "Imprinting evolution and the price of silence." *Bioessays* **25**(6): 577–588.

Nahar, M. S., C. Liao et al. (2013). "Fetal liver bisphenol A concentrations and biotransformation gene expression reveal variable exposure and altered capacity for metabolism in humans." *Journal of Biochemical and Molecular Toxicology* **27**(2): 116–123.

Nguyen, C. P. and L. H. Goodman (2012). "Fetal risk in diagnostic radiology." *Seminars in Ultrasound, CT and MRI* **33**(1): 4–10.

Nilsson, E., G. Larsen et al. (2012). "Environmentally induced epigenetic transgenerational inheritance of ovarian disease." *PLoS One* **7**(5): e36129.

Niwa, T., T. Toyoda et al. (2013). "Prevention of helicobacter pylori-induced gastric cancers in gerbils by a DNA demethylating agent." *Cancer Prevention Research* **6**: 263–270.

Niwa, T., T. Tsukamoto et al. (2010). "Inflammatory processes triggered by Helicobacter pylori infection cause aberrant DNA methylation in gastric epithelial cells." *Cancer Research* **70**: 1430–1440.

Paschos, K. and M. J. Allday (2010). "Epigenetic reprogramming of host genes in viral and microbial pathogenesis." *Trends in Microbiology* **18**: 439–447.

Paschos, K., P. Smith et al. (2009). "Epstein-Barr virus latency in B cells leads to epigenetic repression and CpG methylation of the tumour suppressor gene Bim." *PLoS Pathogens* **5**: e1000492.

Perng, W., L. S. Rozek et al. (2012). "Micronutrient status and global DNA methylation in school-age children." *Epigenetics* **7**(10): 1133–1141.

Petronis, A. (2004). "The origin of schizophrenia: Genetic thesis, epigenetic antithesis, and resolving synthesis." *Biological Psychiatry* **55**(10): 965–970.

Qian, X., C. Huang et al. (2008). "E-cadherin promoter hypermethylation induced by interleukin-1beta treatment or H. pylori infection in human gastric cancer cell lines." *Cancer Letters* **263**: 107–113.

Queen, K. J., M. Shi et al. (2013). "Epstein-Barr virus-induced epigenetic alterations following transient infection." *International Journal of Cancer. Journal International du Cancer* **132**: 2076–2086.

Reik, W., W. Dean et al. (2001). "Epigenetic reprogramming in mammalian development." *Science* **293**(5532): 1089–1093.

Rennoll-Bankert, K. E. and J. S. Dumler (2012). "Lessons from Anaplasma phagocytophilum: chromatin remodeling by bacterial effectors." *Infectious Disorders Drug Targets* **12**: 380–387.

Rhodehamel, E. J. (1995). Overview of biological, chemical, and physical hazards. In *HACCP*, M. Pierson and D. Corlett, Jr. eds. Springer, US, 8–28.

Rogers, J. A., L. Metz et al. (2013). "Review: Endocrine disrupting chemicals and immune responses: a focus on bisphenol-A and its potential mechanisms." *Molecular Immunology* **53**(4): 421–430.

Rojas-Squella, X., L. Santos et al. (2013). "Presence of organochlorine pesticides in breast milk samples from Colombian women." *Chemosphere* **91**(6): 733–739.

Ross, J. A., C. F. Blackman et al. (2010). "A potential microRNA signature for tumorigenic conazoles in mouse liver." *Molecular Carcinogenesis* **49**(4): 320–323.

Roth, T. L., P. R. Zoladz et al. (2011). "Epigenetic modification of hippocampal Bdnf DNA in adult rats in an animal model of post-traumatic stress disorder." *Journal of Psychiatric Research* **45**(7): 919–926.

Rugo, R. E., J. T. Mutamba et al. (2011). "Methyltransferases mediate cell memory of a genotoxic insult." *Oncogene* **30**(6): 751–756.

Sakakibara, S., K. Ueda et al. (2004). "Accumulation of heterochromatin components on the terminal repeat sequence of Kaposi's sarcoma-associated herpesvirus mediated by the latency-associated nuclear antigen." *J Virol* **78**: 7299–7310.

Satarug, S. (2010). "Cadmium, environmental exposure, and health outcomes." *Environmental Health Perspectives* **118**(2): 182–190.

Sato, H., T. Oka et al. (2010). "Multi-step aberrant CpG island hyper-methylation is associated with the progression of adult T-cell leukemia/lymphoma." *The American Journal of Pathology* **176**: 402–415.

Schneider, J. S., S. K. Kidd et al. (2013). "Influence of developmental lead exposure on expression of DNA methyltransferases and methyl cytosine-binding proteins in hippocampus." *Toxicology Letters* **217**(1): 75–81.

Schwartz, S., I. Friedberg et al. (2012). "A metagenomic study of diet-dependent interaction between gut microbiota and host in infants reveals differences in immune response." *Genome Biology* **13**: r32.

Selhub, J. (1999). "Homocysteine metabolism." *Annual Review of Nutrition* **19**: 217–246.

Shamay, M., A. Krithivas et al. (2006). "Recruitment of the de novo DNA methyltransferase Dnmt3a by Kaposi's sarcoma-associated herpesvirus LANA." *Proc Natl Acad Sci* **103**(39): 14554–14559.

Shivapurkar, N., T. Takahashi et al. (2004). "Presence of simian virus 40 DNA sequences in human lymphoid and hematopoietic malignancies and their relationship to aberrant promoter methylation of multiple genes." *Cancer Res* **64**(11): 3757–3760.

Silva Lima, B. and J. W. Van der Laan (2000). "Mechanisms of nongenotoxic carcinogenesis and assessment of the human hazard." *Regulatory Toxicology and Pharmacology* **32**(2): 135–143.

Silva, M. J. (2004). "Urinary levels of seven phthalate metabolites in the U.S. population from the National Health and Nutrition Examination Survey (NHANES) 1999–2000." *Environmental Health Perspectives* **112**(3): 331–338.

Skalska, L., R. E. White et al. (2010). "Epigenetic repression of p16(INK4A) by latent Epstein-Barr virus requires the interaction of EBNA3A and EBNA3C with CtBP." *PLoS Pathogens* **6**: e1000951.

Skinner, M. K., M. Manikkam et al. (2011). "Epigenetic transgenerational actions of endocrine disruptors." *Reproductive Toxicology* **31**(3): 337–343.

Slack, A. (1999). "DNA methyltransferase is a downstream effector of cellular transformation triggered by simian virus 40 large T antigen." *Journal of Biological Chemistry* **274**: 10105–10112.

Sohi, G., K. Marchand et al. (2011). "Maternal protein restriction elevates cholesterol in adult rat offspring due to repressive changes in histone modifications at the cholesterol 7alpha-hydroxylase promoter." *Molecular Endocrinology* **25**(5): 785–798.

Soto, A., C. Brisken et al. (2013). "Does cancer start in the womb? Altered mammary gland development and predisposition to breast cancer due to in utero exposure to endocrine disruptors." *Journal of Mammary Gland Biology and Neoplasia* **18**(2):199–208.

Steegers-Theunissen, R., S. Obermann-Borst et al. (2009). "Periconceptional maternal folic acid use of 400ug per day is related to increased methylation of the *IGF2* gene in the very young child." *PLoS One* **4**(11): e7805.

Stølevik, S. B., U. C. Nygaard et al. (2013). "Prenatal exposure to polychlorinated biphenyls and dioxins from the maternal diet may be associated with immunosuppressive effects that persist into early childhood." *Food and Chemical Toxicology* **51**(0): 165–172.

Strakovsky, R. S., X. Zhang et al. (2011). "Gestational high fat diet programs hepatic phosphoenolpyruvate carboxykinase gene expression and histone modification in neonatal offspring rats." *The Journal of Physiology* **589**(Pt 11): 2707–2717.

Sun, Y., Z. Guo et al. (2012). "Diethyl phthalate enhances expression of SIRT1 and DNMT3a during apoptosis in PC12 cells." *Journal of Applied Toxicology* 1484–1492.

Susiarjo, M., I. Sasson et al. (2013). "Bisphenol A exposure disrupts genomic imprinting in the mouse." *PLoS Genet* **9**(4): e1003401.

Tawa, R., Y. Kimura et al. (1998). "Effects of x-ray irradiation on genomic DNA methylation levels in mouse tissues." *Journal of Radiation Research* **39**: 271–278.

Temple, I. K. and J. P. H. Shield (2010). "6Q24 transient neonatal diabetes." *Reviews in Endocrine & Metabolic Disorders* **11**: 199–204.

Trasande, L., T. M. Attina et al. (2012). "Association between urinary bisphenol A concentration and obesity prevalence in children and adolescents." *JAMA: The Journal of the American Medical Association* **308**(11): 1113–1121.

Tsai, C.-N., C.-L. Tsai et al. (2002). "The Epstein—Barr virus oncogene product, latent membrane protein 1, induces the down-regulation of E-cadherin gene expression via activation of DNA methyltransferases." *Proceedings of the National Academy of Sciences of the United States of America* **99**: 10084–10089.

Tsang, V., R. C. Fry et al. (2012). "The epigenetic effects of a high prenatal folate intake in male mouse fetuses exposed in utero to arsenic." *Toxicology and Applied Pharmacology* **264**(3): 439–450.

Uriarte, G., L. Paternain et al. (2013). "Shifting to a control diet after a high-fat, high-sucrose diet intake induces epigenetic changes in retroperitoneal adipocytes of Wistar rats." *Journal of Physiology and Biochemistry* **69**(3): 601–611.

Van Dyck, F., J. Declercq et al. (2007). "PLAG1, the prototype of the PLAG gene family: Versatility in tumour development (review)." *International Journal of Oncology* **30**: 765–774.

Vandenberg, L. N., I. Chahoud et al. (2010). "Urinary, circulating, and tissue biomonitoring studies indicate widespread exposure to bisphenol A." *Environmental Health Perspectives* **118**(8): 1055–1070.

Vandenberg, L. N., T. Colborn et al. (2012). "Hormones and endocrine-disrupting chemicals: Low-dose effects and nonmonotonic dose responses." *Endocrine Reviews* **33**(3): 378–455.

Veiga-Lopez, A., L. J. Luense et al. (2013). "Developmental programming: Gestational bisphenol-A treatment alters trajectory of fetal ovarian gene expression." *Endocrinology* **154**(5): 1873–1884.

Vercelli, D. (2004). "Genetics, epigenetics, and the environment: switching, buffering, releasing." *Journal of Allergy and Clinical Immunology* **113**(3): 381–386.

Vidal, A. C., S. K. Murphy et al. (2013). "Associations between antibiotic exposure during pregnancy, birth weight and aberrant methylation at imprinted genes among offspring." *International Journal of Obesity (2005)* **37**: 907–913.

Vivekanandan, P., H. D.-J. Daniel et al. (2010). "Hepatitis B virus replication induces methylation of both host and viral DNA." *Journal of Virology* **84**: 4321–4329.

Vo, T. X., A. Revesz et al. (2013). "Maternal protein restriction leads to enhanced hepatic gluconeogenic gene expression in adult male rat offspring due to impaired expression of the liver X receptor." *Journal of Endocrinology* **218**(1): 85–97.

Wallace, A. D. (2012). Toxic endpoints in the study of human exposure to environmental chemicals. *Prog Mol Biol Transl Sci.* **112**: 89–115.

Wang, X. (2010). "Effect of triazophos, fipronil and their mixture on miRNA expression in adult zebrafish." *Journal of Environmental Science and Health. Part B, Pesticides, Food Contaminants, and Agricultural Wastes* **45**(7): 648–657.

Warning, J. C., S. A. McCracken et al. (2011). "A balancing act: Mechanisms by which the fetus avoids rejection by the maternal immune system." *Reproduction (Cambridge, England)* **141**: 715–724.

Waterland, R., M. Travisano et al. (2008). "Methyl donor supplementation prevents transgenerational amplification of obesity." *International Journal of Obesity* **32**: 1373–1379.

Weaver, I. C., N. Cervoni et al. (2004). "Epigenetic programming by maternal behavior." *Nature Neuroscience* **7**(8): 847–854.

Weinhouse, C., O. S. Anderson et al. (2011). "An expression microarray approach for the identification of metastable epialleles in the mouse genome." *Epigenetics* **6**(9): 1105–1113.

Weiss, B. (2012). "The intersection of neurotoxicology and endocrine disruption." *NeuroToxicology* **33**(6): 1410–1419.

Wigle, D. T., T. E. Arbuckle et al. (2007). "Environmental hazards: Evidence for effects on child health." *Journal of Toxicology and Environmental Health, Part B* **10**(1–2): 3–39.

Witkin, S., I. Linhares et al. (2011). "Unique alterations in infection-induced immune activation during pregnancy." *BJOG: An International Journal of Obstetrics & Gynaecology* **118**: 145–153.

Wolstenholme, J. T., J. A. Taylor et al. (2011). "Gestational exposure to low dose bisphenol A alters social behavior in juvenile mice." *PLoS One* **6**(9): e25448.

Woods, R., R. O. Vallero et al. (2012). "Long-lived epigenetic interactions between perinatal PBDE exposure and Mecp2308 mutation." *Human Molecular Genetics* **21**(11): 2399–2411.

Wu, S. (2010). "Dynamic effect of di-2-(ethylhexyl) phthalate on testicular toxicity: Epigenetic changes and their impact on gene expression." *International Journal of Toxicology* **29**(2): 193–200.

Yeo, M., K. Berglund et al. (2013). "Bisphenol A delays the perinatal chloride shift in cortical neurons by epigenetic effects on the Kcc2 promoter." *Proceedings of the National Academy of Sciences of the United States of America* **110**(11): 4315–4320.

Zhang, X.-F., L.-J. Zhang et al. (2012). "Bisphenol A exposure modifies DNA methylation of imprint genes in mouse fetal germ cells." *Molecular Biology Reports* **39**(9): 8621–8628.

Zheng, D.-L., L. Zhang et al. (2009). "Epigenetic modification induced by hepatitis B virus X protein via interaction with de novo DNA methyltransferase DNMT3A." *Journal of Hepatology* **50**: 377–387.

2 Maternal Diet and Exercise

Influences on Obesity and Insulin Resistance

Lindsay G. Carter, Kellie L. K. Tamashiro, and Kevin J. Pearson

CONTENTS

2.1 DEVELOPMENTAL PROGRAMMING

It has been known for decades that birth weight is a predictor of later-life diseases (Barker et al. 1989a, b). The thrifty phenotype hypothesis was presented in the early 1990s and suggests that low birth weight offspring have a higher incidence of multiple diseases later in life (Hales and Barker 1992). The developmental origins of health and disease (DOHaD) hypothesis has grown out of this and the work of many others (Gluckman and Hanson 2004; Gluckman et al. 2005). The DOHaD hypothesis suggests that fetal programming is likely playing a role in the alarming increase in obesity of children and adolescents and will result in increased diabetes, cardiovascular disease, and early death during adulthood (Gluckman and Hanson 2008; Gluckman et al. 2008). Pregnancy is a critical time for both the mother and developing fetus. Events that occur during pregnancy can have a permanent effect on the health of the offspring. Developmental programming is a theory that the stimuli received from the intrauterine environment can result in long-term changes in an organism that can predispose or protect it from diseases later in life. It is well known that changes to the intrauterine environment caused by things such as drug use (for example, tobacco and alcohol) can cause changes in fetal development that result in physical and mental dysfunction in offspring that last a lifetime. Less studied is how maternal diet and physical activity can affect offspring, particularly offspring metabolic health. Over the past two decades, research has emerged showing that maternal diet, whether it is undernutrition or overnutrition, can alter offspring insulin sensitivity and increase offspring incidence of type 2 diabetes. This chapter will define and discuss diabetes as well as outline developmental programming of this disease in humans and animal models. Lastly, this section will introduce the benefits of exercise and the potential use of exercise during pregnancy as an intervention to positively impact offspring health.

2.2 TYPE 2 DIABETES

Type 2 diabetes affects approximately 8%–9% of the population in the United States, and 300 million people are projected to have the disease by 2020 (O'Rahilly 1997). Consequences of the disease include heart disease, neuropathy, and kidney disease (Szepietowska et al. 2004). Type 2 diabetes is characterized by insulin resistance, hyperglycemia, and progressive β-cell failure (the pancreatic islet cells responsible for the formation and secretion of insulin) (Stumvoll et al. 2008). In response to glucose stimulation, β cells release insulin, which in turn acts primarily on white adipose tissue and skeletal muscle to promote glucose uptake into the cells (Muoio and Newgard 2008). A signaling cascade within the cell is initiated upon insulin binding to its receptor on the cell surface in skeletal muscle and adipose tissue (DeFronzo et al. 1981). Insulin binds to the extracellular α subunits of the insulin receptor (IR), causing autophosphorylation of the transmembrane β subunits. Phosphorylated IR then recruits and phosphorylates a tyrosine residue on IR substrate 1 (IRS-1). Tyrosine phosphorylated IRS-1 in turn activates phosphatidylinositol 3-kinase (PI3K), which can catalyze the conversion of phosphatidylinositol-4,5-bisphosphate (PIP2) to phosphatidylinositol-3,4,5-triphosphate (PIP3). PIP3 activates the 3-phosphoinositide-dependent protein kinase 1 (PDK1). PDK1 can then activate other kinases including protein kinase B (PKB or Akt1) and atypical protein kinase C (PKCλ/ζ), leading to the translocation of glucose transporter type 4 (GLUT4) from the cytosol to the membrane for glucose uptake (DeFronzo et al. 1981, 1985).

Obesity is a risk factor associated with type 2 diabetes. Adipose tissue plays a major role in the development of insulin resistance because it secretes adipokines, a group of cytokines and hormones that can deregulate the activation and activity of proteins involved in the insulin signaling cascade. For example, pro-inflammatory cytokines such as interleukin 6 (IL-6) and tumor necrosis factor α (TNF-α) released from adipose can lead to serine/threonine phosphorylation of IRS-1, causing inactivation of the protein and decreased translocation of GLUT4 to the cell surface for glucose uptake into the cell (Trujillo and Scherer 2006). Free fatty acids (FFA) can also work in a similar manner to TNF-α, contributing to insulin resistance. Elevated FFA levels have been shown to produce diacylglycerol (DAG) accumulation in skeletal muscle. This accumulation of DAG can cause serine/threonine phosphorylation of IRS-1, therefore inhibiting insulin stimulated glucose uptake into the cell (Schmitz-Peiffer 2000; Kahn et al. 2006). Meanwhile, adiponectin, a peptide hormone also secreted by adipose tissue (circulating levels are negatively correlated with adiposity) can actually enhance glucose uptake by activating AMP-activated protein kinase (AMPK), a protein that stimulates insulin independent glucose uptake in response to a fall in cellular ATP levels (Trujillo and Scherer 2006). Of particular interest is how levels of these adipokines and peptide hormones are affected by obesity and/or exercise during pregnancy.

It is important to mention that both β-cell dysfunction and hepatic insulin resistance can contribute to the development of type 2 diabetes. In β cells, deficits in glucose stimulated insulin secretion can occur when the cells are exposed to excess nutrients such as lipids and glucose (Porte and Kahn 2001; Zhang et al. 2001). This results in an overall decrease in circulating insulin release in response to rising glucose levels, meaning less glucose is taken up into peripheral tissues. Reduction of hepatic insulin sensitivity by factors such as consumption of a high-fat diet can further promote a diabetic state (Cai et al. 2005; Petersen et al. 2005). Hepatic insulin stimulation inhibits gluconeogenesis, the process of endogenous glucose production from glycogen stored in the liver (Claus and Pilkis 1976; Dentin et al. 2007). When the liver becomes insulin resistant, endogenous glucose production is not decreased in response to circulating insulin and glucose levels, which contributes to hyperglycemia (Bahl et al. 1997).

2.3 DEVELOPMENTAL PROGRAMMING OF DIABETES: HUMAN MATERNAL MALNUTRITION

Geographical studies conducted in England and Wales were the first to show a link between a poor intrauterine environment, caused by inadequate nutrient intake during gestation, to disease in aged offspring (Barker et al. 1989a, b, c). In areas where there was maternal malnutrition, David Barker found a correlation between low birth weight, due to poor maternal nutrition, and ischemic heart disease in the adult offspring who had been exposed to undernutrition during their fetal development (Barker et al. 1989b). A subsequent study looked at individual cases in these areas and found men with the lowest birth weights had approximately a 3-fold higher death rate from ischemic heart disease (Barker et al. 1989a, b). Other studies in this cohort of offspring also showed links between low birth weight and hypertension and fibrinogen in adulthood (Barker et al. 1990, 1992). Barker, his colleague C. Nicholas Hales, and others then looked at glucose regulation and type 2 diabetes in these adults and found that lower birth weights were associated with impaired glucose tolerance and type 2 diabetes (Hales et al. 1991). Later, they went on to show that adults who had a low birth weight had higher incidence of insulin resistance and type 2 diabetes that was independent of body mass and adiposity (Phillips et al. 1994). Other groups were also able to demonstrate insulin resistance and type 2 diabetes in later life related to low birth weight (McCance et al. 1994; Valdez et al. 1994; Yajnik et al. 1995; Lithell et al. 1996). Another well-studied population of offspring exposed to maternal malnutrition comes from those who were born during the Dutch Famine between 1944 and 1945. In adulthood, this population showed higher rates of obesity, coronary heart disease, and impaired glucose tolerance and insulin secretion compared with those born before or after the famine period (Ravelli et al. 1999; Roseboom et al. 2000; de Rooij et al. 2006). Barker and Hales proposed the "thrifty phenotype" hypothesis to summarize the findings from maternal malnutrition research, which states that poor fetal nutrition is detrimental to development and consequently predisposes an individual to type 2 diabetes in adulthood (Hales and Barker 1992). An individual can be "programmed" during fetal development to have an advantage after birth in an environment that is predicted to have low nutrient availability, but this programming is disadvantageous when there is sufficient or excess nutrient availability. Since the development of the thrifty phenotype hypothesis, numerous animal models have been used to study the effects of intrauterine growth restriction (IUGR) leading to low birth weight and the effects on offspring insulin sensitivity and glucose regulation.

2.4 DEVELOPMENTAL PROGRAMMING OF DIABETES: ANIMAL MODELS OF MATERNAL MALNUTRITION

Uterine artery ligation and maternal nutrient/protein restriction are two of the most commonly used models to study the effects of IUGR in animals. Uterine artery ligation creates poor nutrient flow to the fetus, a common cause of reduced fetal growth in humans. In rats, ligation results in low birth weight and insulin resistance and glucose intolerance in aged offspring (Simmons et al. 2001). Offspring also have reduced β-cell mass and higher fasting glucose compared with normal offspring (De Prins and Van Assche 1982; Simmons et al. 2001). In rodents, maternal nutrient restriction causes restricted fetal growth and low birth weight, which is accompanied by impaired β-cell function and impaired glucose tolerance, and hyperinsulinemia in adult life (4 and 12 months of age) compared with offspring from normal fed dams (Garofano et al. 1999; Vickers et al. 2000). Other studies using maternal protein restriction in rodents have found similar results. Offspring have decreased β-cell mass as well as reduced insulin content, and in response to glucose stimulation, have decreased insulin secretion compared with offspring from normal fed dams (Snoeck et al. 1990; Boujendar et al. 2002; Merezak et al. 2004; Fernandez-Twinn et al. 2005). Adult offspring at 15 and 17 months of age also have impaired glucose tolerance (Petry et al. 2001; Ozanne et al. 2003). Maternal nutrient restriction has also been studied using sheep. Offspring born to nutrient-restricted mothers show decreased insulin sensitivity, increased insulin secretion, along with increased weight gain compared with normal offspring (Murphy et al. 2003).

Using these animal models, researchers have been able to attempt to determine the mechanisms behind the observed insulin resistance and impaired glucose tolerance in offspring exposed to IUGR. Research has shown that rat offspring exposed to IUGR have reduced expression of the p110β subunit of PI3-kinase, as well as decreased association of the p110β with the p85 subunit in adipose tissue compared with control offspring, which inhibits GLUT4 translocation, decreasing glucose uptake into the cell (Ozanne et al. 2001; Fernandez-Twinn et al. 2005). In muscle samples from growth-restricted offspring, there is decreased expression of PCKζ, another important protein in the insulin signaling cascade involved with GLUT4 translocation, compared with expression in control animals (Ozanne et al. 2003). Hepatic insulin resistance has also been observed in IUGR offspring. After insulin stimulation, hepatic glucose production in IUGR is not decreased as in control animals and there is decreased phosphorylation of Akt compared with control animals (Vuguin et al. 2004). These decreases in insulin pathway protein expression and activation could make the IR signaling cascade less sensitive to insulin stimulation and activation.

Other studies have found that IUGR can result in epigenetic changes to genes involved in metabolism and growth. Insulin-like growth factor (IGF) 1 is a protein produced mainly by the liver that plays an important role in growth and cellular proliferation by activating the Akt pathway. Circulating levels of this protein are reduced in models of IUGR (El-Khattabi et al. 2003). Decreased IGF-1 has been linked to epigenetic modifications along the hepatic *IGF-1* gene, specifically histone modifications that result in downregulation of gene transcription (Fu et al. 2009). Other epigenetic changes have been associated with the dysfunctional pancreatic β cells found in offspring exposed to IUGR. Pancreatic and duodenal homeobox 1 (Pdx1) is a transcription factor that regulates pancreas growth and differentiation of β cells (Jonsson et al. 1994). Pdx1 expression is reduced in offspring that experience IUGR compared with normal offspring, predisposing offspring to β-cell dysfunction and type 2 diabetes (Park et al. 2008). This reduction in expression is due to histone modifications caused by IUGR that result in silencing of the gene.

IUGR restriction affects many tissues and proteins involved with insulin sensitivity and glucose homeostasis. Clearly, poor nutrient supply to the developing fetus has many long-term consequences that can predispose offspring to insulin resistance and diabetes. Interestingly, maternal overnutrition causes many negative effects in the mother, as well as fetal overgrowth that results in similar long-term offspring disease incidence.

2.5 DEVELOPMENTAL PROGRAMMING OF DIABETES: MATERNAL OBESITY AND GESTATIONAL DIABETES

Obesity rates are on the rise, and almost 30% of women of child-bearing age are considered obese (Hedley et al. 2004). Obesity during pregnancy can be detrimental to both mother and child. High body mass index and consumption of a high calorie diet are risk factors for the development of gestational diabetes, defined as glucose handling impairment first recognized during pregnancy (Chu et al. 2007; Radesky et al. 2008; Kim et al. 2010). Although a natural insulin resistance develops as all pregnancies progress to ensure adequate nutrient flow to the fetus, in some women, especially those who are obese, this natural decrease in insulin sensitivity can develop into diabetes (McIntyre et al. 2010). Excess adipose tissue in obese women may play a role in inducing gestational diabetes by causing an increase in whole-body inflammation. Inflammatory cytokine levels, such as TNF-α, are positively correlated with adiposity, and can contribute to insulin resistance by inactivating proteins, like IRS-1, in the insulin signaling cascade (Kahn et al. 2006). Obese pregnant women with gestational diabetes do indeed have increased circulating levels of these pro-inflammatory cytokines along with downregulated IRS-1 in skeletal muscle (Friedman et al. 1999; Wolf et al. 2004; Barbour et al. 2007). Obesity is also associated with decreased adiponectin levels, a protein that can contribute to increased glucose uptake by promoting AMPK phosphorylation and therefore increase glucose uptake in skeletal muscle (Yamauchi et al. 2002). In women with gestational diabetes, adiponectin levels are significantly decreased compared with women experiencing a normal pregnancy, and this too may contribute to the impaired glucose tolerance in these women (Worda et al. 2004). Finally, increased fat mass is also linked to high levels of circulating triglycerides and free fatty acids, which both have insulin-desensitizing effects, and these levels are higher in women with gestational diabetes than those with a normal pregnancy (Kahn and Flier 2000; Kahn et al. 2006; McIntyre et al. 2010). Similar to inflammatory cytokines, these molecules affect proteins in the insulin signaling cascade and can deactivate them (Kahn et al. 2006; Barbour et al. 2007).

Due to all these changes in protein expression, circulation, and activation, women who develop gestational diabetes are at a substantially higher risk than other women for type 2 diabetes in the years after their pregnancies (Kim et al. 2002). Therefore, treatment or prevention of gestational diabetes is of high importance. Management of gestational diabetes, however, is also critical as high levels of maternal glucose can be detrimental to the developing fetus. Similar to poor nutrient availability during development, excess nutrient and glucose exposure can also "program" offspring to be predisposed to insulin resistance and diabetes.

Fetal growth increases as a result of high maternal glucose levels (Freinkel 1980). This is due to higher levels of fetal insulin being secreted in response to maternal glucose levels because, although maternal glucose crosses the placenta, maternal insulin does not (Freinkel 1980). Children from gestational diabetic mothers, therefore, have excess fat mass and are heavier at birth than children from normal women (Silverman et al. 1991). This increase in body weight and adiposity persists past birth and throughout childhood (Silverman et al. 1991; Pettitt et al. 1993; Chandler-Laney et al. 2011). Even in women with only mild glucose intolerance, as opposed to women with gestational diabetes, infant fat mass, and body weight are increased compared with infants of a normal pregnancy (Catalano et al. 2003). In a well-studied population of women with a high incidence of gestational diabetes, the Pima Indian cohort, it has been found that gestational diabetes results in a high risk of type 2 diabetes in offspring compared with those born to nondiabetic women (Dabelea et al. 2000; Franks et al. 2006). Of the children who become diabetic in this population, almost all of them are from diabetic mothers (Franks et al. 2006). Interestingly, siblings born to the same mothers prior to the development of diabetes or gestational diabetes do not show the same incidence of type 2 diabetes as siblings born to the mothers once diabetes has developed (Pettitt et al. 1993).

Animal models of excess calorie intake before and during pregnancy have shown similar results to human studies. However, animal models have allowed researchers to follow health outcomes in offspring through adulthood. In rodents, offspring from high calorie fed dams have increased adiposity and elevated blood insulin and glucose levels in adult life (between 3 and 6 months of age) compared with offspring from normal diet fed dams (Samuelsson et al. 2008; Volpato et al. 2012). One-year-old offspring from dams fed a high-fat diet also exhibit decreased whole-body insulin sensitivity (as measured by hyperinsulinemic–euglycemic clamp) when compared with offspring from dams fed a normal diet (Taylor et al. 2005). These offspring demonstrate β-cell dysfunction; they have decreased glucose stimulated insulin release from pancreatic β cells compared with control offspring (Taylor et al. 2005). Offspring from diet induced obese dams have also been shown to develop diabetes by 26 weeks of age (Boloker et al. 2002). Using animal models, researchers have also been able to identify a number of other negative health outcomes in offspring caused by maternal high calorie feeding; these outcomes include offspring hypertension, fatty streaks on aortas, fatty liver, and hyperphagia (Khan et al. 2005; Samuelsson et al. 2008; Ashino et al. 2012).

To date, relatively few animal studies have investigated the epigenetic influences maternal high-fat feeding can have on offspring. In a mouse model, DNA methylation has been implicated in increased IGF-2 expression in fetal livers from dams fed a high-fat diet (Zhang et al. 2009). Another group has shown that adiponectin and leptin receptor genes of 20-week-old mice exposed to maternal high-fat feeding have increased methylation, coinciding with decreases in leptin gene methylation compared with offspring from standard diet fed dams, suggesting that these offspring are predisposed to leptin- and insulin-resistant states (Khalyfa et al. 2013). This area of fetal programming is still poorly understood, and many more studies will be needed to elucidate epigenetic mechanism involved in maternal overnutrition and offspring outcomes.

Effective management or prevention of maternal glucose levels and gestational diabetes is extremely important for the health of mother and child. To date, there is evidence that changes to diet and weight loss prior to conception can decrease risk of gestational diabetes. Once gestational diabetes has developed, diet and lifestyle modifications are recommended, and insulin can be used to maintain normal maternal blood glucose levels if needed (Denison and Chiswick 2011). Surprisingly, exercise has received very little attention in the treatment and prevention of gestational diabetes (Denison and Chiswick 2011). It is widely accepted that exercise has innumerable health benefits, including improving whole-body glucose regulation, and may have the same health benefits in pregnant women as in nonpregnant individuals. More recently it has come to light that exercise during pregnancy may also have benefits for offspring. The remainder of this chapter will focus on the effects of exercise during normal and diabetic pregnancy as well as potential benefits of maternal exercise for the next generation.

2.6 EXERCISE DURING PREGNANCY: MATERNAL EFFECTS IN NORMAL PREGNANCIES

There are countless health benefits of physical activity, including but not limited to enhanced mood and cognition, weight management, improved cardiovascular health, and decreased risk of disease such as cancer, heart disease, and type 2 diabetes (Thompson et al. 2003; Snowling and Hopkins 2006; Friedenreich et al. 2010; Saeed et al. 2010). Many physiological adaptations occur in a woman in response to pregnancy as well as exercise, and these can work synergistically to maintain the best health possible for mother and baby. In pregnancy, the cardiovascular system changes to support the developing child. Hormones released by the placenta increase blood vessel elasticity and cause water and sodium retention (Longo 1983; Islami et al. 2003). Blood volume is thereby increased, which in turn can raise cardiac output (Longo 1983). Overall cardiac output can be increased by almost 40% compared with the nonpregnant state, allowing for proper blood flow and glucose and oxygen delivery for fetal development (Longo 1983; Duvekot and Peeters 1994). The cardiovascular adaptations that occur in pregnancy are very similar to those that happen in response to exercise training. Regular exercise also results in increased blood volume and cardiac output (Duncker and Bache 2008). Consistent exercise training and pregnancy can have additive effects on cardiovascular structure and function. Women who exercise during pregnancy have up to 15% higher plasma volume compared with sedentary pregnant women (Clapp 1996a, 2000). Stroke volume can also be raised as much as 50% in exercising pregnant women compared with sedentary pregnant women because exercise training and pregnancy both increase left ventricular volume (Clapp 1994; Clapp and Capeless 1997).

There has been concern that exercise during pregnancy could be harmful for the fetus due to higher maternal body temperature during physical activity bouts (Mudd et al. 2009). However, greater blood circulation due to exercise and pregnancy also allows the mother to maintain core body temperature since more blood is circulating to the surface/skin, allowing heat to dissipate (Clapp 1996a). Another concern has been the changes in blood flow caused by exercise. During an exercise bout, blood flow is shifted away from digestive and reproductive systems and other organs and redirected toward muscle and skin (Kenney and Johnson 1992; Duncker and Bache 2008). This raised fear that exercise would deprive the fetus of blood and therefore oxygen and glucose. In pregnancy, exercise causes a change in blood flow from the fetus to maternal skin and muscle, and at high intensities, this change in flow can be as great as 50% (Clapp 1996a). However, once the exercise bout has ended, there is a rapid return to normal blood flow. Also, women who have trained from the start of their pregnancy actually have increased placental size and improved placental function, meaning that even during physical activity, there is still adequate blood flow to the fetus despite shifts in distribution of flow (Clapp 2003).

There are also several metabolic changes that occur in during pregnancy. Adipose tissue is formed during early pregnancy since it is able to store high amounts of energy. Muscle and fat also become more insulin resistant as pregnancy progresses (Ryan et al. 1985), which causes more maternal fat deposition and decreases maternal glucose uptake and utilization, leaving more glucose for the fetus. Opposite from the pregnancy response, exercise decreases fat mass and increases muscle and bone mass (Ballor and Keesey 1991). Exercise also enhances insulin sensitivity in skeletal muscle so that muscle can take up glucose and store it for energy during an exercise bout (Holloszy 2005). Since exercise can counteract the insulin-desensitizing effects of pregnancy, it is important that during a healthy pregnancy, a woman eats regularly to maintain glucose levels reaching the fetus (Clapp and Little 1995). The metabolic effects of exercise training are particularly important in pregnancies complicated by gestational diabetes, which will be discussed in greater detail the next section.

There are many reported maternal benefits of exercising during healthy pregnancy. There are multiple types of exercise including endurance and resistance training. Examples of endurance training are activities such as running, cycling, and swimming while resistance training includes exercises like weight lifting that focus on increasing muscle size and/or strength (Farup et al. 2012). Pregnant women who participate in continuous weight-bearing endurance training (such as running or aerobics) are leaner with decreased fat mass and weight gain compared with sedentary women. Body fat, as measured by skin-fold thickness, was much lower in women who maintained an exercise routine of at least 3 to 5 times a week (Clapp and Little 1995). They also return to prepregnancy weight faster (Clapp and Little 1995). Women who exercise also have less fluid retention and decreased lower back pain and blood pressure, decreasing the risk of preeclampsia (high blood pressure in pregnancy) (Melzer et al. 2010). Labor time is shorter; in one study 65% of women who remained active throughout pregnancy had a labor of 4 h or less versus only 31% of control women being in labor for 4 h or less (Clapp and Little 1995). Women also complain less of discomfort during pregnancy and labor when exercise was maintained throughout pregnancy. There is up to a 35% decrease in request for pain relievers and 75% decrease in exhaustion during labor compared with sedentary pregnant women (Clapp and Little 1995). A more subtle benefit of physical activity during pregnancy is improved mood; several studies have reported that, particularly in the third trimester, women who exercise have lower anxiety and are more stable in their moods compared with sedentary women (Da Costa et al. 2003; Poudevigne and O'Connor 2005).

Clearly, maintaining high levels of physical activity and conditioning can contribute to healthy pregnancy outcomes. Therefore, the American College of Obstetricians and Gynecologists recommends that women "engage in regular, moderate intensity physical activity to continue to derive the same associated health benefits during their pregnancies as they did prior to pregnancy" (ACOG Committee opinion. Number 267, January 2002: Exercise during pregnancy and the postpartum period 2002). Women who were inactive prior to pregnancy are encouraged to slowly and progressively work up to 30 min of exercise (ACOG Committee opinion, Number 267, January 2002: Exercise during pregnancy and the postpartum period, 2002). The Society of Obstetricians and Gynecologists of Canada also recommends that all pregnant women without complications should perform some form of endurance and/or strength training to promote a healthy pregnancy (Davies et al. 2003). The American College of Sports Medicine (2011) also suggests that healthy women continue a regular exercise regimen throughout pregnancy. All groups advocate safe physical activities such as swimming, jogging, or stationary biking and warn against activities such as contact sports or those that require special balance, for example, cycling, that could result in maternal or fetal injury.

2.7 EXERCISE DURING PREGNANCY: MATERNAL EFFECTS IN DIABETIC PREGNANCIES

There are several maternal and offspring risks of gestational diabetes, including increased risk of type 2 diabetes in mother and offspring; therefore, physical activity throughout pregnancy could be particularly important if a mother has gestational diabetes. Since exercise has metabolic effects that include enhancing insulin sensitivity, it may be an effective method of preventing or treating gestational diabetes. Given the metabolic impact exercise training can have in humans, it is surprising that there have been relatively few studies looking at the physical activity effects on gestational diabetes, in particular the intensity and frequency of exercise required to improve diabetic outcomes.

As mentioned earlier, studies have shown that exercise can improve insulin sensitivity in muscle (Wojtaszewski et al. 2000; Yu et al. 2001). Endurance training has been shown in several studies to increase GLUT4 levels in skeletal muscle as well as increase levels of muscle glycogen synthase and hexokinase, thereby promoting glucose uptake and phosphorylation in response to insulin stimulation (Hofmann and Pette 1994; Yu et al. 2001; Holloszy 2005; Wasserman and Ayala 2005). The anti-inflammatory effects of exercise also play a role in improving insulin sensitivity. Research looking at exercise has found that decreases in TNF-α with training, which could contribute to insulin-sensitizing effects by preventing inactivation of proteins in the insulin signaling cascade (Tsukui et al. 2000; Petersen and Pedersen 2005). The insulin sensitivity enhancing effects of resistance training have been less studied, but it is thought that it is able to improve overall insulin sensitivity by increasing muscle mass (Miller et al. 1984; Fenicchia et al. 2004). Exercise can help regulate blood glucose levels by stimulating insulin independent glucose uptake in skeletal muscle. Muscle contraction stimulates AMPK, which leads to GLUT4 translocation to the cell membrane and glucose uptake into the cell (Vavvas et al. 1997; Derave et al. 1999; Winder and Hardie 1999). AMPK is a fuel sensor that is stimulated in response to a decrease in the ATP/AMP ratio; therefore, when muscles contract and ATP levels decrease and AMP levels increase, AMPK is activated (Winder and Hardie 1999; Hayashi et al. 2000).

Exercise is clearly able to affect glucose disposal in several different ways, and so it has been studied, although not extensively, in the prevention and management of gestational diabetes. Studies looking at the effects of exercise on risk of developing gestational diabetes have yielded mixed results. Observational studies have found that risk of gestational diabetes can be decreased with various levels of physical activity. Women who reported any level of activity early in their pregnancy showed a 56% decrease in gestational diabetes risk in one prospective study (Dempsey et al. 2004b). Dempsey et al. (2004a) also found that, in a separate cohort of women, daily stair-climbing during pregnancy was associated with a 50%–75% decrease in gestational diabetes risk. This study also showed that intensity and duration of exercise bouts were negatively correlated with incidence of gestational diabetes in the study cohort. Other studies have found that women who reported vigorous exercise before pregnancy, or even low intensity physical activity such as brisk walking, have reduced incidence of gestational diabetes compared to women who were not physically active (Zhang et al. 2006; George et al. 2012). In a meta-analysis, Tobias et al. (2011) concluded that high levels of exercise before or during early pregnancy decreased gestational diabetes incidence. Using data from the National Maternal and Infant Health Survey, another group found that women who began physical activity at the onset of pregnancy had 57% lower adjusted odds of developing gestational diabetes (Liu et al. 2008).

Despite these promising findings, other studies have found no clear evidence that physical activity can reduce the risk of gestational diabetes. Han et al. (2012) pooled data from multiple clinical trials and concluded that there was no evidence that any level of physical activity before or during pregnancy had any effects on development of gestational diabetes. In a recent randomized control trial, pregnant women who were assigned to moderate intensity exercise (combination of endurance, muscle strength, and flexibility training) for 3 days a week were not protected from gestational diabetes development (Barakat et al. 2013). It is important to note that although there was no effect on the development of gestational diabetes, this exercise regimen did result in lower incidence of macrosomia in the infants of diabetic mothers compared with those from control diabetic mothers. In another randomized control trial, overweight and obese women were put on a moderate intensity aerobic and resistance exercise program consisting of two 60-min sessions a week (Oostdam et al. 2012). Exercise training was unable to reduce fasting glucose levels or improve insulin sensitivity as defined by the homeostatic model assessment index, which estimates insulin sensitivity and β-cell function.

Little research has been conducted investigating the effects of exercise on pregnant women who have already developed gestational diabetes; however, some yielded positive results. In one, an arm exercise program in which an arm ergometer was used to monitor and maintain heart rates (20 min of arm exercise at ~50% maximum capacity; 3 times a week) in women diagnosed with gestational diabetes was able to normalize fasting glucose levels after 6 weeks (Jovanovic-Peterson et al. 1989). In another study, gestational diabetic women were assigned to diet or diet plus exercise consisting of resistance circuit-type training (i.e., squats, knee extensions, seated row, and lateral pull-down) 3 times a week, and the need for insulin to control blood glucose levels was monitored (Brankston et al. 2004). In normal weight women, diet and exercise was no different than diet alone at preventing the need for insulin to control blood glucose; however, in overweight women, diet plus exercise reduced the need for insulin compared with diet therapy alone. Additionally, another group was able to show that resistance training of moderate intensity with an elastic band 3 times a week was effective in reducing the number of patients with gestational diabetes requiring insulin therapy (de Barros et al. 2010).

Exercise has potential as a treatment or preventative therapy for gestational diabetes, but there is a lack of evidence based studies that can be used to look at the physiological effects of exercise and the type and intensity of exercise required to improve gestational diabetes outcomes. Animal models will be essential in the study of physical activity effects on gestational diabetes.

2.8 EXERCISE DURING PREGNANCY: EFFECTS ON OFFSPRING

There are numerous maternal health benefits of physical activity during healthy and diabetic pregnancies, and more recently, attention has turned to the offspring benefits of maternal exercise. In humans, the offspring effects of maternal exercise can be apparent from birth. In women who exercise before and during pregnancy, rigorous exercise can decrease birth weight, but not to the level associated with IUGR or small for gestational age (SGA) babies. This decrease in birth weight is associated with reduced fat mass and no changes in lean mass in the infants (Clapp and Capeless 1990; Clapp 1994). Beginning moderate weight-bearing exercise in pregnancy can actually increase birth weight due to increased placental growth and volume; however, the increased birth weight is only due to increased lean mass with no changes in fat mass (Clapp et al. 2000a). Other studies have shown that non–weight-bearing, yet fairly intense exercise, such as cycling and swimming had no effects on birth weight (Collings et al. 1983; Clapp 2000). A more recent study, however, found that a non–weight-bearing stationary cycling exercise program started mid gestation was effective in reducing birth weight (Hopkins et al. 2010). Others have shown that even weight-bearing exercise, when at reduced intensities compared with what women exercised prior to pregnancy, did not have effects on birth weight (Bell et al. 1995; Kardel 2005). Interestingly, when intense exercise is stopped at some point during pregnancy, because placental growth has already been increased due to the physical activity, infants are born with a higher birth weight than others from mothers who were sedentary throughout the entire pregnancy, and this increase in birth weight is the result of higher levels of fat mass (Clapp et al. 2002). The type of exercise and timing of exercise during pregnancy can all have different impacts on offspring birth weight, making the amount, type, and timing of exercise during pregnancy important factors to consider in future studies. Currently, it is thought that these differences in birth weight with difference exercise regimens during pregnancy are the result of placental growth changes and oxygen and glucose levels reaching the fetus in response to exercise (Clapp 2003). Again, it is important to note that although exercise can result in lower birth weights; it does not increase risk of low birth weight (less than 2500 g) and that infant body length and size are not altered by maternal exercise.

In humans, there is still very limited research-based information on the long-term outcomes in offspring resulting from maternal exercise during pregnancy. Clinician James Clapp has conducted the most extensive research on maternal and offspring effects of exercise during pregnancy. When evaluating the behavior of infants 5 days after birth using the Brazelton Neonatal Behavioral Assessment Scales, he found that offspring from women who exercised throughout pregnancy (an average of 4 times per week and 60% maximum capacity) performed better on orientation and state level skills than offspring from less active mothers, suggesting that maternal exercise had an impact on neurological development (Clapp et al. 1999). In another cohort of women, Clapp looked at the effects of exercise (stair-climbing, aerobics, or running) on offspring morphometric and neurodevelopmental outcomes at 1 year of age (Clapp et al. 1998). Women in the exercise group exercised at least 3 times a week for 20 min per exercise bout, reaching 60%–90% maximum capacity at their highest intensities. At birth, the infants from exercise mothers had lower body weight and fat mass compared with those from less active mothers. However, at 1 year, they had comparable body weight and body composition to children from control mothers. There were also no differences in neurodevelopmental outcomes as measured by the Bayley Pyschomotor Scales at that age. Using yet another group of exercising women, Clapp measured morphometric and neurodevelopmental outcomes in 5-year-old children and compared these outcomes to children from less active, control mothers (Clapp 1996b). To be included in the exercise group in this study, women had to exercise (running, aerobics, cross country skiing, or a combination of these) 3 or more times per week for at least 30 min a bout and reach at least 55% maximum capacity intensity. As expected, at birth the infants from exercise mothers had lower body weight and reduced fat mass compared with those from control mothers. Interestingly, at 5 years old, children from exercise mothers continued to weigh less and have reduced fat mass compared to children from control mothers. When neurodevelopmental outcomes were measured, Clapp found that although children from both experimental groups performed similarly in motor and integrative skills tasks, children from exercise mothers performed significantly better in oral language skills and on the Wechsler Scales Intelligence Test compared with children from control mothers. The mechanisms behind the improved neurological outcomes and continued reduction in fat mass in this study are unclear, but this does suggest that exercise during pregnancy can have positive impacts on offspring many years after birth.

Using animal models, researchers have been able to start to evaluate the cellular changes associated with the observed effects of maternal exercise in offspring, as well as follow offspring as they age to assess the more long-term effects of exercise during pregnancy. In rodent models, many have looked at the impact of maternal exercise on the neurological outcomes in offspring. Lee et al. (2006) studied the effects of maternal swimming in rats on hippocampal neurogenesis and memory in offspring. Starting at time of conception, pregnant rats in the exercise cohort were forced to swim 10 min/day throughout pregnancy. Offspring were tested on postnatal day 21 in a step-down avoidance task for changes in short-term memory function and on postnatal day 29 offspring were euthanized and hippocampal neurogenesis and brain-derived neurotrophic factor (BDNF) mRNA levels were evaluated. Offspring from exercise dams had increased latency to step-down the avoidance task indicating enhanced short-term memory. On postnatal day 29, offspring from exercise dams showed increased expression of BDNF and increased neurogenesis in the hippocampus, the area of the brain associated with learning and memory. Another group also used forced exercise in a rat model of exercise during pregnancy to evaluate the effects of maternal exercise on offspring anxiety and neurogenesis in the prefrontal cortex (Aksu et al. 2012). Dams in the exercise group were acclimated to treadmill running for 1 week prior to pregnancy and during pregnancy dams ran at 8 m/min for 30 min/day, 5 days a week. Offspring from exercise and control dams were tested on postnatal day 26 and at 4 months of age in an open-field anxiety test. At both of these age points, offspring expression of BDNF and vascular endothelial growth factor (VEGF) (decreases in both are associated with anxiety levels) was measured. At both ages, offspring from exercise dams performed better in the open field task, indicating lower levels of anxiety compared with offspring from control dams. In the prefrontal cortex, offspring from exercise dams also had increased expression of BDNF and VEGF at 26 days and 4 months of age, indicating increased growth and vascularization in this area of the brain compared with offspring from control dams.

In an identical model of maternal exercise in rats, offspring were tested for changes in spatial memory in response to maternal exercise (Dayi et al. 2012). At postnatal days 21 and 120, offspring from control and exercise dams were tested for spatial learning and memory using a water maze. Animals were trained to find a platform and then tested for latency to find the platform in a water maze the next day. At both ages, offspring from exercise dams had decreased latency to find the platform in the maze following training compared with offspring from control dams, demonstrating enhanced spatial memory. Offspring from exercise dams also had increased neurogenesis in the hippocampus at both ages compared with those from control dams. Using mice, Park et al. (2013) used maternal treadmill running during pregnancy (40 min/day at 12 m/min for the 3 weeks of pregnancy) to look at BDNF in the hippocampus and memory in offspring. On postnatal day 3, offspring from exercise dams had enhanced BDNF expression in the hippocampus compared with those from control dams, which could functionally improve learning and memory in the offspring. At postnatal day 30, offspring from exercise dams also performed better on a Y-maze task, signifying improved memory compared with offspring from sedentary dams. Finally, Herring et al. (2012) evaluated the effects voluntary maternal exercise from the start and throughout the duration of pregnancy on offspring Alzheimer's disease outcomes in transgenic mice predisposed for the disease. On postnatal day 150, female offspring from control and exercise dams were evaluated for β-amyloid plaque formation and angiogenesis in the brain. Offspring from exercise dams had significantly decreased β-amyloid plaque burden and increased angiogenesis, as well as an improvement in several other outcomes associated with Alzheimer's disease compared with offspring from control dams. All of these studies suggest that at early and later ages, exercise during pregnancy can positively influence offspring brain development and function and potentially protect offspring from age-related disease.

Given the effects exercise during pregnancy can have on placental blood flow and glucose and oxygen reaching the fetus, surprisingly few studies have investigated the metabolic outcomes of maternal exercise in offspring. Almost two decades ago, Vanheest and Rodgers assessed offspring metabolic outcomes of maternal treadmill running in diabetic rat dams (Vanheest and Rodgers 1997). Female rats were made diabetic via streptozotocin injection and then divided into control and exercise groups. Exercise rats ran on a treadmill at a pace of 20 m/min for 1 h/day, 5 days a week immediately before and during pregnancy. Although there were no changes in maternal glucose regulation as a result of the treadmill training, offspring from exercise dams had significantly lower blood glucose levels during glucose tolerance testing following a glucose challenge at postnatal day 28 compared with those from control diabetic dams. Glucose tolerance in the offspring from exercise diabetic dams was significantly improved even when compared with offspring from control, non-diabetic dams. Despite improvements in glucose disposal, there were no changes in offspring fasting insulin levels as a result of maternal exercise. A more recent study evaluated the ability of maternal exercise to improve offspring metabolic dysfunction programmed by maternal protein nutrient restriction (Fidalgo et al. 2013). As described earlier, maternal protein restriction can cause insulin resistance and glucose intolerance in offspring. In this study, dams were given a low-protein diet, which caused hyperglycemia, hypercholesterolemia, and glucose intolerance in offspring by 150 days of age. These parameters were reversed, however, in offspring from dams that ran on a treadmill for 60 min/day 5 days a week at 65% maximum capacity for 4 weeks prior to pregnancy and throughout pregnancy. The results from this study suggest that exercise during pregnancy can improve glucose regulation in offspring predisposed to glucose intolerance.

Using rats, two separate studies have demonstrated that maternal exercise in unhealthy pregnancies (gestational diabetes and maternal nutrient restriction) can improve offspring glucose metabolism compared with offspring from diabetic or nutrient-restricted sedentary dams (Vanheest and Rodgers 1997; Fidalgo et al. 2013). In humans, offspring exposed to maternal exercise have lower fat mass and body weight at birth (Jackson et al. 1995; Clapp et al. 1998; Clapp 2003; Hopkins et al. 2010). In one cohort of women, 5-year-old children from exercising mothers had higher intelligence scores and lower fat mass compared with children from sedentary mothers (Clapp 1996b).

No studies in humans or animals have looked at the lifelong impact of exercise during a normal pregnancy on offspring insulin sensitivity or glucose disposal. Therefore, our group hypothesized that maternal exercise during a healthy pregnancy would provide metabolic benefits to offspring. We tested this hypothesis in both Institute for Cancer (ICR) mice and Sprague-Dawley rats (Carter et al. 2012, 2013). The main findings of these studies were that maternal voluntary exercise could improve offspring insulin sensitivity and glucose regulation in both mouse and rat models. These results suggest that maternal exercise could potentially be used as a method for preventing or decreasing insulin resistance and type 2 diabetes in future generations. In the mouse model (Carter et al. 2012), exercise during pregnancy did not change maternal body weight; however, it did cause an increase in maternal food intake. Exercise did not affect litter size or offspring body weights during nursing or throughout lifespan. At approximately 7 months of age (Carter et al. 2012) and up through 16–17 months of age when the study was ended (Figure 2.1a and b), male and female offspring from exercise dams had enhanced glucose disposal compared with offspring from control dams as measured by glucose tolerance tests. Offspring from exercise dams also showed enhanced glucose uptake into peripheral tissues following injection of exogenous insulin, indicating enhanced insulin sensitivity in adipose and skeletal muscle compared with offspring from control dams. In female offspring, insulin stimulated glucose uptake was measured ex vivo in an adipose depot and soleus muscle. Unfortunately, tissues from male offspring were not tested. Offspring from exercise dams had increased glucose uptake in both tissues in response to insulin treatment compared with tissues from offspring born to control dams. Finally, body composition in female offspring was unaffected by maternal

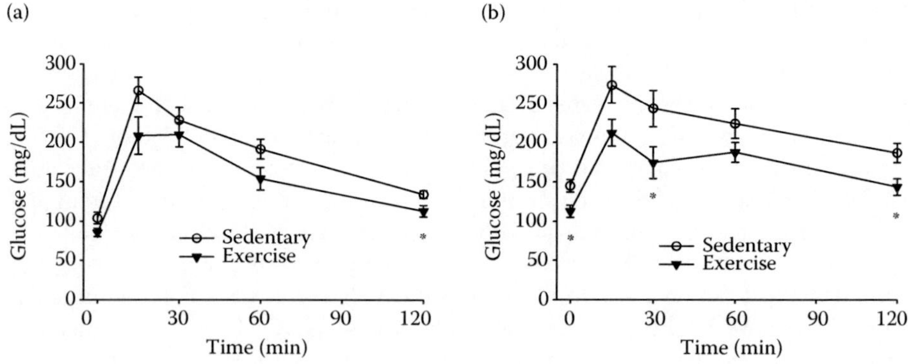

FIGURE 2.1 Seventy-one- and seventy-two-week-old female and male offspring born to exercised dams had improved glucose disposal. Using ICR mice, glucose tolerance was measured in aged female (a) and male (b) offspring born to sedentary or exercise dams. Male but not female offspring from exercised dams had significantly decreased fasting blood glucose levels compared with offspring from sedentary dams. Following oral glucose administration, blood glucose levels were significantly lower in offspring from exercised dams at 120 min in females and at 30 and 120 min in males. Overall glucose disposal as measured by repeated measures analysis of variance was also significantly improved in both female and male offspring born to exercise compared with sedentary dams. *$p < 0.05$ compared with offspring born to sedentary dams; n = 9–15 for (a) and (b). Error bars indicate SEM.

exercise, but males from exercise dams had decreased fat mass and increased lean mass at several age points (starting at 9 months) compared with males from control dams (Carter et al. 2012). Figure 2.2a and b shows the lean to fat mass ratio of male and female offspring born sedentary and exercised offspring at ~16 months of age. The offspring born to exercised dams showed a trend toward increased lean to fat mass ratio while male offspring born to exercised dams had significantly higher lean to fat mass ratio compared with offspring from sedentary dams.

Similar results were found in the rat model. Maternal exercise did not affect food intake but did cause a small decrease in maternal body weight. Again, litter size and offspring body weights during nursing were unchanged with the exercise intervention. Female offspring from exercise dams showed enhanced glucose disposal compared with offspring from control dams at approximately 10 months of age, and these improvements were seen again at 12 and 14 months of age. At 17 months, female offspring from exercise dams had enhanced whole-body insulin sensitivity compared with those from control dams as measured by hyperinsulinemic–euglycemic clamp. These females required a higher glucose infusion rate to maintain target blood glucose levels during insulin infusion and had higher glucose uptake into skeletal muscle compared with females from control dams. Again body composition was unchanged in the female offspring born to exercising dams during adulthood. A timed breeding in the rats also revealed that voluntary exercise did not change maternal glucose tolerance midgestation; however, exercise did reduce maternal fasting insulin and decreased fat pad mass compared with control dams. Unfortunately, glucose tolerance, hyperinsulinemic clamp, and body composition were not measured in the male

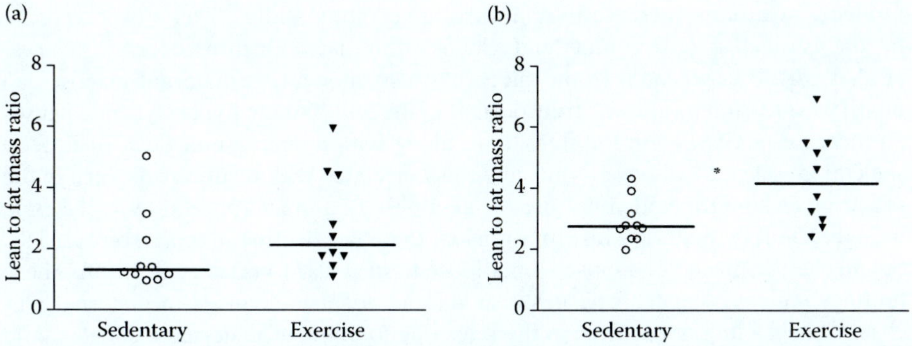

FIGURE 2.2 Maternal exercise significantly impacts body composition in 68-week-old male but not female offspring. Total body fat mass and lean mass were analyzed in aged female and male ICR mouse offspring using EchoMRI. Lean to fat mass ratio was calculated and is shown for female (a) and male (b) offspring born to sedentary and exercised dams. There was a trend observed in female offspring ($p = 0.076$), but male offspring born to exercised dams had significantly higher lean to fat mass ratio compared with offspring from sedentary dams. *$p < 0.05$ compared with offspring born to sedentary dams; n = 10 for (a) and (b). Data were not normally distributed; horizontal line indicates median.

offspring as part of this experiment. Recently, Vega et al. (2013) explored the effects of maternal exercise in normal and obese Wistar rats and found that female and male offspring born to exercising control fed dams had significantly lower fasting glucose levels at 36 days of age compared with those born to sedentary dams. Exercise also attenuated many of the maternal and offspring outcomes associated with high-fat diet consumption before and during pregnancy and nursing (Vega et al. 2013). These promising studies suggest that work in this area should be intensified.

2.9 DISCUSSION AND FUTURE DIRECTIONS

Our results are the first to show that exercise during healthy pregnancy can impact long-term offspring insulin sensitivity and glucose regulation; however, the mechanisms behind these improvements are still unknown (Carter et al. 2012, 2013). The main question is what is happening with maternal exercise that could be influencing fetal development that results in long-term metabolic changes in offspring. To begin, sustained, weight-bearing endurance exercise, such as running, increases placental growth, size, and blood flow in early pregnancy (Jackson et al. 1995; Clapp et al. 2000a, 2002), although in our mice and rats, we did not measure differences in placental weight with exercise toward the end of gestation (data not shown). These changes in placenta size and blood flow can increase glucose and oxygen levels reaching the fetus; however, bouts of sustained weight-bearing endurance exercises cause intermittent reduction in blood flow to placenta in favor of maternal muscle and skin (Rowell and Blackmon 1987; Clapp et al. 2000b). These alterations in maternal glucose flow could be important for several reasons. In the sheep model, maternal glucose signals to the fetus to release insulin; acute increases in glucose cause an increase in fetal insulin secretion while chronic overexposure to glucose can actually blunt fetal glucose stimulated insulin secretion (Carver et al. 1996). Past research in the sheep has also shown that maternal glucose can modify expression of glucose transporters in the fetus. Acute hyperglycemia causes an increase in GLUT1 in the fetal brain along with a decrease of GLUT4 in fetal skeletal muscle and adipose tissue, and this correlates with insulin resistance in the offspring (Aldoretta et al. 1994; Das et al. 1999). In a model of fetal hypoglycemia brought about by maternal insulin infusion, there is a decline in fetal liver GLUT1, but no changes in skeletal muscle or adipose tissue GLUT4 (Das et al. 1999). These findings indicate that fetal tissues respond and adapt to changes in maternal glucose. Changes in glucose flow to the fetus due to bouts of maternal exercise could be a potential programming mechanism behind the observed results in our studies. Intermittent reductions in glucose reaching the fetus could be altering fetal insulin secretion patterns and/or altering how the pancreas will respond to glucose stimulation, thereby modifying adipose tissue and skeletal muscle expression of insulin signaling proteins. In mouse and rat models, changes in fetal exposure to glucose levels would be particularly important late in gestation when the rodent fetal pancreas exhibits increased glucose stimulated insulin release (Kervran and Girard 1974; Freinkel et al. 1984). Insulin signaling proteins may be upregulated in these tissues and therefore making the tissues more sensitive to insulin stimulation. A next step should be to look more extensively at insulin signaling proteins in the skeletal muscle and adipose tissue.

In our studies (Carter et al. 2012, 2013), differences in offspring insulin sensitivity and glucose regulation were not detected until the animals were aged. In mice, differences were first observed around 7 months of age and in rats, differences were detected starting at 10 months of age. This is not necessarily surprising given that much of the developmental programming research investigating glucose metabolism in offspring do not see differences until animals are aged (Ozanne et al. 2001; Fernandez-Twinn et al. 2005). It may be that modifications in insulin sensitivity created by maternal malnutrition, or in the case of this project, maternal exercise, are subtle and cannot be detected until an "insult" such as age, occurs. It has been well documented that aging, even in healthy subjects, can lead to impaired glucose tolerance and peripheral insulin resistance (Davidson 1979; DeFronzo 1979, 1981; Robert et al. 1982; Fink et al. 1983). In elderly subjects with impaired glucose uptake into skeletal muscle and adipose tissue, there is normal binding of insulin to IR in these tissues, indicating impaired postreceptor signaling (Fink et al. 1983). Changes induced in offspring by maternal exercise could be protecting the offspring from this age related decline in insulin sensitivity. It will be important to look at insulin signaling protein expression and activation both before and after there are noticeable differences in glucose regulation (at young and older ages) in offspring. A goal of a future study may also be to examine whether or not maternal exercise can protect offspring from the harmful effects of high-fat feeding, similar to its effects seen in the aging offspring.

Timing and intensity of exercise required to produce beneficial effects in offspring should also be a focus of upcoming studies. In the current rodent model, mice and rats had voluntary access to running wheels before and during pregnancy and nursing. Once acclimated to the wheels, both mice and rats ran up to an average of 10 km/day. To reduce and control the amount of running, a controlled exercise paradigm could be used in which mice or rats are removed from the home cage and placed on a treadmill or wheel for a prescribed amount of time per day.

Rodents can be offered an exercise intervention before and during pregnancy, or start exercise right at conception as a means to investigate the critical window of exercise exposure. It is most likely that exercise before and during pregnancy will produce the most robust results in offspring since, in humans, the greatest reduction in fat mass at birth occurred when women exercise before and during pregnancy (Clapp and Capeless 1990). This particular timing of exercise may bring about the greatest changes in placental blood flow and therefore the greatest change in glucose flow to fetus, leading to the biggest changes in fetal insulin secretion and possibly programming of insulin sensitivity. Exercise training can increase the growth and density of blood vessels in skeletal muscle (Saltin and Rowell 1980; Honig et al. 1982), and perhaps in women who are already trained this contributes to greater changes in blood flow during exercise bouts during pregnancy than in women who are only starting an exercise regimen at the onset of pregnancy. The evidence from humans also suggests that it may be detrimental to stop exercise during pregnancy because of the impact exercise can have on placental growth. Again, physical activity in early pregnancy can increase placenta size and blood flow (Jackson et al. 1995; Clapp et al. 2000a, 2002), so stopping exercise after this period can actually cause excess nutrient flow to the fetus resulting in higher fat mass and weight at birth compared with infants from mothers who were not active throughout their entire pregnancy (Jackson et al. 1995). Therefore, it will be most interesting to see if exercise initiated at the beginning of pregnancy as opposed to before can have the same effects in offspring as seen in our studies. This might also be more relevant in the human population since many women who become pregnant are not actively trying to become pregnant and would not have an opportunity to start an exercise regimen before conception (Singh et al. 2010).

Cross-fostering can also be used to evaluate the importance of maternal exercise during nursing on offspring. Pups from control dams would be cross-fostered to exercise dams and vice versa. Offspring would then be tested for differences between those exposed to maternal exercise during only gestation and those exposed only during the nursing period. In humans, moderate to intense endurance training during lactation does not impact lipid or lactose concentrations of energy density in breastmilk but can significantly increase protein content (Dewey et al. 1994). This level of exercise can also increase milk volume compared with volume from control mothers (Lovelady et al. 1990). In rats, swim training does not change milk protein or fat content but can lower lactose concentrations (Treadway and Lederman 1986). Since exercise can somewhat alter milk composition, it will important to assess its influence on offspring in upcoming studies.

Given that data from both human and animal studies seems to suggest that even fairly intense weight-bearing exercises (such as running and aerobics) are not harmful to the fetus (Lotgering et al. 1985; Clapp et al. 1995; Clapp 2009), and that many women of childbearing age already exercise regularly (Bureau of Labor Statistics 2008), a feasible and important future goal of this project is to investigate the offspring metabolic effects of exercise during pregnancy in humans (Hatoum et al. 1997; Clapp et al. 2003). There would be several factors to consider when designing a human trial for this project such as subject characteristics and exercise type/regimen. Since the initial goal of a study like this would be to follow metabolic outcomes in offspring as opposed to improving maternal health, it would be easiest to choose a population of healthy women who are already performing some type of regular endurance exercise. Recruiting women who already exercise on a regular basis could also help with compliance throughout the study. Endurance exercise would be preferable over only resistance training since the evidence thus far in humans indicates that this type of physical activity result in some offspring differences like reduced fat mass at birth and reduced IGF-1 in cord blood (Clapp and Capeless 1990; Clapp et al. 1998; Hopkins et al. 2010). Resistance training may be more desirable in pregnancies complicated by obesity and/or diabetes.

Additionally, there has been little investigation into exercise as an intervention for maternal obesity and gestational diabetes. Obesity is a major risk factor for the development of gestational diabetes, which is detrimental to the health of both mother and offspring (Chu et al. 2007). Gestational diabetes heightens the risk of developing type 2 diabetes after pregnancy and can lead to obesity and diabetes in offspring (Ben-Haroush et al. 2004; Franks et al. 2006). Although a few studies have shown that resistance training can decrease the need for insulin to control blood glucose in pregnant women with gestational diabetes (Brankston et al. 2004; de Barros et al. 2010), there is a general lack of knowledge on the type, timing, and intensity of exercise best used as a treatment in pregnancies complicated by obesity and diabetes.

Decreases in fat mass could reduce inflammation-induced insulin resistance. Inflammatory cytokines, such as TNF-α, are increased with obesity and pregnancy and can inactivate proteins in the insulin signaling pathway, and these levels can be even higher in obese pregnant women (Hotamisligil et al. 1993; Clapp and Kiess 2000; Lappas et al. 2004), while exercise training can decrease circulating cytokine levels (Tsukui et al. 2000). For some overweight women with infertility issues, diet and exercise have been effective in helping to conceive (Clark et al. 1995). It is perhaps more relevant to study exercise at the onset of pregnancy when targeting maternal obesity and gestational diabetes. Certain types of exercise may not be suitable for overweight and/or diabetic women and their developing child. In humans, the intense weight-bearing exercises that have resulted in offspring benefits were performed by healthy women (Clapp 1996a; Clapp et al. 1999). For previously sedentary and overweight women, high levels of endurance training could be stressful and potentially harmful for the fetus (Blaza and Garrow 1983; Hulens et al. 2001). If the goal is to maintain normal glucose homeostasis, less rigorous cardiovascular exercises like resistance training that still promote muscle contraction and thereby stimulate glucose uptake into muscle, may be more reasonable and less stressful. Previous studies have shown that resistance training, such as use of resistance bands, can be effective in helping to control hyperglycemia associated with gestational diabetes (Brankston et al. 2004).

Epigenetic processes have gained increasing attention as molecular mediators through which gene expression can be controlled over the long term, independent of genetic sequences. Nutritional stimuli at critical stages during development have lasting influences on gene expression (Waterland and Jirtle 2003, 2004; Gallou-Kabani and Junien 2005). Alterations in prenatal or early postnatal diet, such as high-fat diet or low-protein diet, have significant influences on epigenetic modifications, such as DNA methylation and chromatin remodeling, in the brain including DNA methylation differences in the IR promoter and proopiomelanocortin in the hypothalamus (Plagemann et al. 2009, 2010; Vucetic et al. 2010a, b, 2011). Mice maintained on chronic high-fat diet for 3 months have offspring that display a preference for palatable foods and this behavior is concurrent with global and gene-specific (dopamine transporter, μ-opioid receptor, preproenkephalin) promoter DNA hypomethylation in brain circuits associated with reward (Vucetic et al. 2010a). Protein restriction alters hepatic methylation and gene expression of the glucocorticoid receptor, peroxisome proliferator–activated receptor (PPAR) α, and PPAR-γ (Lillycrop et al. 2005, 2007, 2008; Fujiki et al. 2009). Additional genes involved in energy homeostasis that show in vivo epigenetic regulation by DNA methylation include leptin, 11β-hydroxysteroid dehydrogenase 2, and corticotrophin-releasing hormone (Alikhani-Koopaei et al. 2004; McGill et al. 2006; Friso et al. 2008; Mueller and Bale 2008; Milagro et al. 2009). Studies in vitro have demonstrated that DNA methylation regulates the expression of leptin, SOCS3, and GLUT4 (Yokomori et al. 1999; Melzner et al. 2002; Stoger 2008; Campion et al. 2009). Finally, human studies, such as those that have followed offspring whose mothers experienced severe famine during the Dutch Hunger Winter, indicate that DNA methylation patterns can be influenced by maternal diet. These epigenetic modifications can be long-lasting and may be passed on to future generations (Stein et al. 2007; Painter et al. 2008; Ahmed 2010; Lange and Schneider 2010). These data further support the notion that epigenetic mechanisms may contribute to the long-lasting effects of altered perinatal nutrition, maternal obesity, or gestational diabetes and could represent novel therapeutic targets.

To date, little is known about the possible epigenetic changes exercise during pregnancy may exert on offspring. Some studies have, however, looked at the epigenetic changes that occur with exercise, and these findings may be an important starting point for determining epigenetic changes that may contribute to improvements in offspring insulin sensitivity and glucose homeostasis. Although the majority of exercise and epigenetic studies have focused on skeletal muscle adaptations, exercise is linked to genome-wide alterations in DNA methylation in both skeletal muscle and adipose tissue (Barres et al. 2012; Nitert et al. 2012; Ronn et al. 2013). Of particular interest, after muscle contraction, methylation at the promoter region of genes involved in glucose homeostasis such as PPAR-δ, PPAR-γ coactivator 1α (PGC-1α), and PDK4 is significantly decreased (Barres et al. 2012). Nitert et al. (2012) have shown that DNA methylation and expression of genes involved in muscle metabolism, such as adiponectin receptors 1 and 2, change in response to exercise. Also, in adipose tissue, exercise training resulted in alterations in DNA methylation on several obesity and type 2 diabetes candidate genes including IGF-binding protein 2 and transcription factor 7-like 2 (TCF7L2) (Ronn et al. 2013). Exercise has also been linked to histone modifications in skeletal muscle (McGee et al. 2009; Egan et al. 2010; McGee and Hargreaves 2011), and McGee and Hargreaves (2011) provide a thorough review of exercise-associated histone changes. Regardless, many more experiments must be done before any final conclusions can be made about the epigenetic mechanisms involved in the programming of obesity and metabolic disease.

2.10 CONCLUSION

It is clear that the intrauterine and early postnatal environments are critical in supporting proper development and long-term health of offspring. Even slight alterations during these critical windows can have detrimental effects on the embryo/fetus/neonate that can lead to numerous pathologies in adolescence and adulthood. Determining these long-term effects in offspring will provide insight into the mechanisms through which altered maternal diet, maternal obesity, or gestational diabetes during the perinatal period can lead to offspring obesity and insulin resistance. The development of interventions, such as exercise regimens during pregnancy, to prevent long-term metabolic dysfunction would elevate the field to the next level and provide critical information to facilitate translational steps to improve human public health. Once the benefits of maternal exercise are realized, physicians may be better able to advise women about participation in exercise regimens throughout pregnancy. Women may also be more willing to start a healthy diet and training program if they are educated on the benefits it could have for their children. Exercise during pregnancy is a potential short-term intervention that may decrease susceptibility and incidence of insulin resistance and diabetes in future generations.

ACKNOWLEDGMENTS

Data collection in this chapter's figures was supported by U.S. National Institutes of Health (NIH) grants (R01 DK090460 to K.J.P.) and the research core of the COBRE on Obesity and Cardiovascular Diseases from the National Institute of General Medical Sciences (8P20 GM103527). L.G.C. was supported by an NIH training grant (DK07778). K.L.K.T. was supported by an NIH grant from NICHD (HD055030).

REFERENCES

ACOG Committee opinion. Number 267, January 2002: Exercise during pregnancy and the postpartum period. 2002. *Obstet Gynecol* 99 (1):171–3.

Ahmed, F. 2010. Epigenetics: Tales of adversity. *Nature* 468 (7327):S20.

Aksu, I., B. Baykara, S. Ozbal et al. 2012. Maternal treadmill exercise during pregnancy decreases anxiety and increases prefrontal cortex VEGF and BDNF levels of rat pups in early and late periods of life. *Neurosci Lett* 516 (2):221–5.

Aldoretta, P. W., T. D. Carver, and W. W. Hay Jr. 1994. Ovine uteroplacental glucose and oxygen metabolism in relation to chronic changes in maternal and fetal glucose concentrations. *Placenta* 15 (7):753–64.

Alikhani-Koopaei, R., F. Fouladkou, F. J. Frey, and B. M. Frey. 2004. Epigenetic regulation of 11 beta-hydroxysteroid dehydrogenase type 2 expression. *J Clin Invest* 114 (8):1146–57.

American College of Sports Medicine. 2011. ACSM current comment—Exercise during pregnancy. Available at http://www.acsm.org/docs/current-comments.

Ashino, N. G., K. N. Saito, F. D. Souza et al. 2012. Maternal high-fat feeding through pregnancy and lactation predisposes mouse offspring to molecular insulin resistance and fatty liver. *J Nutr Biochem* 23 (4):341–8.

Bahl, J. J., M. Matsuda, R. A. DeFronzo, and R. Bressler. 1997. In vitro and in vivo suppression of gluconeogenesis by inhibition of pyruvate carboxylase. *Biochem Pharmacol* 53 (1):67–74.

Ballor, D. L., and R. E. Keesey. 1991. A meta-analysis of the factors affecting exercise-induced changes in body mass, fat mass and fat-free mass in males and females. *Int J Obes* 15 (11):717–26.

Barakat, R., M. Pelaez, C. Lopez, A. Lucia, and J. R. Ruiz. 2013. Exercise during pregnancy and gestational diabetes-related adverse effects: A randomised controlled trial. *Br J Sports Med* 47 (10):630–6.

Barbour, L. A., C. E. McCurdy, T. L. Hernandez, J. P. Kirwan, P. M. Catalano, and J. E. Friedman. 2007. Cellular mechanisms for insulin resistance in normal pregnancy and gestational diabetes. *Diabetes Care* 30 Suppl 2:S112–9.

Barker, D. J., A. R. Bull, C. Osmond, and S. J. Simmonds. 1990. Fetal and placental size and risk of hypertension in adult life. *BMJ* 301 (6746):259–62.

Barker, D. J., T. W. Meade, and C. H. Fall. 1992. Relation of fetal and infant growth to plasma fibrinogen and factor VII concentrations in adult life. *BMJ* 304 (6820):148–52.

Barker, D. J., C. Osmond, J. Golding, D. Kuh, and M. E. Wadsworth. 1989a. Growth in utero, blood pressure in childhood and adult life, and mortality from cardiovascular disease. *BMJ* 298 (6673):564–7.

Barker, D. J., C. Osmond, and C. M. Law. 1989c. The intrauterine and early postnatal origins of cardiovascular disease and chronic bronchitis. *J Epidemiol Community Health* 43 (3):237–40.

Barker, D. J., P. D. Winter, C. Osmond, B. Margetts, and S. J. Simmonds. 1989b. Weight in infancy and death from ischaemic heart disease. *Lancet* 2 (8663):577–80.

Barres, R., J. Yan, B. Egan et al. 2012. Acute exercise remodels promoter methylation in human skeletal muscle. *Cell Metab* 15 (3):405–11.

Bell, R. J., S. M. Palma, and J. M. Lumley. 1995. The effect of vigorous exercise during pregnancy on birth-weight. *Aust N Z J Obstet Gynaecol* 35 (1):46–51.

Ben-Haroush, A., Y. Yogev, and M. Hod. 2004. Epidemiology of gestational diabetes mellitus and its association with Type 2 diabetes. *Diabet Med* 21 (2):103–13.

Blaza, S., and J. S. Garrow. 1983. Thermogenic response to temperature, exercise and food stimuli in lean and obese women, studied by 24 h direct calorimetry. *Br J Nutr* 49 (2):171–80.

Boloker, J., S. J. Gertz, and R. A. Simmons. 2002. Gestational diabetes leads to the development of diabetes in adulthood in the rat. *Diabetes* 51 (5):1499–506.

Boujendar, S., B. Reusens, S. Merezak et al. 2002. Taurine supplementation to a low protein diet during foetal and early postnatal life restores a normal proliferation and apoptosis of rat pancreatic islets. *Diabetologia* 45 (6):856–66.

Brankston, G. N., B. F. Mitchell, E. A. Ryan, and N. B. Okun. 2004. Resistance exercise decreases the need for insulin in overweight women with gestational diabetes mellitus. *Am J Obstet Gynecol* 190 (1):188–93.

Bureau of Labor Statistics. 2008. Spotlight on statistics—Sports and exercise. Available at http://www.bls.gov/.

Cai, D., M. Yuan, D. F. Frantz et al. 2005. Local and systemic insulin resistance resulting from hepatic activation of IKK-beta and NF-kappaB. *Nat Med* 11 (2):183–90.

Campion, J., F. I. Milagro, and J. A. Martinez. 2009. Individuality and epigenetics in obesity. *Obes Rev* 10 (4):383–92.

Carter, L. G., K. N. Lewis, D. C. Wilkerson et al. 2012. Perinatal exercise improves glucose homeostasis in adult offspring. *Am J Physiol Endocrinol Metab* 303 (8):E1061–8.

Carter, L. G., N. R. Qi, R. de Cabo, and K. J. Pearson. 2013. Maternal exercise improves insulin sensitivity in mature rat offspring. *Med Sci Sports Exerc* 45 (5):832–40.

Carver, T. D., S. M. Anderson, P. W. Aldoretta, and W. W. Hay Jr. 1996. Effect of low-level basal plus marked "pulsatile" hyperglycemia on insulin secretion in fetal sheep. *Am J Physiol* 271 (5 Pt 1):E865–71.

Catalano, P. M., A. Thomas, L. Huston-Presley, and S. B. Amini. 2003. Increased fetal adiposity: a very sensitive marker of abnormal in utero development. *Am J Obstet Gynecol* 189 (6):1698–704.

Chandler-Laney, P. C., N. C. Bush, D. J. Rouse, M. S. Mancuso, and B. A. Gower. 2011. Maternal glucose concentration during pregnancy predicts fat and lean mass of prepubertal offspring. *Diabetes Care* 34 (3):741–5.

Chu, S. Y., W. M. Callaghan, S. Y. Kim et al. 2007. Maternal obesity and risk of gestational diabetes mellitus. *Diabetes Care* 30 (8):2070–6.

Clapp, J. F. III. 1994. A clinical approach to exercise during pregnancy. *Clin Sports Med* 13 (2):443–58.

Clapp, J. F. III. 1996a. The effect of continuing regular endurance exercise on the physiologic adaptations to pregnancy and pregnancy outcome. *Am J Sports Med* 24 (6 Suppl):S28–9.

Clapp, J. F. III. 1996b. Morphometric and neurodevelopmental outcome at age five years of the offspring of women who continued to exercise regularly throughout pregnancy. *J Pediatr* 129 (6):856–63.

Clapp, J. F. III. 2000. Exercise during pregnancy. A clinical update. *Clin Sports Med* 19 (2):273–86.

Clapp, J. F. III. 2003. The effects of maternal exercise on fetal oxygenation and feto-placental growth. *Eur J Obstet Gynecol Reprod Biol* 110 Suppl 1:S80–5.

Clapp, J. F. III. 2009. Is exercise during pregnancy related to preterm birth? *Clin J Sport Med* 19 (3):241–3.

Clapp, J. F. III, and E. Capeless. 1997. Cardiovascular function before, during, and after the first and subsequent pregnancies. *Am J Cardiol* 80 (11):1469–73.

Clapp, J. F. III, and E. L. Capeless. 1990. Neonatal morphometrics after endurance exercise during pregnancy. *Am J Obstet Gynecol* 163 (6 Pt 1):1805–11.

Clapp, J. F. III, and W. Kiess. 2000. Effects of pregnancy and exercise on concentrations of the metabolic markers tumor necrosis factor alpha and leptin. *Am J Obstet Gynecol* 182 (2):300–6.

Clapp, J. F. III, H. Kim, B. Burciu, and B. Lopez. 2000a. Beginning regular exercise in early pregnancy: Effect on fetoplacental growth. *Am J Obstet Gynecol* 183 (6):1484–8.

Clapp, J. F. III, H. Kim, B. Burciu, S. Schmidt, K. Petry, and B. Lopez. 2002. Continuing regular exercise during pregnancy: Effect of exercise volume on fetoplacental growth. *Am J Obstet Gynecol* 186 (1):142–7.

Clapp, J. F. III, and K. D. Little. 1995. The interaction between regular exercise and selected aspects of women's health. *Am J Obstet Gynecol* 173 (1):2–9.

Clapp, J. F. III, K. D. Little, S. K. Appleby-Wineberg, and J. A. Widness. 1995. The effect of regular maternal exercise on erythropoietin in cord blood and amniotic fluid. *Am J Obstet Gynecol* 172 (5):1445–51.

Clapp, J. F. III, K. D. Little, and J. A. Widness. 2003. Effect of maternal exercise and fetoplacental growth rate on serum erythropoietin concentrations. *Am J Obstet Gynecol* 188 (4):1021–5.

Clapp, J. F. III, B. Lopez, and R. Harcar-Sevcik. 1999. Neonatal behavioral profile of the offspring of women who continued to exercise regularly throughout pregnancy. *Am J Obstet Gynecol* 180 (1 Pt 1):91–4.

Clapp, J. F. III, S. Simonian, B. Lopez, S. Appleby-Wineberg, and R. Harcar-Sevcik. 1998. The one-year morphometric and neurodevelopmental outcome of the offspring of women who continued to exercise regularly throughout pregnancy. *Am J Obstet Gynecol* 178 (3):594–9.

Clapp, J. F. III, W. Stepanchak, J. Tomaselli, M. Kortan, and S. Faneslow. 2000b. Portal vein blood flow-effects of pregnancy, gravity, and exercise. *Am J Obstet Gynecol* 183 (1):167–72.

Clark, A. M., W. Ledger, C. Galletly et al. 1995. Weight loss results in significant improvement in pregnancy and ovulation rates in anovulatory obese women. *Hum Reprod* 10 (10):2705–12.

Claus, T. H., and S. J. Pilkis. 1976. Regulation by insulin of gluconeogenesis in isolated rat hepatocytes. *Biochim Biophys Acta* 421 (2):246–62.

Collings, C. A., L. B. Curet, and J. P. Mullin. 1983. Maternal and fetal responses to a maternal aerobic exercise program. *Am J Obstet Gynecol* 145 (6):702–7.

Da Costa, D., N. Rippen, M. Dritsa, and A. Ring. 2003. Self-reported leisure-time physical activity during pregnancy and relationship to psychological well-being. *J Psychosom Obstet Gynecol* 24 (2):111–9.

Dabelea, D., R. L. Hanson, R. S. Lindsay et al. 2000. Intrauterine exposure to diabetes conveys risks for type 2 diabetes and obesity: A study of discordant sibships. *Diabetes* 49 (12):2208–11.

Das, U. G., R. E. Schroeder, W. W. Hay Jr., and S. U. Devaskar. 1999. Time-dependent and tissue-specific effects of circulating glucose on fetal ovine glucose transporters. *Am J Physiol* 276 (3 Pt 2):R809–17.

Davidson, M. B. 1979. The effect of aging on carbohydrate metabolism: A review of the English literature and a practical approach to the diagnosis of diabetes mellitus in the elderly. *Metabolism* 28 (6):688–705.

Davies, G. A., L. A. Wolfe, M. F. Mottola, and C. MacKinnon. 2003. Joint SOGC/CSEP clinical practice guideline: Exercise in pregnancy and the postpartum period. *Can J Appl Physiol* 28 (3):330–41.

Dayi, A., S. Agilkaya, S. Ozbal et al. 2012. Maternal aerobic exercise during pregnancy can increase spatial learning by affecting leptin expression on offspring's early and late period in life depending on gender. *Sci World J* 2012:429803.

de Barros, M. C., M. A. Lopes, R. P. Francisco, A. D. Sapienza, and M. Zugaib. 2010. Resistance exercise and glycemic control in women with gestational diabetes mellitus. *Am J Obstet Gynecol* 203 (6):556.e1–6.

De Prins, F. A., and F. A. Van Assche. 1982. Intrauterine growth retardation and development of endocrine pancreas in the experimental rat. *Biol Neonate* 41 (1–2):16–21.

de Rooij, S. R., R. C. Painter, D. I. Phillips et al. 2006. Impaired insulin secretion after prenatal exposure to the Dutch famine. *Diabetes Care* 29 (8):1897–901.

DeFronzo, R. A. 1979. Glucose intolerance and aging: Evidence for tissue insensitivity to insulin. *Diabetes* 28 (12):1095–101.

DeFronzo, R. A. 1981. Glucose intolerance and aging. *Diabetes Care* 4 (4):493–501.

DeFronzo, R. A., R. Gunnarsson, O. Bjorkman, M. Olsson, and J. Wahren. 1985. Effects of insulin on peripheral and splanchnic glucose metabolism in noninsulin-dependent (type II) diabetes mellitus. *J Clin Invest* 76 (1):149–55.

DeFronzo, R. A., E. Jacot, E. Jequier, E. Maeder, J. Wahren, and J. P. Felber. 1981. The effect of insulin on the disposal of intravenous glucose. Results from indirect calorimetry and hepatic and femoral venous catheterization. *Diabetes* 30 (12):1000–7.

Dempsey, J. C., C. L. Butler, T. K. Sorensen et al. 2004a. A case-control study of maternal recreational physical activity and risk of gestational diabetes mellitus. *Diabetes Res Clin Pract* 66 (2):203–15.

Dempsey, J. C., T. K. Sorensen, M. A. Williams et al. 2004b. Prospective study of gestational diabetes mellitus risk in relation to maternal recreational physical activity before and during pregnancy. *Am J Epidemiol* 159 (7):663–70.

Denison, F. C., and C. Chiswick. 2011. Improving pregnancy outcome in obese women. *Proc Nutr Soc* 70 (4):457–64.

Dentin, R., Y. Liu, S. H. Koo et al. 2007. Insulin modulates gluconeogenesis by inhibition of the coactivator TORC2. *Nature* 449 (7160):366–9.

Derave, W., S. Lund, G. D. Holman, J. Wojtaszewski, O. Pedersen, and E. A. Richter. 1999. Contraction-stimulated muscle glucose transport and GLUT-4 surface content are dependent on glycogen content. *Am J Physiol* 277 (6 Pt 1):E1103–10.

Dewey, K. G., C. A. Lovelady, L. A. Nommsen-Rivers, M. A. McCrory, and B. Lonnerdal. 1994. A randomized study of the effects of aerobic exercise by lactating women on breast-milk volume and composition. *N Engl J Med* 330 (7):449–53.

Duncker, D. J., and R. J. Bache. 2008. Regulation of coronary blood flow during exercise. *Physiol Rev* 88 (3):1009–86.

Duvekot, J. J., and L. L. Peeters. 1994. Maternal cardiovascular hemodynamic adaptation to pregnancy. *Obstet Gynecol Surv* 49 (12 Suppl):S1–14.

Egan, B., B. P. Carson, P. M. Garcia-Roves et al. 2010. Exercise intensity-dependent regulation of peroxisome proliferator–activated receptor coactivator-1 mRNA abundance is associated with differential activation of upstream signalling kinases in human skeletal muscle. *J Physiol* 588 (Pt 10):1779–90.

El-Khattabi, I., F. Gregoire, C. Remacle, and B. Reusens. 2003. Isocaloric maternal low-protein diet alters IGF-I, IGFBPs, and hepatocyte proliferation in the fetal rat. *Am J Physiol Endocrinol Metab* 285 (5):E991–1000.

Farup, J., T. Kjolhede, H. Sorensen et al. 2012. Muscle morphological and strength adaptations to endurance vs. resistance training. *J Strength Cond Res* 26 (2):398–407.

Fenicchia, L. M., J. A. Kanaley, J. L. Azevedo Jr. et al. 2004. Influence of resistance exercise training on glucose control in women with type 2 diabetes. *Metabolism* 53 (3):284–9.

Fernandez-Twinn, D. S., A. Wayman, S. Ekizoglou, M. S. Martin, C. N. Hales, and S. E. Ozanne. 2005. Maternal protein restriction leads to hyperinsulinemia and reduced insulin-signaling protein expression in 21-mo-old female rat offspring. *Am J Physiol Regul Integr Comp Physiol* 288 (2):R368–73.

Fidalgo, M., F. Falcao-Tebas, A. Bento-Santos et al. 2013. Programmed changes in the adult rat offspring caused by maternal protein restriction during gestation and lactation are attenuated by maternal moderate-low physical training. *Br J Nutr* 109 (3):449–56.

Fink, R. I., O. G. Kolterman, J. Griffin, and J. M. Olefsky. 1983. Mechanisms of insulin resistance in aging. *J Clin Invest* 71 (6):1523–35.

Franks, P. W., H. C. Looker, S. Kobes et al. 2006. Gestational glucose tolerance and risk of type 2 diabetes in young Pima Indian offspring. *Diabetes* 55 (2):460–5.

Freinkel, N. 1980. Banting Lecture 1980. Of pregnancy and progeny. *Diabetes* 29 (12):1023–35.

Freinkel, N., N. J. Lewis, R. Johnson, I. Swenne, A. Bone, and C. Hellerstrom. 1984. Differential effects of age versus glycemic stimulation on the maturation of insulin stimulus-secretion coupling during culture of fetal rat islets. *Diabetes* 33 (11):1028–38.

Friedenreich, C. M., H. K. Neilson, and B. M. Lynch. 2010. State of the epidemiological evidence on physical activity and cancer prevention. *Eur J Cancer* 46 (14):2593–604.

Friedman, J. E., T. Ishizuka, J. Shao, L. Huston, T. Highman, and P. Catalano. 1999. Impaired glucose transport and insulin receptor tyrosine phosphorylation in skeletal muscle from obese women with gestational diabetes. *Diabetes* 48 (9):1807–14.

Friso, S., F. Pizzolo, S. W. Choi et al. 2008. Epigenetic control of 11 beta-hydroxysteroid dehydrogenase 2 gene promoter is related to human hypertension. *Atherosclerosis* 199 (2):323–7.

Fu, Q., X. Yu, C. W. Callaway, R. H. Lane, and R. A. McKnight. 2009. Epigenetics: Intrauterine growth retardation (IUGR) modifies the histone code along the rat hepatic IGF-1 gene. *FASEB J* 23 (8):2438–49.

Fujiki, K., F. Kano, K. Shiota, and M. Murata. 2009. Expression of the peroxisome proliferator activated receptor gamma gene is repressed by DNA methylation in visceral adipose tissue of mouse models of diabetes. *BMC Biol* 7:38.

Gallou-Kabani, C., and C. Junien. 2005. Nutritional epigenomics of metabolic syndrome: New perspective against the epidemic. *Diabetes* 54 (7):1899–906.

Garofano, A., P. Czernichow, and B. Breant. 1999. Effect of ageing on beta-cell mass and function in rats malnourished during the perinatal period. *Diabetologia* 42 (6):711–8.

George, L. A., L. Zhang, N. Tuersunjiang et al. 2012. Early maternal undernutrition programs increased feed intake, altered glucose metabolism and insulin secretion, and liver function in aged female offspring. *Am J Physiol Regul Integr Comp Physiol* 302 (7):R795–804.

Gluckman, P. D., and M. A. Hanson. 2004. Developmental origins of disease paradigm: A mechanistic and evolutionary perspective. *Pediatr Res* 56 (3):311–7.

Gluckman, P. D., and M. A. Hanson. 2008. Developmental and epigenetic pathways to obesity: An evolutionary-developmental perspective. *Int J Obes (Lond)* 32 Suppl 7: S62–71.

Gluckman, P. D., M. A. Hanson, A. S. Beedle, and D. Raubenheimer. 2008. Fetal and neonatal pathways to obesity. *Front Horm Res* 36:61–72.

Gluckman, P. D., M. A. Hanson, and C. Pinal. 2005. The developmental origins of adult disease. *Matern Child Nutr* 1 (3):130–41.

Hales, C. N., and D. J. Barker. 1992. Type 2 (non-insulin-dependent) diabetes mellitus: The thrifty phenotype hypothesis. *Diabetologia* 35 (7):595–601.

Hales, C. N., D. J. Barker, P. M. Clark et al. 1991. Fetal and infant growth and impaired glucose tolerance at age 64. *BMJ* 303 (6809):1019–22.

Han, S., P. Middleton, and C. A. Crowther. 2012. Exercise for pregnant women for preventing gestational diabetes mellitus. *Cochrane Database Syst Rev* 7:CD009021.

Hatoum, N., J. F. Clapp III, M. R. Newman, N. Dajani, and S. B. Amini. 1997. Effects of maternal exercise on fetal activity in late gestation. *J Matern Fetal Med* 6 (3):134–9.

Hayashi, T., M. F. Hirshman, N. Fujii, S. A. Habinowski, L. A. Witters, and L. J. Goodyear. 2000. Metabolic stress and altered glucose transport: Activation of AMP-activated protein kinase as a unifying coupling mechanism. *Diabetes* 49 (4):527–31.

Hedley, A. A., C. L. Ogden, C. L. Johnson, M. D. Carroll, L. R. Curtin, and K. M. Flegal. 2004. Prevalence of overweight and obesity among US children, adolescents, and adults, 1999–2002. *JAMA* 291 (23):2847–50.

Herring, A., A. Donath, M. Yarmolenko et al. 2012. Exercise during pregnancy mitigates Alzheimer-like pathology in mouse offspring. *FASEB J* 26 (1):117–28.

Hofmann, S., and D. Pette. 1994. Low-frequency stimulation of rat fast-twitch muscle enhances the expression of hexokinase II and both the translocation and expression of glucose transporter 4 (GLUT-4). *Eur J Biochem* 219 (1–2):307–15.

Holloszy, J. O. 2005. Exercise-induced increase in muscle insulin sensitivity. *J Appl Physiol* 99 (1):338–43.

Honig, C. R., C. L. Odoroff, and J. L. Frierson. 1982. Active and passive capillary control in red muscle at rest and in exercise. *Am J Physiol* 243 (2):H196–206.

Hopkins, S. A., J. C. Baldi, W. S. Cutfield, L. McCowan, and P. L. Hofman. 2010. Exercise training in pregnancy reduces offspring size without changes in maternal insulin sensitivity. *J Clin Endocrinol Metab* 95 (5):2080–8.

Hotamisligil, G. S., N. S. Shargill, and B. M. Spiegelman. 1993. Adipose expression of tumor necrosis factor-alpha: Direct role in obesity-linked insulin resistance. *Science* 259 (5091):87–91.

Hulens, M., G. Vansant, R. Lysens, A. L. Claessens, and E. Muls. 2001. Exercise capacity in lean versus obese women. *Scand J Med Sci Sports* 11 (5):305–9.

Islami, D., P. Bischof, and D. Chardonnens. 2003. Modulation of placental vascular endothelial growth factor by leptin and hCG. *Mol Hum Reprod* 9 (7):395–8.

Jackson, M. R., P. Gott, S. J. Lye, J. W. Ritchie, and J. F. Clapp III. 1995. The effects of maternal aerobic exercise on human placental development: Placental volumetric composition and surface areas. *Placenta* 16 (2):179–91.

Jonsson, J., L. Carlsson, T. Edlund, and H. Edlund. 1994. Insulin-promoter-factor 1 is required for pancreas development in mice. *Nature* 371 (6498):606–9.

Jovanovic-Peterson, L., E. P. Durak, and C. M. Peterson. 1989. Randomized trial of diet versus diet plus cardiovascular conditioning on glucose levels in gestational diabetes. *Am J Obstet Gynecol* 161 (2):415–9.

Kahn, B. B., and J. S. Flier. 2000. Obesity and insulin resistance. *J Clin Invest* 106 (4):473–81.

Kahn, S. E., R. L. Hull, and K. M. Utzschneider. 2006. Mechanisms linking obesity to insulin resistance and type 2 diabetes. *Nature* 444 (7121):840–6.

Kardel, K. R. 2005. Effects of intense training during and after pregnancy in top-level athletes. *Scand J Med Sci Sports* 15 (2):79–86.

Kenney, W. L., and J. M. Johnson. 1992. Control of skin blood flow during exercise. *Med Sci Sports Exerc* 24 (3):303–12.

Kervran, A., and J. R. Girard. 1974. Glucose-induced increase of plasma insulin in the rat foetus in utero. *J Endocrinol* 62 (3):545–51.

Khalyfa, A., A. Carreras, F. Hakim, J. M. Cunningham, Y. Wang, and D. Gozal. 2013. Effects of late gestational high-fat diet on body weight, metabolic regulation and adipokine expression in offspring. *Int J Obes (Lond)* 37 (11):1481–9.

Khan, I. Y., V. Dekou, G. Douglas et al. 2005. A high-fat diet during rat pregnancy or suckling induces cardiovascular dysfunction in adult offspring. *Am J Physiol Regul Integr Comp Physiol* 288 (1):R127–33.

Kim, C., K. M. Newton, and R. H. Knopp. 2002. Gestational diabetes and the incidence of type 2 diabetes: A systematic review. *Diabetes Care* 25 (10):1862–8.

Kim, S. Y., L. England, H. G. Wilson, C. Bish, G. A. Satten, and P. Dietz. 2010. Percentage of gestational diabetes mellitus attributable to overweight and obesity. *Am J Public Health* 100 (6):1047–52.

Lange, U. C., and R. Schneider. 2010. What an epigenome remembers. *Bioessays* 32 (8): 659–68.

Lappas, M., M. Permezel, and G. E. Rice. 2004. Release of proinflammatory cytokines and 8-isoprostane from placenta, adipose tissue, and skeletal muscle from normal pregnant women and women with gestational diabetes mellitus. *J Clin Endocrinol Metab* 89 (11):5627–33.

Lee, H. H., H. Kim, J. W. Lee et al. 2006. Maternal swimming during pregnancy enhances short-term memory and neurogenesis in the hippocampus of rat pups. *Brain Dev* 28 (3):147–54.

Lillycrop, K. A., E. S. Phillips, A. A. Jackson, M. A. Hanson, and G. C. Burdge. 2005. Dietary protein restriction of pregnant rats induces and folic acid supplementation prevents epigenetic modification of hepatic gene expression in the offspring. *J Nutr* 135 (6):1382–6.

Lillycrop, K. A., E. S. Phillips, C. Torrens, M. A. Hanson, A. A. Jackson, and G. C. Burdge. 2008. Feeding pregnant rats a protein-restricted diet persistently alters the methylation of specific cytosines in the hepatic PPAR alpha promoter of the offspring. *Br J Nutr* 100 (2):278–82.

Lillycrop, K. A., J. L. Slater-Jefferies, M. A. Hanson, K. M. Godfrey, A. A. Jackson, and G. C. Burdge. 2007. Induction of altered epigenetic regulation of the hepatic glucocorticoid receptor in the offspring of rats fed a protein-restricted diet during pregnancy suggests that reduced DNA methyltransferase-1 expression is involved in impaired DNA methylation and changes in histone modifications. *Br J Nutr* 97 (6):1064–73.

Lithell, H. O., P. M. McKeigue, L. Berglund, R. Mohsen, U. B. Lithell, and D. A. Leon. 1996. Relation of size at birth to non-insulin dependent diabetes and insulin concentrations in men aged 50–60 years. *BMJ* 312 (7028):406–10.

Liu, J., J. N. Laditka, E. J. Mayer-Davis, and R. R. Pate. 2008. Does physical activity during pregnancy reduce the risk of gestational diabetes among previously inactive women? *Birth* 35 (3):188–95.

Longo, L. D. 1983. Maternal blood volume and cardiac output during pregnancy: A hypothesis of endocrinologic control. *Am J Physiol* 245 (5 Pt 1):R720–9.

Lotgering, F. K., R. D. Gilbert, and L. D. Longo. 1985. Maternal and fetal responses to exercise during pregnancy. *Physiol Rev* 65 (1):1–36.

Lovelady, C. A., B. Lonnerdal, and K. G. Dewey. 1990. Lactation performance of exercising women. *Am J Clin Nutr* 52 (1):103–9.

McCance, D. R., D. J. Pettitt, R. L. Hanson, L. T. Jacobsson, W. C. Knowler, and P. H. Bennett. 1994. Birth weight and non-insulin dependent diabetes: Thrifty genotype, thrifty phenotype, or surviving small baby genotype? *BMJ* 308 (6934):942–5.

McGee, S. L., E. Fairlie, A. P. Garnham, and M. Hargreaves. 2009. Exercise-induced histone modifications in human skeletal muscle. *J Physiol* 587 (Pt 24):5951–8.

McGee, S. L., and M. Hargreaves. 2011. Histone modifications and exercise adaptations. *J Appl Physiol (1985)* 110 (1):258–63.

McGill, B. E., S. F. Bundle, M. B. Yaylaoglu, J. P. Carson, C. Thaller, and H. Y. Zoghbi. 2006. Enhanced anxiety and stress-induced corticosterone release are associated with increased Crh expression in a mouse model of Rett syndrome. *Proc Natl Acad Sci U S A* 103 (48):18267–72.

McIntyre, H. D., A. M. Chang, L. K. Callaway et al. 2010. Hormonal and metabolic factors associated with variations in insulin sensitivity in human pregnancy. *Diabetes Care* 33 (2):356–60.

Melzer, K., Y. Schutz, M. Boulvain, and B. Kayser. 2010. Physical activity and pregnancy: Cardiovascular adaptations, recommendations and pregnancy outcomes. *Sports Med* 40 (6):493–507.

Melzner, I., V. Scott, K. Dorsch et al. 2002. Leptin gene expression in human preadipocytes is switched on by maturation-induced demethylation of distinct CpGs in its proximal promoter. *J Biol Chem* 277 (47):45420–7.

Merezak, S., B. Reusens, A. Renard et al. 2004. Effect of maternal low-protein diet and taurine on the vulnerability of adult Wistar rat islets to cytokines. *Diabetologia* 47 (4):669–75.

Milagro, F. I., J. Campion, D. F. Garcia-Diaz, E. Goyenechea, L. Paternain, and J. A. Martinez. 2009. High fat diet-induced obesity modifies the methylation pattern of leptin promoter in rats. *J Physiol Biochem* 65 (1):1–9.

Miller, W. J., W. M. Sherman, and J. L. Ivy. 1984. Effect of strength training on glucose tolerance and post-glucose insulin response. *Med Sci Sports Exerc* 16 (6):539–43.

Mudd, L. M., S. Nechuta, J. M. Pivarnik, and N. Paneth. 2009. Factors associated with women's perceptions of physical activity safety during pregnancy. *Prev Med* 49 (2–3):194–9.

Mueller, B. R., and T. L. Bale. 2008. Sex-specific programming of offspring emotionality after stress early in pregnancy. *J Neurosci* 28 (36):9055–65.

Muoio, D. M., and C. B. Newgard. 2008. Mechanisms of disease: Molecular and metabolic mechanisms of insulin resistance and beta-cell failure in type 2 diabetes. *Nat Rev Mol Cell Biol* 9 (3):193–205.

Murphy, H. C., G. Regan, I. G. Bogdarina et al. 2003. Fetal programming of perivenous glucose uptake reveals a regulatory mechanism governing hepatic glucose output during refeeding. *Diabetes* 52 (6):1326–32.

Nitert, M. D., T. Dayeh, P. Volkov et al. 2012. Impact of an exercise intervention on DNA methylation in skeletal muscle from first-degree relatives of patients with type 2 diabetes. *Diabetes* 61 (12):3322–32.

O'Rahilly, S. 1997. Science, medicine, and the future. Non-insulin dependent diabetes mellitus: The gathering storm. *BMJ* 314 (7085):955–9.

Oostdam, N., M. N. M. van Poppel, M. G. A. J. Wouters et al. 2012. No effect of the FitFor2 exercise programme on blood glucose, insulin sensitivity, and birthweight in pregnant women who were overweight and at risk for gestational diabetes: Results of a randomised controlled trial. *BJOG-an International Journal of Obstetrics and Gynaecology* 119 (9):1098–107.

Ozanne, S. E., M. W. Dorling, C. L. Wang, and B. T. Nave. 2001. Impaired PI 3-kinase activation in adipocytes from early growth-restricted male rats. *Am J Physiol Endocrinol Metab* 280 (3):E534–9.

Ozanne, S. E., G. S. Olsen, L. L. Hansen et al. 2003. Early growth restriction leads to down regulation of protein kinase C zeta and insulin resistance in skeletal muscle. *J Endocrinol* 177 (2):235–41.

Painter, R. C., C. Osmond, P. Gluckman, M. Hanson, D. I. Phillips, and T. J. Roseboom. 2008. Transgenerational effects of prenatal exposure to the Dutch famine on neonatal adiposity and health in later life. *BJOG* 115 (10):1243–9.

Park, J. H., D. A. Stoffers, R. D. Nicholls, and R. A. Simmons. 2008. Development of type 2 diabetes following intrauterine growth retardation in rats is associated with progressive epigenetic silencing of Pdx1. *J Clin Invest* 118 (6):2316–24.

Park, J. W., M. H. Kim, S. J. Eo et al. 2013. Maternal exercise during pregnancy affects mitochondrial enzymatic activity and biogenesis in offspring brain. *Int J Neurosci* 123 (4):253–64.

Petersen, A. M., and B. K. Pedersen. 2005. The anti-inflammatory effect of exercise. *J Appl Physiol* 98 (4):1154–62.

Petersen, K. F., S. Dufour, D. Befroy, M. Lehrke, R. E. Hendler, and G. I. Shulman. 2005. Reversal of nonalcoholic hepatic steatosis, hepatic insulin resistance, and hyperglycemia by moderate weight reduction in patients with type 2 diabetes. *Diabetes* 54 (3):603–8.

Petry, C. J., M. W. Dorling, D. B. Pawlak, S. E. Ozanne, and C. N. Hales. 2001. Diabetes in old male offspring of rat dams fed a reduced protein diet. *Int J Exp Diabetes Res* 2 (2):139–43.

Pettitt, D. J., R. G. Nelson, M. F. Saad, P. H. Bennett, and W. C. Knowler. 1993. Diabetes and obesity in the offspring of Pima Indian women with diabetes during pregnancy. *Diabetes Care* 16 (1):310–4.

Phillips, D. I., D. J. Barker, C. N. Hales, S. Hirst, and C. Osmond. 1994. Thinness at birth and insulin resistance in adult life. *Diabetologia* 37 (2):150–4.

Plagemann, A., T. Harder, M. Brunn et al. 2009. Hypothalamic proopiomelanocortin promoter methylation becomes altered by early overfeeding: An epigenetic model of obesity and the metabolic syndrome. *J Physiol* 587 (Pt 20):4963–76.

Plagemann, A., K. Roepke, T. Harder et al. 2010. Epigenetic malprogramming of the insulin receptor promoter due to developmental overfeeding. *J Perinat Med* 38 (4):393–400.

Porte, D. Jr., and S. E. Kahn. 2001. beta-cell dysfunction and failure in type 2 diabetes: Potential mechanisms. *Diabetes* 50 Suppl 1:S160–3.

Poudevigne, M. S., and P. J. O'Connor. 2005. Physical activity and mood during pregnancy. *Med Sci Sports Exerc* 37 (8):1374–80.

Radesky, J. S., E. Oken, S. L. Rifas-Shiman, K. P. Kleinman, J. W. Rich-Edwards, and M. W. Gillman. 2008. Diet during early pregnancy and development of gestational diabetes. *Paediatr Perinat Epidemiol* 22 (1):47–59.

Ravelli, A. C. J., J. H. P. van der Meulen, C. Osmond, D. J. P. Barker, and O. P. Bleker. 1999. Obesity at the age of 50 y in men and women exposed to famine prenatally. *Am J Clin Nutr* 70 (5):811–6.

Robert, J. J., J. C. Cummins, R. R. Wolfe et al. 1982. Quantitative aspects of glucose production and metabolism in healthy elderly subjects. *Diabetes* 31 (3):203–11.

Ronn, T., P. Volkov, C. Davegardh et al. 2013. A six months exercise intervention influences the genome-wide DNA methylation pattern in human adipose tissue. *PLoS Genet* 9 (6):e1003572.

Roseboom, T. J., J. H. P. van der Meulen, C. Osmond et al. 2000. Coronary heart disease after prenatal exposure to the Dutch famine, 1944–1945. *Heart* 84 (6):595–8.

Rowell, L. B., and J. R. Blackmon. 1987. Human cardiovascular adjustments to acute hypoxemia. *Clin Physiol* 7 (5):349–76.

Ryan, E. A., M. J. O'Sullivan, and J. S. Skyler. 1985. Insulin action during pregnancy. Studies with the euglycemic clamp technique. *Diabetes* 34 (4):380–9.

Saeed, S. A., D. J. Antonacci, and R. M. Bloch. 2010. Exercise, yoga, and meditation for depressive and anxiety disorders. *Am Fam Physician* 81 (8):981–6.

Saltin, B., and L. B. Rowell. 1980. Functional adaptations to physical-activity and inactivity. *Fed Proc* 39 (5):1506–13.

Samuelsson, A. M., P. A. Matthews, M. Argenton et al. 2008. Diet-induced obesity in female mice leads to offspring hyperphagia, adiposity, hypertension, and insulin resistance: A novel murine model of developmental programming. *Hypertension* 51 (2):383–92.

Schmitz-Peiffer, C. 2000. Signalling aspects of insulin resistance in skeletal muscle: Mechanisms induced by lipid oversupply. *Cell Signal* 12 (9–10):583–94.

Silverman, B. L., T. Rizzo, O. C. Green et al. 1991. Long-term prospective evaluation of offspring of diabetic mothers. *Diabetes* 40 Suppl 2:121–5.

Simmons, R. A., L. J. Templeton, and S. J. Gertz. 2001. Intrauterine growth retardation leads to the development of type 2 diabetes in the rat. *Diabetes* 50 (10):2279–86.

Singh, S., G. Sedgh, and R. Hussain. 2010. Unintended pregnancy: Worldwide levels, trends, and outcomes. *Stud Fam Plann* 41 (4):241–50.

Snoeck, A., C. Remacle, B. Reusens, and J. J. Hoet. 1990. Effect of a low protein diet during pregnancy on the fetal rat endocrine pancreas. *Biol Neonate* 57 (2):107–18.

Snowling, N. J., and W. G. Hopkins. 2006. Effects of different modes of exercise training on glucose control and risk factors for complications in type 2 diabetic patients: A meta-analysis. *Diabetes Care* 29 (11):2518–27.

Stein, A. D., H. S. Kahn, A. Rundle, P. A. Zybert, K. van der Pal-de Bruin, and L. H. Lumey. 2007. Anthropometric measures in middle age after exposure to famine during gestation: Evidence from the Dutch famine. *Am J Clin Nutr* 85 (3):869–76.

Stoger, R. 2008. Epigenetics and obesity. *Pharmacogenomics* 9 (12):1851–60.

Stumvoll, M., B. J. Goldstein, and T. W. van Haeften. 2008. Type 2 diabetes: Pathogenesis and treatment. *Lancet* 371 (9631):2153–6.

Szepietowska, B., M. Szelachowska, M. Gorska, D. Jakubczyk, and I. Kinalska. 2004. Chronic complications in adult patients with newly diagnosed diabetes mellitus in relation to the presence of humoral autoimmune markers against pancreatic islet cells. *Pol Arch Med Wewn* 111 (5):563–9.

Taylor, P. D., J. McConnell, I. Y. Khan et al. 2005. Impaired glucose homeostasis and mitochondrial abnormalities in offspring of rats fed a fat-rich diet in pregnancy. *Am J Physiol Regul Integr Comp Physiol* 288 (1):R134–9.

Thompson, P. D., D. Buchner, I. L. Pina et al. 2003. Exercise and physical activity in the prevention and treatment of atherosclerotic cardiovascular disease: A statement from the Council on Clinical Cardiology (Subcommittee on Exercise, Rehabilitation, and Prevention) and the Council on Nutrition, Physical Activity, and Metabolism (Subcommittee on Physical Activity). *Circulation* 107 (24):3109–16.

Tobias, D. K., C. L. Zhang, R. M. van Dam, K. Bowers, and F. B. Hu. 2011. Physical activity before and during pregnancy and risk of gestational diabetes mellitus a meta-analysis. *Diabetes Care* 34 (1):223–9.

Treadway, J. L., and S. A. Lederman. 1986. The effects of exercise on milk yield, milk composition, and offspring growth in rats. *Am J Clin Nutr* 44 (4):481–8.

Trujillo, M. E., and P. E. Scherer. 2006. Adipose tissue-derived factors: Impact on health and disease. *Endocr Rev* 27 (7):762–78.

Tsukui, S., T. Kanda, M. Nara, M. Nishino, T. Kondo, and I. Kobayashi. 2000. Moderate-intensity regular exercise decreases serum tumor necrosis factor-alpha and HbA1c levels in healthy women. *Int J Obes Relat Metab Disord* 24 (9):1207–11.

Valdez, R., M. A. Athens, G. H. Thompson, B. S. Bradshaw, and M. P. Stern. 1994. Birthweight and adult health outcomes in a biethnic population in the USA. *Diabetologia* 37 (6):624–31.

Vanheest, J. L., and C. D. Rodgers. 1997. Effects of exercise in diabetic rats before and during gestation on maternal and neonatal outcomes. *Am J Physiol* 273 (4 Pt 1):E727–33.

Vavvas, D., A. Apazidis, A. K. Saha et al. 1997. Contraction-induced changes in acetyl-CoA carboxylase and 5'-AMP-activated kinase in skeletal muscle. *J Biol Chem* 272 (20):13255–61.

Vega, C. C., L. A. Reyes-Castro, C. J. Bautista, F. Larrea, P. W. Nathanielsz, and E. Zambrano. 2013. Exercise in obese female rats has beneficial effects on maternal and male and female offspring metabolism. *Int J Obes (Lond).* doi:10.1038/ijo.2013.150

Vickers, M. H., B. H. Breier, W. S. Cutfield, P. L. Hofman, and P. D. Gluckman. 2000. Fetal origins of hyperphagia, obesity, and hypertension and postnatal amplification by hypercaloric nutrition. *Am J Physiol Endocrinol Metab* 279 (1):E83–7.

Volpato, A. M., A. Schultz, E. Magalhaes-da-Costa, M. L. Correia, M. B. Aguila, and C. A. Mandarim-de-Lacerda. 2012. Maternal high-fat diet programs for metabolic disturbances in offspring despite leptin sensitivity. *Neuroendocrinology* 96 (4):272–84.

Vucetic, Z., J. Kimmel, and T. M. Reyes. 2011. Chronic high-fat diet drives postnatal epigenetic regulation of mu-opioid receptor in the brain. *Neuropsychopharmacology* 36 (6):1199–206.

Vucetic, Z., J. Kimmel, K. Totoki, E. Hollenbeck, and T. M. Reyes. 2010a. Maternal high-fat diet alters methylation and gene expression of dopamine and opioid-related genes. *Endocrinology* 151 (10):4756–64.

Vucetic, Z., K. Totoki, H. Schoch et al. 2010b. Early life protein restriction alters dopamine circuitry. *Neuroscience* 168 (2):359–70.

Vuguin, P., E. Raab, B. Liu, N. Barzilai, and R. Simmons. 2004. Hepatic insulin resistance precedes the development of diabetes in a model of intrauterine growth retardation. *Diabetes* 53 (10):2617–22.

Wasserman, D. H., and J. E. Ayala. 2005. Interaction of physiological mechanisms in control of muscle glucose uptake. *Clin Exp Pharmacol Physiol* 32 (4):319–23.

Waterland, R. A., and R. L. Jirtle. 2003. Transposable elements: Targets for early nutritional effects on epigenetic gene regulation. *Mol Cell Biol* 23 (15):5293–300.

Waterland, R. A., and R. L. Jirtle. 2004. Early nutrition, epigenetic changes at transposons and imprinted genes, and enhanced susceptibility to adult chronic diseases. *Nutrition* 20 (1):63–8.

Winder, W. W., and D. G. Hardie. 1999. AMP-activated protein kinase, a metabolic master switch: Possible roles in type 2 diabetes. *Am J Physiol* 277 (1 Pt 1):E1–10.

Wojtaszewski, J. F., B. F. Hansen, Gade et al. 2000. Insulin signaling and insulin sensitivity after exercise in human skeletal muscle. *Diabetes* 49 (3):325–31.

Wolf, M., J. Sauk, A. Shah et al. 2004. Inflammation and glucose intolerance: A prospective study of gestational diabetes mellitus. *Diabetes Care* 27 (1):21–7.

Worda, C., H. Leipold, C. Gruber, A. Kautzky-Willer, M. Knofler, and D. Bancher-Todesca. 2004. Decreased plasma adiponectin concentrations in women with gestational diabetes mellitus. *Am J Obstet Gynecol* 191 (6):2120–4.

Yajnik, C. S., C. H. Fall, U. Vaidya et al. 1995. Fetal growth and glucose and insulin metabolism in four-year-old Indian children. *Diabet Med* 12 (4):330–6.

Yamauchi, T., J. Kamon, Y. Minokoshi et al. 2002. Adiponectin stimulates glucose utilization and fatty-acid oxidation by activating AMP-activated protein kinase. *Nat Med* 8 (11):1288–95.

Yokomori, N., M. Tawata, and T. Onaya. 1999. DNA demethylation during the differentiation of 3T3–L1 cells affects the expression of the mouse GLUT4 gene. *Diabetes* 48 (4):685–90.

Yu, M., E. Blomstrand, A. V. Chibalin, H. Wallberg-Henriksson, J. R. Zierath, and A. Krook. 2001. Exercise-associated differences in an array of proteins involved in signal transduction and glucose transport. *J Appl Physiol* 90 (1):29–34.

Zhang, C. Y., G. Baffy, P. Perret et al. 2001. Uncoupling protein-2 negatively regulates insulin secretion and is a major link between obesity, beta cell dysfunction, and type 2 diabetes. *Cell* 105 (6):745–55.

Zhang, C., C. G. Solomon, J. E. Manson, and F. B. Hu. 2006. A prospective study of pregravid physical activity and sedentary behaviors in relation to the risk for gestational diabetes mellitus. *Arch Intern Med* 166 (5):543–8.

Zhang, J., F. Zhang, X. Didelot et al. 2009. Maternal high fat diet during pregnancy and lactation alters hepatic expression of insulin like growth factor-2 and key microRNAs in the adult offspring. *BMC Genomics* 10:478.

3 Maternal Protein and Fat Intake

Epigenetic Consequences on Fetal Development

Yuan-Xiang Pan, Rita S. Strakovsky, Dan Zhou, Huan Wang, and Hong Chen

CONTENTS

3.1 INTRODUCTION

Maternal nutrition during gestation has long been acknowledged as one of the most important factors contributing to pregnancy outcomes. We are only beginning to understand how maternal nutrition and intrauterine environment may impact not only immediate pregnancy outcomes, but also health throughout the course of the offspring's life. Associations between maternal nutrition and infant growth and development suggest that improving the diets of women of child-bearing age might be an important component of public health strategies aimed at improving the health, nutrition, and well-being of women themselves as well as reducing the burden of chronic disease in their offspring.

During pregnancy, the mother's basal energy metabolism changes to adapt to the needs of the developing embryo.[1] After delivery, major organs, including liver, mammary gland, and gut further adapt for milk production by altering their metabolism.[2] Therefore, any dietary disturbance during gestation or lactation plays a pivotal role in offspring development. Poor maternal nutrition leads to long-term physiological changes in offspring, including dysregulation of appetite,[3,4] adult obesity,[5] increased risk of diabetes, and reduced longevity.[6] These observations in animals have been supported by epidemiology showing nutrient deficits during pregnancy relate to increased incidence of chronic diseases in adults.[7] Recently, it has been demonstrated that fetal epigenetic changes that last into adulthood may originate from dietary restriction during gestation.[8]

Intrauterine growth restriction (IUGR) is a condition in which a fetus in the womb fails to grow at the expected rate during the pregnancy. The most common definition of IUGR is a fetal weight that is below the 10th percentile for gestational age, which is also called small-for-gestational age (SGA) or fetal growth restriction (FGR). IUGR is a common cause of perinatal morbidity and mortality and is associated with increased risk of adult disease, such as diabetes and cardiovascular disease. Fetal undernutrition can result in IUGR and maternal protein restriction is a common approach to induce fetal undernutrition and one of the best characterized developmental programming models. A marginal protein restriction diet (PR, 8% protein/dry matter) during pregnancy has physiological impacts both on mother and offspring.[9] Protein restriction causes alterations in the fetal hepatic DNA methylation and further regulates the expression of important metabolic genes in offspring, including peroxisome proliferator–activated receptor (Ppar) α and glucocorticoid receptor.[10] Furthermore, accumulating evidence shows that maternal protein restriction can cause fetal epigenetic changes to genes involved in various diseases, such as type 2 diabetes and coronary heart diseases.[11]

In the meantime, the intake of high-fat (HF), energy-dense foods is on the rise in industrialized as well as developing nations,[12,13] which may have devastating effects on many populations. A total daily fat intake more than 30 percent of total calories will be considered as an HF diet. Because maternal nutrition programs both short- and long-term health outcomes in offspring, the increased intake of HF may be especially detrimental during pregnancy. A maternal HF diet has adverse effects on fetal liver development in nonhuman primates,[14] which has the potential to lead to altered glucose metabolism in adulthood. Prolonged HF consumption was shown to disturb metabolic functions and results in obesity, hyperglycemia, and systemic insulin resistance.[15–17] More importantly, such phenomenon not only affects the fat-craved individuals, but also predisposes subsequent generations to similar health concerns. Perinatal exposure to HF diet profoundly alters the intrauterine environment and leads to permanent phenotypic alterations with adverse outcomes that persist through the adulthood of progeny. These outcomes are manifested at certain physiological levels including growth, learning ability, and susceptibility to chronic diseases.[18,19] They are revealed or even aggravated in the face of postnatal HF/carbohydrate challenge[20–22] or upon the addition of drug treatment.[23]

Despite of considerable evidence from animal models to support a role for maternal nutrition in fetal programming such that alterations in metabolism during gestation or lactation result in metabolic changes in the offspring in adult life, the mechanistic link between maternal diets and the adverse outcomes in offspring remains the subject of debate. As the prevalence of disease increases,[24] determining the mechanisms behind its in utero development in response to maternal diet and physiology may become the key to its prevention.

3.2 PHYSIOLOGICAL CONSEQUENCES OF MATERNAL DIETS

Many adult diseases have fetal origins. Links among maternal nutrition, intrauterine environment of the fetus, and susceptibility to these adult diseases have attracted numerous attentions. A large number of studies has demonstrated that maternal diets typically affect fetal growth and programs the fetus to develop hypertension, insulin resistance, obesity, and cancer later in life.

3.2.1 Maternal Protein Restriction

Maternal undernutrition during the pregnancy has been demonstrated to result in IUGR and low birth weight. To adapt to the maternal undernutrition environment and increase the chance of postnatal survival, the fetus responds with a number of adaptations, such as changing its metabolic rate, altering the production of hormones, storing nutrients as fat, and redistribution of fetal blood flow to protect brain at the expenses of other tissues such as muscle. These factors lead to a slower fetal growth and low birth weight. Data have shown that IUGR is associated with a late life increased prevalence of metabolic syndrome, obesity, impaired glucose tolerance, type 2 diabetes.[25] A maternal protein restriction (PR) diet, when compared with postnatal protein restriction, was shown to increase renal albuminuria in young mice, and this was accompanied by altered expression of genes that function to protect the kidney against oxidative stress.[26] Our group has shown that gestational protein restriction in rats has severe consequences for the growth capacity of offspring,[27] and Ozanne et al. have repeatedly demonstrated that protein restriction during gestation or lactation influences oxidative balance in the offspring. Additionally, a perinatal PR diet resulted in an increase in markers of fibrosis in pancreatic islets of adult offspring, and this was accompanied by increased islet expression of xanthine oxidase, and decreased manganese superoxide dismutase (MnSOD), copper- and zinc-containing superoxide dismutase (Cu/Zn-SOD), and heme oxygenase 1,[28] implying increased oxidative stress and diminished anti-oxidative capacity in pancreatic cells of offspring exposed to a gestational PR diet. Using the same model, aortas of offspring exposed to a gestational PR diet had lower MnSOD protein content than the postnatally PR-fed group, and there was a significant decrease of MnSOD between 3 and 12 months of age in the gestational PR group, suggesting a decrease in anti-oxidative capacity in these animals.[29] How gestational protein restriction modulates offspring development remains unclear. Potentially, protein malnutrition can be sensed by several molecular signaling pathways as a cellular stressor or amino acid sensor, which activates the adaptation response. Alternatively, inadequate protein supply during pregnancy may increase the metabolism of other substrates, such as lipids and carbohydrates, by both the mother and fetus, therefore increasing free fatty acids, ketones, glucose, and specific amino acid in circulation, all of which have been shown to be capable of dysregulating metabolism and development.

3.2.2 Maternal HF Diet

HF diet, which is high in n-6 polyunsaturated fatty acids, increases the susceptibility of female offspring to develop carcinogen induced mammary tumors as adults.[30] It also shows mammary tumorigenesis in out bred mice and mice that overexpress the c-neu oncogene.[31] These phenomena may be caused by increased exposure of fetus to estrogens because of HF diet.[32]

Maternal HF diets demonstrate altered hepatic fatty acid profiles, including saturated fatty acids, monounsaturated fatty acids, polyunsaturated fatty acids, indicating increased hepatic lipogenesis,[33] and also increased activity for liver alanine aminotransferase, aspartate aminotransferase, and key regulating proteins for fat metabolism such as sterol regulatory element-binding proteins (sREBP) 1 and Ppar-α.[34] HF diets during gestation and lactation also contribute to the development of liver steatosis and later steatohepatitis in adult life.[14,35,36] More importantly, in both mice,[36] and nonhuman primates,[14] maternal HF diets affect liver gene expression of key factors involved in gluconeogenesis, lipogenesis, oxidative stress, and inflammation, thus greatly increasing the risk of fatty liver development in the offspring, "priming" them to develop severe steatosis.[14,36] An HF diet is thought to alter oxidative balance by increasing lipid peroxidation and the production of reactive oxygen species (ROS), and our group recently showed that a gestational HF in rats repressed the expression of antioxidant genes in livers of adult offspring. Despite being on a standard control diet for 12 weeks after birth, offspring exposed to a gestational HF diet had increased plasma triglycerides, and their livers had increased thiobarbituric acid reactive substances (TBARS), as well as decreased transcription of *Gpx-1*, *Sod1*, *Pon1*, *Pon2*, and *Pon3*.[37] Our data demonstrate that the maternal environment is a critical programmer of the antioxidant capacity and that molecular events programmed in utero can last into adulthood. While the fetus initially responds to a maternal HF diet by increasing its anti-oxidative capacity, this response may be short-lived. Under a chronic stress, the anti-oxidative system no longer functions with time, which eventually leads to oxidative stress-induced tissue damage. Additionally, what remains to be clarified is whether the dysregulation in oxidative balance in the offspring is due to the increased in utero supply of maternal lipids itself, or the innate increase in the production of lipids or ROS due to the offspring's programmed altered lipid handling and metabolism.

When pregnant rat dams were fed an HF diet supplemented with dietary antioxidants, the offspring of these animals had significantly improved outcomes when compared with offspring of dams fed the HF diet without anti-oxidative supplementation. These beneficial changes included decreased adiposity at 2 weeks and 2 months of age, decreased ROS, and increased glutathione (GSH) in the preimplantation embryos, and decreased TBARS in the fetus and neonate.[38] A high-fiber diet fed to pregnant mice that were also consuming an HF diet improved the hydroxyl radical scavenging capacity in the placenta as well as maternal plasma when compared with the HF diet alone. Additionally, the livers of offspring whose mothers were supplemented with fiber, in conjunction with an HF diet, had increased total SOD, Cu/ZnSOD and MnSOD protein, as well as increased mRNA expression of these genes, *Hif-1α*, *Trx1*, *Trx2*, and *Gpx1*. A similar increase of *Hif-1α*, *Trx1*, Cu/ZnSOD, and MnSOD mRNA was also observed in fetal hearts when compared with offspring of dams consuming an HF diet alone without fiber supplementation.[39] In a rat model of gestational high-saturated fat feeding, fetuses of HF-fed dams had lower total mineralized tissues, crown–rump length, and total bone volume than controls, and quercetin supplementation improved these outcomes. The authors proposed that the HF diet increased oxidative stress, and quercetin, which has been shown to have strong antioxidant capacity, decreased it, thus improving fetal outcomes.[40]

3.2.3 Molecular Pathways Link Maternal Diets to Fetal Development

3.2.3.1 Maternal Protein Restriction

Both single and multicellular organisms respond to immediate and long-term exposures to an amino acid-poor environment.[41,42] Short-term dietary amino acid restriction activates the amino acid response (AAR) pathway in animals.[43] In vitro studies illustrated that under amino acid limitation, the AAR pathway becomes activated after the accumulation of uncharged transfer RNAs (tRNAs) is sensed and eukaryotic translation initiation factor 2α (eIF2α) is phosphorylated by GCN2 kinase. Initiation of the AAR pathway by phosphorylation of eIF2α leading to the decrease of global protein synthesis,[44] while the expression and translation of selected genes, including activating transcription factor (ATF) 4, is induced. ATF4 is a regulator directly impacting downstream genes including Asns,[45] ATF3,[46] and C/EBP-homologous protein (Chop).[47,48] At the same time, ATF4 protein synthesis increases by the modulation of short upstream open reading frames (uORFs) within the 5′-leader of ATF4 mRNA.[44,49,50] As a consequence of elevated ATF4 protein expression, multiple stress-induced genes are activated, including genes containing an amino acid response element (AARE).[51]

The AAR pathway is associated with protein metabolism and reflects the availability of protein in organisms. Activation of the AAR pathway reflects the limitation of dietary protein. The mammalian AAR pathway in placenta is upregulated by a gestational protein restriction diet, and the activation of the AAR may act as a cue for the fetus to develop an adaptive response suited to its predicted postnatal environment.[27] Studies of Sprague-Dawley rats during gestation and lactation have revealed the biphasic nature of protein metabolism. By dividing the pregnancy period into two stages, the first 2 weeks and the last 1 week, Naismith found that protein redistribution occurred in these two stages.[52] During the first 2 weeks of pregnancy, rats start to reserve protein and calories when the competition from the fetus is minimal. The magnitude of the reserve is determined by the protein content of the diet. During the last week of pregnancy, the period of rapid fetal growth, the protein reserve is diminished, which occurs regardless of the level of protein intake. Regardless of stage, it has been concluded that lean tissues have the greatest role during this active protein metabolism.[52–54] Starvation causes muscle catabolism produced by decreased protein synthesis and increased protein degradation. Autophagy, primarily known as macroautophagy in cellular metabolism, involves cell degradation of unnecessary or dysfunctional cellular components through the lysosomal machinery, and has been highly associated with myopathy.[55,56] Protein degradation in skeletal muscle is associated with two highly conserved pathways, the ubiquitin-proteasomal pathway and the autophagic/lysosomal pathway.[57] Forkhead box O3 (FoxO3) has been reported to link muscle atrophy with increased autophagy in skeletal muscle, which can be caused by nutrient starvation.[58–60] A previous study showed that the mammalian AAR pathway in placenta is upregulated by a gestational protein restriction diet and the activation of the AAR pathway may act as a cue for the fetus to develop an adaptive response suited to their predicted postnatal environment.[27] Therefore, maternal sensing of protein deficiency may occur in muscle via increased the AAR pathway and ATF4 bindings at target genes providing insight into the transduction of the nutrient signaling in the mother and the subsequent prediction of adult-onset diseases in the offspring due to maternal protein deficiency (Figure 3.1).

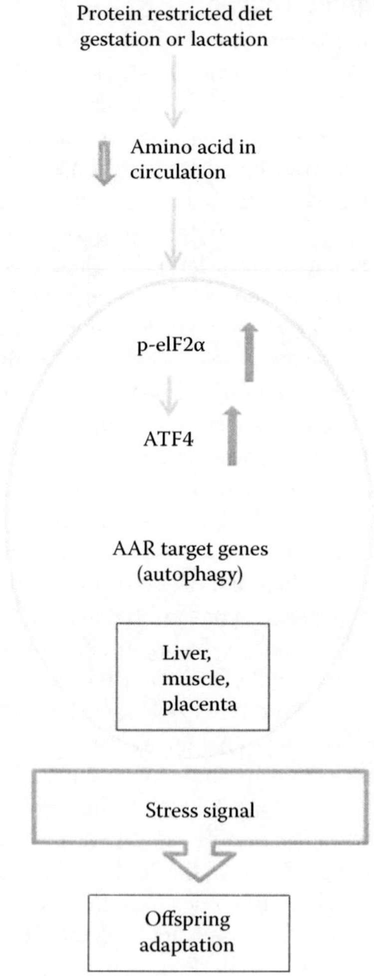

FIGURE 3.1 Proposed nutrient signaling pathways linking maternal protein deficiency to dysfunction in the offspring.

3.2.3.2 Maternal HF Diet

While HF diets alter oxidative balance by increasing lipid peroxidation and the production of ROS, they also increase chronic inflammation. Inflammation is a part of the body's natural defense system against injury and disease. As a response to the environmental factors/stimuli, inflammation triggers a series of body reactions and nonspecific immune response, such as increased blood flow and cellular metabolism, vasodilatation, and fluid influx.[61] Conversely, chronic inflammation is also linked to disease and long-term health concerns. It plays a pivotal role in the pathogenesis of non-alcoholic fatty liver disease (NAFLD)/nonalcoholic steatohepatitis (NASH),[62] cardiovascular diseases,[63] renal failure,[64] nervous system disorders,[65] aging,[66] diabetes,[67] and several types of cancer.[68] Low-grade chronic inflammatory response is commonly observed in obese individuals and sustained HF eating populations. Inflammation is a pathogenic component of various types of liver diseases and progressive liver damage[69] and precedes NAFLD-associated hepatic lipid accumulation. The overload of energy substrates from diet, including fatty acids from an HF diet, can directly induce inflammatory signaling in the liver by activating JNK and IκB kinase (IKK)[70,71] and repressing AMPK[72,73] signaling in hepatocytes, which also increases inflammatory macrophage activation. These changes in hepatocytes impair insulin signaling by increasing the phosphorylation of inhibitory serine residues of insulin receptor substrate 1 (IRS1)[74,75] and/or by decreasing Akt phosphorylation.[76] The resulting hepatic insulin resistance leads to increased expression of genes for lipogenic enzymes such as acetyl-CoA carboxylase (ACC)[77] and fatty acid synthase (FAS)[78] by increasing activities of transcription factors such as SREBP1c[79] and/or carbohydrate-responsive element-binding protein (ChREBP)[80,81] and decreased fatty acid oxidation by inhibiting carnitine palmitoyltransferase 1 (CPT1) activity.[82] As a result, hepatic steatosis develops. In macrophages, overnutrition-induced increase in JNK1, IKKβ, Nuclear factor kappa-light-chain-enhancer of activated B cells (NF-κB), and/or decrease in AMPK signaling stimulate the production of proinflammatory cytokines, which in turn exacerbate liver inflammatory response and hepatic lipid metabolic dysregulation through impairing hepatocyte insulin signaling. Expression of inflammation related genes, such as cyclooxygenase 2 (also known as prostaglandin-endoperoxide synthase, Cox-2),[83–85] interleukin (IL) 1β,[86,87] IL-6,[88–96] and IL-12[97,98] are associated with fatty liver disease. Overall, maternal HF dietary feeding regulates broad ranges of genes in the progeny, including activation of oxidative stress and proinflammatory genes. These changes of gene expression further initiate a series of undesirable physiological imbalance that become pathogenic and either leads to the predisposition or increased vulnerability to chronic diseases. Chronic inflammation and oxidative stress form one of the physiological bases to the compromised health in the offspring exposed to maternal HF diet (Figure 3.2). In an attempt to discover the molecular mechanisms behind the in utero programming of disease, recent studies have focused on connecting the physiological response to maternal diet and any epigenetically induced changes in gene transcription.

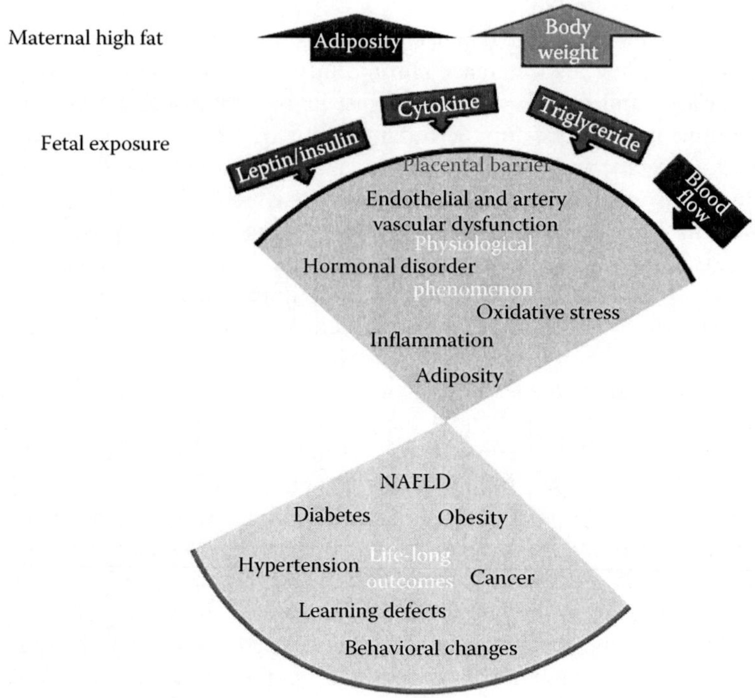

FIGURE 3.2 Proposed impact of maternal HF diet and adiposity on physiological response and health outcomes in the offspring.

3.3 ROLE OF EPIGENETICS IN FETAL DEVELOPMENT

Epigenetics is a mechanism for storing and preserving a "molecular memory" at the cellular level. Epigenetic modifications (epigenome) are heritable changes in gene expression that are not coded in the DNA sequence itself (genome). Epigenetic mechanisms represent a mechanistic layer that sits at the interface of genes and environmental experiences[99] and can be either transient or persist throughout the lifetime of the organism.[100] There are two basic molecular epigenetic mechanisms that are widely studied – covalent modification of DNA through DNA methylation, and regulation of chromatin structure through histone posttranslational modifications. Other epigenetic molecular mechanisms such as the regulation of gene expression through noncoding RNA also exist, but much less studied.[101] Nutritional epigenetics, especially during the critical prenatal period, has become a vast new frontier for studying the way in which our environment influences not only our immediate health but also how it may leave a mark on the way we in which we respond to our future surroundings.

3.3.1 Molecular Basis of Epigenetics

DNA methylation and the covalent modification of histone tails are both epigenetic modifications that have been shown to be powerful regulators of gene expression.

The effect of DNA methylation is dependent on the location of the methylated phosphodiester bond between cytosine and guanine (CpG) sites.[102] When the CpGs are located in promoter regions, methylation of DNA is associated with suppression of gene transcription and broad DNA methylation in many cases causes complete silencing of the associated gene.[103] DNA methylation at the promoter prevents binding of transcription factors and attracts methylated-CpG-binding domain proteins (MBDs), such as MeCP2, MBD1, MBD2, MBD4, and Kaiso.[104–109] These MBD proteins also contain a transcription-regulatory domain (TRD), which recruits scaffolding proteins. In turn, scaffolding proteins attract MBD repressive complexes, such as the SIN3 transcription regulator family member A (Sin3a)[110] and *Drosophila* nucleosome remodeling factor (NurF)[111] complexes, containing histone deacetylases (HDACs), to the methylated site. The HDACs modify chromatin structure locally by removing acetyl groups from histone core proteins, leading to compaction of chromatin and transcriptional suppression.[112] Under normal conditions, the genome undergoes several regulated bouts of methylation and demethylation during the earliest stages of development;[113] however, in adulthood, any considerable change in the methylation of DNA is associated primarily with abnormal pathologies, such as carcinogenesis.[114] Clinical, animal, and cell studies have demonstrated the role of DNA methylation in the development and progression of numerous cancers, including colon, breast, hepatic, ovarian, prostate, and others.

While DNA methylation is relatively permanent and results in the silencing of genes, the modification of histone tails has been thought to be less static and more responsive to diet and the environment. Histones are basic proteins whose function is to interact with DNA through posttranslational modifications.[115] The degree to which nucleosomes are condensed is mediated by posttranslational modifications of the N-terminal tails of histone proteins protruding from the nucleosome, which in turn determine the transcriptional activity of the associated gene (DNA). Histone modification can occur because of DNA methylation[116] or can occur through mechanisms controlled by signaling pathways that are independent of DNA methylation.[117] There are four major posttranslational modifications of histone tails that serve as epigenetic marks: acetylation, methylation, phosphorylation, and ubiquitination.[118] Histone modifications act by modifying the physical configuration of the DNA, either loosening or tightening the chromatin, and these arrangements either activate or inhibit gene transcription by regulating the binding of transcription factors and the transcription-initiation machinery.[119] Additionally, the complicated interplay or cross-talk between the various histone modifications has made it more difficult to classify these as absolute activators or inhibitors of transcription. Methylation of histone H3 at lysine 4 (H3K4Me2) or lysine 9 (H3K9Me3) play opposite roles in regulating the accessibility of chromatin domains. H3K4Me2 at the promoter is positively associated with gene transcription[120] In higher eukaryotic cells, a ChIP-on-chip assay even demonstrated that H3K4Me2 was an epigenetic marker for gene activation in 97% of the 9328 promoters tested.[121] Both of H3K9Me3 and H3K4Me2 are regulated by the recruitment and binding of sequence-specific transcription factors and associated chromatin modifying complexes.[122] Furthermore, it has been suggested that H3K9Me3 is a primary signal in the silencing of gene transcription, regardless of other modifications.[123,124] In addition to the complicated interplay between the many histone modifications, it has also been shown that DNA methylation and histone modifications can regulate each other and/or be regulated by the same factors in response to various stimuli.[125] Therefore, while epigenetics has become a new frontier for the study of gene transcription, the context within which these modifications occur should be considered when making any concrete conclusions about their regulation of specific genes.

3.3.2 Maternal Programming of Fetal Gene Expression

Maternal physiology and diet are potent regulators of fetal outcomes, and there is mounting evidence suggesting that these changes occur in response to the in utero programming of gene expression. Numerous studies have shown that maternal diet results in alterations in fetal birth outcomes and may predict adult disease development.[126] Additionally, studies that manipulate gestational diet have confirmed that these physiological outcomes in the offspring are accompanied by molecular alterations to vital pathways that regulate almost all chronic diseases.[127–132]

Studies in A^{vy} mice were shown to have epigenetic modifications to the *Agouti* gene that altered coat color, as well as increased their susceptibility towards obesity and tumor development.[133] The DNA methylation that led to the altered expression of the *Agouti* gene was shown to be modified by the addition of methyl donors to the maternal diet, showing a direct link among genotype, phenotype, and maternal intake. In later studies, the Jirtle group showed that the supplementation of the maternal diet with genistein increased the methylation of the *Agouti* gene, which not only changed the coat color in A^{vy} mice, but also protected them from developing obesity, and these effects persisted into adulthood.[134] Consequently, further studies of these agouti mice (those carrying the viable yellow agouti A^{vy} epiallele) showed that supplementation with other dietary methyl donors, which are known to influence DNA methylation, induced the pseudoagouti phenotype.[135] These initial discoveries paved the way for the study of the developmental origins of disease from the epigenetic perspective, specifically focusing on the interaction between maternal nutrition and the fetal epigenome in programming adult-onset diseases. Several other animal models of maternal nutritional manipulation connect gestational dietary intake and alterations within the fetal epigenome. Lillycrop et al. have demonstrated that maternal protein restriction results in considerable gene-specific DNA methylation in the offspring, and affects the transcription of genes critical for hepatic metabolic processes, such as glucocorticoid receptor[136] and *Ppar*-α.[137] Our group also showed that a maternal protein restriction diet resulted in a significant increase of hepatic insulin-like growth factor (*Igf2*) transcription, and this was correlated with the methylation of the gene.[138] This maternal protein restriction diet resulted in gender-specific histone modifications (histone acetylation at lysine 4 and lysine 9 residues) within the CCAAT-enhancer-binding protein (*CEBP/*β) gene in skeletal muscle of adult offspring that corresponded to the transcription of this critical metabolic regulator.[139] The same treatment also resulted in histone modifications and the repression of cyclin-dependent kinase inhibitor 2 (*p16/Cdkn2a*), a key cell-cycle control gene, in mammary glands of adult female offspring, an event that has been linked to breast carcinogenesis.[140] We also reported the distinct regulation of the ATF3 and Asns genes by maternal protein restriction, which involves both histone modification and transcription factor binding that leads to these differing transcriptional responses.[141] While ATF4 binding was increased in response to the gestational PR diet at both the ATF3 and Asns proximal promoters at the C/EBP-ATF response element (CARE) region, the transcriptional responses were not the same. At the Asns promoter, ATF4, which belongs to the cAMP response element-binding protein (CREB) family, may recruit CREB-binding protein (CBP) that acts as a histone acetyltransferase (HAT)[142] and results in elevated acetylation of histone H4 in the PR group. Increased H4Ac at the Asns promoter may indicate a more open chromatin state, resulting in a significantly increased binding of ATF3, C/EBPβ, and possibly CHOP at the CARE region of the Asns promoter. ATF3, together with C/EBPβ and CHOP, may function as a repressive complex, leading to a significant increase of trimethylation at lysine 9 on histone H3 (H3K9Me3), a marker of silenced chromatin. Therefore, in the current study, we propose that the complex histone modifications and transcription factor associations at the Asns promoter resulted in unchanged Asns transcription rate and mRNA levels in response to amino acid or protein deprivation.

A maternal HF diet not only increases adiposity,[38,143] adipocyte hypertrophy,[144] accompanied with decreased lean mass[145] in the offspring, but also changes gene expression as well as the phenotype through epigenetic modifications. Inflammation programmed cell differentiation via promotion of adipogenesis and simultaneous repression of myogenesis,[146,147] thus increasing adiposity that prepares extra space for obesity formation. Particularly, IL-6, monocyte chemoattractant protein 1 (MCP-1), and tumor necrosis factor α (TNF-α) are closely associated with body fat deposition.[148–151] A gestational HF diet in mice was shown to lead to the hypomethylation of dopamine and opioid-related genes in brains of offspring that also had a preference for sucrose and fat.[152] Either supplementation or deficiency of folic acid and other methyl donors during pregnancy has also been implicated in regulating methylation events in the offspring.[152] Gestational choline supplementation led to an increase in H3K9Me2 and H3K27Me2, whereas choline-deficiency increased the levels of H3K4Me2 in offspring rats.[153] In another study in rats, gestational food restriction led to the dimethylation of histone H3 at lysine residue 4 (H3K4Me2) within the *Igf1* promoter, which affected the mRNA expression of the gene,[154] and a gestational low-protein diet in mice resulted in a decrease of H3K4Me3 and H4K20Me3 and an increase of H3K9Me3 and H3K27Me3 within *Igf2*, which was accompanied by a decrease in the mRNA expression of the gene.[155] Although it is clear that a maternal diet that is high in fat is deleterious to fetal development, the consequences of a gestational HF diet on the histone code have not been thoroughly elucidated. In primates, the chronic consumption of a maternal HF diet led to an increase in acetylation of histone H3 at lysine 14 (H3K14Ac), a trend of increase in acetylation of histone H3 at lysine 9 (H3K9Ac), lysine 18 (H3K18Ac), dimethylation of histone H3 at lysine 9 (H3K9Me2), trimethylation of histone H3 at lysine 9 (H3K9Me3), and trimethylation of histone H4 at lysine 27 (H3K27Me3), and these fetal hepatic tissues had an increase in triglycerides and NAFLD.[156] In rodents, a maternal HF diet results in a decreased association of H3Ac, H3K4Me2, H3K9Me3, and H3K27Me3 within the promoter of the hepatic gluconeogenic *Pck1* gene. Additionally, a maternal HF diet led to a decrease in H3K9Me3, and an increase in H4Ac and H3K4Me2, and that these modifications were associated with an increased transcription and transcriptional rate of the *Pck1* gene.[157] The studies above suggest that maternal diets alter fetal chromatin structure in rodents as well as in primates via covalent histone modifications, and this contributes to changes to gene expression as well as to the phenotype of the offspring.

3.3.3 Epigenetic Mechanisms Involving Maternal Dietary Effect

There is a strong support from human, animal, and cell studies showing that epigenetic events are critical for the regulation of fetal development. Increased fat intake before or during pregnancy results in suboptimal concentrations of energy, hormones, and blood supply,[14,18,144,158,159] which promotes abnormal development and growth of fetus in such an unfriendly intrauterine environment. The fetal consequences caused by increased maternal circulating cytokines and placenta inflammation are still under debate. Increased inflammation was observed in fetal circulation, liver, and brain following the maternal HF diet.[14,160] However, bidirectional transfer of proinflammatory cytokines is barely detected in a normal term placenta.[161,162] There is little knowledge available to establish the connection between placenta inflammation and initiation of a fetal inflammatory response. Indeed, there is evidence showing that induced cytokines in the placenta did not necessarily cross the placenta barrier to impact the fetal inflammatory status,[163] despite the altered permeability due to placental inflammation.[164] There is increasing evidence suggesting the reciprocal interaction between inflammation and epigenetic modifications. On the one hand, the capacity for epigenetic regulation of inflammatory genes that eventually result in multiple physiological consequences was shown in multiple studies.[165] DNA methylation was suggested to participate in the gene regulation of Cox-2,[166,167] IL-6,[168] and IL-1β[169] under multiple pathological conditions in liver. Histone acetylation was also shown to be associated with inflammation, and is tightly controlled by histone acetyltransferase (HAT) and histone deacetylases (HDACs).[170–172] Inhibition of HDAC reduced the production of IL-6 and thus attenuated liver injury.[173] Meanwhile, the pathogenesis of NASH is potentially attributed to the recruitment of chromatin-remodeling proteins related gene Brahma-related gene 1 (Brg1, also known as ATP-dependent helicase SMARCA4) and Brahma (Brm) to the promoter of proinflammatory mediators IL-1, IL-6, and MCP-1, which induces histone modifications and enhances gene transcription.[174] The newly identified epigenetic regulator ten-eleven translocation methylcytosine dioxygenase 1 (TET1) also appears to suppress the transcription of IL-1β.[175] On the other hand, inflammation also influences the epigenomic status. The proinflammatory signaling networks of IL-1β, PRS, and TNF-α-induced pathways have been shown to activate HDAC1 ubiquitination and proteosomal degradation,[176,177] thereby potentially affecting the downstream target genes that are subjected to HDAC1 regulation. IL-6 was also found to regulate its responsive genes through histone methylation by acting on the association of signal transducer and activator of transcription 3 (STAT3) to the promoter regions.[178] However, a complete documentation of epigenetic modification is yet to be established for inflammatory genes.

Although little data are available regarding the epigenetic modifications occurring within the anti-oxidative system in response to maternal diet, the response of anti-oxidative genes to maternal nutrition is likely similar to that of other vital pathways, in that they are regulated in utero, which likely occurs at the epigenetic level. Few studies are available relating nutrition and epigenetics in humans, but a study of mucosal swabs showed that MnSOD was differentially expressed in vegetarians versus meat-eaters and that this was associated with differential global DNA methylation as well as methylation of CpG islands within the MnSOD gene.[179] Observations from our laboratory have shown that a maternal HF diet impacts the expression of the Pon genes in liver, and these changes are gender-specific and may occur through the epigenetic modifications of these genes. We showed that *Pon1*, *Pon2*, and *Pon3* were increased following a maternal HF diet in livers of male but not female neonates. Histone modifications within the *Pon1* gene promoter corresponded to the mRNA expression of the gene, with increased H4Ac and H3K4Me2 and decreased H3K9Me3. These modifications were associated with increased oxidative stress in the liver, suggesting that the fetal anti-oxidative system responds to maternal nutrition, which occurs at the epigenetic level. The early-life environment results in lifelong epigenetically programmed differences in gene expression;[180] therefore, future efforts to address maternal nutrition programming of offspring gene expression may help to develop more effective personal nutritional recommendations to prevent or alleviate adult-onset diseases.

3.4 PERSPECTIVES AND FUTURE DIRECTIONS

Although it is valuable to determine specific epigenetic modifications governing the regulation of fetal development, it is more imperative to clarify how these changes contribute to the overall health of offspring. The activation of the mammalian AAR may act as a cue for the fetus to develop an adaptive response suited to its predicted postnatal environment but could in turn increase their vulnerability to developing disease and also influence their responsiveness to therapy. A clear understanding of the molecular mechanisms at the epigenetic level for inflammatory genes will provide novel knowledge that is critical for preventing and/or treating a variety of inflammatory gene-related pathological conditions through epigenetic therapies. The overall balance of prooxidants and antioxidants is the determinant of oxidative balance, and it remains unclear whether maternal diet regulates one or both of these. Additionally, more research is needed to clearly establish how maternal physiology or the nutrients themselves are responsible for altered physiology observed in offspring. While recent advancements in the field make it possible to conclude that epigenetic modifications play a critical role in regulating fetal development in response to maternal diet, few studies have addressed whether these modifications can be reversed or whether the resulting adult diseases can be ameliorated. Because inflammation and the anti-oxidative defense system are critical for mediating the relationship between the environment and human disease, it is essential to determine whether these systems can be modified epigenetically and whether reversing these marks can lead to improved health outcomes. Additionally, because studies have been shown that maternal nutrition is a potent regulator of fetal development, future epigenetic research in the field will be crucial to demonstrate the promise of disease prevention or therapy during the earliest developmental periods.

REFERENCES

1. Rieger D. Relationships between energy metabolism and development of early mammalian embryos. *Theriogenology* 1992; 37:75–93.
2. Widdowson EM. Changes in the body and its organs during lactation: Nutritional implications. *Ciba Found Symp* 1976; (45):103–18.
3. Ozanne SE. The long term effects of early postnatal diet on adult health. *Adv Exp Med Biol* 2009; 639:135–44.
4. Winick M, Noble A. Cellular response in rats during malnutrition at various ages. *J Nutr* 1966; 89:300–6.
5. Petry CJ, Ozanne SE, Wang CL, Hales CN. Early protein restriction and obesity independently induce hypertension in 1-year-old rats. *Clin Sci (Lond)* 1997; 93:147–52.
6. Hales CN, Desai M, Ozanne SE, Crowther NJ. Fishing in the stream of diabetes: From measuring insulin to the control of fetal organogenesis. *Biochem Soc Trans* 1996; 24:341–50.
7. Godfrey KM, Barker DJ. Fetal nutrition and adult disease. *Am J Clin Nutr* 2000; 71:1344S–52S.
8. Waterland RA, Michels KB. Epigenetic epidemiology of the developmental origins hypothesis. *Annu Rev Nutr* 2007; 27:363–88.
9. Snoeck A, Remacle C, Reusens B, Hoet JJ. Effect of a low protein diet during pregnancy on the fetal rat endocrine pancreas. *Biol Neonate* 1990; 57:107–18.
10. Lillycrop KA, Phillips ES, Jackson AA, Hanson MA, Burdge GC. Dietary protein restriction of pregnant rats induces and folic acid supplementation prevents epigenetic modification of hepatic gene expression in the offspring. *J Nutr* 2005; 135:1382–6.
11. Ozanne SE, Constancia M. Mechanisms of disease: The developmental origins of disease and the role of the epigenotype. *Nat Clin Pract Endocrinol Metab* 2007; 3:539–46.
12. Misra A, Singhal N, Khurana L. Obesity, the metabolic syndrome, and type 2 diabetes in developing countries: Role of dietary fats and oils. *J Am Coll Nutr* 2010; 29:289S–301S.
13. Zhai F, Wang H, Du S, He Y, Wang Z, Ge K et al. Prospective study on nutrition transition in China. *Nutr Rev* 2009; 67 Suppl 1:S56–S61.
14. McCurdy CE, Bishop JM, Williams SM, Grayson BE, Smith MS, Friedman JE et al. Maternal high-fat diet triggers lipotoxicity in the fetal livers of nonhuman primates. *J Clin Invest* 2009; 119:323–35.
15. Tschop M, Heiman ML. Rodent obesity models: An overview. *Exp Clin Endocrinol Diabetes* 2001; 109:307–19.
16. Oakes ND, Cooney GJ, Camilleri S, Chisholm DJ, Kraegen EW. Mechanisms of liver and muscle insulin resistance induced by chronic high-fat feeding. *Diabetes* 1997; 46:1768–74.
17. Buettner R, Scholmerich J, Bollheimer LC. High-fat diets: Modeling the metabolic disorders of human obesity in rodents. *Obesity (Silver Spring)* 2007; 15:798–808.
18. Srinivasan M, Katewa SD, Palaniyappan A, Pandya JD, Patel MS. Maternal high-fat diet consumption results in fetal map programming predisposing to the onset of metabolic syndrome-like phenotype in adulthood. *Am J Physiol Endocrinol Metab* 2006; 291:E792–9.
19. Elahi MM, Cagampang FR, Mukhtar D, Anthony FW, Ohri SK, Hanson MA. Long-term maternal high-fat feeding from weaning through pregnancy and lactation predisposes offspring to hypertension, raised plasma lipids and fatty liver in mice. *Br J Nutr* 2009; 102:514–9.
20. Shankar K, Harrell A, Liu X, Gilchrist JM, Ronis MJ, Badger TM. Maternal obesity at conception programs obesity in the offspring. *Am J Physiol Regul Integr Comp Physiol* 2008; 294:R528–38.
21. Plagemann A, Harder T, Rake A, Voits M, Fink H, Rohde W et al. Perinatal elevation of hypothalamic insulin, acquired malformation of hypothalamic galaninergic neurons,

and syndrome x-like alterations in adulthood of neonatally overfed rats. *Brain Res* 1999; 836:146–55.
22. Levin BE, Govek E. Gestational obesity accentuates obesity in obesity-prone progeny. *Am J Physiol* 1998; 275:R1374–9.
23. Karnik HB, Sonawane BR, Adkins JS, Mohla S. High dietary fat feeding during perinatal development of rats alters hepatic drug metabolism of progeny. *Dev Pharmacol Ther* 1989; 14:135–40.
24. Zhang X, Saaddine JB, Chou CF, Cotch MF, Cheng YJ, Geiss LS et al. Prevalence of diabetic retinopathy in the United States, 2005–2008. *JAMA* 2010; 304:649–56.
25. Martin-Gronert MS, Ozanne SE. Mechanisms linking suboptimal early nutrition and increased risk of type 2 diabetes and obesity. *J Nutr* 2010; 140:662–6.
26. Chen JH, Tarry-Adkins JL, Matharu K, Yeo GS, Ozanne SE. Maternal protein restriction affects gene expression profiles in the kidney at weaning with implications for the regulation of renal function and lifespan. *Clin Sci (Lond)* 2010; 119:373–84.
27. Strakovsky RS, Zhou D, Pan YX. A low-protein diet during gestation in rats activates the placental mammalian amino acid response pathway and programs the growth capacity of offspring. *J Nutr* 2010; 140(12):2116–20.
28. Tarry-Adkins JL, Chen JH, Jones RH, Smith NH, Ozanne SE. Poor maternal nutrition leads to alterations in oxidative stress, antioxidant defense capacity, and markers of fibrosis in rat islets: Potential underlying mechanisms for development of the diabetic phenotype in later life. *FASEB J* 2010; 24:2762–71.
29. Tarry-Adkins JL, Martin-Gronert MS, Chen JH, Cripps RL, Ozanne SE. Maternal diet influences DNA damage, aortic telomere length, oxidative stress, and antioxidant defense capacity in rats. *FASEB J* 2008; 22:2037–44.
30. de AS, Khan G, Hilakivi-Clarke L. High birth weight increases mammary tumorigenesis in rats. *Int J Cancer* 2006; 119:1537–46.
31. Luijten M, Thomsen AR, van den Berg JA, Wester PW, Verhoef A, Nagelkerke NJ et al. Effects of soy-derived isoflavones and a high-fat diet on spontaneous mammary tumor development in Tg.NK (MMTV/c-neu) mice. *Nutr Cancer* 2004; 50:46–54.
32. Hilakivi-Clarke L, Clarke R, Onojafe I, Raygada M, Cho E, Lippman M. A maternal diet high in n - 6 polyunsaturated fats alters mammary gland development, puberty onset, and breast cancer risk among female rat offspring. *Proc Natl Acad Sci USA* 1997; 94:9372–7.
33. Bouanane S, Merzouk H, Benkalfat NB, Soulimane N, Merzouk SA, Gresti J et al. Hepatic and very low-density lipoprotein fatty acids in obese offspring of overfed dams. *Metabolism* 2010; 59(12):1701–9.
34. Gregorio BM, Souza-Mello V, Carvalho JJ, Mandarim-de-Lacerda CA, Aguila MB. Maternal high-fat intake predisposes nonalcoholic fatty liver disease in C57BL/6 offspring. *Am J Obstet Gynecol* 2010; 203(5):495.e1–8.
35. Burgueño AL, Carabelli J, Sookoian S, Pirola CJ. The impact of maternal high-fat feeding on liver and abdominal fat accumulation in adult offspring under a long-term high-fat diet. *Hepatology* 2010; 51:2234–5.
36. Bruce KD, Cagampang FR, Argenton M, Zhang J, Ethirajan PL, Burdge GC et al. Maternal high-fat feeding primes steatohepatitis in adult mice offspring, involving mitochondrial dysfunction and altered lipogenesis gene expression. *Hepatology* 2009; 50:1796–808.
37. Zhang X, Strakovsky R, Zhou D, Zhang Y, Pan YX. A maternal high-fat diet represses the expression of antioxidant defense genes and induces the cellular senescence pathway in the liver of male offspring rats. *J Nutr* 2011; 141:1254–9.
38. Sen S, Simmons RA. Maternal antioxidant supplementation prevents adiposity in the offspring of Western diet-fed rats. *Diabetes* 2010; 59:3058–65.
39. Lin Y, Han XF, Fang ZF, Che LQ, Nelson J, Yan TH et al. Beneficial effects of dietary fibre supplementation of a high-fat diet on fetal development in rats. *Br J Nutr* 2011; 106(4):510–8.

40. Liang C, Decourcy K, Prater MR. High-saturated-fat diet induces gestational diabetes and placental vasculopathy in C57BL/6 mice. *Metabolism* 2010; 59(7):943–50.
41. Endo Y, Fu Z, Abe K, Arai S, Kato H. Dietary protein quantity and quality affect rat hepatic gene expression. *J Nutr* 2002; 132:3632–7.
42. Roseboom TJ, van der Meulen JH, Ravelli AC, Osmond C, Barker DJ, Bleker OP. Effects of prenatal exposure to the Dutch famine on adult disease in later life: An overview. *Mol Cell Endocrinol* 2001; 185:93–8.
43. Thiaville MM, Dudenhausen EE, Zhong C, Pan YX, Kilberg MS. Deprivation of protein or amino acid induces C/EBPβ synthesis and binding to amino acid response elements, but its action is not an absolute requirement for enhanced transcription. *Biochem J* 2008; 410:473–84.
44. Lu PD, Harding HP, Ron D. Translation reinitiation at alternative open reading frames regulates gene expression in an integrated stress response. *J Cell Biol* 2004; 167: 27–33.
45. Barbosa-Tessmann IP, Chen C, Zhong C, Siu F, Schuster SM, Nick HS et al. Activation of the human asparagine synthetase gene by the amino acid response and the endoplasmic reticulum stress response pathways occurs by common genomic elements. *J Biol Chem* 2000; 275:26976–85.
46. Pan Y, Chen H, Siu F, Kilberg MS. Amino acid deprivation and endoplasmic reticulum stress induce expression of multiple activating transcription factor-3 mRNA species that, when overexpressed in HepG2 cells, modulate transcription by the human asparagine synthetase promoter. *J Biol Chem* 2003; 278:38402–12.
47. Bruhat A, Jousse C, Wang XZ, Ron D, Ferrara M, Fafournoux P. Amino acid limitation induces expression of CHOP, a CCAAT/enhancer binding protein-related gene, at both transcriptional and post-transcriptional levels. *J Biol Chem* 1997; 272:17588–93.
48. Abcouwer SF, Schwarz C, Meguid RA. Glutamine deprivation induces the expression of GADD45 and GADD153 primarily by mRNA stabilization. *J Biol Chem* 1999; 274:28645–51.
49. Harding HP, Novoa I, Zhang Y, Zeng H, Wek R, Schapira M et al. Regulated translation initiation controls stress-induced gene expression in mammalian cells. *Mol Cell* 2000; 6:1099–108.
50. Vattem KM, Wek RC. Reinitiation involving upstream ORFs regulates ATF4 mRNA translation in mammalian cells. *Proc Natl Acad Sci USA* 2004; 101:11269–74.
51. Harding HP, Zhang Y, Zeng H, Novoa I, Lu PD, Calfon M et al. An integrated stress response regulates amino acid metabolism and resistance to oxidative stress. *Mol Cell* 2003; 11:619–33.
52. Naismith DJ, Morgan BL. The biphasic nature of protein metabolism during pregnancy in the rat. *Br J Nutr* 1976; 36:563–6.
53. Naismith DJ. The requirement for protein, and the utilization of protein and calcium during pregnancy. *Metabolism* 1966; 15:582–95.
54. Naismith DJ. The foetus as a parasite. *Proc Nutr Soc* 1969; 28:25–31.
55. Nishino I. Autophagic vacuolar myopathies. *Curr Neurol Neurosci Rep* 2003; 3:64–9.
56. Mizushima N, Levine B, Cuervo AM, Klionsky DJ. Autophagy fights disease through cellular self-digestion. *Nature* 2008; 451:1069–75.
57. Sandri M, Lin J, Handschin C, Yang W, Arany ZP, Lecker SH et al. PGC-1alpha protects skeletal muscle from atrophy by suppressing FoxO3 action and atrophy-specific gene transcription. *Proc Natl Acad Sci USA* 2006; 103:16260–5.
58. Mammucari C, Milan G, Romanello V, Masiero E, Rudolf R, Del PP et al. FoxO3 controls autophagy in skeletal muscle in vivo. *Cell Metab* 2007; 6:458–71.
59. Zhao J, Brault JJ, Schild A, Cao P, Sandri M, Schiaffino S et al. FoxO3 coordinately activates protein degradation by the autophagic/lysosomal and proteasomal pathways in atrophying muscle cells. *Cell Metab* 2007; 6:472–83.

60. Sandri M, Sandri C, Gilbert A, Skurk C, Calabria E, Picard A et al. Foxo transcription factors induce the atrophy-related ubiquitin ligase atrogin-1 and cause skeletal muscle atrophy. *Cell* 2004; 117:399–412.
61. Ferrero-Miliani L, Nielsen OH, Andersen PS, Girardin SE. Chronic inflammation: Importance of NOD2 and NALP3 in interleukin-1beta generation. *Clin Exp Immunol* 2007; 147:227–35.
62. Andreozzi P, Viscogliosi G, Colella F, Subic M, Cipriani E, Marigliano B et al. Predictors of liver fibrosis in patients with non-alcoholic fatty liver disease. The role of metabolic syndrome, insulin-resistance and inflammation. *Recenti Prog Med* 2012; 103:570–4.
63. Taube A, Schlich R, Sell H, Eckardt K, Eckel J. Inflammation and metabolic dysfunction: Links to cardiovascular diseases. *Am J Physiol Heart Circ Physiol* 2012; 302:H2148–65.
64. Manabe I. Chronic inflammation links cardiovascular, metabolic and renal diseases. *Circ J* 2011; 75:2739–48.
65. Sartori AC, Vance DE, Slater LZ, Crowe M. The impact of inflammation on cognitive function in older adults: Implications for healthcare practice and research. *J Neurosci Nurs* 2012; 44:206–17.
66. Jenny NS. Inflammation in aging: Cause, effect, or both? *Discov Med* 2012; 13:451–60.
67. Romeo GR, Lee J, Shoelson SE. Metabolic syndrome, insulin resistance, and roles of inflammation—mechanisms and therapeutic targets. *Arterioscler Thromb Vasc Biol* 2012; 32:1771–6.
68. Chaturvedi AK, Moore SC, Hildesheim A. Invited commentary: Circulating inflammation markers and cancer risk—implications for epidemiologic studies. *Am J Epidemiol* 2013; 177:14–9.
69. Medzhitov R. Origin and physiological roles of inflammation. *Nature* 2008; 454:428–35.
70. Kumashiro N, Erion DM, Zhang D, Kahn M, Beddow SA, Chu X et al. Cellular mechanism of insulin resistance in nonalcoholic fatty liver disease. *Proc Natl Acad Sci USA* 2011; 108:16381–5.
71. Perseghin G, Petersen K, Shulman GI. Cellular mechanism of insulin resistance: Potential links with inflammation. *Int J Obes Relat Metab Disord* 2003; 27 Suppl 3:S6–11.
72. Lian Z, Li Y, Gao J, Qu K, Li J, Hao L et al. A novel AMPK activator, WS070117, improves lipid metabolism discords in hamsters and HepG2 cells. *Lipids Health Dis* 2011; 10:67.
73. Barroso E, Rodriguez-Calvo R, Serrano-Marco L, Astudillo AM, Balsinde J, Palomer X et al. The PPARbeta/delta activator GW501516 prevents the down-regulation of AMPK caused by a high-fat diet in liver and amplifies the PGC-1alpha-lipin 1-PPARalpha pathway leading to increased fatty acid oxidation. *Endocrinology* 2011; 152:1848–59.
74. Samuel VT, Liu ZX, Qu X, Elder BD, Bilz S, Befroy D et al. Mechanism of hepatic insulin resistance in non-alcoholic fatty liver disease. *J Biol Chem* 2004; 279:32345–53.
75. Solinas G, Naugler W, Galimi F, Lee MS, Karin M. Saturated fatty acids inhibit induction of insulin gene transcription by JNK-mediated phosphorylation of insulin-receptor substrates. *Proc Natl Acad Sci USA* 2006; 103:16454–9.
76. Schattenberg JM, Wang Y, Singh R, Rigoli RM, Czaja MJ. Hepatocyte CYP2E1 overexpression and steatohepatitis lead to impaired hepatic insulin signaling. *J Biol Chem* 2005; 280:9887–94.
77. Abu-Elheiga L, Wu H, Gu Z, Bressler R, Wakil SJ. Acetyl-CoA carboxylase 2–/– mutant mice are protected against fatty liver under high-fat, high-carbohydrate dietary and de novo lipogenic conditions. *J Biol Chem* 2012; 287:12578–88.
78. Morgan K, Uyuni A, Nandgiri G, Mao L, Castaneda L, Kathirvel E et al. Altered expression of transcription factors and genes regulating lipogenesis in liver and adipose tissue of mice with high fat diet-induced obesity and nonalcoholic fatty liver disease. *Eur J Gastroenterol Hepatol* 2008; 20:843–54.

79. Chisholm JW, Hong J, Mills SA, Lawn RM. The LXR ligand T0901317 induces severe lipogenesis in the db/db diabetic mouse. *J Lipid Res* 2003; 44:2039–48.
80. Denechaud PD, Dentin R, Girard J, Postic C. Role of ChREBP in hepatic steatosis and insulin resistance. *FEBS Lett* 2008; 582:68–73.
81. Dentin R, Benhamed F, Hainault I, Fauveau V, Foufelle F, Dyck JR et al. Liver-specific inhibition of ChREBP improves hepatic steatosis and insulin resistance in ob/ob mice. *Diabetes* 2006; 55:2159–70.
82. Schreurs M, Kuipers F, van der Leij FR. Regulatory enzymes of mitochondrial beta-oxidation as targets for treatment of the metabolic syndrome. *Obes Rev* 2010; 11:380–8.
83. Yu J, Ip E, Dela PA, Hou JY, Sesha J, Pera N et al. COX-2 induction in mice with experimental nutritional steatohepatitis: Role as pro-inflammatory mediator. *Hepatology* 2006; 43:826–36.
84. Peng Y, Rideout D, Rakita S, Lee J, Murr M. Diet-induced obesity associated with steatosis, oxidative stress, and inflammation in liver. *Surg Obes Relat Dis* 2012; 8:73–81.
85. Hsieh PS, Jin JS, Chiang CF, Chan PC, Chen CH, Shih KC. COX-2-mediated inflammation in fat is crucial for obesity-linked insulin resistance and fatty liver. *Obesity (Silver Spring)* 2009; 17:1150–7.
86. Stanton MC, Chen SC, Jackson JV, Rojas-Triana A, Kinsley D, Cui L et al. Inflammatory signals shift from adipose to liver during high fat feeding and influence the development of steatohepatitis in mice. *J Inflamm (Lond)* 2011; 8:8. doi: 10.1186/1476-9255-8-8.
87. Fang X, Zou S, Zhao Y, Cui R, Zhang W, Hu J et al. Kupffer cells suppress perfluorononanoic acid-induced hepatic peroxisome proliferator–activated receptor alpha expression by releasing cytokines. *Arch Toxicol* 2012; 86:1515–25.
88. Haukeland JW, Damas JK, Konopski Z, Loberg EM, Haaland T, Goverud I et al. Systemic inflammation in nonalcoholic fatty liver disease is characterized by elevated levels of CCL2. *J Hepatol* 2006; 44:1167–74.
89. Hill DB, Marsano L, Cohen D, Allen J, Shedlofsky S, McClain CJ. Increased plasma interleukin-6 concentrations in alcoholic hepatitis. *J Lab Clin Med* 1992; 119:547–52.
90. Kugelmas M, Hill DB, Vivian B, Marsano L, McClain CJ. Cytokines and NASH: A pilot study of the effects of lifestyle modification and vitamin E. *Hepatology* 2003; 38:413–9.
91. Kang JH, Goto T, Han IS, Kawada T, Kim YM, Yu R. Dietary capsaicin reduces obesity-induced insulin resistance and hepatic steatosis in obese mice fed a high-fat diet. *Obesity (Silver Spring)* 2010; 18:780–7.
92. Wieckowska A, Papouchado BG, Li Z, Lopez R, Zein NN, Feldstein AE. Increased hepatic and circulating interleukin-6 levels in human nonalcoholic steatohepatitis. *Am J Gastroenterol* 2008; 103:1372–9.
93. Cai D, Yuan M, Frantz DF, Melendez PA, Hansen L, Lee J et al. Local and systemic insulin resistance resulting from hepatic activation of IKK-beta and NF-kappaB. *Nat Med* 2005; 11:183–90.
94. Wang Y, Zhang L, Wu X, Gurley EC, Kennedy E, Hylemon PB et al. The role of CCAAT enhancer-binding protein homologous protein in human immunodeficiency virus protease-inhibitor-induced hepatic lipotoxicity in mice. *Hepatology* 2013; 57:1005–16.
95. Chen L, Jarujaron S, Wu X, Sun L, Zha W, Liang G et al. HIV protease inhibitor lopinavir-induced TNF-alpha and IL-6 expression is coupled to the unfolded protein response and ERK signaling pathways in macrophages. *Biochem Pharmacol* 2009; 78:70–7.
96. Zhou H, Jarujaron S, Gurley EC, Chen L, Ding H, Studer E et al. HIV protease inhibitors increase TNF-alpha and IL-6 expression in macrophages: Involvement of the RNA-binding protein HuR. *Atherosclerosis* 2007; 195:e134–43.
97. Li Z, Soloski MJ, Diehl AM. Dietary factors alter hepatic innate immune system in mice with nonalcoholic fatty liver disease. *Hepatology* 2005; 42:880–5.
98. Nam H, Ferguson BS, Stephens JM, Morrison RF. Impact of obesity on IL-12 family gene expression in insulin responsive tissues. *Biochim Biophys Acta* 2013; 1832(1): 11–9.

99. Marx V. Epigenetics: Reading the second genomic code. *Nature* 2012; 491:143–7.
100. Margueron R, Reinberg D. Chromatin structure and the inheritance of epigenetic information. *Nat Rev Genet* 2010; 11:285–96.
101. Esteller M. Cancer, epigenetics and the Nobel Prizes. *Mol Oncol* 2012; 6:565–6.
102. Jones PA, Liang G. Rethinking how DNA methylation patterns are maintained. *Nat Rev Genet* 2009; 10:805–11.
103. Deaton AM, Bird A. CpG islands and the regulation of transcription. *Genes Dev* 2011; 25:1010–22.
104. Hendrich B, Bird A. Identification and characterization of a family of mammalian methyl-CpG binding proteins. *Mol Cell Biol* 1998; 18:6538–47.
105. Nakao M, Matsui S, Yamamoto S, Okumura K, Shirakawa M, Fujita N. Regulation of transcription and chromatin by methyl-CpG binding protein MBD1. *Brain Dev* 2001; 23 Suppl 1:S174–6.
106. Kudo S. Methyl-CpG-binding protein MeCP2 represses Sp1-activated transcription of the human leukosialin gene when the promoter is methylated. *Mol Cell Biol* 1998; 18:5492–9.
107. Boyes J, Bird A. DNA methylation inhibits transcription indirectly via a methyl-CpG binding protein. *Cell* 1991; 64:1123–34.
108. Chatagnon A, Bougel S, Perriaud L, Lachuer J, Benhattar J, Dante R. Specific association between the methyl-CpG-binding domain protein 2 and the hypermethylated region of the human telomerase reverse transcriptase promoter in cancer cells. *Carcinogenesis* 2009; 30:28–34.
109. Lopez-Serra L, Ballestar E, Fraga MF, Alaminos M, Setien F, Esteller M. A profile of methyl-CpG binding domain protein occupancy of hypermethylated promoter CpG islands of tumor suppressor genes in human cancer. *Cancer Res* 2006; 66:8342–6.
110. McDonel P, Demmers J, Tan DW, Watt F, Hendrich BD. Sin3a is essential for the genome integrity and viability of pluripotent cells. *Dev Biol* 2012; 363:62–73.
111. Badenhorst P, Voas M, Rebay I, Wu C. Biological functions of the ISWI chromatin remodeling complex NURF. *Genes Dev* 2002; 16:3186–98.
112. Klose RJ, Bird AP. Genomic DNA methylation: The mark and its mediators. *Trends Biochem Sci* 2006; 31:89–97.
113. Liang P, Song F, Ghosh S, Morien E, Qin M, Mahmood S et al. Genome-wide survey reveals dynamic widespread tissue-specific changes in DNA methylation during development. *BMC Genomics* 2011; 12:231.
114. Maunakea AK, Chepelev I, Zhao K. Epigenome mapping in normal and disease States. *Circ Res* 2010; 107:327–39.
115. Strahl BD, Allis CD. The language of covalent histone modifications. *Nature* 2000; 403:41–5.
116. Gilbert N, Thomson I, Boyle S, Allan J, Ramsahoye B, Bickmore WA. DNA methylation affects nuclear organization, histone modifications, and linker histone binding but not chromatin compaction. *J Cell Biol* 2007; 177:401–11.
117. Cedar H, Bergman Y. Linking DNA methylation and histone modification: Patterns and paradigms. *Nat Rev Genet* 2009; 10:295–304.
118. Bartova E, Krejci J, Harnicarova A, Galiova G, Kozubek S. Histone modifications and nuclear architecture: A review. *J Histochem Cytochem* 2008; 56:711–21.
119. Kouzarides T. Chromatin modifications and their function. *Cell* 2007; 128:693–705.
120. Bernstein BE, Kamal M, Lindblad-Toh K, Bekiranov S, Bailey DK, Huebert DJ et al. Genomic maps and comparative analysis of histone modifications in human and mouse. *Cell* 2005; 120:169–81.
121. Schubeler D, Macaprine DM, Scalzo D, Wirbelauer C, Kooperberg C, van LF et al. The histone modification pattern of active genes revealed through genome-wide chromatin analysis of a higher eukaryote. *Genes Dev* 2004; 18:1263–71.

122. Lachner M, Jenuwein T. The many faces of histone lysine methylation. *Curr Opin Cell Biol* 2002; 14:286–98.
123. Snowden AW, Gregory PD, Case CC, Pabo CO. Gene-specific targeting of H3K9 methylation is sufficient for initiating repression in vivo. *Curr Biol* 2002; 12:2159–66.
124. Szyf M, Weaver I, Meaney M. Maternal care, the epigenome and phenotypic differences in behavior. *Reprod Toxicol* 2007; 24:9–19.
125. Murr R. Interplay between different epigenetic modifications and mechanisms. *Adv Genet* 2010; 70:101–41.
126. Fall C. Maternal nutrition: Effects on health in the next generation. *Indian J Med Res* 2009; 130:593–9.
127. Metges CC. Early nutrition and later obesity: Animal models provide insights into mechanisms. *Adv Exp Med Biol* 2009; 646:105–12.
128. Akyol A, Langley-Evans SC, McMullen S. Obesity induced by cafeteria feeding and pregnancy outcome in the rat. *Br J Nutr* 2009; 102:1601–10.
129. Bayol SA, Simbi BH, Stickland NC. A maternal cafeteria diet during gestation and lactation promotes adiposity and impairs skeletal muscle development and metabolism in rat offspring at weaning. *J Physiol* 2005; 567:951–61.
130. Caluwaerts S, Lambin S, van Bree R, Peeters H, Vergote I, Verhaeghe J. Diet-induced obesity in gravid rats engenders early hyperadiposity in the offspring. *Metabolism* 2007; 56:1431–8.
131. Howie GJ, Sloboda DM, Kamal T, Vickers MH. Maternal nutritional history predicts obesity in adult offspring independent of postnatal diet. *J Physiol* 2009; 587:905–15.
132. Jones HN, Woollett LA, Barbour N, Prasad PD, Powell TL, Jansson T. High-fat diet before and during pregnancy causes marked up-regulation of placental nutrient transport and fetal overgrowth in C57/BL6 mice. *FASEB J* 2009; 23:271–8.
133. Wolff GL, Kodell RL, Moore SR, Cooney CA. Maternal epigenetics and methyl supplements affect agouti gene expression in Avy/a mice. *FASEB J* 1998; 12:949–57.
134. Dolinoy DC, Weidman JR, Waterland RA, Jirtle RL. Maternal genistein alters coat color and protects Avy mouse offspring from obesity by modifying the fetal epigenome. *Environ Health Perspect* 2006; 114:567–72.
135. Waterland RA, Travisano M, Tahiliani KG, Rached MT, Mirza S. Methyl donor supplementation prevents transgenerational amplification of obesity. *Int J Obes (Lond)* 2008; 32:1373–9.
136. Lillycrop KA, Slater-Jefferies JL, Hanson MA, Godfrey KM, Jackson AA, Burdge GC. Induction of altered epigenetic regulation of the hepatic glucocorticoid receptor in the offspring of rats fed a protein-restricted diet during pregnancy suggests that reduced DNA methyltransferase-1 expression is involved in impaired DNA methylation and changes in histone modifications. *Br J Nutr* 2007; 97:1064–73.
137. Lillycrop KA, Phillips ES, Torrens C, Hanson MA, Jackson AA, Burdge GC. Feeding pregnant rats a protein-restricted diet persistently alters the methylation of specific cytosines in the hepatic PPAR alpha promoter of the offspring. *Br J Nutr* 2008; 100: 278–82.
138. Gong L, Pan YX, Chen H. Gestational low protein diet in the rat mediates Igf2 gene expression in male offspring via altered hepatic DNA methylation. *Epigenetics* 2010; 5(7):619–26.
139. Zheng S, Rollet M, Pan YX. Maternal protein restriction during pregnancy induces CCAAT/enhancer-binding protein (C/EBPbeta) expression through the regulation of histone modification at its promoter region in female offspring rat skeletal muscle. *Epigenetics* 2011; 6:161–70.
140. Zheng S, Pan YX. Histone modifications, not DNA methylation, cause transcriptional repression of p16 (CDKN2A) in the mammary glands of offspring of protein-restricted rats. *J Nutr Biochem* 2011; 22:567–73.

141. Zhou D, Pan YX. Gestational low protein diet selectively induces the amino acid response pathway target genes in the liver of offspring rats through transcription factor binding and histone modifications. *Biochim Biophys Acta* 2011; 1809(10):549–56.
142. Ogryzko VV, Schiltz RL, Russanova V, Howard BH, Nakatani Y. The transcriptional coactivators p300 and CBP are histone acetyltransferases. *Cell* 1996; 87:953–9.
143. Taylor PD, McConnell J, Khan IY, Holemans K, Lawrence KM, Asare-Anane H et al. Impaired glucose homeostasis and mitochondrial abnormalities in offspring of rats fed a fat-rich diet in pregnancy. *Am J Physiol Regul Integr Comp Physiol* 2005; 288:R134–9.
144. Samuelsson AM, Matthews PA, Argenton M, Christie MR, McConnell JM, Jansen EH et al. Diet-induced obesity in female mice leads to offspring hyperphagia, adiposity, hypertension, and insulin resistance: A novel murine model of developmental programming. *Hypertension* 2008; 51:383–92.
145. Grant WF, Gillingham MB, Batra AK, Fewkes NM, Comstock SM, Takahashi D et al. Maternal high fat diet is associated with decreased plasma n-3 fatty acids and fetal hepatic apoptosis in nonhuman primates. *PLoS One* 2011; 6:e17261.
146. Tong JF, Yan X, Zhu MJ, Ford SP, Nathanielsz PW, Du M. Maternal obesity downregulates myogenesis and beta-catenin signaling in fetal skeletal muscle. *Am J Physiol Endocrinol Metab* 2009; 296:E917–24.
147. Bayol SA, Simbi BH, Bertrand JA, Stickland NC. Offspring from mothers fed a 'junk food' diet in pregnancy and lactation exhibit exacerbated adiposity that is more pronounced in females. *J Physiol* 2008; 586:3219–30.
148. Kubaszek A, Pihlajamaki J, Komarovski V, Lindi V, Lindstrom J, Eriksson J et al. Promoter polymorphisms of the TNF-alpha (G-308A) and IL-6 (C-174G) genes predict the conversion from impaired glucose tolerance to type 2 diabetes: The Finnish Diabetes Prevention Study. *Diabetes* 2003; 52:1872–6.
149. Yudkin JS. Adipose tissue, insulin action and vascular disease: Inflammatory signals. *Int J Obes Relat Metab Disord* 2003; 27 Suppl 3:S25–8.
150. Kirwan JP, Hauguel-De Mouzon S, Lepercq J, Challier JC, Huston-Presley L, Friedman JE et al. TNF-alpha is a predictor of insulin resistance in human pregnancy. *Diabetes* 2002; 51:2207–13.
151. Sell H, Eckel J. Monocyte chemotactic protein-1 and its role in insulin resistance. *Curr Opin Lipidol* 2007; 18:258–62.
152. Vucetic Z, Kimmel J, Totoki K, Hollenbeck E, Reyes TM. Maternal high-fat diet alters methylation and gene expression of dopamine and opioid-related genes. *Endocrinology* 2010; 151:4756–64.
153. Davison JM, Mellott TJ, Kovacheva VP, Blusztajn JK. Gestational choline supply regulates methylation of histone H3, expression of histone methyltransferases G9a (Kmt1c) and Suv39h1 (Kmt1a), and DNA methylation of their genes in rat fetal liver and brain. *J Biol Chem* 2009; 284:1982–9.
154. Tosh DN, Fu Q, Callaway CW, McKnight RA, McMillen IC, Ross MG et al. Epigenetics of programmed obesity: Alteration in IUGR rat hepatic IGF1 mRNA expression and histone structure in rapid versus delayed postnatal catch-up growth. *Am J Physiol Gastrointest Liver Physiol* 2010; 299(5):G1023–9.
155. Sharif J, Nakamura M, Ito T, Kimura Y, Nagamune T, Mitsuya K et al. Food restriction in pregnant mice can induce changes in histone modifications and suppress gene expression in fetus. *Nucleic Acids Symp Ser (Oxf)* 2007; (51):125–6.
156. Aagaard-Tillery KM, Grove K, Bishop J, Ke X, Fu Q, McKnight R et al. Developmental origins of disease and determinants of chromatin structure: Maternal diet modifies the primate fetal epigenome. *J Mol Endocrinol* 2008; 41:91–102.
157. Strakovsky RS, Zhang X, Zhou D, Pan YX. Gestational high fat diet programs hepatic phosphoenopryruvate carboxykinase (Pck) expression and histone modification in neonatal offspring rats. *J Physiol* 2011; 589(Pt 11):2707–17.

158. Chen H, Simar D, Lambert K, Mercier J, Morris MJ. Maternal and postnatal overnutrition differentially impact appetite regulators and fuel metabolism. *Endocrinology* 2008; 149:5348–56.
159. Bilbo SD, Tsang V. Enduring consequences of maternal obesity for brain inflammation and behavior of offspring. *FASEB J* 2010; 24:2104–15.
160. Grayson BE, Levasseur PR, Williams SM, Smith MS, Marks DL, Grove KL. Changes in melanocortin expression and inflammatory pathways in fetal offspring of nonhuman primates fed a high-fat diet. *Endocrinology* 2010; 151:1622–32.
161. Zaretsky MV, Alexander JM, Byrd W, Bawdon RE. Transfer of inflammatory cytokines across the placenta. *Obstet Gynecol* 2004; 103:546–50.
162. Aaltonen R, Heikkinen T, Hakala K, Laine K, Alanen A. Transfer of proinflammatory cytokines across term placenta. *Obstet Gynecol* 2005; 106:802–7.
163. Ashdown H, Dumont Y, Ng M, Poole S, Boksa P, Luheshi GN. The role of cytokines in mediating effects of prenatal infection on the fetus: Implications for schizophrenia. *Mol Psychiatry* 2006; 11:47–55.
164. Assadi F. Neonatal nephrotic syndrome associated with placental transmission of proinflammatory cytokines. *Pediatr Nephrol* 2011; 26:469–71.
165. Shanmugam MK, Sethi G. Role of epigenetics in inflammation-associated diseases. *Subcell Biochem* 2012; 61:627–57.
166. Maffei L, Murata Y, Rochira V, Tubert G, Aranda C, Vazquez M et al. Dysmetabolic syndrome in a man with a novel mutation of the aromatase gene: Effects of testosterone, alendronate, and estradiol treatment. *J Clin Endocrinol Metab* 2004; 89:61–70.
167. Cheng F, Shen J, Peng B, Pan Y, Tao Z, Chen J. Rapid room-temperature synthesis of nanocrystalline spinels as oxygen reduction and evolution electrocatalysts. *Nat Chem* 2011; 3:79–84.
168. Kang SC, Lee BM. DNA methylation of estrogen receptor alpha gene by phthalates. *J Toxicol Environ Health A* 2005; 68:1995–2003.
169. Vanhees K, Coort S, Ruijters EJ, Godschalk RW, van Schooten FJ, Barjesteh van Waalwijk van Doorn-Khosrovani S. Epigenetics: Prenatal exposure to genistein leaves a permanent signature on the hematopoietic lineage. *FASEB J* 2011; 25:797–807.
170. Villagra A, Sotomayor EM, Seto E. Histone deacetylases and the immunological network: Implications in cancer and inflammation. *Oncogene* 2010; 29:157–73.
171. Enya K, Hayashi H, Takii T, Ohoka N, Kanata S, Okamoto T et al. The interaction with Sp1 and reduction in the activity of histone deacetylase 1 are critical for the constitutive gene expression of IL-1 alpha in human melanoma cells. *J Leukoc Biol* 2008; 83:190–9.
172. Lu J, Sun H, Wang X, Liu C, Xu X, Li F et al. Interleukin-12 p40 promoter activity is regulated by the reversible acetylation mediated by HDAC1 and p300. *Cytokine* 2005; 31:46–51.
173. Zhang L, Wan J, Jiang R, Wang W, Deng H, Shen Y et al. Protective effects of trichostatin A on liver injury in septic mice. *Hepatol Res* 2009; 39:931–8.
174. Tian W, Xu H, Fang F, Chen Q, Xu Y, Shen A. Brahma related gene 1 (Brg1) bridges epigenetic regulation of pro-inflammatory cytokine production to steatohepatitis. *Hepatology* 2013; 58(2):576–88.
175. Neves-Costa A, Moita LF. TET1 is a negative transcriptional regulator of IL-1beta in the THP-1 cell line. *Mol Immunol* 2013; 54:264–70.
176. Gopal YN, Van Dyke MW. Depletion of histone deacetylase protein: A common consequence of inflammatory cytokine signaling? *Cell Cycle* 2006; 5:2738–43.
177. Aung HT, Schroder K, Himes SR, Brion K, van ZW, Trieu A et al. LPS regulates proinflammatory gene expression in macrophages by altering histone deacetylase expression. *FASEB J* 2006; 20:1315–27.

178. Young TY, Jerome D, Gupta S. Hyperimmunoglobulinemia E syndrome associated with coronary artery aneurysms: Deficiency of central memory CD4+ T cells and expansion of effector memory CD4+ T cells. *Ann Allergy Asthma Immunol* 2007; 98:389–92.
179. Thaler R, Karlic H, Rust P, Haslberger AG. Epigenetic regulation of human buccal mucosa mitochondrial superoxide dismutase gene expression by diet. *Br J Nutr* 2009; 101:743–9.
180. Szyf M. The early life environment and the epigenome. *Biochim Biophys Acta* 2009; 1790(9); 878–85.

Section II

Methyl Donors

4 Folate, DNA Methylation, and Colorectal Cancer

Sung-Eun Kim, Shannon Masih, and Young-In Kim

CONTENTS

4.1 INTRODUCTION

The overwhelming body of evidence suggests that environmental factors, including diet, influence our susceptibility to disease including cancer. Individual variations in disease susceptibility to environmental factors cannot entirely be accounted for by conventional genetic mechanisms. In this regard, epigenetics modifications including DNA methylation, histone modifications, chromatin remodeling, and RNA interference have emerged as a promising mechanism by which environmental factors including diet can influence disease susceptibility. In contrast to genetic changes in human diseases, epigenetic changes are gradual in onset and progression and are potentially reversible by dietary and pharmacological manipulations.[1] Aberrant patterns and dysregulation of epigenetics play a critical role in the development and progression of cancer including colorectal cancer (CRC).[1] Because of the critical role of folate in biological methylation reactions including DNA methylation and a purported relationship between folate status and CRC risk, there has been considerable interest in investigating whether DNA methylation might be a link between folate status and CRC risk.[1,2] Given this consideration, the aim of this chapter is to review the currently available evidence for the purported association between folate status and CRC risk and to review whether the modulatory effect of folate on CRC risk is mediated in part through epigenetic modifications, in particular alterations in DNA methylation.

4.2 FOLATE

Folate is a water-soluble B vitamin that is present naturally in foods such as green leafy vegetables, asparagus, broccoli, brussels sprouts, citrus fruit, legumes, dry cereals, whole grain, yeast, lima beans, liver, and other organ meats.[3] Folate is the generic term referring to compounds that have similar chemical structures and nutritional properties. Folic acid is the fully oxidized monoglutamyl form of this vitamin that is used commercially in supplements and in fortified foods. The main structure of folate and folic acid consists of a pteridine moiety attached to *para*-aminobenzoic acid with one or more glutamate residues attached via γ-peptide bonds (Figure 4.1).[4] Dietary folates have reduced pteridine rings and are characterized by their polyglutamylated tails, whereas folic acid is monoglutamylated and fully oxidized (Figure 4.1). Naturally occurring folates are very unstable and rapidly lose their activity in foods and are easily oxidized under low pH.[3] Up to 50%–75% of the original folate values are lost through food harvesting, storage, processing, and preparation.[3] Folate bioavailability varies widely depending on the food source and food preparation method.[3] In contrast, folic acid is highly stable for months or even years and has higher bioavailability compared with naturally occurring folates.[3] Except for the de novo synthesis by intestinal microflora, some of which are absorbed across the large intestine and incorporated into the host's tissues,[5–7] mammals are generally unable to synthesize folate and therefore must obtain folate from dietary or supplemental sources.[4]

Folic acid is not found in nature nor is it a normal metabolite. It must be reduced, first to dihydrofolate (DHF) and then to tetrahydrofolate (THF) by dihydrofolate

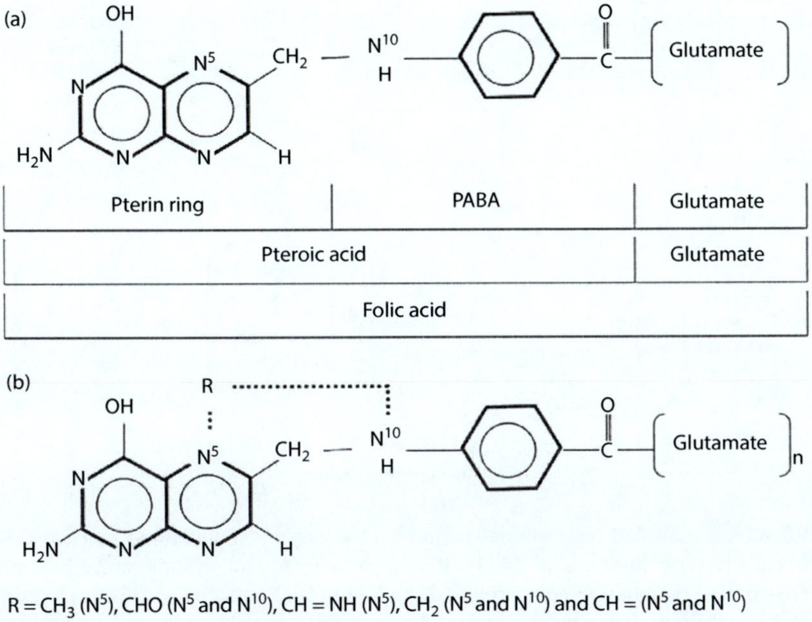

FIGURE 4.1 Chemical structures of folic acid (a) and folate (b). Folic acid consists of three moieties: the pterin (or pteridine) ring, which is conjugated to *para*-aminobenzoic acid (PABA) by a methylene bridge, which is in turn joined to a glutamic residue via a peptide bond. Folic acid is the fully oxidized monoglutamyl form of this vitamin that is commercially used in supplements and in fortified foods. Folate is the generic term referring to compounds that have similar chemical structures and nutritional properties. All naturally occurring folates found in food differ from the oxidized folic acid in the oxidation state of the pteridine ring and are typically reduced. Furthermore, one-carbon units (R) can be linked to tetrahydrofolate (THF) at the N-5 and N-10 positions, which confers folate the role of mediating the transfer of one-carbon units. In addition, multiple glutamate residues of varying numbers (up to nine) can be added via a γ-peptide linkage.

reductase (DHFR) and methylated to 5-methylTHF (predominant folate found in blood) in the liver and to a lesser degree in the intestine[8] before it can enter the folate cycle that is critical for folate's metabolism and function. Because the capacity of this conversion is limited, excessive intake of folic acid saturates this capacity and results in its appearance unaltered in circulation.[9]

4.2.1 Biochemical Functions of Folate

Folate mediates the transfer of one-carbon units involved in de novo purine and thymidylate synthesis, methionine cycle, and biological methylation reactions (Figure 4.2).[4] As an essential cofactor for the de novo biosynthesis of nucleotides, folate plays an important role in DNA synthesis, stability and integrity, and repair.[4,10] In the methionine cycle, 5-methylTHF transfers one methyl group to homocysteine to synthesize methionine catalyzed by methionine synthase (MS), a vitamin B_{12}

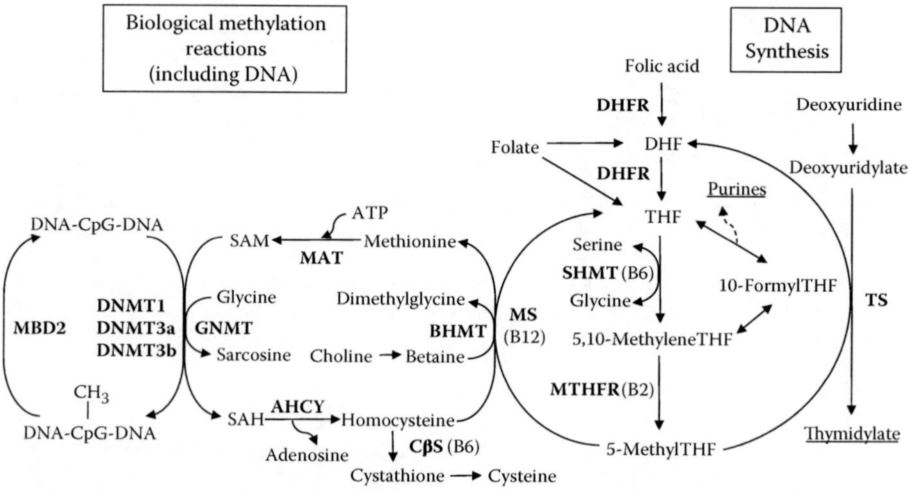

FIGURE 4.2 Simplified biochemical pathway of folate metabolism and reactions. SAM is both an allosteric inhibitor of MTHFR and an activator of cystathionine β-synthase. AHCY, *S*-adenosylhomocysteine hydrolase; B2, vitamin B_2; B6, vitamin B_6; B12, vitamin B_{12}; BHMT, betaine-homocysteine methyltransferase; CβS, cystathionine β-synthase; CH_3, methyl group; CpG, cytosine–guanine dinucleotide sequence; DHF, dihydrofolate; DHFR, dihydrofolate reductase; DNMT, DNA methyltransferase; GNMT, glycine *N*-methyltransferase; MAT, methionine adenosyltransferase; MBD2, methyl-CpG-binding domain protein 2, DNA demethylase; MS, methionine synthase; MTHFR, methylenetetrahydrofolate reductase; SAH, *S*-adenosylhomocysteine; SAM, *S*-adenosylmethionine; SHMT, serine hydroxymethyltransferase; THF, tetrahydrofolate. Enzymes are shown in bold.

(cobalamin)-dependent enzyme.[11,12] Methionine can then be activated by ATP and methionine adenosyltransferase (MAT) to produce *S*-adenosylmethionine (SAM), the primary methyl group donor for most biological methylation reactions.[1,2,13] MAT, also known as SAM synthase, is encoded by *MAT1A* and *MAT2A*.[14] Glycine *N*-methyltransferase (GNMT) catalyzes the SAM-dependent methylation of glycine to yield sarcosine, thus optimizing the ratio of SAM to *S*-adenosylhomocysteine (SAH).[15] GNMT is a major depot for methyl groups and regulates the ratio of SAM to SAH; its regulation and activity when upregulated can lead to insufficient SAM for other reactions.[16] SAM decarboxylase is a SAM depletion enzyme involved in the early steps for polyamine biosynthesis and serves as a way to deplete SMA, thereby limiting its bioavailability for methylation reactions and under certain developmental and pathological conditions.[17] *S*-Methyl-5-thioadenosine phosphorylase, another enzyme that participates in methionine metabolism, removes one of the products of SAM decarboxylase and is often affected in certain pathological conditions including cancer.[18] After donating the methyl group, 5-methylTHF is converted to THF and then to 5-methylenetetrahydrofolate (5,10-methyleneTHF) by serine hydroxymethyltransferase (SHMT).[19,20] SHMT catalyzes the reversible interconversion of serine and THF to glycine and 5,10-methyleneTHF and serves as a major entry point for one-carbon units into the folate pathway.[19,20] 5,10-MethyleneTHF is a key substrate in folate metabolism, which can be directed toward nucleotide biosynthesis or toward methionine regeneration.[21] Methylenetetrahydrofolate reductase (MTHFR) catalyzes the irreversible conversion of 5,10-methyleneTHF to 5-methylTHF.[22]

4.2.2 Folate in Health and Disease

Due to its role in purine and thymidylate synthesis and methylation reactions, folate is essential for proper cell division, normal growth, and development, and therefore plays a crucial role in human health and disease. Folate deficiency has been linked to a number of human diseases such as anemia, neural tube and other congenital defects, adverse pregnancy outcomes, coronary heart disease, stroke, neuropsychiatric disorders, cognitive impairments, osteoporosis, and certain cancers.[3,23] Although observational epidemiological studies have generally supported the relationship between folate status and the risk of these diseases, results from several large folic acid intervention trials in humans have been inconsistent and generally have not been supportive of the protective effect of folic acid supplementation on coronary heart disease[24–30] and cognitive function.[31,32] However, folic acid supplementation appears to reduce the risk of stroke[29,33] and, in more recent studies, to delay age-related cognitive decline[34] and slow and/or prevent the rate of brain atrophy.[35,36]

4.2.3 Folic Acid Fortification and Supplementation

In Canada and the United States, an overwhelming body of evidence for a protective effect of periconceptional folic acid supplementation on neural tube defects led to mandatory folic acid fortification of white flour, cornmeal, and enriched pasta in 1998, providing a daily average of 100–200 μg.[37,38] Mandatory folic acid fortification has significantly increased dietary intake and blood measurements of folate/folic acid[39–44] and has significantly reduced neural tube defect rates by 15%–50%.[45–50] In addition, up to 40% of the North American population are taking folic acid supplementation for several possible but as yet unproven health benefits including the prevention of coronary heart disease, cancer, and cognitive decline.[51,52] Women of child-bearing age are also routinely advised to take 0.4–1.0 mg folic acid for the prevention of neural tube defects.[53] Furthermore, vitamins and supplements are commonly used among patients newly diagnosed with cancer and cancer survivors with a much higher frequency of use (up to 80%) compared with the general population.[54–59] The highest prevalence of use is among Caucasian female breast cancer patients and survivors who are of middle to upper socioeconomic status.[59] A recent study of women with early-stage breast cancer has shown that 54% and 72% consume multivitamins pre-cancer and post-cancer diagnosis, respectively,[58] higher than the reported figure of 35% and 55% pre- and post-CRC diagnosis.[57]

The most recent National Health and Nutrition Examination Survey 1988–2010 survey has reported post-fortification geometric mean serum and red blood cell (RBC) folate concentrations to be 2.5× and 1.5× pre-fortification values and they were ~30 and ~750 nM higher in supplement users than in nonusers, respectively.[42] Serum folate concentrations are indicative of short-term folate intake and can be influenced by recent folate and/or folic acid consumption prior to blood draw, whereas RBC folate concentrations are indicative of long-term folate intake, and its measurement is considered the gold standard for determining folate status.[60] A recent study of the Canadian Health Measures Survey has reported that folate deficiency (RBC folate <305 nM) is present in <1% of the Canadian population and >40% have high RBC folate concentrations (>1360 nM; a level reflecting the 97% percentile from the NHANES 1999–2004 data) in the post-fortification era.[40] Higher levels of circulating unmetabolized folic acid (UMFA) in blood have been reported in the post-fortification era, likely attributable to a high folic acid (FA) intake from both supplementation and fortification.[39] Studies have found oral doses of FA as little as 280 μg to overwhelm the capacity of DHFR, an enzyme that reduces folic acid.[9,61] Even in countries without mandatory folic acid fortification, some members of the population, including infants, have detectable UMFA in their blood, probably because of the voluntary fortification[61] or intake of supplemental folic acid.[62] After mandatory folic acid fortification in the United States, UMFA was found in the blood of 78% of postmenopausal women.[63] A recent study from the Framingham Offspring Cohort reported that fortification doubled the median levels of UMFA in plasma in both supplement and nonsupplement users compared with the pre-fortification era and that the prevalence of high circulating UMFA (≥85th percentile) increased from 9.4% to 19.1% in nonsupplement users and from 15.9% to 24.3% in supplement users after fortification.[64] Whether or no high circulating UMFA poses a health risk is a hotly debated topic as some of the purported adverse effects of high intakes and levels of folate/FA have been attributed to UMFA.[8,23]

4.2.4 Potential Adverse Effects of Folic Acid Supplementation

In addition to the potential masking effect of folate on vitamin B_{12} deficiency, primarily in elderly,[3] an emerging body of evidence has recently warned of new potentially harmful effects of FA supplementation including resistance or tolerance to antifolate drugs used against arthritis and cancer, decreased natural killer cell cytotoxicity, genetic selection of disease alleles, an increased risk of insulin resistance and obesity and asthma in children born to mothers supplemented with high levels of FA, and an increased risk of cognitive impairment in elderly in combination with low vitamin B_{12} status.[23]

4.3 FOLATE AND CRC

4.3.1 Folate and CRC Risk: Epidemiological Evidence

Perhaps one of the most speculative and provocative new medical applications of folate nutrition is the potential role of folate as a cancer preventive agent.[65,66] The concept that folate deficiency enhances, whereas folate supplementation reduces, the risk of neoplastic transformation appears counterintuitive and contradictory to our conventional understanding of folate biochemistry. Folate is an essential cofactor for the de novo biosynthesis of purines and thymidylate, and in this role, folate plays an important role in DNA synthesis and replication.[4,10] Consequently, folate deficiency in tissues with rapidly replicating cells results in ineffective DNA synthesis.

In neoplastic cells where DNA replication and cell division are occurring at an accelerated rate, interruption of folate metabolism causes ineffective DNA synthesis, resulting in inhibition of tumor growth.[65,66] Indeed, this has been the basis for cancer chemotherapy using a number of antifolate agents (e.g., methotrexate) and 5-fluorouracil.[65,66] Furthermore, folate deficiency has been shown to induce regression and suppress progression of pre-existing neoplasms in experimental models.[65,66] In contrast to the inhibitory and promoting effect of folate deficiency and supplementation, respectively, on established neoplasms, however, folate status appears to have an opposite effect in normal tissues. An accumulating body of epidemiological, clinical, and experimental evidence suggests that folate deficiency in normal tissues appears to predispose them to neoplastic transformation, and folic acid supplementation suppresses the development of tumors in normal tissues.[65,66]

Counterintuitive to the traditional role of folate in cancer (i.e., antifolate in cancer chemotherapy),[67] epidemiological studies have suggested that folate insufficiency may increase the risk of several human malignancies including cancer of the colorectum, oropharynx, esophagus, stomach, pancreas, lungs, cervix, ovary, and breast and neuroblastoma and leukemia and that folic acid supplementation may reduce this risk.[66,68] The precise nature and magnitude of the inverse relationship between folate status and the risk of these malignancies, however, have not been clearly established because of inconsistent results.[66,68] The relationship between folate status and cancer risk has been best studied for CRC, and there is a reasonable body of evidence that supports the purported inverse association.[69–71] Although not uniformly consistent, the portfolio of retrospectively and prospectively conducted epidemiological studies suggest a 20% to 40% reduction in the risk of CRC or its precursor, adenomas, in subjects with the highest folate status compared with those with the lowest status.[69–72] Some, and not all, molecular epidemiological studies have reported that certain genetic polymorphisms in the folate metabolic pathway (e.g., *MTHFR* C677T) modify CRC risk and interact with folate and other nutritional and environmental factors involved in the folate metabolic and one-carbon transfer pathways to further modulate the risk.[73–75]

Some studies have suggested that folate and the synthetic folic acid might have differential effects on colorectal carcinogenesis. For example, epidemiological studies have shown dietary folate, not synthetic folic acid, to be protective against CRC.[72] Although folate and folic acid are ultimately metabolized to the same metabolite (5-methylTHF) in the body, there is probable cause to believe folate and folic acid may exert different effects on tissues. Unlike reduced folate, folic acid requires additional reduction steps by DHFR to be converted to THF so it can then enter the folate metabolism pathway (Figure 4.2).[8] DHFR in humans has been shown to be easily saturated and enzyme activity varies widely among individuals.[76] In this regard, there is concern that high circulating and tissue levels of UMFA could interfere with the metabolism, cellular transport, and regulatory functions of natural folates by competing with the reduced forms for binding with enzymes, carrier proteins, and binding proteins, and this could influence metabolic flux in the nucleotide synthesis and biological methylation pathways involving one-carbon units.[8,23,77] There is evidence that folic acid supplementation may preferentially shuttle the flux of one-carbon units to the nucleotide synthesis over the methionine cycle necessary for biological methylatioin reactions.[21] Although folic acid is an inhibitor of DHFR,[76] folic acid has also been shown to upregulate DHFR in vitro.[78] This upregulation may increase thymidylate synthase (TS) activity because the transcription of both of the genes encoding DFHR and TS is coregulated by the same E2F-1 transcription factor.[79,80] This would increase thymidylate production, thereby increasing cellular proliferation, at the expense of biological methylation reactions.[81] Furthermore, excess DHF resulting from folic acid supplementation inhibits MTHFR, thereby reducing the formation of 5-methylTHF and consequent DNA hypomethylation.[82]

4.3.2 Folate and CRC Risk: Clinical Trials

Small human intervention studies have reported that folic acid supplementation (0.4–10 mg/day for 3 months to 2 years) can improve or reverse several functional biomarkers of folate and one-carbon metabolisms and CRC.[66] However, these studies were limited by the small number of subjects, short duration of intervention, and uncertain validity of the selected biomarkers to reflect CRC risk.[66] Four randomized placebo-controlled clinical trials have shown a null effect[83] or limited support for a potential protective effect of folic acid supplementation on the recurrence of colorectal adenomas, the well-established precursor of CRC.[84–86] In contrast to either the protective or null effect of folic acid supplementation on surrogate end point biomarkers of CRC in the aforementioned clinical trials, the Aspirin/Folate Polyp Prevention Study reported that folic acid supplementation at 1 mg/day for 6 years in individuals likely harboring (pre)neoplastic lesions significantly increased the risk of recurrence of advanced adenomas by 67% and the risk of multiple adenomas by 2.3-fold,[87] thereby suggesting a potential tumor promoting effect of folic acid supplementation. Furthermore, folic acid supplementation significantly increased the risk of prostate cancer after 10 years.[88] A recent combined analysis of three large randomized clinical trials of folic acid supplementation on the prevention of colorectal adenomas among patients with a history of adenomas found no relationship between folic acid supplementation and the risk of recurrence of adenomas.[89] In fact, there was some evidence, albeit nonsignificant, for a decreased risk among subjects with low baseline plasma folate and those with high alcohol intake.[89]

Combined or meta-analyses of several large randomized clinical trials that investigated the effect of FA supplementation with or without other B vitamins on cardiovascular disease outcomes as the primary end point reported either a tumor promoting[90] or null[28] effect on cancer risk as the secondary end point. Two most recent systemic review and meta-analysis of folic acid trials with cancer risk as either the primary or secondary end point again reported either an increased[91] or null[92] effect associated with folic acid supplementation. A meta-analysis of 6 clinical trials that examined the effect of folic acid supplementation on the recurrence of colorectal adenomas in those with a history of adenomas and on the risk of CRC in general populations found no significant effect of folic acid supplementation on CRC risk.[93] Two ecologic studies that examined a temporal post-fortification trend of CRC incidence in the United States, Canada, and Chile reported increased CRC rates in these countries following fortification, suggesting that FA fortification may have been wholly or partly responsible for this trend.[94,95] However, two recent large population-based cohort studies conducted after folic acid fortification have reported a decreasing trend of CRC incidence post-fortification in the United States and have not demonstrated a CRC-promoting effect of folic acid supplementation.[96,97] At present, the potential tumor promoting effect on the progression of (pre)neoplastic lesions remains the most controversial and potentially most serious adverse effect of folic acid supplementation and fortification.

4.3.3 Folate and CRC Risk: Dual Modulatory Effects of Folate on CRC

The potential tumor-promoting effect of folic acid supplementation observed in some of the above-mentioned clinical trials and inconsistent results from epidemiological studies are not entirely surprising as animal studies from our and other laboratories have suggested that folate possesses dual modulatory effects on cancer development and progression depending on the dose and the stage of cell transformation at the time of folate intervention (Figure 4.3).[66,68,98,99] Folate deficiency has an inhibitory effect, whereas folic acid supplementation has a promoting effect on the progression of established (pre)neoplastic lesions.[100–103] In contrast, in normal tissues, folate deficiency predisposes them to neoplastic transformation, and modest levels of folic acid supplementation suppress, whereas supraphysiological supplemental doses enhance, the development and progression of (pre)neoplastic lesions.[102–105]

Several biologically plausible mechanisms relating to folate's essential role in DNA synthesis, integrity, and repair and methylation that explain the dual effects of folate on cancer development and progression (Figure 4.3).[66] The accumulating

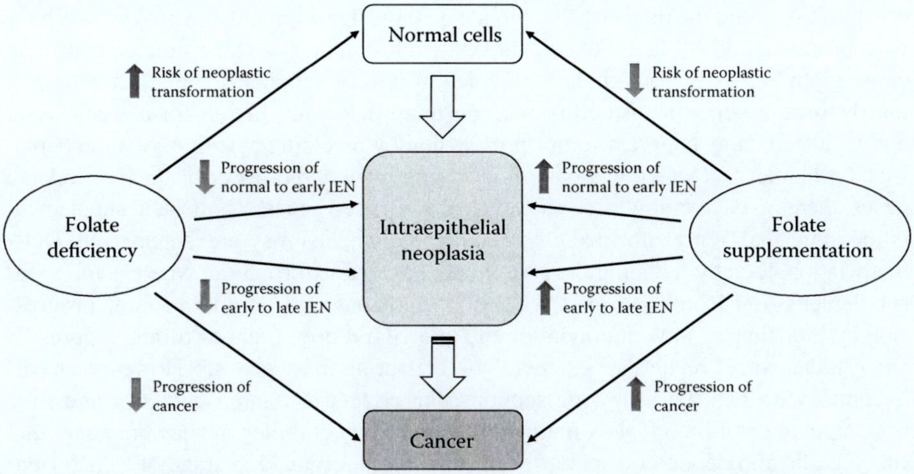

FIGURE 4.3 Dual modulatory role of folate in carcinogenesis. Folate deficiency in normal tissues predispose them to neoplastic transformation, and modest supplemental levels suppress, whereas supraphysiological doses of supplementation enhances, the development of tumors in normal tissues. In contrast, folate deficiency has an inhibitory effect whereas folate supplementation has a promoting effect on the progression of established neoplasms. The mechanisms by which folate exerts dual modulatory effects on carcinogenesis depending on the timing and dose of folate intervention relate to its essential role in one-carbon transfer reactions involved in DNA synthesis and biological methylation reactions. IEN, intraepithelial neoplasia.

body of evidence of in vivo and in vitro evidence indicates that in normal colorectal epithelial cells, folate deficiency induces DNA strand breaks, chromosomal and genomic instability, uracil misincorporation, impaired DNA repair, and increased mutations, thereby increasing the risk of neoplastic transformation.[65,66,106–111] The portfolio of in vitro and in vivo evidence indicates that folic acid supplementation can correct some of these defects in colorectal epithelial cells, thereby reducing the risk of neoplastic transformation.[65,66,106–111] In contrast, in preneoplastic or neoplastic colorectal epithelial cells where DNA replication and cell division are occurring at an accelerated rate, folate deficiency causes ineffective DNA synthesis, resulting in inhibition of the growth and progression of these (pre)neoplastic lesions, which is the basis for cancer chemotherapy using antifolate agents and 5-fluorouracil.[66,68] Mechanistically, the most likely mechanism by which folic acid supplementation may promote the progression of established (pre)neoplastic lesion in the colorectum is the provision of nucleotide precursors to rapidly replicating cells for accelerated proliferation and progression.[66,68]

Several other potential biological mechanisms for the dual effects of folate on colorectal carcinogenesis have not yet been well established. However, because of the essential role of folate in the provision of SAM (Figure 4.2), there has been intense interest in determining whether aberrant patterns and dysregulation of DNA methylation might be a mechanistic link between folate status and CRC risk.[1,2] As will be discussed in the next section of this chapter, several biologically plausible epigenetic mechanisms pertaining to the effects of folate deficiency and folic acid supplementation on global and gene-specific DNA methylation in the normal or transformed colorectum to explain the dual effects of folate on CRC development and progression have been proposed and studied. Folate deficiency in normal colorectal epithelial cells may cause global DNA hypomethylation, which promotes neoplastic transformation primarily through genomic instability, whereas folate deficiency in transformed colorectal epithelial cells may suppress tumor progression by reversing cytosine-guanine (CpG) island promoter DNA methylation of tumor suppressor and other critical cancer related genes, thereby reactivating these epigenetically silenced genes.[66] Folic acid supplementation in normal, untransformed colorectal epithelial cells may prevent aberrant DNA methylation, thereby reducing the risk of neoplastic transformation, whereas folic acid supplementation in transformed colorectal epithelial may promote tumor progression by inducing de novo methylation of CpG-island promoters of tumor suppressor and critical cancer related genes, thereby inactivating these genes.[66] However, as will become evident in the following sections, the effects of folate deficiency and folic acid supplementation on DNA methylation are highly complex as they are gene- and site-specific and depend on species, cell type, target organ, and stage of transformation as well as on the timing, degree, and duration of folate intervention.[1,2,13,66,112]

4.4 EPIGENETICS: DNA METHYLATION

4.4.1 Epigenetics

Epigenetics refers to the study of changes in the regulation of gene activity and expression that are not dependent on gene sequence.[113,114] Epigenetic mechanisms include DNA methylation, histone modifications, chromatin remodeling, and RNA interference, all of which alter gene expression and function.[115] In contrast to genetic changes in human diseases, epigenetic changes are gradual in onset and progression and are potentially reversible by dietary and pharmacological manipulations.[1,116]

4.4.2 DNA Methylation

One of the main epigenetic DNA modifications in mammals is the methylation of cytosine located within the CpG sequences, which is heritable and tissue- and species-specific.[116,117] DNA methylation is an important epigenetic determinant in gene expression (an inverse relation except for few exceptions), in the maintenance of DNA integrity and stability, in chromatin modifications, and in the development of mutations.[114,117] Up to 80% of all CpG sites in human DNA are normally methylated.[114,117] However, this methylation occurs primarily in the bulk of the genome where CpG density is low, including exons, noncoding regions, and repeat DNA sites, and allows correct organization of chromatin in active and inactive states[118] (Figure 4.4). By contrast, most CpG rich areas clustered in small stretches of DNA termed "CpG islands," which span the 5′-end of approximately half of the human genes including the promoter, untranslated region, and exon 1, are unmethylated in normal cells, thereby allowing transcription[114,117,119] (Figure 4.4). When methylated, CpG islands cause stable heritable transcriptional silencing,[114,117] which is mediated by the transcriptional repressor, methyl-CpG binding protein 2 (MeCP2), which

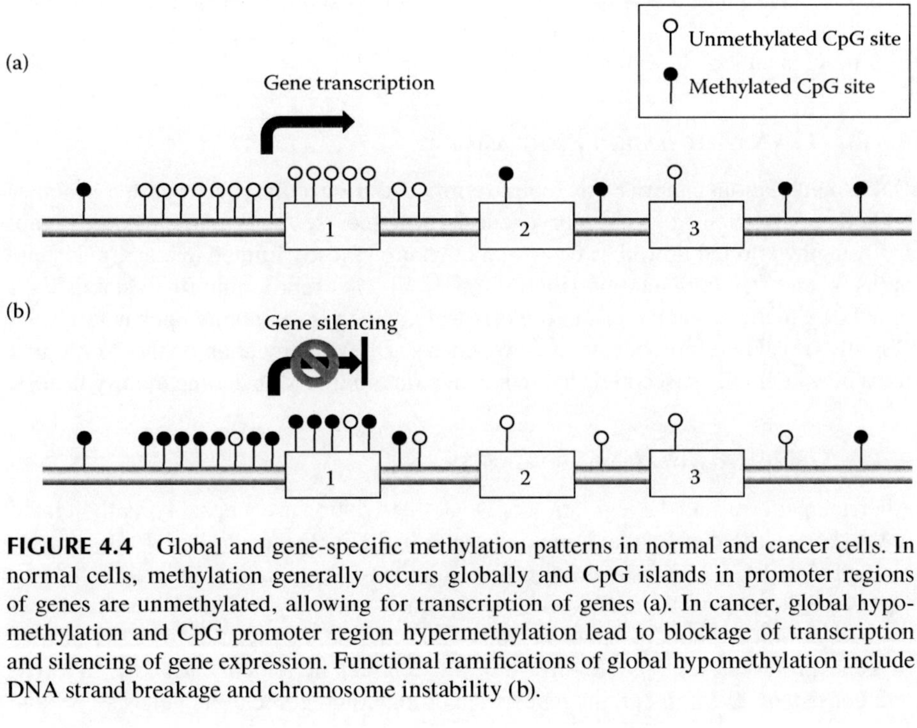

FIGURE 4.4 Global and gene-specific methylation patterns in normal and cancer cells. In normal cells, methylation generally occurs globally and CpG islands in promoter regions of genes are unmethylated, allowing for transcription of genes (a). In cancer, global hypomethylation and CpG promoter region hypermethylation lead to blockage of transcription and silencing of gene expression. Functional ramifications of global hypomethylation include DNA strand breakage and chromosome instability (b).

binds methylated CpG islands and recruits a complex containing a transcriptional corepressor and a histone deacetylase (HDAC).[120,121] Deacetylation of histones suppresses transcription by allowing tighter nucleosomal packaging and thus rendering an inactive chromatin conformation.[122,123]

DNA methylation is a dynamic process between active methylation, mediated by CpG methyltransferases (DNMT1, 3a, 3b)[124–128] using SAM,[129] and removal of methyl groups from 5-methylcytosine (5-mC) residues by several mechanisms,[130] including active demethylation by a purported demethylase, methyl DNA-binding domain protein 2 (MBD2).[131]

Most recently, hydroxymethyl cytosine was found. The conversion of 5-mC to 5-hydroxymethylcytosine (5-hmC) in mammalian DNA is mediated by methylcytosine oxygenase TET1.[132] In addition, 5-hmC might be produced by the addition of formaldehyde to cytosines in DNA by DNMT proteins.[133] The exact function of 5-hmC is yet to be determined but it appears that 5-hmC might serve biologically important roles by itself or it might serve as an intermediate in DNA demethylation. 5-hmC may be involved in gene regulation; while 5-mC in the CpG promoter region of genes is generally linked to gene silencing, 5-hmC has been shown to be associated with gene activation.[134] 5-hmC is present in mammalian DNA at a significant level in a tissue-specific manner.[135] 5-hmC may play an important role in neurodevelopment[136] as it is found in very high levels in the central nervous system.[137] Furthermore, 5-hmC is highly enriched in primordial germ cells, where it appears to play a role in global DNA demethylation.[138] 5-hmC is also found in embryonic stem cells[139] and self-renewal in embryonic stem cells is impaired with decreased 5-hmC levels.[135] The levels of 5-hmC are profoundly reduced in cancers of prostate, breast, gliolma, and colon compared with normal tissues, suggesting a possible role of 5-hmC in cancer development in these tissues.[140]

4.4.3 DNA Methylation Programming

DNA methylation patterns are reprogrammed during embryogenesis by genome-wide demethylation early in embryogenesis, which erases large parts of the parental DNA methylation, followed by de novo methylation mostly limited to non-CpG island areas, soon after implantation (Figure 4.5).[128,141] This reprogramming establishes a new DNA methylation pattern in the offspring, which is maintained postnatally[128,141] (Figure 4.5). Therefore, during embryogenesis, DNA methylation of the developing fetus may be highly susceptible to environmental modifiers, including dietary factors.

4.4.4 DNA Methylation and Cancer

Aberrant patterns and dysregulation of DNA methylation are mechanistically related to the development of several human diseases including cancer.[114,117,142,143] In contrast to methylated CpG sites in the CpG-poor bulk of the genome and unmethylated CpG islands in normal cells, cancer cells simultaneously harbor widespread loss of methylation in the CpG-poor regions and gains in methylation of CpG islands in gene promoter regions (Figure 4.3).[142,143] Global hypomethylation is an early, and consistent, event in carcinogenesis including colorectal carcinogenesis[142,143] and

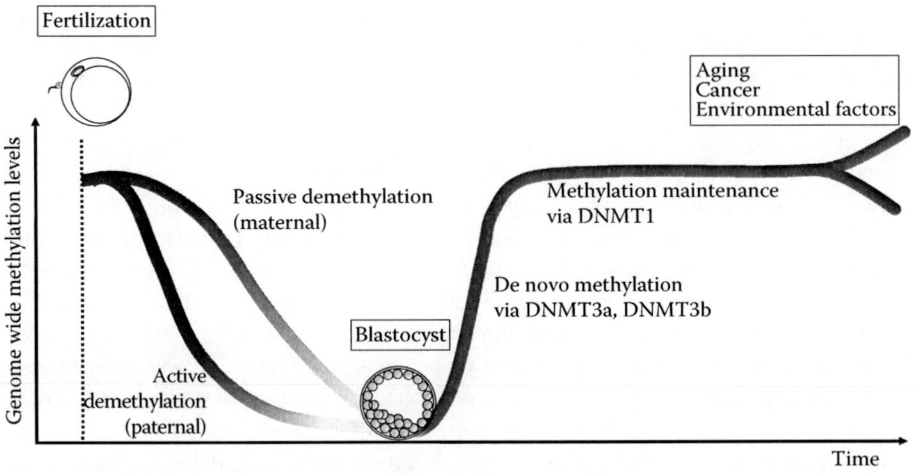

FIGURE 4.5 DNA methylation throughout the life cycle. DNA methylation patterns are reprogrammed during embryogenesis by genome-wide demethylation after fertlization. Active and passive demethylation erases significant parts of the parental DNA methylation patterns, followed by de novo methylation, which establishes a new DNA methylation pattern soon after implantation when the blastocyst is formed. Maintenance methyltransferase (DNMT1) uses hemimethylated sites to ensure DNA methylation patterns, whereas de novo methyltransferases (DNMT3a, 3b) do not require preexisting methylation patterns. Later in life, factors such as aging, cancer, and environmental exposures can cause epigenetic divergence with increased or decreased methylation levels.

contributes to the development of cancer primarily via chromosomal instability, loss of heterozygosity, rearrangements, aneuploidy, loss of imprinting, upregulation of protooncogenes, and reactivation of intragenomic parasitic sequences.[142,143] DNA methylation at promoter CpG islands silences transcription and hence inactivates the function of a wide array of genes that have classic tumor-suppressor function or play critical roles in cancer development and progression.[142,143] DNA hypermethylation at promoter CpG islands of tumor suppressor and mismatch repair genes and their inactivation has been shown to be mechanistically involved in colorectal carcinogenesis.[118] Increased frequencies of promoter region CpG methylation in specific groups of genes gave rise to the CpG island methylator phenotype (CIMP) in CRC.[144]

4.5 FOLATE AND DNA METHYLATION

4.5.1 In Vitro Studies

A number of in vitro studies utilizing various untransformed human cell lines have demonstrated the potential of folic acid to modulate DNA methylation. The results of these studies show that the effects of folic acid concentrations in in vitro systems on global and gene-specific DNA methylation vary vastly by cell line. One of the earlier studies demonstrated that immortalized but untransformed human colonocytes grown in a folate-deficient medium were associated with significant global DNA hypomethylation.[145] In another study, folate deficiency was shown to induce significant global DNA hypomethylation in untransformed mouse NIH/3TE fibroblast cell line and Chinese hamster ovarian cell line, CHO-K1.[146] In contrast, high concentrations of folic acid (2.3–9.1 μM), but not physiologically relevant doses, were shown to induce long interspersed nuclear element 1 (LINE-1) hypomethylation and gene-specific CpG island hypermethylation in normal human WI-38 fibroblasts and FHC colon epithelial cells.[147] These studies collectively suggest that both high concentrations of folic acid and folate depletion can induce global DNA hypomethylation in untransformed cells and that the direction of change in global DNA methylation and CpG island promoter methylation is not always in the same direction in response to folate manipulation.

Studies using human cancer cell lines have also demonstrated similar effects of folate depletion or folic acid supplementation on DNA methylation. In one study, human nasopharyngeal carcinoma KB cells grown in folate delete medium was associated with paradoxical CpG island promoter DNA methylation and consequent downregulation of the *H-cadherin* gene.[148] In SW620 human colon cancer cells, folate depletion was associated with significant global and region-specific *p53* gene DNA hypomethylation, which was reversed by folic acid supplementation.[149] In contrast, folate depletion, even a severe degree, was not able to alter global DNA methylation in HCT116 and Caco2 human colon cancer.[146] Interestingly, folate depletion in this study was associated with significant CpG island promoter DNA hypermethylation in the *ER* gene in both HCT116 and Caco-2 cells.[146] However, this CpG promoter *ER* hypermethylation in these cell lines was not associated with a significant change in *ER* gene expression.[146] A recent study investigated the effect of folate deprivation on invasion in HCT116 cell line considering that activation of the Sonic hedgehog (Shh) signal coupled with the transcription nuclear factor κB (NF-κB) pathway is important in cancer aggressiveness. This study found that HCT116 cells cultured in a folate deplete medium exhibited enhanced invasiveness as a result of CpG promoter DNA hypomethylation of the *Shh* gene and consequent increased *Shh* gene expression, leading to increased activation of the Shh signaling in interaction with increased binding with NF-κB pathway.[150] In one study, supraphysiologically high concentrations of folic acid in culture medium (20 μM) increased CpG promoter DNA methylation of tumor suppressor genes *ESR1*, *p16INK4a*, and *p15INK4b* and led to a significant reduction in *ESR1* gene expression in Caco2 cells.[151] These in vitro studies conducted in human cancer cell lines collectively suggest that effects of folate deficiency and folic acid supplementation on DNA methylation are cell-, site-, and gene-specific and that the direction of DNA methylation changes may not be the same between global and gene- or site-specific DNA methylation.

The major limitation of the in vitro system to study the effect of folate on DNA methylation is that the degree of folate deficiency and folic acid supplementation used in this system is neither physiologically nor clinically relevant and therefore is not applicable to in vivo situations. Nevertheless, these in vitro studies collectively provide proof of principle that supports the ability of folate to modulate DNA methylation.

4.5.2 Animal Studies

The liver is the major storage organ for folate and the development of hepatocellular carcinoma has been shown to be influenced by the status of folate and other methyl group donors in part via epigenetic dysregulation. As such, the liver is an excellent model to study the effect of folate and other methyl group donors on epigenetics and cancer development. Diets deficiency in methyl group donors (choline, folate, methionine, and vitamin B_{12}) are associated with spontaneous and chemically induced development of hepatocellular carcinoma in rodents.[152] Diets deficient in different combinations of methyl group donors have been consistently observed to induced global and protooncogene (*c-myc*, *c-fos*, *c-Ha-ras*) DNA hypomethylation and elevated steady-state levels of corresponding mRNAs[153–158] and site-specific *p53* hypomethylation[155,159,160] in rat liver. Methyl group donor deficiency has also been shown to upregulate DNMT in rat liver.[156,157,159–161] However, another study reported that a diet deficiency in choline, methionine, and folate did not induce a significant degree of global DNA hypomethylation in rat colon, suggesting that the colorectum may be resistant to the hypomethylating effect of methyl group deficiency.[162]

Although isolated folate deficiency has been shown to reduce SAM levels and SAM to SAH ratios and increase SAH concentrations in rat liver,[163–167] conflicting data exist for the effect of isolated folate deficiency on DNA methylation in rodent liver. Collectively, the results from animals studies suggest that folate deficiency of a moderate degree or short duration appears to induce global DNA hypermethylation in rodent liver, likely due to compensatory upregulation of Dnmt, and that the effect of sustained or severe folate deficiency on global DNA methylation in rodent liver is not consistent.[103,166–168] One intriguing observation from one of these animal studies was that severe folate deficiency produced significant hypomethylation (by 40%) within mutation hot spot (exons 6–7), but not in exon 8, of the *p53* tumor suppressor gene despite a 56% increase in global DNA methylation in rat liver.[167] This observation supports observations from in vitro studies and suggests that the effect of folate deficiency on DNA methylation may be site- and gene-specific and suggests that the changes in global and site-specific DNA methylation in response to folate deficiency may not be in the same direction.

The effect of isolated folate deficiency on DNA methylation in the colorectum, the primary target tissue that is particularly susceptible to the folate deficiency-induced carcinogenic effect, has not yet been clearly elucidated. It is likely that other players are involved to account for inconsistent or nonuniform effects such as histone modifications, protein–protein interactions, and targeting of Dnmts to specific gene regions. A moderate degree of folate deficiency alone or in conjunction with colorectal carcinogens (dimethylhydrazine [DMH] and azoxymethane [AOM]) for 10–26 weeks failed to induce significant global and gene-specific DNA hypomethylation in rat colon.[105,162,166,169–171] In contrast, significant *p53* hypomethylation in exon 8, but not in exons 6–7, was observed in the DMH-treated rat colon in conjunction with folate deficiency, although it remains unclear whether this was due to the DMH, the folate deficiency, or the combination of the two, and this was effectively overcome in a dose-dependent manner by increasing levels of dietary folate.[172] In another study, severe folate deficiency with a marked reduction in colonic folate concentrations and a progressive decrease in colonic SAM to SAH ratio did not induce significant global DNA hypomethylation in the colon.[173] Paradoxically, severe folate deficiency significantly increased (by 30%) the extent of global DNA methylation in the colon at an intermediate time point,[173] which is consistent with prior observations made in rodent liver.[103,167] Folic acid supplementation did not modulate colonic SAM, SAH, and SAM to SAH ratios, and global DNA methylation at any time point.[173] In this study, the extent of *p53* methylation in the promoter and exons 6–7 was variable over time at each of the CpG sites examined, and no associations with time or dietary folate were observed at any CpG site.[173] Steady-state levels of *p53* mRNA did not significantly correlated with either global or *p53* methylation within the promoter region and exons 6–7.[173] These observations suggest that isolated folate deficiency does not induce consistent and predictable changes in *p53* methylation in rat colon whereas it may produce *p53* hypomethylation in specific exons in rat liver and in rat colon in conjunction with alkylating agents. The functional significance of *p53* hypomethylation in the colon in exons 6–7 and 8 in response to folate deficiency has not yet been clearly elucidated but it may be related to DNA strand breaks within these regions.[174] In contrast, dietary depletion of combined methyl donors predictably induces *p53* hypomethyation within exons 6–7 of the *p53* gene in rat liver.[155,159,160,173] These observations suggest that *p53* methylation changes likely depend on the degree of methyl donor supply and consequent levels of methylation intermediates that are predictably and consistently achieved by combined methyl deficiency and not by isolated folate deficiency. Interestingly, a recent study has suggested that the effect of folate deficiency on DNA methylation in the colon may be species-specific. In this study, folate deficiency in mice induced significant global, but not gene-specific, DNA hypomethylation in the colon. However, this observed low-folate diet-induced global DNA methylation changes were insufficient to promote tumor development, even in the presence of increased point mutations resulting from uracil DNA glycosylate deficiency.[175] Whether or not DNA methylation changes—either global or site/gene-specific—in response to isolated folate or combined methyl group donor manipulations in diet in rodent lead to significant functional ramifications including changes in gene expression or gene stability have not yet been clearly established. It is likely that folate influences other epigenetic mechanisms such as histone modifications, chromatin remodeling, and RNA interference, and it may be difficult to determine the effect of folate exclusively on DNA methylation and its functional ramification given complex interactions among these epigenetic mechanisms.

One critical factor that may explain the inability of folate deficiency to modulate global and site- and gene-specific DNA methylation in the colorectum may be related to the fact that modulation of SAM and SAH in the colonic mucosa is particularly resistant to the level of dietary folate.[105,166,173] The reason for this tenacious resistance to altered SAM and SAH levels in the colorectum compared with other tissues is unclear at present. There is a possibility that in response to folate deficiency, the choline- and betaine-dependent transmethylation pathway might be upregulated to maintain critical SAM and SAH concentrations in the colorectum (Figure 4.2).[163,176,177]

An emerging body of evidence in rodent studies has suggested that the aging colon might be more susceptible to folate manipulation. A recent animal study utilizing a moderate degree of dietary folate deficiency has suggested that the colon of old rats (12 months) is more susceptible to changes in SAM and SAH in response to folate deficiency as compared with that of young rats (weanling), although these changes in SAM and SAH do not appear to induce global DNA hypomethylation.[178] However, the effect of folate deficiency on global DNA methylation in the aging colon appears to be species-specific. In a study using mice, old mice (18 months) had lower global DNA methylation in the colon compared with corresponding young mice (4 months) in each dietary folate (deficient, control, and supplemented) group.[179] In the old mice, global DNA methylation increased in a manner that was directly related to dietary folate, whereas this association was not at all evident in the young mice.[179] At each level of dietary folate, old mice demonstrated a markedly greater degree of *p16* promoter methylation compared with the corresponding young groups.[179] As was the case with global DNA methylation, *p16* methylation in the young mice was not altered by dietary folate levels, whereas old mice showed a stepwise increase in *p16* promoter methylation across the spectrum of dietary folate.[179] Interestingly, counterintuitive to the expected *p16* gene silencing resulting from *p16* promoter methylation, old mice fed the control and folic acid–supplemented diets had greater *p16* expression compared with young mice fed the same diets.[179] This observation suggests that promoter methylation is not the sole determinant of *p16* expression and other factors associated aging and other epigenetic mechanisms might have been responsible for the observed increased *p16* expression associated with folic acid supplementation in old mice.

Aging has been shown to alter DNA methylation in a complex fashion, which has the potential to promote age-related conditions, including cancer.[180,181] Studies in humans and rodents have consistently observed incremental decrease in genomic DNA methylation with aging in a tissue-specific manner.[179,182–185] Aging is also an important contributor to the promoter CpG island hypermethylation of several genes including *ER*, *IGF2*, and *p16* with consequent gene inactivation observed in several cancers, including CRC and atherosclerosis.[179,186–193] Aging is a major determinant of cancer risk, including CRC.[194] It is therefore possible that age-related alterations and dysregulation of DNA methylation serve as a functional link between aging and cancer.[188] The causes and mechanisms of age-related DNA methylation alterations are largely unknown, although some evidence suggests the role of environmental exposures or modifier genes.[189,195] Whether or not folate deficiency and folic acid supplementation can influence aging-related DNA methylation changes and how this may promote age-related conditions, including CRC, is worthy of further investigation.

The available evidence from animal studies indicates that the effects of folate deficiency and folic acid supplementation on DNA methylation are highly complex as they are gene- and site-specific and depends on species, cell type, target organ, and stage of transformation as well as on the timing, degree, and duration of folate manipulation.[1,2,13,66,112] Data from these animal studies may not be entirely applicable to humans as there are significant inherent difference in folate metabolism and epigenetics between animal models and humans. Nevertheless, these studies provide experimental evidence to support the ability of isolated folate deficiency or folic acid supplementation alone to modulate global and gene-specific DNA methylation.

4.5.3 Human Studies

4.5.3.1 Effects of Folate on Global DNA Methylation

A growing body of evidence from observation studies and clinical trials in humans suggests that altered folate status can affect DNA methylation (see Table 4.1). Folate

TABLE 4.1
The Effects of Folate Depletion and Folic Acid Supplementation on Global DNA Methylation in Human Clinical Trials

Study (Reference)	Subjects	Dose	Design	DNA Methylation	Effect
		Folate Deficiency			
Jacob et al. 1998[196]	Women: 49–63 years	56–111 μg/day	9 weeks	Lymphocytes	120% decrease ($p < 0.05$)
Rampersaud et al. 2000[197]	Women: 60–85 years	118 μg/day	7 weeks	Leukocytes	10% decrease ($p = 0.0012$)
Shelnutt et al. 2004[198]	Woman: 20–30 years with *MTHFR* 677TT or CC	115 μg/day, then 400 μg/day	7 weeks 7 weeks	Leukocytes	5% decrease ($p = 0.08$) 8% increase ($p = 0.04$) in woman with *MTHFR* 677 TT after repletion
Abratte et al. 2009[199]	Woman: 18–46 years with *MTHFR* 677TT or CC	133 μg/day then 400–800 μg/day and/or choline (344–486 mg/day), betaine (122–349 mg/day)	2 weeks 12 weeks	Leukocytes	No effect
		Folic Acid Supplementation			
Cravo et al. 1994[200]	Patients with colon cancer and adenoma	10 mg/day	6 months	Rectal mucosa	93% increase ($p < 0.002$)
Cravo et al. 1998[201]	Patients with colonic adenoma	5 mg/day	3 months	Rectal mucosa	37% increase in patients with 1 adenoma ($p = 0.05$) and no change in those with >1 adenomas

Kim et al. 2001[202]	Patients with colonic adenoma	5 mg/day	6 months	Rectal mucosa	57% increase ($p = 0.001$)
			1 year		No effect
Cravo et al. 1995[203]	Patients with inflammatory bowel disease	5 mg/day	6 months	Rectal mucosa	No effect
Fenech et al. 1998[204]	Normal subjects	2 mg/day	12 weeks	Lymphocytes	No effect
Ingrosso et al. 2003[205]	Uremic patients with hyperhomocysteinemia and preexisting DNA hypomethylation	15 mg/day 5-methylTHF	8 weeks	Lymphocytes	Restored to normal levels
Pufulete et al. 2005[206]	Patients with colonic adenoma	400 µg/day	10 weeks	Lymphocytes	31% increase ($p = 0.05$)
				Rectal mucosa	25% increase ($p = 0.09$)
Figueiredo et al. 2009[207]	Patients with colonic adenoma	1 mg/day	6–8 years	Colonic mucosa	No effect

depletion in healthy human volunteers in a metabolic unit setting has been observed to diminish global DNA methylation in leukocytes. Significant decreases in lymphocyte global DNA methylation were observed after intervention with a low folate diet for 3 months in postmenopausal (49–63 years)[196] and 7 weeks in elderly women (60–85 years).[197] However, after folate repletion, global DNA methylation levels were restored in postmenopausal,[196] but not in elderly,[197] women. In another controlled feeding study, a nonsignificant decreased trend in leukocyte global DNA methylation in young women was observed with low intake of dietary folate for 7 weeks.[198] After an additional 7 weeks of repletion, global DNA methylation increased, but only in the *MTHFR* 677TT genotype.[198] In another study, dietary intakes of choline, betaine, and/or folate for 12 weeks did not influence global DNA methylation in leukocytes in healthy premenopausal women.[199]

Several human studies have investigated correlations between DNA methylation and folate status in easily accessible blood cells (see Table 4.2). In human subjects with normal folate status, no significant correlations between lymphocyte global DNA methylation and RBC folate and plasma homocysteine concentrations were observed.[204] In contrast, folate status was directly correlated with peripheral blood mononuclear cell (PBMC) global DNA methylation, which was specific for individuals with the *MTHFR* 677TT genotype.[208] In a cross-sectional study, postmenopausal women exhibited an inverse relationship between lymphocyte LINE-1 methylation and sex hormone concentrations among women with low serum folate levels, whereas LINE-1 methylation was positively associated with immune markers in those with high serum folate levels.[209]

Few studies have investigated a relationship between folate status and global and gene-specific DNA methylation in normal-appearing colorectal mucosa in humans. Two studies have shown that global DNA methylation in normal-appearing colorectal mucosa was positively correlated with serum and RBC folate concentrations and negatively with plasma homocysteine concentrations in individuals with colonic adenomas and adenocarcinomas[210] and in those without these lesions.[211] The relationship between global DNA methylation and blood levels of folate has also been examined in other tissues. In one human study, serum and cervical tissue folate concentrations correlated inversely, albeit weakly, with cervical global DNA methylation.[212] The interpretation of this study is, however, limited by the measurement of folate concentrations and DNA methylation in (pre)malignant cervical tissues instead of normal tissue as was the case in studies examining this association in the colorectum.[210,211] Additional evidence lends support to a positive relationship between folate status and global DNA methylation. In a combined analysis of CRCs from participants from the Nurses' Health Study and the Health Professionals Follow-up Study, subjects with a high total daily folate intake were 43% less likely to develop CRCs with significant global DNA hypomethylation (define as $< 55\%$ LINE-1 methylation) compared with those with a low total daily folate intake.[213]

In some human intervention studies, folic acid supplementation at 12.5–25 times the daily requirement for 3–12 months significantly increased the extent of global DNA methylation in normal-appearing colorectal mucosa of subjects with resected colorectal adenoma or cancer,[200–202] whereas no such effect was observed in patients with chronic ulcerative colitis who were given folic acid supplementation at 12.5 times

TABLE 4.2
Correlations between Folate Status and DNA Methylation in Human Epidemiological Studies

Study (Reference)	Design	Subjects	Folate Source and Levels	DNA Methylation	Effect
			Global DNA Methylation		
Fenech et al. 1998[204]	Cross-sectional	Normal subjects	RBC (nM) 363.8–440.0	Lymphocytes	No correlations
Friso et al. 2002[208]	Cross-sectional	Patients with and without coronary artery disease with *MTHFR* 677TT or CC	Plasma (nM) <12 versus ≥12 RBC (nM/g Hb) <1.1 versus ≥1.1	Peripheral blood mononuclear cell	61% decrease ($p < 0.0001$) and 65% decrease ($p < 0.0001$) in subjects with *MTHFR* 677TT
Ulrich et al. 2012[209]	Cross-sectional	Overweight/obese postmenopausal woman	Serum (ng/mL) 15.2–49.9 (low) versus 50.5–258.7 (high)	Lymphocytes	Inverse relation with sex hormone concentrations among women with low serum folate levels ($p < 0.05$) Positive relation with immune markers in those with high serum folate levels ($p < 0.07$)
Pufulete et al. 2003[210]	Case-control	Patients with colon cancer or adenoma	Serum (nM) 12.2 versus 18.1	Colon mucosa Leukocytes	26% decrease ($p = 0.009$) 14% decrease ($p < 0.001$)
Pufulete et al. 2005[211]	Cross-sectional	Patients without colon cancer or adenoma	Serum (nM) 18.6 RBC (nM) 648.1 Hcy (μM) 9.9	Colon mucosa	$r = -0.311$ ($p = 0.01$) $r = -0.356$ ($p = 0.03$) $r = 0.256$ ($p = 0.04$)
Fowler et al. 1998[212]	Prospective	Patients with cervical dysplasia	Serum (nM) 5.2 Cervical tissue (ng/mg protein) 1.2	Cervical tissue	$r = -0.239$ ($p = 0.06$) $r = -0.279$ ($p = 0.02$)
Schernhammer et al. 2010[213]	Prospective	Patients with colon cancer	Dietary intake (μg/day) ≥400 versus <200	Colon cancer	43% decreased risk of hypomethylation ($p = 0.05$)

(continued)

TABLE 4.2 (Continued)
Correlations between Folate Status and DNA Methylation in Human Epidemiological Studies

Study (Reference)	Design	Subjects	Folate Source and Levels	DNA Methylation	Effect
			Gene-Specific DNA Methylation		
van Engeland et al. 2003[214]	Prospective	Patients with colon cancer	Dietary intake (μg/day)/alcohol <215/high versus ≥ 215/low	Colon cancer: *APC-1A*, *p14*^{ARF}, *p16*^{INK4A}, *hMLH1*, *O*^{6}*-MGMT*, *RASSFIA*	Increased prevalence for all genes ($p > 0.05$)
de Vogel et al. 2008[215]	Prospective	Patients with colon cancer	Dietary intake (μg/day) 142.4–163.2 versus 247–279.9	Colon cancer: *MLH1*	No effect
Al-Ghnaniem et al. 2007[216]	Case-control	Patients with colon cancer or adenoma	Serum (nM) 12.3 versus 17.9	Colon mucosa: *ERα*, *MLH1*	No effects
Al-Ghnaniem et al. 2012[217]	Case-control	Patients with colorectal adenoma	Dietary inke (μg/day) 400 versus 0	Colon mucosa: *ESR1*, *MLH1*	No effects
Wallace et al. 2010[203]	Case-control	Patients with a history of colorectal adenoma	Dietary inke (μg/day) 1,000 versus 0	Normal colorectal mucosa: *ERα*, *SFRP1*	Higher RBC folate levels were associated with higher levels of methylation
van Guelpen et al. 2010[205]	Case-control	Patients with colon cancer	Plasma (nM) ≥ 6.8 versus < 6.8	Colon cancer: *CDKN2A*, *MLH1*, *CACNA1G*, *NEUROG1*, *RUNX3*, *SOCS1*, *IGF2*, *CRABP1*	~3-fold increase ($p < 0.05$)
Kawakami et al. 2003[206]	Cross-sectional	Patients with colon cancer	5,10-MethyleneTHF (pmol/g tissue) 2.95 versus 1.53 THF (pmol/g tissue) 3.71 versus 1.99	Colon cancer, *hMLH1*, *p16*, *TIMP3*, *ARF*, *MINT2*, *DAPK*, *APC*	Tumors with methylated promoters had higher tissue concentration of folate intermediates

the daily requirement for 6 months[203] (see Table 4.1). Folic acid supplementation at five times the daily requirement, which was sufficient to correct a marker of DNA damage, failed to modulate genomic DNA methylation in lymphocytes in healthy volunteers.[204] In another study, folic acid supplementation with 15 mg 5-methylTHF a day for 8 weeks restored global DNA methylation in lymphocytes to normal levels in 32 men with uremia, hyperhomocysteinemia, and pre-existing global DNA hypomethylation.[205] Another study demonstrated that a physiological dose of folic acid equivalent to the daily requirement (400 μg/day) for 10 weeks increased global DNA methylation in lymphocytes (by 31%; $p = 0.05$) and in colonic mucosa (by 25%; $p = 0.09$) compared with placebo in patients with colorectal adenomas.[206] In contrast, folic acid at 2.5 times the daily requirement for 3 years did not influence LINE-1 methylation in normal-appearing colonic mucosal in participants of the Aspirin/Folate Polyp Prevention Study.[207]

Although not uniformly consistent, the data from these human studies render stronger support for the folate's ability to modulate global DNA methylation in PBMC and in the colorectum, the target tissue of interest. Most of these studies indicate a positive correlation between global DNA methylation in PBMC and in normal-appearing colorectal mucosa. Generally, folate deficiency appears to induce PBMC global DNA hypomethylation but there is no conclusive data suggesting that folate deficiency of a physiologically and clinical relevant degree induces significant global DNA hypomethylation in the colorectum. In contrast, folic acid supplementation, even at the modest supplemental levels, appears to be able to increase global DNA methylation in normal-appearing colorectal mucosa in certain situations. Overall, folic acid supplementation, even at modest supplemental levels, appears to be able to reverse preexisitng global DNA hypomethylation and to increase the extent of global DNA methylation above the preexisting level in PBMC and the colorectum.

4.5.3.2 Effects of Folate on Gene-Specific DNA Methylation

Aberrant CpG island methylation is characteristic of cancer development and specific promoter CpG islands are frequently and simultaneously methylated in sporadic CRC, leading to transcriptional silencing.[142,143] Several human studies have investigated the effects of folate status on gene-specific DNA methylation (see Table 4.2). In the Netherlands Cohort Study on Diet and Cancer, the prevalence of CpG island promoter hypermethylation was higher, albeit nonsignificant, in CRCs derived from patients with low folate/high alcohol intake compared with CRCs from patients with high folate/low alcohol intake for each of the six tested genes (*APC*, *p14*, *p16*, *hMLH1*, *O^6-MGMT*, and *RASSF1A*).[214] The number of CRCs with at least one gene methylated was higher (84%) in the low folate intake/high alcohol intake group compared with the high folate intake/low alcohol intake group (70%; $p = 0.085$).[214] A follow-up analysis in a subcohort of this population did not show a significant correlation between dietary intake of folate and the risk of CRC with the mismatch repair gene *hMLH1* hypermethylation.[215] However, a significant positive association between dietary folate intake and *BRAF*-mutated CRC was observed among men and a nonsignificant positive association among women.[215] Sporadic CRCs with microsatellite instability present a distinct molecular phenotype and may develop predominantly as a consequence of *hMLH1* hypermethylation.[218] Furthermore, mutations in the *BRAF* oncogene are strongly associated with *hMLH1* promoter methylation and microsatellite instability[219] and are an additional type of molecular alteration characterizing this phenotype. In contrast, another study found no significant relationship between CpG promoter methylation of the *hMLH1* and *ERα* genes and serum or RBC folate concentrations in normal-appearing colorectal mucosa in individuals with and without colorectal neoplasia.[216] However, in this study, *ERα* promoter DNA methylation was significantly inversely correlated with serum vitamin B_{12} concentrations.[216] Furthermore, a recent clinical trial found 400-μg/day folic acid supplementation for 10 weeks had no effect on *ESR1* and *hMLH1* methylation in colonic mucosa.[217] In contrast, CpG island promoter methylation of the *ERα* and *SFRP1* genes in normal-appearing colorectal mucosa was significantly positively associated with RBC folate levels in participants of the Aspirin/Folate Polyp Prevention Study.[220] In another study, the odds of CpG island promoter methylation of *CDKN2A*, *MLH1*, *CACNA1G*, *NEUROG1*, *RUNX3*, *SOCS1*, *IGF2*, and *CRABP1* in CRC was found to be significantly greater in patients with high plasma folate concentrations.[221] In another study, CRCs with CpG promoter hypermethylation of the *hMLH1*, *TIMP3*, and *ARF* genes had higher tumor concentrations of 5,10-methyleneTHF and THF compared with those without.[222]

Taken together, the effect of folate on gene-specific DNA methylation is not well established. Most of these studies used a candidate gene approach to interrogate a limited number of genes. A comprehensive survey of the effect of folate status on gene- and site-specific DNA methylation using an epigenomic approach is warranted.

4.5.3.3 Folic Acid Fortification, DNA Methylation, and CRC Risk

There has been a concern that the significantly increased dietary intake and blood levels of folate and folic acid resulting from mandatory fortification and widespread supplemental use may increase the rate of CRC in North America and this may be mediated in part via altered DNA methylation.[68,99] Two ecologic studies that examined a temporal post-fortification trend of CRC incidence in the United States, Canada, and Chile reported increased CRC rates in these countries following fortification, suggesting that FA fortification may have been wholly or partly responsible for this trend.[94,95] However, two recent large population-based cohort studies conducted after folic acid fortification have reported a decreasing trend of CRC incidence post-fortification in the United States and have not demonstrated a CRC-promoting effect of folic acid supplementation.[96,97] At present, no studies have examined the effect of the high folate/folic acid intake and blood levels on DNA methylation in the colorectum in the post-fortification era. Recent studies have reported that folic acid fortification seems to have resulted in increased expression of DNMT 1 in the cervix when archived specimens of cervical intraepithelial neoplasia diagnosed before (1990–1992) and after (2000–2002) mandatory folic acid fortification were compared.[223] However, the observed increased DNMT 1 expression was not associated with a significant change in the degree or pattern of global DNA methylation in post-fortificaiton compared with pre-fortification cervical samples.[224] Whether or not folic acid fortification and widespread supplemental use of folic acid can alter the degree and pattern of DNA methylation in the colorectum or in other tissues is an important public health issue as aberrant patterns and dysregulation of DNA methylation are mechanistically related to the development of cancer in the colorectum and other organs.

4.6 *MTHFR* POLYMORPHISMS AND DNA METHYLATION

4.6.1 *MTHFR* Polymorphisms and Functional Ramifications

There is evidence that folate status influences DNA methylation through an interaction with the *MTHFR* polymorphisms. Intracellular folate homeostasis depends on MTHFR, a critical enzyme in folate metabolism that catalyzes the irreversible conversion of 5,10-methyleneTHF to 5-methylTHF (Figure 4.2).[225] The substrate 5,10-methyleneTHF, an intracellular coenzymatic form of folate, is the methyl donor for the nonreversible methylation, catalyzed by TS of deoxyuridine-5-monophosphate (dUMP) to deoxythymidine-5-monophosphate (dTMP; thymidylate), a precursor for DNA synthesis. 5,10-MethyleneTHF can also be oxidized to 10-formylTHF for de novo purine synthesis. Therefore, 5,10-methyleneTHF is critical in maintaining the balance of the nucleotide pool for DNA synthesis. 5-MethylTHF, the product of MTHFR, is the predominant form of folate in blood and provides the methyl group for remethylation of homocysteine to methionine, thereby ensuring the provision of SAM necessary for biological methylation reactions (Figure 4.2).

A mutation (C→T at nucleotide position 677) in the *MTHFR* gene, which results in an alanine-to-valine substitution, has been identified.[226,227] The *MTHFR* C677T polymorphism is associate with increased thermolability and reduced MTHFR activity, leading to lower levels of 5-methylTHF, an accumulation of 5,10-methyleneTHF, increased plasma homocysteine levels, and changes in cellular composition of one-carbon folate derivatives.[225] Furthermore, the *MTHFR* C677T polymorphism has been shown to be associated with global DNA hypomethylation[198,208,228–230] in lymphocytes only in individuals with low folate or riboflavin status. The *MTHFR* C677T is a common mutation, with an allele frequency of about 35% in the general North American population.[227,231,232] This variant occurs frequently among Caucasian and Asian populations, with rates of ~12%–15% for individuals who are homozygous (TT) for the variant and up to 50% for individuals who are heterozygous (CT).[22,233]

A second common polymorphism in the *MTHFR* gene (A→C at nucleotide position 1298), which results in a glutamine to alanine substitution, has also been identified.[234,235] Similar to the C667T polymorphism, the A1298C polymorphism results in decreased *MTHFR* activity, but does not result in a thermolabile protein.[234,235] Allele frequencies up to 33% have been previous reported for this polymorphism.[234,235] The CC prevalence in North American studies, which included mainly Caucasian subjects, was 7%–12%.[236] Unlike the C677T polymorphism, the effect of the A1298C polymorphism on plasma homocysteine and folate levels is inconsistent, and its functional significance has not yet been well characterized.[235,237,238] However, there is preliminary evidence that under conditions of extreme folate depletion, this polymorphism may become clinically relevant.[239] Studies have also indicated that the *MTHFR* A1298C polymorphism may also modulate global DNA methylation in human lymphocytes, although the degree and direction of change have not been clearly established.[229,240]

4.6.2 Folate, *MTHFR* Polymorphisms, and DNA Methylation

Although not uniformly consistent, the *MTHFR* C677T polymorphism has been shown to modify risk for several cancers in a site-specific manner.[22,241] It appears to decrease the risk of CRC, hepatocellular carcinoma, cervical cancer, and certain leukemias and lymphomas.[22,241] In contrast, this polymorphism seems to increase the risk of cancer of the breast, endometrium, esophagus, stomach, pancreas, and bladder.[22,241] For some cancers, the cancer risk modification conferred by this polymorphism is further modified by the status of folate and nutrients involved in one-carbon and folate metabolism.[22,75,236] For CRC, epidemiological evidence indicates a protective effect of this polymorphism in individuals with adequate or high status of folate and nutrients involved in one-carbon metabolism.[73–75] In those with an inadequate status of folate and related nutrients, the protective effect conferred by this polymorphism was diminished, and in some cases, an increased risk of CRC was observed.[75] However, this perceived diminished protection conferred by this polymorphism with increasingly low folate status has not been unequivocally confirmed, and it has been suggested that this may be an analytical artefact.[242]

The mechanisms by which the *MTHFR* C677T polymorphism could modulate the risk of CRC have not yet been elucidated. This polymorphism is associated with reduced MTHFR activity[227] and increased thermolability of *MTHFR*[226,227] in vitro, which lead to decreased 5-methylTHF and an accumulation of 5,10-methyleneTHF in red blood cells.[243] Recent biochemical and structural studies of *Escherichia coli* and human MTHFR have revealed that the *MTHFR* C677T polymorphism allows a faster dissociation of a critical stabilizing cofactor, FAD (flavin adenine dinucleotide or a coenzyme form of riboflavin), from the variant MTHFR compared with the wild-type MTHFR, resulting in thermolability and significantly reduced MTHFR activity.[244,245] Adequate folate or riboflavin protected MTHFR from the loss of FAD cofactor, thereby ensuring functional MTHFR activity.[244,245] Under the conditions of high folate or riboflavin, the enzyme kinetics of the variant MTHFR were similar to those of the wild-type enzyme.[244,245] This infers that only at low folate or riboflavin concentrations will the functional impact of the *MTHFR* C677T variant become significant. Indeed, the *MTHFR* C677T polymorphism was associated with increased plasma homocysteine (a sensitive inverse indicator of folate status) concentrations[231,246–248] and genomic DNA hypomethylation[198,208,228–230] in lymphocytes only in individuals with low folate or riboflavin status.

Based on these biochemical consequences, one proposed mechanism suggests that when the dietary supply of folate and related nutrients is high, individuals with this polymorphism might be at reduced risk of cancer because higher intracellular levels of 5,10-methyleneTHF might prevent imbalances of the nucleotide pool during DNA synthesis, thereby ensuring DNA replication with a high fidelity.[232] Furthermore, with high intakes of folate and related cofactors, the flux of 5,10-methyleneTHF to 5-methylTHF would function at its full capacity, and hence, individuals with this polymorphism would have adequate levels of SAM for optimal DNA methylation.[232] When intakes of folate and related nutrients are low, the reduced stability of the variant MTHFR results in deactivation of the MTFHR enzyme and therefore decreased flux of 5,10-methyleneTHF toward the methionine cycle pathway at a higher threshold of folate availability. This would maintain the availability of 5,10-methyleneTHF and reduced the likelihood of compromised DNA synthesis and consequent nucleotide pool imbalance.[232] In this instance, however, DNA methylation might be affected because of reduced levels of 5-methylTHF resulting from an insufficient supply from the diet and reduced flux of 5,10-methyleneTHF to the methionine cycle due to the decreased stability of the variant MTHFR enzyme.[232]

Although not uniformly consistent,[249] evidence suggests that the *MTHFR* C677T polymorphism induces genomic DNA hypomethylation in human lymphocytes in conjunction with low folate status.[198,208,228,229] However, these observations made in lymphocytes cannot be extrapolated to target organs of interest. Published studies in cancer cell lines and human colorectal biopsies have reported no effect of the *MTHFR* C677T polymorphism on CpG islands methylation of a small number (9–15) of tumor suppressor genes.[250–252] *Mthfr*$^{-/-}$ and *Mthfr*$^{+/-}$ mice had significantly decreased SAM and increased SAH concentrations, and decreased genomic DNA methylation compared with *Mthfr*$^{+/+}$ mice in a organ-specific manner.[253,254] However, these functional effects of MTHFR deficiency were not determined in the colorectum. In an in vitro model of the *MTHFR* 677T mutation in HCT116 human colon adenocarcinoma cells, the *MTHFR* 677T mutation was associated with significantly increased global DNA methylation when folate supply was adequate or high; however, in the setting of folate insufficiency, this mutation was associated with significantly decreased global DNA methylation.[255]

More recent investigations of folate status and DNA methylation in humans have included analysis of common *MTHFR* polymorphisms. In a study investigating the combined effects of folic acid and vitamin B_{12} supplementation for 6 months on CpG promoter DNA methylation of 6 tumor suppressor and DNA repair genes frequently reported to be aberrantly methylated in CRC, folic acid, and vitamin B_{12} supplementation was associated with a nonsignificant trend toward a 67% increase in promoter DNA hypermethylation in the rectal mucosa of patients with resected colorectal adenomas.[256] In a subsequent study, these investigators demonstrated that dietary folate intake interacted with the *MTHFR* C677T polymorphism to influence CpG promoter DNA methylation such that among individuals homozygous for this variant, the risk of promoter DNA methylation was inversely related to dietary folate intake, but statistical significance was only observed for the *O*^{6}*-MGMT* DNA-methyltransferase gene.[257] Slattery et al. initially failed to identify a significant association between dietary folate intake and colon tumor CpG island methylation of *p16*, *hMLH1* and *MINT-1*, *MINT-2*, and *MINT-3 loci*,[258] but in their follow-up analysis, subjects heterozygous or homozygous for the *MTHFR* A1298C genotype with low folate/low methionine/high alcohol intake had a >2-fold greater odds of developing tumors presenting CpG island DNA hypermethylation compared with subjects with the wild-type genotype and high folate/high methionine/low alcohol intake.[259] In contrast, one study reported that the odds of promoter DNA methylation in CRC were significantly higher in patients with high serum folate levels, and that this positive association was further modified by the *MTHFR* C677T polymorphism, reaching significance only in subjects heterozygous or homozygous for the *MTHFR* C677T polymorphism.[260]

MTHFR is a critical enzyme for intracellular folate homeostasis as it balances the shuttling of one-carbon units to the nucleotide synthesis and the biological methylation pathway.[225] The functionally significant *MTHFR* polymorphisms, in particular the *MTHFR* C677T polymorphism is capable of modulating DNA methylation. The aforementioned studies collectively suggest that the common *MTHFR* polymorphisms can further modify the effect of folate status on DNA methylation. These studies emphasize the importance of taking into account interactions between folate status and critical genes in the folate and one-carbon metabolic pathways when investigating the effect of folate nutrition on DNA methylation.

4.7 MATERNAL AND EARLY NUTRITION AND DNA METHYLATION AND CANCER RISK IN THE OFFSPRING

The potential role of maternal and early life nutrition on cancer risk in the offspring has been recently investigated. Epigenetic modifications in utero and in early postnatal period have been postulated as one of the underlying mechanisms by which maternal and early postnatal nutrition may modulate cancer risk in the offspring. In this regard, DNA methylation patterns are reprogrammed during embryogenesis by active and passive genome-wide demethylation, which erases large parts of the parental DNA methylation pattern (Figure 4.5). After implantation of the blastocyst into the uterus, de novo methylation occurs, which establishes a new DNA methylation pattern for the offspring, with methylation mostly limited to non-CpG island areas, which is maintained postnatally (Figure 4.5).[261] Therefore, DNA methylation during the embryogenic stage, when DNA methylation patterns are reprogrammed, may be highly susceptible to the modulating effects of nutritional, hormonal, and metabolic factors. In particular, the one-carbon nutrient pool including folate available during this critical time period of development can have a significant effect on the DNA methylation patterns in the offspring, and subsequently on chronic disease risk, including cancer, later in life.

Proof-of-principle studies have shown that maternal methyl donor supplementation can significantly change DNA methylation and associated phenotypes in the offspring.[262–264] For example, the pioneering research using the agouti viable yellow (A^{vy}/a) mouse showed that maternal supplementation of high levels of one-carbon nutrients, including folic acid, B_{12}, choline, and methionine, during pregnancy led to CpG DNA hypermethylation at the regulatory region of the A^{vy} gene, which resulted in a change in offspring's coat color from yellow to brown.[262,263] Furthermore, the brown coat mice were found to have lower obesity, hypertension, and insulin resistance,[264] and when challenged with a chemical carcinogen, tumor multiplicity in the liver and tumor incidence in the lungs were reduced[265] compared with their yellow coat color siblings. In addition, supplementation with one-carbon nutrients during pregnancy led to a "straight-tail" phenotype compared with the "kinky-tailed" phenotype in the offspring in the *Axin Fused* (*Axin^{FU}*) mouse model.[266] The "straight-tail" phenotype was correlated with increased CpG promoter DNA methylation at the *Axin* gene.[266] These observations demonstrate that the maternal supplementation of one-carbon nutrients can influence the phenotype and disease susceptibility in the offspring via DNA methylation changes.

4.7.1 Animal Studies

Evidence for the potential cancer modulating effects of maternal environment in the offspring is largely demonstrated in animal studies, since cancer is an age-related disease and requires a lengthy follow-up period for diagnosis. Two animal studies have demonstrated the protective effects of maternal folic acid supplementation alone[267] or in combination with vitamins B_2, B_6, and B_{12}[268] on colorectal and small intestinal tumorigenesis, respectively, in the offspring. In contrast, McKay et al. reported no effect of maternal folate depletion on small intestinal tumorigenesis.[269] One recent study has reported that maternal and postweaning folic acid supplementation significantly increased the incidence of mammary tumors and accelerated the development and progression of mammary tumors in the offspring.[270]

A recent study from our laboratory using rats showed that maternal folic acid supplementation at 2.5× the control (5-mg folic acid/kg diet; The control diet contains 2-mg folic acid/kg diet, which is the basal dietary requirement for rodents; The 2.5× folic acid supplementation approximates 1 mg of folic acid/day or 2.5× the recommended daily allowance [0.4 mg dietary folate equivalent/day] in humans significantly, albeit very modestly by 3%, increased global DNA methylation of the colon of the weanling offspring.[267] Unexpectedly, in this study, postweaning, but no maternal, folic acid supplementation significantly decreased global DNA methylation of the colon of the offspring by 3.3%–6.4%, at 14 weeks of age.[267] In contrast, at weaning, maternal folic acid supplementation significantly decreased global (by 25%) and site-specific DNA methylation of the *Ppar-γ*, *ERα*, *p53*, and *Apc* genes in the liver.[271] At 14 weeks of age, postweaning, but not maternal, folic acid supplementation significantly decreased global DNA methylation by 10%–16%.[271] At 14 weeks of age, both maternal and postweaning folic acid supplementation significantly increased DNA methylation of the *Ppar-γ*, *p53*, and *p16* genes whereas only postweaning folic acid supplementation significantly increased DNA methylation of the *ERα* and *Apc* genes.[271] *Apc*, *p16*, and *p53* are tumor suppressor genes and mutations and/or aberrant DNA methylation of these genes have been implicated in CRC.[272] Therefore, aberrant DNA methylation of the *Apc*, *p16*, and *p53* could contribute to colorectal carcinogenesis via gene silencing as well as gene instability. In another study using a chemical carcinogen model of mammary tumors in the same strain of rats, maternal, but not postweaning, folic acid supplementation at 2.5× the control significantly decreased global DNA methylation of the mammary glands in the offspring by 7% at 28 weeks of age.[270] These observations indicate for the first time that maternal folic acid supplementation at 2.5× the control can modulate global and gene-specific DNA methylation in the rat offspring in an organ- and gene-specific manner.

The observed unexpected decrease in global and gene- and site-specific DNA methylation in the offspring associated with maternal and postweaning folic acid supplementation in some of these studies may be partly explained by the preferential shuttling of the flux of one-carbon units to the nucleotide synthesis pathway over the biological methylation pathway in response to folic acid supplementation. Although folic acid is an inhibitor of DHFR,[76] it can upregulate DHFR in certain situations.[78] This upregulation may increase thymidylate synthase activity because the transcription of these genes is coregulated by several transcription factors,[273] and this would increase thymidylate production at the expense of methylation reactions.[81] Furthermore, excess DHF resulting from folic acid supplementation inhibits MTHFR, thereby reducing the formation of 5-methylTHF and consequent DNA hypomethylation.[82] Another mechanism by which maternal folic acid supplementation induced both global and gene- and site-specific DNA hypomethylation in the offspring is the observed downregulation of DNMT.[270,271]

Several other animal studies have also investigated the effect of maternal and early life folate manipulation on DNA methylation in the offspring with conflicting results. Lillycrop et al. demonstrated that maternal folic acid supplementation (5× the control) prevented the promoter hypomethylation of the *Ppar*-α and *Gr* genes and the decreased Dnmt1 expression induced by protein restriction in maternal diet in the liver of adult rat offspring.[274,275] However, in a mouse model, daily administration of folinic acid by gavage during the periconceptional period until day 15.5 of gestation significantly decreased global DNA methylation in both the liver and brain of the embryo.[276] Generally, folate depletion imposed during the in utero and lactation stages, but not during the postweaning stage, appears to decrease global DNA methylation in the offspring in an organ-specific manner.[277–280] In contrast, a maternal diet deplete of folate fed to *Apc+/Min* and wild-type mice had no effect on gene-specific DNA methylation in weaning mice from both groups.[281] However, decreased *p53*-specific DNA methylation in the small intestine of adult mice was seen with the depleted maternal folate diet.[281] Furthermore, a low folate diet in the postweaning period increased *Apc*-specific DNA methylation in the *Apc+/Min* mice but not in the wild-type mice.[281] These studies used different protocols of folate manipulation in several different models, and hence, definitive conclusions cannot be drawn.[274–281]

A recent study using pregnant CD1 mice found maternal folic acid supplementation followed by gestational arsenic exposure altered DNA methylation in CpG islands located in nearly 3000 genes, including differentially methylated regions (DMRs) of imprinted genes involved in embryogenesis and cancer.[282] After identification of various gene networks, the most affected pathway was the Wnt-signaling pathway, which plays a role in fetal development and regeneration of tissues in adults.[283] Over 50 genes associated with this pathway were found to have altered DNA methylation patterns (e.g., *Wnt3*, *Fzd8*, *Fzd10*, *Dvl1*, *Dvl3*, *Axin2*, *Ctnnb1*, *Cdkn2a*, *Rara*, *Tgf*-β).[282] Alterations in the Wnt signaling pathway have been implicated in the development of CRC;[283] thus, these DNA methylation changes could have implications for the development of CRC in the offspring.

4.7.2 Human Studies

Epidemiological studies have reported a protective effect of periconceptional maternal folic acid supplementation on several pediatric cancers in the offspring.[284–287] However, some studies have not confirmed this purported protective effect[288,289] and have even reported an increased risk.[286] Furthermore, most of these epidemiological studies could not delineate the effect on cancer risk specific to folic acid from other vitamins in the supplements.[290] At present, no information concerning the effect of maternal folic acid supplementation on CRC risk in the offspring exists in humans.

Recent small pilot human studies have provided limited support, albeit equivocal, for the modulatory effect of periconceptional folic acid supplementation on global DNA methylation in the offspring.[291–293]

Several studies that have investigated DNA methylation patterns of specific candidate genes have focused on the *IGF2* gene, likely because IGF2 acts as a growth-promoting hormone during fetal development. *IGF2* is an imprinted gene requiring DNA methylation to suppress either paternal or maternal allelic expression. Aberrant biallelic *IGF2* expression has been linked to the development of certain cancers including CRC.[294] Four studies investigating the effect of maternal folic acid supplementation on *IGF2*-specific DNA methylation in the infant found conflicting results. In two studies, *IGF2* DNA methylation was increased with maternal folic acid supplementation[295,296] and was positively associated with RBC folate concentrations in cord blood.[295] The other two studies, however, demonstrated that DNA methylation of *IGF2* was decreased[297] or had no association[298] with maternal folic acid supplementation. Furthermore, maternal *MTHFR* 677T variant allele was associated with an increase in *IGF2* methylation in the infant.[299] Another candidate gene of interest is the imprinted gene *paternally expressed 3* (*PEG3*). Haggarty et al. found that women reporting folic acid supplementation at 12 weeks of gestation had offspring with lower *PEG3* DNA methylation compared with those stopping folic acid supplementation at 12 weeks of gestation.[295] Similar to *p53* and *APC*, *PEG3* is a tumor suppressor gene, which inhibits the Wnt signaling pathway and regulates tumor growth.[300] Therefore, the lower *PEG3* DNA methylation observed with folic acid supplementation at 12 weeks of gestation could cause improper gene imprinting required for normal *PEG3* function and tumor growth regulation.

Overall, the available body of evidence from animal and human studies, although not uniformly consistent, provide considerable evidence that global and gene-specific DNA methylation can be modulated by maternal dietary and supplemental intakes of folate/folic acid as well as maternal genotype of critical enzymes involved in the folate pathway. Whether these DNA methylation changes resulting from maternal folate status play a mechanistic role in cancer development and progression in the offspring needs further exploration.

4.7.3 Future Studies

A gap in the knowledge still exist with the lack of information on how changes in the maternal diet can influence epigenetics and/or disease risk in the mothers themselves. Studies to date have generally focused on the health outcomes of the offspring but we need to still consider these effects on the mother. Furthermore, the effect of paternal dietary status on epigenetic changes in the offspring is an emerging important area of research as some studies have suggested that paternal diet may contribute to epigenetic alterations and modulate health outcomes in the offspring, independent of the maternal diet.[301–303] Kim et al. reported fetuses from male mice fed folate-deficient diets before and during mating were associated with a lower percentage of fetal whole brain DNA methylation and Igf2 protein expression compared with fetuses from folic acid–supplemented males.[304] Therefore, in future studies, both paternal and maternal diets need to be considered in the design and interpretation of studies investigating the role of in utero nutrition on DNA methylation and health outcomes in the offspring.

Whether or not DNA methylation changes are transmitted to future generations (i.e., transgenerational epigenetic inheritance) is a highly controversial topic.[305,306] Overall interpretation of the available evidence does not provide unequivocal support for the transmission of epigenetic changes in the F1 generation induced by maternal or paternal diets to the F3–4 generations. Elucidating the mechanisms of epigenetic inheritance is particularly challenging, especially due to the phenomenon of passive genome-wide demethylation that occurs postconception in mammalian embryogenesis. This is an important factor that needs to be taken into consideration when determining the heritability of parental DNA methylation patterns.[307] Epigenetic inheritance of histone modification, chromatin remodeling, and DNA methylation is an area of growing interest. To date, preliminary research has been conducted to establish proof of principle of the mechanisms behind transgenerational epigenetic inheritance in plants, worms, and flies.[307]

4.8 CONCLUSIONS

Genetic changes in cancer are abrupt in onset, their effects are often all or nothing, the loss of function occurs at a fixed level, and they are not reversible in most cases. In contrast, epigenetic changes are gradual in onset and progressive, their effects are dose-dependent and are potentially reversible. These observations present new opportunities in cancer risk modification and prevention using dietary and lifestyle factors and potential chemopreventive drugs. In this regard, folate has been a focus of intense interest due to its potential modulatory effect on cancer development observed in animal studies, epidemiological observations and intervention trials in humans and its potential ability to modulate DNA methylation. However, folate will likely influence other epigenetic mechanisms such as histone modifications, chromatin remodeling, and RNA interference, although we have exclusively focused on DNA methylation in this chapter. Indeed, studies investigating these complex interactions have begun to emerge. Furthermore, investigating the effects of folate on different methylation products beyond 5-mC such as 5-hmC are of great interest.

The portfolio of evidence from in vitro, animal, and human studies collectively suggests that the effects of folate deficiency and folic acid supplementation on DNA methylation are highly complex as they are gene- and site-specific and depends on species, cell type, target organ, and stage of transformation as well as on the timing, degree, and duration of folate manipulation. These studies also suggest that changes in DNA methylation also depend on interactions with other methyl group donors and with other dietary and environmental factors and on genetic variants of critical genes in the folate metabolic and one-carbon transfer pathways. Furthermore, the pattern of site- and gene-specific DNA methylation may not be in concert with the direction of changes in global DNA methylation. Consequently, linking DNA methylation changes resulting from folate manipulations to cancer development and progression in specific organ sites remains a very difficult task. Nevertheless, a considerable amount of evidence supports that folate may modulate the risk of cancer in certain sites via aberrant global and gene-specific DNA methylation. An emerging body of evidence suggests that the folate status during pregnancy and early postnatal life may play a critical role in the development of cancer of the offspring later in adult life through its potential modulatory effects on DNA methylation reprogramming during this highly critical and susceptible period of epigenetic programming. Also, it appears that folate status can influence DNA methylation more readily in the elderly. Whether or not folate deficiency and folic acid supplementation can modulate aging-related DNA methylation changes and how this may promote age-related conditions, including CRC, is worthy of further investigation. Although the jury is still out, the potential for folate to modulate DNA methylation and thus modify the risk of chronic diseases including cancer in humans across the lifespan remains provocative and is worthy of further studies.

ACKNOWLEDGMENT

This work was supported in part by the Canadian Institutes of Health Research (grant 14126).

REFERENCES

1. Kim YI. Nutritional epigenetics: Impact of folate deficiency on DNA methylation and colon cancer susceptibility. *J Nutr.* Nov 2005;135(11):2703–2709.
2. Kim YI. Folate and DNA methylation: A mechanistic link between folate deficiency and colorectal cancer? *Cancer Epidemiol Biomarkers Prev.* 2004;13(4):511–519.
3. Institute of Medicine. Folate. *Dietary Reference Intakes for Thiamin, Riboflavin, Niacin, Vitamin B6, Folate, Vitamin B12, Pantothenic Acid, Biotin, and Choline*. Washington, DC: National Academy Press; 1998:196–305.
4. Shane B. Folate chemistry and metabolism. In: Bailey LB, ed. *Folate in Health and Disease*. Boca Raton, FL: CRC Press; 2010:1–24.
5. Aufreiter S, Gregory JF III, Pfeiffer CM et al. Folate is absorbed across the colon of adults: Evidence from cecal infusion of (13)C-labeled [6S]–5-formyltetrahydrofolic acid. *Am J Clin Nutr.* Jul 2009;90(1):116–123.
6. Camilo E, Zimmerman J, Mason JB et al. Folate synthesized by bacteria in the human upper small intestine is assimilated by the host. *Gastroenterology.* Apr 1996;110(4):991–998.
7. Rong N, Selhub J, Goldin BR, Rosenberg IH. Bacterially synthesized folate in rat large intestine is incorporated into host tissue folyl polyglutamates. *J Nutr.* Dec 1991;121(12):1955–1959.
8. Wright AJ, Dainty JR, Finglas PM. Folic acid metabolism in human subjects revisited: Potential implications for proposed mandatory folic acid fortification in the UK. *Br J Nutr.* Oct 2007;98(4):667–675.
9. Kelly P, McPartlin J, Goggins M, Weir DG, Scott JM. Unmetabolized folic acid in serum: Acute studies in subjects consuming fortified food and supplements. *Am J Clin Nutr.* Jun 1997;65(6):1790–1795.
10. Lucock M. Folic acid: Nutritional biochemistry, molecular biology, and role in disease processes. *Mol Genet Metab.* Sep–Oct 2000;71(1–2):121–138.
11. Chen LH, Liu ML, Hwang HY, Chen LS, Korenberg J, Shane B. Human methionine synthase. cDNA cloning, gene localization, and expression. *J Biol Chem.* Feb 7 1997;272(6):3628–3634.

12. Leclerc D, Wilson A, Dumas R et al. Cloning and mapping of a cDNA for methionine synthase reductase, a flavoprotein defective in patients with homocystinuria. *Proc Natl Acad Sci U S A.* Mar 17 1998;95(6):3059–3064.
13. Ly A, Hoyt L, Crowell J, Kim YI. Folate and DNA methylation. *Antioxid Redox Signal.* Jul 15 2012;17(2):302–326.
14. Mato JM, Alvarez L, Ortiz P, Pajares MA. S-adenosylmethionine synthesis: Molecular mechanisms and clinical implications. *Pharmacol Ther.* 1997;73(3):265–280.
15. Chen YM, Chen LY, Wong FH, Lee CM, Chang TJ, Yang-Feng TL. Genomic structure, expression, and chromosomal localization of the human glycine N-methyltransferase gene. *Genomics.* May 15 2000;66(1):43–47.
16. Luka Z, Cerone R, Phillips JA III, Mudd HS, Wagner C. Mutations in human glycine N-methyltransferase give insights into its role in methionine metabolism. *Hum Genet.* Jan 2002;110(1):68–74.
17. Pegg AE, Xiong H, Feith DJ, Shantz LM. S-adenosylmethionine decarboxylase: Structure, function and regulation by polyamines. *Biochem Soc Trans.* Nov 1998;26(4): 580–586.
18. CarteniFarina M, Oliva A, Romeo G et al. 5′-Methylthioadenosine phosphorylase from Caldariella acidophila. Purification and properties. *Eur J Biochem.* Nov 1979;101(2):317–324.
19. Girgis S, Nasrallah IM, Suh JR et al. Molecular cloning, characterization and alternative splicing of the human cytoplasmic serine hydroxymethyltransferase gene. *Gene.* Apr 14 1998;210(2):315–324.
20. Stover PJ, Chen LH, Suh JR, Stover DM, Keyomarsi K, Shane B. Molecular cloning, characterization, and regulation of the human mitochondrial serine hydroxymethyltransferase gene. *J Biol Chem.* Jan 17 1997;272(3):1842–1848.
21. Green JM, MacKenzie RE, Matthews RG. Substrate flux through methylenetetrahydrofolate dehydrogenase: Predicted effects of the concentration of methylenetetrahydrofolate on its partitioning into pathways leading to nucleotide biosynthesis or methionine regeneration. *Biochemistry.* Oct 18 1988;27(21):8014–8022.
22. Ueland PM, Hustad S, Schneede J, Refsum H, Vollset SE. Biological and clinical implications of the MTHFR C677T polymorphism. *Trends Pharmacol Sci.* Apr 2001;22(4):195–201.
23. Smith AD, Kim YI, Refsum H. Is folic acid good for everyone? *Am J Clin Nutr.* Mar 2008;87(3):517–533.
24. Albert CM, Cook NR, Gaziano JM et al. Effect of folic acid and B vitamins on risk of cardiovascular events and total mortality among women at high risk for cardiovascular disease: A randomized trial. *JAMA.* May 7 2008;299(17):2027–2036.
25. Armitage JM, Bowman L, Clarke RJ et al. Effects of homocysteine-lowering with folic acid plus vitamin B12 vs placebo on mortality and major morbidity in myocardial infarction survivors: A randomized trial. *JAMA.* Jun 23 2010;303(24):2486–2494.
26. Bonaa KH, Njolstad I, Ueland PM et al. Homocysteine lowering and cardiovascular events after acute myocardial infarction. *N Engl J Med.* Apr 13 2006;354(15): 1578–1588.
27. Clarke R, Halsey J, Bennett D, Lewington S. Homocysteine and vascular disease: Review of published results of the homocysteine-lowering trials. *J Inherit Metab Dis.* Feb 2011;34(1):83–91.
28. Clarke R, Halsey J, Lewington S et al. Effects of lowering homocysteine levels with B vitamins on cardiovascular disease, cancer, and cause-specific mortality: Meta-analysis of 8 randomized trials involving 37 485 individuals. *Arch Intern Med.* Oct 11 2010;170(18):1622–1631.
29. Lonn E, Yusuf S, Arnold MJ et al. Homocysteine lowering with folic acid and B vitamins in vascular disease. *N Engl J Med.* Apr 13 2006;354(15):1567–1577.

30. Toole JF, Malinow MR, Chambless LE et al. Lowering homocysteine in patients with ischemic stroke to prevent recurrent stroke, myocardial infarction, and death: The Vitamin Intervention for Stroke Prevention (VISP) randomized controlled trial. *JAMA.* Feb 4 2004;291(5):565–575.
31. Balk EM, Raman G, Tatsioni A, Chung M, Lau J, Rosenberg IH. Vitamin B6, B12, and folic acid supplementation and cognitive function: A systematic review of randomized trials. *Arch Intern Med.* Jan 8 2007;167(1):21–30.
32. McMahon JA, Green TJ, Skeaff CM, Knight RG, Mann JI, Williams SM. A controlled trial of homocysteine lowering and cognitive performance. *N Engl J Med.* Jun 29 2006;354(26):2764–2772.
33. Wang X, Qin X, Demirtas H et al. Efficacy of folic acid supplementation in stroke prevention: A meta-analysis. *Lancet.* Jun 2 2007;369(9576):1876–1882.
34. Durga J, van Boxtel MP, Schouten EG et al. Effect of 3-year folic acid supplementation on cognitive function in older adults in the FACIT trial: A randomised, double blind, controlled trial. *Lancet.* Jan 20 2007;369(9557):208–216.
35. Douaud G, Refsum H, de Jager CA et al. Preventing Alzheimer's disease-related gray matter atrophy by B-vitamin treatment. *Proc Natl Acad Sci USA.* Jun 4 2013;110(23):9523–9528.
36. Smith AD, Smith SM, de Jager CA et al. Homocysteine-lowering by B vitamins slows the rate of accelerated brain atrophy in mild cognitive impairment: A randomized controlled trial. *PLoS One.* 2010;5(9):e12244.
37. Administration FaD. Food standards: Amendment of standards of identity for enriched grain products to require addition of folic acid. Final rule. 21 CFR Parts 136, 137, and 139. *Fed Reg.* 1996;61(44):8781–8807.
38. Health Canada. Regulations amending the Food and Drug Regulations (1066). *Can Gaz, Part 1.* 1997;131:3702–3737.
39. Bailey RL, Mills JL, Yetley EA et al. Unmetabolized serum folic acid and its relation to folic acid intake from diet and supplements in a nationally representative sample of adults aged > or = 60 y in the United States. *Am J Clin Nutr.* Aug 2010;92(2):383–389.
40. Colapinto CK, O'Connor DL, Tremblay MS. Folate status of the population in the Canadian Health Measures Survey. *CMAJ.* Feb 8 2011;183(2):E100–E106.
41. Pfeiffer CM, Caudill SP, Gunter EW, Osterloh J, Sampson EJ. Biochemical indicators of B vitamin status in the US population after folic acid fortification: Results from the National Health and Nutrition Examination Survey 1999–2000. *Am J Clin Nutr.* Aug 2005;82(2):442–450.
42. Pfeiffer CM, Hughes JP, Lacher DA et al. Estimation of trends in serum and RBC folate in the U.S. population from pre- to postfortification using assay-adjusted data from the NHANES 1988–2010. *J Nutr.* May 2012;142(5):886–893.
43. Pfeiffer CM, Johnson CL, Jain RB et al. Trends in blood folate and vitamin B-12 concentrations in the United States, 1988–2004. *Am J Clin Nutr.* Sep 2007;86(3):718–727.
44. Ray JG, Vermeulen MJ, Boss SC, Cole DE. Increased red cell folate concentrations in women of reproductive age after Canadian folic acid food fortification. *Epidemiology.* Mar 2002;13(2):238–240.
45. De Wals P, Rusen ID, Lee NS, Morin P, Niyonsenga T. Trend in prevalence of neural tube defects in Quebec. *Birth Defects Res Part A Clin Mol Teratol.* Nov 2003;67(11):919–923.
46. Gucciardi E, Pietrusiak MA, Reynolds DL, Rouleau J. Incidence of neural tube defects in Ontario, 1986–1999. *CMAJ.* Aug 6 2002;167(3):237–240.
47. Honein MA, Paulozzi LJ, Mathews TJ, Erickson JD, Wong LY. Impact of folic acid fortification of the US food supply on the occurrence of neural tube defects. *JAMA.* Jun 20 2001;285(23):2981–2986.
48. Persad VL, Van den Hof MC, Dube JM, Zimmer P. Incidence of open neural tube defects in Nova Scotia after folic acid fortification. *CMAJ.* Aug 6 2002;167(3):241–245.

49. Williams LJ, Mai CT, Edmonds LD et al. Prevalence of spina bifida and anencephaly during the transition to mandatory folic acid fortification in the United States. *Teratology.* Jul 2002;66(1):33–39.
50. De Wals P, Tairou F, Van Allen MI et al. Reduction in neural-tube defects after folic acid fortification in Canada. *N Engl J Med.* Jul 12 2007;357(2):135–142.
51. Radimer K, Bindewald B, Hughes J, Ervin B, Swanson C, Picciano MF. Dietary supplement use by US adults: Data from the National Health and Nutrition Examination Survey, 1999–2000. *Am J Epidemiol.* Aug 15 2004;160(4):339–349.
52. Bailey RL, Dodd KW, Gahche JJ et al. Total folate and folic acid intake from foods and dietary supplements in the United States: 2003–2006. *Am J Clin Nutr.* Jan 2010;91(1):231–237.
53. Wilson RD, Johnson JA, Wyatt P et al. Pre-conceptional vitamin/folic acid supplementation 2007: The use of folic acid in combination with a multivitamin supplement for the prevention of neural tube defects and other congenital anomalies. *J Obstet Gynaecol Can.* Dec 2007;29(12):1003–1026.
54. Alamian A, Rouleau I, Simard J, Dorval M. Use of dietary supplements among women at high risk of hereditary breast and ovarian cancer (HBOC) tested for cancer susceptibility. *Nutr Cancer.* 2006;54(2):157–165.
55. Bright-Gbebry M, Makambi KH, Rohan JP et al. Use of multivitamins, folic acid and herbal supplements among breast cancer survivors: The Black Women's Health Study. *BMC Complement Altern Med.* 2011;11:30.
56. Giovannucci E, Chan AT. Role of vitamin and mineral supplementation and aspirin use in cancer survivors. *J Clin Oncol.* Sep 10 2010;28(26):4081–4085.
57. Holmes RS, Zheng Y, Baron JA et al. Use of folic acid-containing supplements after a diagnosis of colorectal cancer in the Colon Cancer Family Registry. *Cancer Epidemiol Biomarkers Prev.* Aug 2010;19(8):2023–2034.
58. Kwan ML, Greenlee H, Lee VS et al. Multivitamin use and breast cancer outcomes in women with early-stage breast cancer: The Life After Cancer Epidemiology Study. *Breast Cancer Res Treat.* Nov 2011;130(1):195–205.
59. Velicer CM, Ulrich CM. Vitamin and mineral supplement use among US adults after cancer diagnosis: A systematic review. *J Clin Oncol.* Feb 1 2008;26(4):665–673.
60. Yetley EA, Coates PM, Johnson CL. Overview of a roundtable on NHANES monitoring of biomarkers of folate and vitamin B-12 status: Measurement procedure issues. *Am J Clin Nutr.* Jul 2011;94(1):297S–302S.
61. Sweeney MR, McPartlin J, Weir DG et al. Evidence of unmetabolised folic acid in cord blood of newborn and serum of 4-day-old infants. *Br J Nutr.* Nov 2005;94(5):727–730.
62. Rock CL. Multivitamin-multimineral supplements: Who uses them? *Am J Clin Nutr.* Jan 2007;85(1):277S–279S.
63. Troen AM, Mitchell B, Sorensen B et al. Unmetabolized folic acid in plasma is associated with reduced natural killer cell cytotoxicity among postmenopausal women. *J Nutr.* Jan 2006;136(1):189–194.
64. Kalmbach RD, Choumenkovitch SF, Troen AM, D'Agostino R, Jacques PF, Selhub J. Circulating folic acid in plasma: Relation to folic acid fortification. *Am J Clin Nutr.* Sep 2008;88(3):763–768.
65. Kim YI. Folate and carcinogenesis: Evidence, mechanisms, and implications. *J Nutr Biochem.* 1999;10:66–88.
66. Kim YI. Folate and colorectal cancer: An evidence-based critical review. *Mol Nutr Food Res.* Mar 2007;51(3):267–292.
67. Kamen B. Folate and antifolate pharmacology. *Semin Oncol.* Oct 1997;24(5 Suppl 18): S18-30–S18-39.
68. Kim YI. Folic acid supplementation and cancer risk: Point. *Cancer Epidemiol Biomarkers Prev.* Sep 2008;17(9):2220–2225.

69. Giovannucci E. Epidemiologic studies of folate and colorectal neoplasia: A review. *J Nutr.* Aug 2002;132(8 Suppl):2350S–2355S.
70. Kim DH, Smith-Warner SA, Spiegelman D et al. Pooled analyses of 13 prospective cohort studies on folate intake and colon cancer. *Cancer Causes Control.* Nov 2010;21(11):1919–1930.
71. Sanjoaquin MA, Allen N, Couto E, Roddam AW, Key TJ. Folate intake and colorectal cancer risk: A meta-analytical approach. *Int J Cancer.* Feb 20 2005;113(5):825–828.
72. Kennedy DA, Stern SJ, Moretti M et al. Folate intake and the risk of colorectal cancer: A systematic review and meta-analysis. *Cancer Epidemiol.* Feb 2011;35(1):2–10.
73. de Jong MM, Nolte IM, te Meerman GJ et al. Low-penetrance genes and their involvement in colorectal cancer susceptibility. *Cancer Epidemiol Biomarkers Prev.* Nov 2002;11(11):1332–1352.
74. Houlston RS, Tomlinson IP. Polymorphisms and colorectal tumor risk. *Gastroenterology.* Aug 2001;121(2):282–301.
75. Sharp L, Little J. Polymorphisms in genes involved in folate metabolism and colorectal neoplasia: A HuGE review. *Am J Epidemiol.* Mar 1 2004;159(5):423–443.
76. Bailey SW, Ayling JE. The extremely slow and variable activity of dihydrofolate reductase in human liver and its implications for high folic acid intake. *Proc Natl Acad Sci U S A.* Sep 8 2009;106(36):15424–15429.
77. Lucock M. Is folic acid the ultimate functional food component for disease prevention? *BMJ.* Jan 24 2004;328(7433):211–214.
78. Kamen BA, Nylen PA, Whitehead VM, Abelson HT, Dolnick BJ, Peterson DW. Lack of dihydrofolate reductase in human tumor and leukemia cells in vivo. *Cancer Drug Deliv.* Spring 1985;2(2):133–138.
79. Obama K, Kanai M, Kawai Y, Fukushima M, Takabayashi A. Role of retinoblastoma protein and E2F-1 transcription factor in the acquisition of 5-fluorouracil resistance by colon cancer cells. *Int J Oncol.* Aug 2002;21(2):309–314.
80. Slansky JE, Li Y, Kaelin WG, Farnham PJ. A protein synthesis-dependent increase in E2F1 mRNA correlates with growth regulation of the dihydrofolate reductase promoter. *Mol Cell Biol.* Mar 1993;13(3):1610–1618.
81. Nijhout HF, Reed MC, Budu P, Ulrich CM. A mathematical model of the folate cycle: New insights into folate homeostasis. *J Biol Chem.* Dec 31 2004;279(53):55008–55016.
82. Matthews RG, Haywood BJ. Inhibition of pig liver methylenetetrahydrofolate reductase by dihydrofolate: Some mechanistic and regulatory implications. *Biochemistry.* Oct 30 1979;18(22):4845–4851.
83. Logan RF, Grainge MJ, Shepherd VC, Armitage NC, Muir KR. Aspirin and folic acid for the prevention of recurrent colorectal adenomas. *Gastroenterology.* Jan 2008; 134(1):29–38.
84. Jaszewski R, Misra S, Tobi M et al. Folic acid supplementation inhibits recurrence of colorectal adenomas: A randomized chemoprevention trial. *World J Gastroenterol.* Jul 28 2008;14(28):4492–4498.
85. Paspatis GA, Karamanolis DG. Folate supplementation and adenomatous colonic polyps. *Dis Colon Rectum.* Dec 1994;37(12):1340–1341.
86. Wu K, Platz EA, Willett WC et al. A randomized trial on folic acid supplementation and risk of recurrent colorectal adenoma. *Am J Clin Nutr.* Dec 2009;90(6):1623–1631.
87. Cole BF, Baron JA, Sandler RS et al. Folic acid for the prevention of colorectal adenomas: A randomized clinical trial. *JAMA.* Jun 6 2007;297(21):2351–2359.
88. Figueiredo JC, Grau MV, Haile RW et al. Folic acid and risk of prostate cancer: Results from A randomized clinical trial. *J Natl Cancer Inst.* Mar 18 2009;101(6):432–435.
89. Figueiredo JC, Mott LA, Giovannucci E et al. Folic acid and prevention of colorectal adenomas: A combined analysis of randomized clinical trials. *Int J Cancer.* Jul 1 2011;129(1):192–203.

90. Ebbing M, Bonaa KH, Nygard O et al. Cancer incidence and mortality after treatment with folic acid and vitamin B12. *JAMA*. Nov 18 2009;302(19):2119–2126.
91. Wien TN, Pike E, Wisloff T, Staff A, Smeland S, Klemp M. Cancer risk with folic acid supplements: A systematic review and meta-analysis. *BMJ Open*. 2012;2(1):e000653.
92. Vollset SE, Clarke R, Lewington S et al. Effects of folic acid supplementation on overall and site-specific cancer incidence during the randomised trials: Meta-analyses of data on 50,000 individuals. *Lancet*. Mar 23 2013;381(9871):1029–1036.
93. Carroll C, Cooper K, Papaioannou D et al. Meta-analysis: Folic acid in the chemoprevention of colorectal adenomas and colorectal cancer. *Aliment Pharmacol Ther.* Apr 2010;31(7):708–718.
94. Mason JB, Dickstein A, Jacques PF et al. A temporal association between folic acid fortification and an increase in colorectal cancer rates may be illuminating important biological principles: A hypothesis. *Cancer Epidemiol Biomarkers Prev.* Jul 2007;16(7):1325–1329.
95. Hirsch S, Sanchez H, Albala C et al. Colon cancer in Chile before and after the start of the flour fortification program with folic acid. *Eur J Gastroenterol Hepatol.* Apr 2009;21(4):436–439.
96. Gibson TM, Weinstein SJ, Pfeiffer RM et al. Pre- and postfortification intake of folate and risk of colorectal cancer in a large prospective cohort study in the United States. *Am J Clin Nutr.* Oct 2011;94(4):1053–1062.
97. Stevens VL, McCullough ML, Sun J, Jacobs EJ, Campbell PT, Gapstur SM. High levels of folate from supplements and fortification are not associated with increased risk of colorectal cancer. *Gastroenterology.* Jul 2011;141(1):98–105, 105.e101.
98. Kim YI. Role of folate in colon cancer development and progression. *J Nutr.* 2003;133:3731S–3739S.
99. Kim YI. Will mandatory folic acid fortification prevent or promote cancer? *Am J Clin Nutr.* Nov 2004;80(5):1123–1128.
100. Deghan Manshadi S, Ishiguro L, Sohn KJ et al. Folic acid supplementation promotes mammary tumorigenesis in a rat model. *Proc Am Assoc Cancer Res.* 2011;71:abst LB-460.
101. Lindzon GM, Medline A, Sohn KJ, Depeint F, Croxford R, Kim YI. Effect of folic acid supplementation on the progression of colorectal aberrant crypt foci. *Carcinogenesis.* Jun 18 2009;30(9):1536–1543.
102. Song J, Medline A, Mason JB, Gallinger S, Kim YI. Effects of dietary folate on intestinal tumorigenesis in the *Apc^{Min}* mouse. *Cancer Res.* Oct 1 2000;60(19):5434–5440.
103. Song J, Sohn KJ, Medline A, Ash C, Gallinger S, Kim YI. Chemopreventive effects of dietary folate on intestinal polyps in Apc+/−Msh2−/− mice. *Cancer Res.* Jun 15 2000;60(12):3191–3199.
104. Cravo ML, Mason JB, Dayal Y et al. Folate deficiency enhances the development of colonic neoplasia in dimethylhydrazine-treated rats. *Cancer Res.* Sep 15 1992; 52(18):5002–5006.
105. Kim YI, Salomon RN, Graeme-Cook F et al. Dietary folate protects against the development of macroscopic colonic neoplasia in a dose responsive manner in rats. *Gut.* Nov 1996;39(5):732–740.
106. Ames BN. DNA damage from micronutrient deficiencies is likely to be a major cause of cancer. *Mutat Res.* Apr 18 2001;475(1–2):7–20.
107. Choi SW, Mason JB. Folate and carcinogenesis: An integrated scheme. *J Nutr.* 2000; 130(2):129–132.
108. Choi SW, Mason JB. Folate status: Effects on pathways of colorectal carcinogenesis. *J Nutr.* Aug 2002;132(8 Suppl):2413S–2418S.
109. Duthie SJ. Folic acid deficiency and cancer: Mechanisms of DNA instability. *Br Med Bull.* 1999;55(3):578–592.

110. Fenech M. The role of folic acid and vitamin B12 in genomic stability of human cells. *Mutat Res.* Apr 18 2001;475(1–2):57–67.
111. Lamprecht SA, Lipkin M. Chemoprevention of colon cancer by calcium, vitamin D and folate: Molecular mechanisms. *Nat Rev Cancer.* Aug 2003;3(8):601–614.
112. Crowell J, Ly A, Kim YI. Folate and DNA methylation. In: Maulik N, Maulik G, eds. *Nutrition, Epigeneic Mechanisms, and Human Disease*. Boca Raton, FL: CRC Press; 2011:31–75.
113. Esteller M. Relevance of DNA methylation in the management of cancer. *Lancet Oncol.* Jun 2003;4(6):351–358.
114. Esteller M. Cancer epigenomics: DNA methylomes and histone-modification maps. *Nat Rev Genet.* Apr 2007;8(4):286–298.
115. Egger G, Liang G, Aparicio A, Jones PA. Epigenetics in human disease and prospects for epigenetic therapy. *Nature.* May 27 2004;429(6990):457–463.
116. Ballestar E, Esteller M. The impact of chromatin in human cancer: Linking DNA methylation to gene silencing. *Carcinogenesis.* Jul 2002;23(7):1103–1109.
117. Jones PA, Baylin SB. The fundamental role of epigenetic events in cancer. *Nat Rev Genet.* Jun 2002;3(6):415–428.
118. Herman JG, Baylin SB. Gene silencing in cancer in association with promoter hypermethylation. *N Engl J Med.* Nov 20 2003;349(21):2042–2054.
119. Robertson KD, Wolffe AP. DNA methylation in health and disease. *Nat Rev Genet.* Oct 2000;1(1):11–19.
120. Jones PL, Veenstra GJ, Wade PA et al. Methylated DNA and MeCP2 recruit histone deacetylase to repress transcription. *Nat Genet.* Jun 1998;19(2):187–191.
121. Canman CE, Lim DS, Cimprich KA et al. Activation of the ATM kinase by ionizing radiation and phosphorylation of p53. *Science.* Sep 11 1998;281(5383):1677–1679.
122. Choi SW, Kim YI, Weitzel JN, Mason JB. Folate depletion impairs DNA excision repair in the colon of the rat. *Gut.* Jul 1998;43(1):93–99.
123. Bird AP, Wolffe AP. Methylation-induced repression—Belts, braces, and chromatin. *Cell.* Nov 24 1999;99(5):451–454.
124. Yen RW, Vertino PM, Nelkin BD et al. Isolation and characterization of the cDNA encoding human DNA methyltransferase. *Nucleic Acids Res.* 1992;20(9):2287–2291.
125. Yoder JA, Yen RW, Vertino PM, Bestor TH, Baylin SB. New 5′ regions of the murine and human genes for DNA (cytosine-5)-methyltransferase. *J Biol Chem.* 1996; 271(49):31092–31097.
126. Okano M, Xie S, Li E. Cloning and characterization of a family of novel mammalian DNA (cytosine-5) methyltransferases. *Nat Genet.* 1998;19(3):219–220.
127. Okano M, Bell DW, Haber DA, Li E. DNA methyltransferases Dnmt3a and Dnmt3b are essential for de novo methylation and mammalian development. *Cell.* 1999; 99(3):247–257.
128. Li E, Jaenisch R. DNA methylation and methyltransferases. In: Ehrlich M, ed. *DNA Alterations in Cancer: Genetic and Epigenetic Changes*. Natick, MA: Eaton Publishing; 2000:351–365.
129. Tajima S, Suetake I. Regulation and function of DNA methylation in vertebrates. *J Biochem (Tokyo).* 1998;123(6):993–999.
130. Vairapandi M, Duker NJ. Enzymic removal of 5-methylcytosine from DNA by a human DNA-glycosylase. *Nucleic Acids Res.* 1993;21(23):5323–5327.
131. Bhattacharya SK, Ramchandani S, Cervoni N, Szyf M. A mammalian protein with specific demethylase activity for mCpG DNA. *Nature.* Feb 18 1999;397(6720):579–583.
132. Tahiliani M, Koh KP, Shen Y et al. Conversion of 5-methylcytosine to 5-hydroxymethylcytosine in mammalian DNA by MLL partner TET1. *Science.* May 15 2009; 324(5929):930–935.

133. Liutkeviciute Z, Lukinavicius G, Masevicius V, Daujotyte D, Klimasauskas S. Cytosine-5-methyltransferases add aldehydes to DNA. *Nat Chem Biol.* Jun 2009;5(6):400–402.
134. Mellen M, Ayata P, Dewell S, Kriaucionis S, Heintz N. MeCP2 binds to 5hmC enriched within active genes and accessible chromatin in the nervous system. *Cell.* Dec 21 2012;151(7):1417–1430.
135. Jin SG, Kadam S, Pfeifer GP. Examination of the specificity of DNA methylation profiling techniques towards 5-methylcytosine and 5-hydroxymethylcytosine. *Nucleic Acids Res.* Jun 2010;38(11):e125.
136. Szulwach KE, Li X, Li Y et al. 5-hmC-mediated epigenetic dynamics during postnatal neurodevelopment and aging. *Nat Neurosci.* Dec 2011;14(12):1607–1616.
137. Globisch D, Munzel M, Muller M et al. Tissue distribution of 5-hydroxymethylcytosine and search for active demethylation intermediates. *PLoS One.* 2010;5(12):e15367.
138. Hackett JA, Sengupta R, Zylicz JJ et al. Germline DNA demethylation dynamics and imprint erasure through 5-hydroxymethylcytosine. *Science.* Jan 25 2013; 339(6118):448–452.
139. Ruzov A, Tsenkina Y, Serio A et al. Lineage-specific distribution of high levels of genomic 5-hydroxymethylcytosine in mammalian development. *Cell Res.* Sep 2011;21(9):1332–1342.
140. Haffner MC, Chaux A, Meeker AK et al. Global 5-hydroxymethylcytosine content is significantly reduced in tissue stem/progenitor cell compartments and in human cancers. *Oncotarget.* Aug 2011;2(8):627–637.
141. Reik W, Dean W, Walter J. Epigenetic reprogramming in mammalian development. *Science.* Aug 10 2001;293(5532):1089–1093.
142. Esteller M. Epigenetics in cancer. *N Engl J Med.* Mar 13 2008;358(11):1148–1159.
143. Goel A, Boland CR. Epigenetics of colorectal cancer. *Gastroenterology.* Dec 2012; 143(6):1442–1460 e1441.
144. Issa JP. CpG island methylator phenotype in cancer. *Nat Rev Cancer.* Dec 2004; 4(12):988–993.
145. Duthie SJ, Narayanan S, Blum S, Pirie L, Brand GM. Folate deficiency in vitro induces uracil misincorporation and DNA hypomethylation and inhibits DNA excision repair in immortalized normal human colon epithelial cells. *Nutr Cancer.* 2000;37(2):245–251.
146. Stempak JM, Sohn KJ, Chiang EP, Shane B, Kim YI. Cell and stage of transformation-specific effects of folate deficiency on methionine cycle intermediates and DNA methylation in an in vitro model. *Carcinogenesis.* Feb 3 2005;26(5):981–990.
147. Charles MA, Johnson IT, Belshaw NJ. Supra-physiological folic acid concentrations induce aberrant DNA methylation in normal human cells in vitro. *Epigenetics.* Jul 2012;7(7):689–694.
148. Jhaveri MS, Wagner C, Trepel JB. Impact of extracellular folate levels on global gene expression. *Mol Pharmacol.* Dec 2001;60(6):1288–1295.
149. Wasson GR, McGlynn AP, McNulty H et al. Global DNA and p53 region-specific hypomethylation in human colonic cells is induced by folate depletion and reversed by folate supplementation. *J Nutr.* Nov 2006;136(11):2748–2753.
150. Wang TP, Hsu SH, Feng HC, Huang RF. Folate deprivation enhances invasiveness of human colon cancer cells mediated by activation of sonic hedgehog signaling through promoter hypomethylation and cross action with transcription nuclear factor-kappa B pathway. *Carcinogenesis.* Jun 2012;33(6):1158–1168.
151. Berner C, Aumuller E, Gnauck A, Nestelberger M, Just A, Haslberger AG. Epigenetic control of estrogen receptor expression and tumor suppressor genes is modulated by bioactive food compounds. *Ann Nutr Metab.* 2010;57(3–4):183–189.
152. Newberne PM, Rogers AE. Labile methyl groups and the promotion of cancer. *Annu Rev Nutr.* 1986;6:407–432.

153. Christman JK, Sheikhnejad G, Dizik M, Abileah S, Wainfan E. Reversibility of changes in nucleic acid methylation and gene expression induced in rat liver by severe dietary methyl deficiency. *Carcinogenesis.* Apr 1993;14(4):551–557.
154. Dizik M, Christman JK, Wainfan E. Alterations in expression and methylation of specific genes in livers of rats fed a cancer promoting methyl-deficient diet. *Carcinogenesis.* Jul 1991;12(7):1307–1312.
155. Pogribny IP, Basnakian AG, Miller BJ, Lopatina NG, Poirier LA, James SJ. Breaks in genomic DNA and within the p53 gene are associated with hypomethylation in livers of folate/methyl-deficient rats. *Cancer Res.* May 1 1995;55(9):1894–1901.
156. Wainfan E, Dizik M, Stender M, Christman JK. Rapid appearance of hypomethylated DNA in livers of rats fed cancer-promoting, methyl-deficient diets. *Cancer Res.* Aug 1 1989;49(15):4094–4097.
157. Wainfan E, Poirier LA. Methyl groups in carcinogenesis: Effects on DNA methylation and gene expression. *Cancer Res.* Apr 1 1992;52(7 Suppl):2071s–2077s.
158. Zapisek WF, Cronin GM, Lyn-Cook BD, Poirier LA. The onset of oncogene hypomethylation in the livers of rats fed methyl-deficient, amino acid-defined diets. *Carcinogenesis.* Oct 1992;13(10):1869–1872.
159. Pogribny IP, Miller BJ, James SJ. Alterations in hepatic p53 gene methylation patterns during tumor progression with folate/methyl deficiency in the rat. *Cancer Lett.* May 1 1997;115(1):31–38.
160. Pogribny IP, Poirier LA, James SJ. Differential sensitivity to loss of cytosine methyl groups within the hepatic p53 gene of folate/methyl deficient rats. *Carcinogenesis.* Nov 1995;16(11):2863–2867.
161. Wainfan E, Kilkenny M, Dizik M. Comparison of methyltransferase activities of pair-fed rats given adequate or methyl-deficient diets. *Carcinogenesis.* May 1988;9(5):861–863.
162. Duthie SJ, Narayanan S, Brand GM, Grant G. DNA stability and genomic methylation status in colonocytes isolated from methyl-donor-deficient rats. *Eur J Nutr.* Jun 2000;39(3):106–111.
163. Kim YI, Miller JW, da Costa KA et al. Severe folate deficiency causes secondary depletion of choline and phosphocholine in rat liver. *J Nutr.* Nov 1994;124(11):2197–2203.
164. Miller JW, Nadeau MR, Smith J, Smith D, Selhub J. Folate-deficiency-induced homocysteinaemia in rats: Disruption of S-adenosylmethionine's co-ordinate regulation of homocysteine metabolism. *Biochem J.* Mar 1 1994;298(Pt 2):415–419.
165. Balaghi M, Horne DW, Wagner C. Hepatic one-carbon metabolism in early folate deficiency in rats. *Biochem J.* Apr 1 1993;291(Pt 1):145–149.
166. Kim YI, Christman JK, Fleet JC et al. Moderate folate deficiency does not cause global hypomethylation of hepatic and colonic DNA or c-myc-specific hypomethylation of colonic DNA in rats. *Am J Clin Nutr.* May 1995;61(5):1083–1090.
167. Kim YI, Pogribny IP, Basnakian AG et al. Folate deficiency in rats induces DNA strand breaks and hypomethylation within the p53 tumor suppressor gene. *Am J Clin Nutr.* Jan 1997;65(1):46–52.
168. Balaghi M, Wagner C. DNA methylation in folate deficiency: Use of CpG methylase. *Biochem Biophys Res Commun.* Jun 30 1993;193(3):1184–1190.
169. Le Leu RK, Young GP, McIntosh GH. Folate deficiency diminishes the occurrence of aberrant crypt foci in the rat colon but does not alter global DNA methylation status. *J Gastroenterol Hepatol.* Oct 2000;15(10):1158–1164.
170. Le Leu RK, Young GP, McIntosh GH. Folate deficiency reduces the development of colorectal cancer in rats. *Carcinogenesis.* Dec 2000;21(12):2261–2265.
171. Davis CD, Uthus EO. Dietary folate and selenium affect dimethylhydrazine-induced aberrant crypt formation, global DNA methylation and one-carbon metabolism in rats. *J Nutr.* Sep 2003;133(9):2907–2914.

172. Kim YI, Pogribny IP, Salomon RN et al. Exon-specific DNA hypomethylation of the p53 gene of rat colon induced by dimethylhydrazine. Modulation by dietary folate. *Am J Pathol.* Oct 1996;149(4):1129–1137.
173. Sohn KJ, Stempak JM, Reid S, Shirwadkar S, Mason JB, Kim YI. The effect of dietary folate on genomic and p53-specific DNA methylation in rat colon. *Carcinogenesis.* Jan 2003;24(1):81–90.
174. Kim YI, Shirwadkar S, Choi SW, Puchyr M, Wang Y, Mason JB. Effects of dietary folate on DNA strand breaks within mutation-prone exons of the p53 gene in rat colon. *Gastroenterology.* Jul 2000;119(1):151–161.
175. Linhart HG, Troen A, Bell GW et al. Folate deficiency induces genomic uracil misincorporation and hypomethylation but does not increase DNA point mutations. *Gastroenterology.* Jan 2009;136(1):227–235.e223.
176. Chiuve SE, Giovannucci EL, Hankinson SE et al. The association between betaine and choline intakes and the plasma concentrations of homocysteine in women. *Am J Clin Nutr.* Oct 2007;86(4):1073–1081.
177. Jacob RA, Jenden DJ, Allman-Farinelli MA, Swendseid ME. Folate nutriture alters choline status of women and men fed low choline diets. *J Nutr.* Mar 1999;129(3):712–717.
178. Choi SW, Friso S, Dolnikowski GG et al. Biochemical and molecular aberrations in the rat colon due to folate depletion are age-specific. *J Nutr.* Apr 2003;133(4):1206–1212.
179. Keyes MK, Jang H, Mason JB et al. Older age and dietary folate are determinants of genomic and p16-specific DNA methylation in mouse colon. *J Nutr.* Jul 2007;137(7):1713–1717.
180. Ahuja N, Issa JP. Aging, methylation and cancer. *Histol Histopathol.* Jul 2000; 15(3):835–842.
181. Issa JP. CpG-island methylation in aging and cancer. *Curr Top Microbiol Immunol.* 2000;249:101–118.
182. Drinkwater RD, Blake TJ, Morley AA, Turner DR. Human lymphocytes aged in vivo have reduced levels of methylation in transcriptionally active and inactive DNA. *Mutat Res.* Jan 1989;219(1):29–37.
183. Mays-Hoopes L, Chao W, Butcher HC, Huang RC. Decreased methylation of the major mouse long interspersed repeated DNA during aging and in myeloma cells. *Dev Genet.* 1986;7(2):65–73.
184. Vanyushin BF, Mazin AL, Vasilyev VK, Belozersky AN. The content of 5-methylcytosine in animal DNA: The species and tissue specificity. *Biochim Biophys Acta.* Mar 28 1973;299(3):397–403.
185. Wilson VL, Smith RA, Ma S, Cutler RG. Genomic 5-methyldeoxycytidine decreases with age. *J Biol Chem.* Jul 25 1987;262(21):9948–9951.
186. Toyota M, Ahuja N, Ohe-Toyota M, Herman JG, Baylin SB, Issa JP. CpG island methylator phenotype in colorectal cancer. *Proc Natl Acad Sci U S A.* 1999;96(15):8681–8686.
187. Santini V, Kantarjian HM, Issa JP. Changes in DNA methylation in neoplasia: Pathophysiology and therapeutic implications. *Ann Intern Med.* Apr 3 2001;134(7):573–586.
188. Issa JP, Ottaviano YL, Celano P, Hamilton SR, Davidson NE, Baylin SB. Methylation of the oestrogen receptor CpG island links ageing and neoplasia in human colon. *Nat Genet.* Aug 1994;7(4):536–540.
189. Belinsky SA, Nikula KJ, Baylin SB, Issa JP. Increased cytosine DNA-methyltransferase activity is target-cell-specific and an early event in lung cancer. *Proc Natl Acad Sci U S A.* 1996;93(9):4045–4050.
190. Shiraishi M, Chuu YH, Sekiya T. Isolation of DNA fragments associated with methylated CpG islands in human adenocarcinomas of the lung using a methylated DNA binding column and denaturing gradient gel electrophoresis. *Proc Natl Acad Sci U S A.* Mar 16 1999;96(6):2913–2918.

191. Liang G, Salem CE, Yu MC et al. DNA methylation differences associated with tumor tissues identified by genome scanning analysis. *Genomics.* Nov 1 1998;53(3):260–268.
192. Huang RF, Ho YH, Lin HL, Wei JS, Liu TZ. Folate deficiency induces a cell cycle-specific apoptosis in HepG2 cells. *J Nutr.* Jan 1999;129(1):25–31.
193. So K, Tamura G, Honda T et al. Multiple tumor suppressor genes are increasingly methylated with age in non-neoplastic gastric epithelia. *Cancer Science.* Nov 2006;97(11):1155–1158.
194. Winawer SJ, Fletcher RH, Miller L et al. Colorectal cancer screening: Clinical guidelines and rationale. *Gastroenterology.* Feb 1997;112(2):594–642.
195. Ahuja N, Li Q, Mohan AL, Baylin SB, Issa JP. Aging and DNA methylation in colorectal mucosa and cancer. *Cancer Res.* Dec 1 1998;58(23):5489–5494.
196. Jacob RA, Gretz DM, Taylor PC et al. Moderate folate depletion increases plasma homocysteine and decreases lymphocyte DNA methylation in postmenopausal women. *J Nutr.* Jul 1998;128(7):1204–1212.
197. Rampersaud GC, Kauwell GP, Hutson AD, Cerda JJ, Bailey LB. Genomic DNA methylation decreases in response to moderate folate depletion in elderly women. *Am J Clin Nutr.* Oct 2000;72(4):998–1003.
198. Shelnutt KP, Kauwell GP, Gregory JF III et al. Methylenetetrahydrofolate reductase 677C→T polymorphism affects DNA methylation in response to controlled folate intake in young women. *J Nutr Biochem.* Sep 2004;15(9):554–560.
199. Abratte CM, Wang W, Li R, Axume J, Moriarty DJ, Caudill MA. Choline status is not a reliable indicator of moderate changes in dietary choline consumption in premenopausal women. *J Nutr Biochem.* Jan 2009;20(1):62–69.
200. Cravo M, Fidalgo P, Pereira AD et al. DNA methylation as an intermediate biomarker in colorectal cancer: Modulation by folic acid supplementation. *Eur J Cancer Prev.* Nov 1994;3(6):473–479.
201. Cravo ML, Pinto AG, Chaves P et al. Effect of folate supplementation on DNA methylation of rectal mucosa in patients with colonic adenomas: Correlation with nutrient intake. *Clin Nutr.* Apr 1998;17(2):45–49.
202. Kim YI, Baik HW, Fawaz K et al. Effects of folate supplementation on two provisional molecular markers of colon cancer: A prospective, randomized trial. *Am J Gastroenterol.* Jan 2001;96(1):184–195.
203. Cravo M, Gloria L, Salazar De Sousa L et al. Folate status, DNA methylation, and colon cancer risk in inflammatory bowel disease. *Clin Nutr.* 1995;14:50–53.
204. Fenech M, Aitken C, Rinaldi J. Folate, vitamin B12, homocysteine status and DNA damage in young Australian adults. *Carcinogenesis.* Jul 1998;19(7):1163–1171.
205. Ingrosso D, Cimmino A, Perna AF et al. Folate treatment and unbalanced methylation and changes of allelic expression induced by hyperhomocysteinaemia in patients with uraemia. *Lancet.* May 17 2003;361(9370):1693–1699.
206. Pufulete M, Al-Ghnaniem R, Khushal A et al. Effect of folic acid supplementation on genomic DNA methylation in patients with colorectal adenoma. *Gut.* May 2005;54(5):648–653.
207. Figueiredo JC, Grau MV, Wallace K et al. Global DNA hypomethylation (LINE-1) in the normal colon and lifestyle characteristics and dietary and genetic factors. *Cancer Epidemiol Biomarkers Prev.* Apr 2009;18(4):1041–1049.
208. Friso S, Choi SW, Girelli D et al. A common mutation in the 5,10-methylenetetrahydrofolate reductase gene affects genomic DNA methylation through an interaction with folate status. *Proc Natl Acad Sci U S A.* Apr 16 2002;99(8):5606–5611.
209. Ulrich CM, Toriola AT, Koepl LM et al. Metabolic, hormonal and immunological associations with global DNA methylation among postmenopausal women. *Epigenetics.* Sep 2012;7(9):1020–1028.

210. Pufulete M, Al-Ghnaniem R, Leather AJ et al. Folate status, genomic DNA hypomethylation, and risk of colorectal adenoma and cancer: A case control study. *Gastroenterology.* May 2003;124(5):1240–1248.
211. Pufulete M, Al-Ghnaniem R, Rennie JA et al. Influence of folate status on genomic DNA methylation in colonic mucosa of subjects without colorectal adenoma or cancer. *Br J Cancer.* Mar 14 2005;92(5):838–842.
212. Fowler BM, Giuliano AR, Piyathilake C, Nour M, Hatch K. Hypomethylation in cervical tissue: Is there a correlation with folate status? *Cancer Epidemiol Biomarkers Prev.* Oct 1998;7(10):901–906.
213. Schernhammer ES, Giovannucci E, Kawasaki T, Rosner B, Fuchs CS, Ogino S. Dietary folate, alcohol and B vitamins in relation to LINE-1 hypomethylation in colon cancer. *Gut.* Jun 2010;59(6):794–799.
214. van Engeland M, Weijenberg MP, Roemen GM et al. Effects of dietary folate and alcohol intake on promoter methylation in sporadic colorectal cancer: The Netherlands cohort study on diet and cancer. *Cancer Res.* Jun 15 2003;63(12):3133–3137.
215. de Vogel S, Bongaerts BW, Wouters KA et al. Associations of dietary methyl donor intake with MLH1 promoter hypermethylation and related molecular phenotypes in sporadic colorectal cancer. *Carcinogenesis.* Sep 2008;29(9):1765–1773.
216. Al-Ghnaniem R, Peters J, Foresti R, Heaton N, Pufulete M. Methylation of estrogen receptor alpha and mutL homolog 1 in normal colonic mucosa: Association with folate and vitamin B-12 status in subjects with and without colorectal neoplasia. *Am J Clin Nutr.* Oct 2007;86(4):1064–1072.
217. Al-Ghnaniem Abbadi R, Emery P, Pufulete M. Short-term folate supplementation in physiological doses has no effect on ESR1 and MLH1 methylation in colonic mucosa of individuals with adenoma. *J Nutrigenet Nutrigenomics.* 2012;5(6):327–338.
218. Arnold CN, Goel A, Compton C et al. Evaluation of microsatellite instability, hMLH1 expression and hMLH1 promoter hypermethylation in defining the MSI phenotype of colorectal cancer. *Cancer Biol Ther.* Jan 2004;3(1):73–78.
219. Deng G, Bell I, Crawley S et al. BRAF mutation is frequently present in sporadic colorectal cancer with methylated hMLH1, but not in hereditary nonpolyposis colorectal cancer. *Clin Cancer Res.* Jan 1 2004;10(1 Pt 1):191–195.
220. Wallace K, Grau MV, Levine AJ et al. Association between folate levels and CpG Island hypermethylation in normal colorectal mucosa. *Cancer Prev Res.* Dec 2010; 3(12):1552–1564.
221. Van Guelpen B, Dahlin AM, Hultdin J et al. One-carbon metabolism and CpG island methylator phenotype status in incident colorectal cancer: A nested case-referent study. *Cancer Causes Control.* Apr 2010;21(4):557–566.
222. Kawakami K, Ruszkiewicz A, Bennett G, Moore J, Watanabe G, Iacopetta B. The folate pool in colorectal cancers is associated with DNA hypermethylation and with a polymorphism in methylenetetrahydrofolate reductase. *Clin Cancer Res.* Dec 1 2003;9(16 Pt1): 5860–5865.
223. Piyathilake CJ, Celedonio JE, Macaluso M, Bell WC, Azrad M, Grizzle WE. Mandatory fortification with folic acid in the United States is associated with increased expression of DNA methyltransferase-1 in the cervix. *Nutrition.* Jan 2008;24(1):94–99.
224. Piyathilake CJ, Azrad M, Jhala D et al. Mandatory fortification with folic acid in the United States is not associated with changes in the degree or the pattern of global DNA methylation in cells involved in cervical carcinogenesis. *Cancer Biomark.* 2006;2(6):259–266.
225. Kim YI. Role of the MTHFR polymorphisms in cancer risk modification and treatment. *Future Oncol.* May 2009;5(4):523–542.
226. Goyette P, Sumner JS, Milos R et al. Human methylenetetrahydrofolate reductase: Isolation of cDNA, mapping and mutation identification. *Nat Genet.* Jun 1994;7(2):195–200.

227. Frosst P, Blom HJ, Milos R et al. A candidate genetic risk factor for vascular disease: A common mutation in methylenetetrahydrofolate reductase. *Nat Genet.* May 1995;10(1):111–113.
228. Stern LL, Mason JB, Selhub J, Choi SW. Genomic DNA hypomethylation, a characteristic of most cancers, is present in peripheral leukocytes of individuals who are homozygous for the C677T polymorphism in the methylenetetrahydrofolate reductase gene. *Cancer Epidemiol Biomarkers Prev.* 2000;9(8):849–853.
229. Castro R, Rivera I, Ravasco P et al. 5,10-methylenetetrahydrofolate reductase (MTHFR) 677C→T and 1298A→C mutations are associated with DNA hypomethylation. *J Med Genet.* Jun 2004;41(6):454–458.
230. McNulty H, McKinley MC, Wilson B et al. Impaired functioning of thermolabile methylenetetrahydrofolate reductase is dependent on riboflavin status: Implications for riboflavin requirements. *Am J Clin Nutr.* Aug 2002;76(2):436–441.
231. Jacques PF, Bostom AG, Williams RR et al. Relation between folate status, a common mutation in methylenetetrahydrofolate reductase, and plasma homocysteine concentrations. *Circulation.* Jan 1 1996;93(1):7–9.
232. Ma J, Stampfer MJ, Giovannucci E et al. Methylenetetrahydrofolate reductase polymorphism, dietary interactions, and risk of colorectal cancer. *Cancer Res.* Mar 15 1997;57(6):1098–1102.
233. Bailey LB. Folate, methyl-related nutrients, alcohol, and the MTHFR 677C→T polymorphism affect cancer risk: Intake recommendations. *J Nutr.* Nov 2003;133(11 Suppl 1): 3748S–3753S.
234. van der Put NM, Gabreels F, Stevens EM et al. A second common mutation in the methylenetetrahydrofolate reductase gene: An additional risk factor for neural-tube defects? *Am J Hum Genet.* May 1998;62(5):1044–1051.
235. Weisberg I, Tran P, Christensen B, Sibani S, Rozen R. A second genetic polymorphism in methylenetetrahydrofolate reductase (MTHFR) associated with decreased enzyme activity. *Mol Genet Metab.* Jul 1998;64(3):169–172.
236. Robien K, Ulrich CM. 5,10-Methylenetetrahydrofolate reductase polymorphisms and leukemia risk: A HuGE minireview. *Am J Epidemiol.* Apr 1 2003;157(7):571–582.
237. Friedman G, Goldschmidt N, Friedlander Y et al. A common mutation A1298C in human methylenetetrahydrofolate reductase gene: Association with plasma total homocysteine and folate concentrations. *J Nutr.* Sep 1999;129(9):1656–1661.
238. Lievers KJ, Boers GH, Verhoef P et al. A second common variant in the methylenetetrahydrofolate reductase (MTHFR) gene and its relationship to MTHFR enzyme activity, homocysteine, and cardiovascular disease risk. *J Mol Med.* Sep 2001;79(9):522–528.
239. Ulrich CM, Robien K, Sparks R. Pharmacogenetics and folate metabolism—A promising direction. *Pharmacogenomics.* May 2002;3(3):299–313.
240. Friso S, Girelli D, Trabetti E et al. The MTHFR 1298A>C polymorphism and genomic DNA methylation in human lymphocytes. *Cancer Epidemiol Biomarkers Prev.* Apr 2005;14(4):938–943.
241. Kim YI. 5,10-Methylenetetrahydrofolate reductase polymorphisms and pharmacogenetics: A new role of single nucleotide polymorphisms in the folate metabolic pathway in human health and disease. *Nutr Rev.* Nov 2005;63(11):398–407.
242. Brockton NT. Localized depletion: The key to colorectal cancer risk mediated by MTHFR genotype and folate? *Cancer Causes Control.* Oct 2006;17(8):1005–1016.
243. Bagley PJ, Selhub J. A common mutation in the methylenetetrahydrofolate reductase gene is associated with an accumulation of formylated tetrahydrofolates in red blood cells. *Proc Natl Acad Sci U S A.* Oct 27 1998;95(22):13217–13220.
244. Guenther BD, Sheppard CA, Tran P, Rozen R, Matthews RG, Ludwig ML. The structure and properties of methylenetetrahydrofolate reductase from Escherichia coli suggest how folate ameliorates human hyperhomocysteinemia. *Nat Struct Biol.* Apr 1999;6(4):359–365.

245. Yamada K, Chen Z, Rozen R, Matthews RG. Effects of common polymorphisms on the properties of recombinant human methylenetetrahydrofolate reductase. *Proc Natl Acad Sci U S A.* Dec 18 2001;98(26):14853–14858.
246. Christensen B, Frosst P, Lussier-Cacan S et al. Correlation of a common mutation in the methylenetetrahydrofolate reductase gene with plasma homocysteine in patients with premature coronary artery disease. *Arterioscler Thromb Vasc Biol.* Mar 1997;17(3):569–573.
247. Girelli D, Friso S, Trabetti E et al. Methylenetetrahydrofolate reductase C677T mutation, plasma homocysteine, and folate in subjects from northern Italy with or without angiographically documented severe coronary atherosclerotic disease: Evidence for an important genetic-environmental interaction. *Blood.* Jun 1 1998;91(11):4158–4163.
248. Verhoef P, Kok FJ, Kluijtmans LA et al. The 677C→T mutation in the methylenetetrahydrofolate reductase gene: Associations with plasma total homocysteine levels and risk of coronary atherosclerotic disease. *Atherosclerosis.* Jul 11 1997;132(1):105–113.
249. Narayanan S, McConnell J, Little J et al. Associations between two common variants C677T and A1298C in the methylenetetrahydrofolate reductase gene and measures of folate metabolism and DNA stability (strand breaks, misincorporated uracil, and DNA methylation status) in human lymphocytes in vivo. *Cancer Epidemiol Biomarkers Prev.* Sep 2004;13(9):1436–1443.
250. Paz MF, Avila S, Fraga MF et al. Germ-line variants in methyl-group metabolism genes and susceptibility to DNA methylation in normal tissues and human primary tumors. *Cancer Res.* Aug 1 2002;62(15):4519–4524.
251. Paz MF, Fraga MF, Avila S et al. A systematic profile of DNA methylation in human cancer cell lines. *Cancer Res.* Mar 1 2003;63(5):1114–1121.
252. Kawakami K, Ruszkiewicz A, Bennett G et al. DNA hypermethylation in the normal colonic mucosa of patients with colorectal cancer. *Br J Cancer.* Feb 27 2006; 94(4):593–598.
253. Chen Z, Karaplis AC, Ackerman SL et al. Mice deficient in methylenetetrahydrofolate reductase exhibit hyperhomocysteinemia and decreased methylation capacity, with neuropathology and aortic lipid deposition. *Hum Mol Genet.* Mar 1 2001;10(5):433–443.
254. Devlin AM, Arning E, Bottiglieri T, Faraci FM, Rozen R, Lentz SR. Effect of Mthfr genotype on diet-induced hyperhomocysteinemia and vascular function in mice. *Blood.* Apr 1 2004;103(7):2624–2629.
255. Sohn KJ, Jang H, Campan M et al. The methylenetetrahydrofolate reductase C677T mutation induces cell-specific changes in genomic DNA methylation and uracil misincorporation: A possible molecular basis for the site-specific cancer risk modification. *Int J Cancer.* May 1 2009;124(9):1999–2005.
256. van den Donk M, Pellis L, Crott JW et al. Folic acid and vitamin B-12 supplementation does not favorably influence uracil incorporation and promoter methylation in rectal mucosa DNA of subjects with previous colorectal adenomas. *J Nutr.* Sep 2007;137(9):2114–2120.
257. van den Donk M, van Engeland M, Pellis L et al. Dietary folate intake in combination with MTHFR C677T genotype and promoter methylation of tumor suppressor and DNA repair genes in sporadic colorectal adenomas. *Cancer Epidemiol Biomarkers Prev.* Feb 2007;16(2):327–333.
258. Slattery ML, Curtin K, Sweeney C et al. Diet and lifestyle factor associations with CpG island methylator phenotype and BRAF mutations in colon cancer. *Int J Cancer.* Feb 1 2007;120(3):656–663.
259. Curtin K, Slattery ML, Ulrich CM et al. Genetic polymorphisms in one-carbon metabolism: Associations with CpG island methylator phenotype (CIMP) in colon cancer and the modifying effects of diet. *Carcinogenesis.* Aug 2007;28(8):1672–1679.

260. Mokarram P, Naghibalhossaini F, Saberi Firoozi M et al. Methylenetetrahydrofolate reductase C677T genotype affects promoter methylation of tumor-specific genes in sporadic colorectal cancer through an interaction with folate/vitamin B12 status. *World J Gastroenterol.* Jun 21 2008;14(23):3662–3671.
261. Nafee TM, Farrell WE, Carroll WD, Fryer AA, Ismail KM. Epigenetic control of fetal gene expression. *BJOG.* Jan 2008;115(2):158–168.
262. Waterland RA, Jirtle RL. Transposable elements: Targets for early nutritional effects on epigenetic gene regulation. *Mol Cell Biol.* Aug 2003;23(15):5293–5300.
263. Wolff GL, Kodell RL, Moore SR, Cooney CA. Maternal epigenetics and methyl supplements affect agouti gene expression in Avy/a mice. *FASEB J.* Aug 1998;12(11):949–957.
264. Yen TT, Gill AM, Frigeri LG, Barsh GS, Wolff GL. Obesity, diabetes, and neoplasia in yellow A(vy)/– mice: Ectopic expression of the agouti gene. *FASEB J.* May 1994;8(8):479–488.
265. Wolff GL, Roberts DW, Morrissey RL et al. Tumorigenic responses to lindane in mice: Potentiation by a dominant mutation. *Carcinogenesis.* Dec 1987;8(12):1889–1897.
266. Waterland RA, Dolinoy DC, Lin JR, Smith CA, Shi X, Tahiliani KG. Maternal methyl supplements increase offspring DNA methylation at Axin Fused. *Genesis.* Sep 2006; 44(9):401–406.
267. Sie KK, Medline A, van Weel J et al. Effect of maternal and postweaning folic acid supplementation on colorectal cancer risk in the offspring. *Gut.* Dec 2011;60(12):1687–1694.
268. Ciappio ED, Liu Z, Brooks RS, Mason JB, Bronson RT, Crott JW. Maternal B vitamin supplementation from preconception through weaning suppresses intestinal tumorigenesis in Apc1638N mouse offspring. *Gut.* Dec 2011;60(12):1695–1702.
269. McKay JA, Williams EA, Mathers JC. Gender-specific modulation of tumorigenesis by folic acid supply in the Apc mouse during early neonatal life. *Br J Nutr.* Mar 2008; 99(3):550–558.
270. Ly A, Lee H, Chen J et al. Effect of maternal and postweaning folic acid supplementation on mammary tumor risk in the offspring. *Cancer Res.* Feb 1 2011;71(3):988–997.
271. Sie KK, Li J, Ly A, Sohn KJ, Croxford R, Kim YI. Effect of maternal and postweaning folic acid supplementation on global and gene-specific DNA methylation in the liver of the rat offspring. *Mol Nutr Food Res.* Apr 2013;57(4):677–685.
272. Fearon ER. Molecular genetics of colorectal cancer. *Annu Rev Pathol.* 2011;6:479–507.
273. Sowers R, Toguchida J, Qin J et al. mRNA expression levels of E2F transcription factors correlate with dihydrofolate reductase, reduced folate carrier, and thymidylate synthase mRNA expression in osteosarcoma. *Mol Cancer Ther.* Jun 2003;2(6):535–541.
274. Lillycrop KA, Phillips ES, Jackson AA, Hanson MA, Burdge GC. Dietary protein restriction of pregnant rats induces and folic acid supplementation prevents epigenetic modification of hepatic gene expression in the offspring. *J Nutr.* Jun 2005;135(6):1382–1386.
275. Lillycrop KA, Slater-Jefferies JL, Hanson MA, Godfrey KM, Jackson AA, Burdge GC. Induction of altered epigenetic regulation of the hepatic glucocorticoid receptor in the offspring of rats fed a protein-restricted diet during pregnancy suggests that reduced DNA methyltransferase-1 expression is involved in impaired DNA methylation and changes in histone modifications. *Br J Nutr.* Jun 2007;97(6):1064–1073.
276. Finnell RH, Spiegelstein O, Wlodarczyk B et al. DNA methylation in Folbp1 knockout mice supplemented with folic acid during gestation. *J Nutr.* Aug 2002;132(8 Suppl):2457S–2461S.
277. McKay JA, Waltham KJ, Williams EA, Mathers JC. Folate depletion during pregnancy and lactation reduces genomic DNA methylation in murine adult offspring. *Genes Nutr.* May 2011;6(2):189–196.
278. McKay JA, Wong YK, Relton CL, Ford D, Mathers JC. Maternal folate supply and sex influence gene-specific DNA methylation in the fetal gut. *Mol Nutr Food Res.* Nov 2011;55(11):1717–1723.

279. Lawrance AK, Deng L, Rozen R. Methylenetetrahydrofolate reductase deficiency and low dietary folate reduce tumorigenesis in Apc min/+ mice. *Gut.* Jun 2009;58(6):805–811.
280. Maloney CA, Hay SM, Rees WD. Folate deficiency during pregnancy impacts on methyl metabolism without affecting global DNA methylation in the rat fetus. *Br J Nutr.* Jun 2007;97(6):1090–1098.
281. McKay JA, Williams EA, Mathers JC. Effect of maternal and post-weaning folate supply on gene-specific DNA methylation in the small intestine of weaning and adult apc and wild type mice. *Front Genet.* 2011;2:23.
282. Tsang V, Fry RC, Niculescu MD et al. The epigenetic effects of a high prenatal folate intake in male mouse fetuses exposed in utero to arsenic. *Toxicol Appl Pharmacol.* Nov 1 2012;264(3):439–450.
283. Klaus A, Birchmeier W. Wnt signalling and its impact on development and cancer. *Nat Rev Cancer.* May 2008;8(5):387–398.
284. Bunin GR, Gallagher PR, Rorke-Adams LB, Robison LL, Cnaan A. Maternal supplement, micronutrient, and cured meat intake during pregnancy and risk of medulloblastoma during childhood: A Children's Oncology Group Study. *Cancer Epidemiol Biomarkers Prev.* Sep 2006;15(9):1660–1667.
285. Bunin GR, Kuijten RR, Buckley JD, Rorke LB, Meadows AT. Relation between maternal diet and subsequent primitive neuroectodermal brain tumors in young children. *N Engl J Med.* Aug 19 1993;329(8):536–541.
286. Schuz J, Weihkopf T, Kaatsch P. Medication use during pregnancy and the risk of childhood cancer in the offspring. *Eur J Pediatr.* May 2007;166(5):433–441.
287. Thompson JR, Gerald PF, Willoughby ML, Armstrong BK. Maternal folate supplementation in pregnancy and protection against acute lymphoblastic leukaemia in childhood: A case-control study. *Lancet.* Dec 8 2001;358(9297):1935–1940.
288. Dockerty JD, Herbison P, Skegg DC, Elwood M. Vitamin and mineral supplements in pregnancy and the risk of childhood acute lymphoblastic leukaemia: A case-control study. *BMC Public Health.* 2007;7:136.
289. Milne E, Royle JA, Miller M et al. Maternal folate and other vitamin supplementation during pregnancy and risk of acute lymphoblastic leukemia in the offspring. *Int J Cancer.* Jun 1 2010;126(11):2690–2699.
290. Goh YI, Bollano E, Einarson TR, Koren G. Prenatal multivitamin supplementation and rates of pediatric cancers: A meta-analysis. *Clin Pharmacol Ther.* May 2007; 81(5):685–691.
291. Fryer AA, Nafee TM, Ismail KM, Carroll WD, Emes RD, Farrell WE. LINE-1 DNA methylation is inversely correlated with cord plasma homocysteine in man: A preliminary study. *Epigenetics.* Aug 16 2009;4(6):394–398.
292. Fryer AA, Emes RD, Ismail KM et al. Quantitative, high-resolution epigenetic profiling of CpG loci identifies associations with cord blood plasma homocysteine and birth weight in humans. *Epigenetics.* Jan 2011;6(1):86–94.
293. Chang H, Zhang T, Zhang Z et al. Tissue-specific distribution of aberrant DNA methylation associated with maternal low-folate status in human neural tube defects. *J Nutr Biochem.* Dec 2011;22(12):1172–1177.
294. Cheng YW, Idrees K, Shattock R et al. Loss of imprinting and marked gene elevation are 2 forms of aberrant IGF2 expression in colorectal cancer. *Int J Cancer.* Aug 1 2010;127(3):568–577.
295. Haggarty P, Hoad G, Campbell DM, Horgan GW, Piyathilake C, McNeill G. Folate in pregnancy and imprinted gene and repeat element methylation in the offspring. *Am J Clin Nutr.* Jan 2013;97(1):94–99.
296. Steegers-Theunissen RP, Obermann-Borst SA, Kremer D et al. Periconceptional maternal folic acid use of 400 microg per day is related to increased methylation of the IGF2 gene in the very young child. *PLoS One.* 2009;4(11):e7845.

297. Hoyo C, Murtha AP, Schildkraut JM et al. Methylation variation at IGF2 differentially methylated regions and maternal folic acid use before and during pregnancy. *Epigenetics.* Jul 2011;6(7):928–936.
298. Ba Y, Yu H, Liu F et al. Relationship of folate, vitamin B12 and methylation of insulin-like growth factor-II in maternal and cord blood. *Eur J Clin Nutr.* Apr 2011;65(4): 480–485.
299. McKay JA, Groom A, Potter C et al. Genetic and non-genetic influences during pregnancy on infant global and site specific DNA methylation: Role for folate gene variants and vitamin B12. *PLoS One.* 2012;7(3):e33290.
300. Otsuka S, Maegawa S, Takamura A et al. Aberrant promoter methylation and expression of the imprinted PEG3 gene in glioma. *Proc Jpn Acad Ser B Phys Biol Sci.* 2009;85(4):157–165.
301. Curley JP, Mashoodh R, Champagne FA. Epigenetics and the origins of paternal effects. *Horm Behav.* Mar 2011;59(3):306–314.
302. DelCurto H, Wu G, Satterfield MC. Nutrition and reproduction: Links to epigenetics and metabolic syndrome in offspring. *Curr Opin Clin Nutr Metab Care.* Jul 2013; 16(4):385–391.
303. Soubry A, Schildkraut JM, Murtha A et al. Paternal obesity is associated with IGF2 hypomethylation in newborns: Results from a Newborn Epigenetics Study (NEST) cohort. *BMC Med.* 2013;11:29.
304. Kim HW, Kim KN, Choi YJ, Chang N. Effects of paternal folate deficiency on the expression of insulin-like growth factor-2 and global DNA methylation in the fetal brain. *Mol Nutr Food Res.* Apr 2013;57(4):671–676.
305. Cropley JE, Suter CM, Beckman KB, Martin DIK. Germ-line epigenetic modification of the murine Avy allele by nutritional supplementation. *Proc Natl Acad Sci U S A.* 2006;103:17308–17312.
306. Waterland RA, Travisano M, Tahiliani KG. Diet-induced hypermethylation at agouti viable yellow is not inherited transgenerationally through the female. *FASEB J.* Oct 2007;21(12):3380–3385.
307. Fisher AG, Brockdorff N. Epigenetic memory and parliamentary privilege combine to evoke discussions on inheritance. *Development.* Nov 2012;139(21):3891–3896.

5 Parental Nutrition, Epigenetics, and Chronic Disease

Julia A. Sabet, Eric D. Ciappio, and Jimmy W. Crott

CONTENTS

5.1 INTRODUCTION

Since the publication of the landmark work of Barker et al. [1], the potential for early life exposures to impact on later disease risk has captured the imagination of researchers. In this early work, low birth weight was associated with increased risk for ischemic heart disease in later life. Since then, very elegant studies of the Overkalix parish in Northern Sweden have demonstrated that nutrition during key periods of development of both parents is associated with mortality, heart disease, and diabetes in children and even grandchildren [2–4].

It is thought that alterations of the epigenome are a key mechanistic route by which maternal exposures impact on offspring health. Because DNA methylation patterns are disrupted in a variety of cancers [5], researchers have begun to question whether an individual's own risk for this cancer might be, to some extent, programmed by exposures to their parents. Because one carbon nutrients including the B vitamins and choline provide the methyl groups required for DNA methylation and because these nutrients have been associated with one's own cancer risk, several studies have focused on these nutrients. In this chapter, we have attempted to summarize recent studies investigating maternal methyl donor intake and risk for tumorigenesis in offspring. In addition, we discuss the potential importance for paternal exposures in dictating offspring disease risk, and although there are no data specifically for cancer, we discuss work that provides proof of principle that paternal nutrition can impact on the epigenome of offspring.

5.2 MATERNAL METHYL DONORS AND OFFSPRING TUMORIGENESIS

Epidemiological studies demonstrate that maternal intake of one-carbon nutrients is associated with the risk of tumorigenesis in offspring, particularly among cancers typically observed in children. For instance, the incidence of acute lymphocytic leukemia (ALL) was lower among children born to mothers who used folic acid–containing multivitamins periconceptionally in most [6–8] but not all case control studies [9]. Three additional case control studies report that maternal use of multivitamins containing folic acid was associated with a significantly reduced risk of pediatric brain tumors in offspring [10–12], while three others reported modest reductions in risk that failed to attain statistical significance [13–15]. Maternal folate intake and maternal folic acid supplement use has been associated with a decreased risk of both retinoblastoma [16–19] and non-Hodgkin lymphoma [20]. Ecologic studies examining the impact of folic acid fortification on childhood cancer incidence indicate that both Wilms tumor [21] and neuroblastoma [22] declined significantly following the introduction of mandatory folic acid fortification in Canada, and in the United States, similar reductions were observed again for Wilms tumor, as well as primitive neuroectodermal tumors in the first years immediately following fortification [23].

Recent meta-analyses suggest that the potential protective effect bestowed upon children born to mothers consuming multivitamins periconceptionally may be substantial [24]. Among children born to mothers supplementing with multivitamins containing folic acid, the odds of having pediatric brain tumors were approximately 30% lower (OR = 0.73; 95% CI = 0.60–0.88), the odds of having ALL were roughly 40% lower (OR = 0.61; 95% CI = 0.50–0.74), and the odds of having neuroblastoma were nearly 50% lower (OR = 0.53; 95% CI = 0.42–0.68) as compared to children born to mothers not using these vitamins. It is important to remember, that many of these studies examined supplements in which folate is only one of many nutrients present; therefore, a role for these additional nutrients in modulating carcinogenesis, either alone or in concert with folate, in offspring cannot be excluded. Observations that polymorphisms in genes involved in folate metabolism such as methylene tetrahydrofolate reductase (MTHFR) [25] and methionine synthase [26] can modulate the risk of developing ALL in adulthood, as well as the increasing appreciation for the role of DNA methylation in the pathogenesis of childhood cancers [27,28] lend support to the notion that alterations in one-carbon metabolism are an important major driver behind the apparent chemopreventive effect of maternal multivitamin intake. MTHFR and MS are central enzymes in one-carbon metabolism; the former regulates the partitioning of the folate pool between that available for DNA methylation and that available for thymidine synthesis while the latter transfers a methyl group from 5-methy tetrahydrofolate to homocysteine where it will subsequently be available for DNA (and other) methylation reactions.

To date, associations between maternal one-carbon intake and cancer in offspring have been limited to cancers that affect children. Epidemiological evidence is not yet available to support or refute the idea that maternal one-carbon nutrient intake can impact the risk of developing cancers that typically develop in adulthood, such as sporadic cancers of the colorectum or breast. This is likely due to the relatively advanced age at which these tumors typically present, and because suitable databases that attempt to link maternal diet to diseases in the latter decades of the offspring's adulthood are not yet available. Nevertheless, the majority of preclinical investigations into the relationship between maternal one-carbon nutrient intake and offspring tumorigenesis have focused on colorectal and breast cancers, as tumorigenesis in both of these sites has repeatedly been shown to be sensitive to one-carbon nutrient status.

Rodent studies investigating the impact of maternal folate supplementation on offspring tumorigenesis have generated mixed results: some studies report a protective effect of supplementation on tumorigenesis [29,30], whereas others observed no effect [31] or even a promotion of tumorigenesis in offspring with maternal folate [32]. Of the studies where a protection against tumorigenesis was noted, it appears that the effect is stronger when the related one-carbon nutrients, vitamins B_2, B_6, and B_{12} are provided with folate, presumably because they act as cofactors for enzymes responsible for the interconversion of folate between its several functional forms [29]. In addition, our own work has demonstrated that, despite an unchanged tumor incidence, a significantly higher proportion of intestinal tumors that developed in pups born to dams fed a diet mildly deficient in folate and vitamins B_2, B_6, and B_{12} were invasive relative to tumors from pups born to replete dams [29]. Importantly, these studies utilized both chemically induced and genetically predisposed models, each with a differing tumor burden and latency. Along with differences in experimental design, the relative propensity of the model toward tumorigenesis may account for the disparate results.

Several investigations into the relationship between maternal one-carbon nutrients and offspring tumorigenesis in the breast have also been performed using animal models. One group reported that providing supplemental folate to mothers from preconception through weaning led to alterations in mammary gland morphology consistent with decreased breast tumor risk, namely an increased prevalence of alveolar buds and a decrease in spontaneously occurring terminal end buds [33]. A second study by the same group however, showed that when a carcinogen is administered to the offspring, maternal supplementation may promote breast tumorigenesis [34]. Another group reported that providing supplemental choline to dams during a specific period of gestation resulted in a reduction in the growth rate of breast tumors in the offspring following administration of the carcinogen DMBA. Furthermore, feeding a choline-deficient diet during gestation resulted in a significant increase in the growth rate of offspring tumors [35]. Most recently, providing supplemental levels of several one-carbon nutrients (folate, methionine, choline, vitamin B_{12}) during either gestation and/or in early adolescence reduced the multiplicity of breast tumors in offspring following treatment with the carcinogen *N*-nitroso-*N*-methylurea (NMU) [36].

In both the intestine and breast, it appears that both the severity of the model used and the length of the dietary intervention are important factors that can modify the tumorigenic response of the offspring to the maternal diet. When we consider studies with dietary interventions limited to the periconceptional and suckling periods, it is apparent that folate acts in a protective fashion, provided that a model with a relatively mild tumor burden, such as AOM [30] and *Apc*$^{+/1638N}$ models [29], is used. In contrast, if dietary interventions continue into the offspring's adolescence, thus exposing developing lesions to varied folate concentrations, there may be an opportunity for the tumor-promoting effect of folate to be expressed. This may explain the findings of Lawrance et al. [32], who reported that it was the offspring of deficient rather than supplemented dams that displayed the lowest tumor burden. Because the relatively severe Apc$^{+/min}$ model was used and the dietary intervention was continued through the life of the offspring, the tumor suppressing capacity of maternal high folate was likely overshadowed by the tumor-promoting effect of high folate fueling nucleotide synthesis in the tumor-prone offspring. The same interpretation may be applicable to mouse studies of breast cancer. For example, in two studies from the same group maternal folate supplementation was protective against spontaneously occurring precursor lesions when there was no driver of tumorigenesis [33], but when the potent carcinogen DMBA was used, maternal supplementation, especially when continued postweaning, significantly accelerated tumorigenesis [34]. In contrast to folate, maternal choline supplementation appears to be protective, even in these strongly tumorigenic models [35,36]. Importantly, unlike folate, choline is a one-carbon nutrient that provides methyl groups for the remethylation of homocysteine to methionine, but not for nucleotide synthesis. It is the latter process that is thought to mediate the cancer-promoting effect, and the inability of choline to support nucleotide synthesis may explain why choline does not appear to promote carcinogenesis in settings where folate might.

One-carbon nutrients play a major mechanistic role in the biological methylation process, providing methyl groups for the regeneration of *S*-adenosylmethionine (SAM), which ultimately serve to methylate DNA. Alterations in the intake of these one-carbon nutrients can therefore impact the degree of biological methylation at both the genome-wide and gene-specific levels, which can affect gene expression and ultimately the phenotype of the organism. During development, the embryonic genome experiences tremendous shifts in DNA methylation and is particularly sensitive to alterations in the availability of one-carbon units. These observations have led to the predominant view in the field that the mechanism by which maternal one-carbon nutrient intake affects offspring tumorigenesis is by affecting patterns of DNA methylation.

For reasons that remain to be determined, the methylation status of insulin-like growth factor 2 (*Igf2*) has repeatedly been observed to be responsive to changes in maternal nutrition. *Igf2* is an imprinted gene for which the maternal allele is silenced by DNA methylation and only the paternal allele is expressed. The methylation density of the imprinted (methylated) maternal allele can diminish resulting in biallelic expression of the gene, through an appropriately named process known as loss of imprinting. Loss of *Igf2* imprinting has been implicated in several conditions, including cancers of the prostate and colorectum in both animal models and humans [37,38]. A prominent example that the methylation of this gene may be sensitive to maternal diet during development comes from a cohort of Dutch citizens who were exposed to famine in utero during World War II. Remarkably, those born to mothers exposed to the famine during periconception had significantly lower methylation of the *Igf2* "differentially methylated region" compared with their same-sex siblings who were not exposed to famine in utero—an effect that persisted some 60 years after the exposure [39,40].

Accumulating evidence supports the concept that maternal one-carbon nutrient intake in particular has an impact on offspring *Igf2* methylation in both humans and animal models. Preclinical studies in rodents have reported that inadequate intakes of one-carbon nutrients such as folate, vitamin B_{12}, methionine, and choline during early postweaning life induce changes in *Igf2* methylation and expression that can persist into adulthood [41], and similar effects of maternal dietary choline depletion on offspring *Igf2* methylation/expression were reported [42]. In human studies, maternal vitamin B_{12} status has been reported to have an inverse relationship with *Igf2* methylation in cord blood [43]. Furthermore, maternal folic acid supplementation has been associated with both an increase [44,45] and a decrease [46] in *Igf2* methylation in the offspring. Data from a recent randomized controlled trial supports the latter finding, wherein women in Gambia were provided with supplemental folate (along with several other nutrients) during pregnancy, and *Igf2* methylation in cord blood was shown to be significantly lower relative to control [47].

Perhaps the most widely recognized demonstration of the potential for maternal exposure to impact on DNA methylation and phenotype in offspring comes from studies with the agouti viable yellow (A^{vy}/a) mouse. In this mouse, feeding dams a diet with supplemented with the methyl donor nutrients folate, vitamin B_{12}, betaine, and choline during gestation shifted the distribution of pups from predominately having a yellow (agouti) coat color to having a brown in color (pseudo-agouti) [48,49]. This increase in the proportion of pups with brown coat color coincided with an elevated methylation of specific sequences within the A^{vy} promoter [49,50]. Furthermore, mice with the agouti phenotype have an elevated propensity toward adult-onset obesity, hypertension, and insulin resistance compared with mice with their pseudo-agouti siblings [51]. In addition, Wolff and colleagues demonstrated that the yellow mice were more susceptible to hepatic tumorigenesis following exposure to phenobarbital [52] and lindane [53] compared with their darker colored littermates.

Similar to the A^{vy}/a model, the potential for maternal methyl group consumption to alter the phenotype of the offspring was redemonstrated in the *Axin Fused* ($Axin^{Fu}$), or "kinky-tail" mouse. $Axin^{Fu}$ dams supplemented with the one-carbon nutrients folate, vitamin B_{12}, choline, and betaine during gestation suppresses the kinky-tail phenotype, which is evident in the offspring of control fed dams [54]. Furthermore, the presence of tail kinks was inversely related to the methylation density of a retrotransposon within exon 6 of the *Axin* gene. Taken together, studies with the A^{vy}/a and $Axin^{Fu}$ provide proof of concept that one-carbon nutrients in the maternal diet can impact DNA methylation at the gene-specific level, and that the ensuing change in gene expression can have significant effects on offspring phenotype and disease susceptibility.

One of the cell signaling pathways relevant to carcinogenesis that appears to be sensitive to methyl donor availability is the Wnt signaling pathway. This pathway is involved in the growth and development of several tissues, including that of the gastrointestinal mucosa [55]. Aberrant Wnt signaling is frequently implicated in colorectal carcinogenesis, as mutations in the Wnt pathway are observed in more than 80% of sporadic colorectal tumors [56]. The loss of the Wnt pathway gene

Adenomatous polyposis coli (*Apc*), for example, is considered to be a major initiating event in the development of colorectal tumors [56,57]. Silencing due to promoter methylation has been documented during colorectal carcinogenesis for negative regulators of Wnt signaling such as *Apc* [58], Axis inhibition protein 2 (*Axin2*), Dickkopf-related protein 1 (*Dkk1*), the Secreted frizzled-related protein (*Sfrp*) family (*Sfrp1*, *2*, *4*, and *5*), and Wnt-inhibitory factor 1 (*Wif1*), among others [59,60]. Our own investigations have demonstrated that this pathway is aberrantly activated during multiple B-vitamin deficiency, resulting in increased intestinal tumorigenesis in mice [61,62]. In our studies on the impact of maternal nutrition on intestinal tumorigenesis in offspring we again focused on this pathway and observed a stepwise reduction in *Apc*, *Sfrp1*, *Wif1*, and *Wnt5a* tumor suppressor expression in the intestinal mucosa and, that the expression of *Sfrp1* was inversely related methylation of its promoter region [29].

Further evidence supporting the role of gene-specific methylation in modulating the impact of the maternal diet on offspring cancer risk comes from Kovacheva et al. [35]. These investigators examined the impact of maternal choline intake on breast tumorigenesis in offspring and identified several genes that were differentially expressed in the tumors of offspring born to dams consuming different quantities of choline during gestation. Among these genes, the expression of Stratifin (*Sfn*) in offspring, a gene frequently silenced by methylation in breast carcinogenesis [63], was inversely related to maternal choline intake as well as to the methylation of its promoter region. It is worth nothing that the expression of *Sfn*, like the aforementioned Wnt signaling genes, is regulated by methylation in human cells as well [64], and it may be that these genes are susceptible to these same effects in humans as well. In sum, the data from animal models supports the hypothesis that maternal one-carbon nutrient intake alters offspring cancer risk by modifying the methylation density of key genes.

In addition to alterations in gene-specific methylation, global cytosine methylation may have an important mechanistic role in modifying the relationship between the maternal diet and offspring carcinogenesis. Genomic hypomethylation is a consistent and early event in colorectal carcinogenesis [65,66] and may result from folate depletion [67]. In humans, both folic acid [45] and choline [68] supplementation during pregnancy has been associated with a decrease in LINE-1 methylation, a surrogate marker for genomic methylation. In animal models, maternal folic acid supplementation has been shown to increase [30], decrease [34,69], or have no effect [29] on offspring genomic methylation while maternal deficiency decreased genomic methylation in the intestine of offspring [29,32,70]. That folate depletion might induce genomic hypomethylation in offspring fits well with the role of folate in supplying methyl groups for DNA methylation. The data for supplementation are less consistent with both elevations and reductions in methylation reported. Deciphering the underlying mechanisms for these changes as well as the functional relevance will require further work.

Although changes in cytosine methylation, especially at the gene level, are undoubtedly an important mediator of disease risk in offspring, additional epigenetic processes likely play a role. There is a functional interdependence between epigenetic processes such as DNA methylation, histone modifications, and chromatin structure [71,72], which serves to regulate cellular transcription. For instance, methylated DNA serves as a binding site for the transcriptional repressor methyl CpG binding protein 2 (Mecp2), which then recruits histone deacetylase (Hdac) proteins that remove acetyl groups from the histone tails thereby facilitating compaction of the chromatin fiber [73]. Indeed, dietary intake of one-carbon nutrients has been shown to impact histone modifications in both the adult mouse [74] and in offspring when consumed in the maternal diet [65,75]. Supplemental intakes of the one-carbon nutrients folate, vitamin B_{12}, choline, and methionine in the maternal diet have also been shown to affect the expression of several genes related to epigenetic processes in offspring, such as *Mecp2* and *Hdac1* [36]. There is also evidence to suggest that histone modifications in the offspring may be sensitive to maternal dietary perturbations, as recent data have demonstrated that maternal choline supplementation during the third trimester resulted in increased H3K9 methylation in the placenta [66]. Considering this interactive relationship between epigenetic processes, it is reasonable to assume that modifications to histones and changes to chromatin dynamics, in addition to DNA methylation, play a functional role in the relationship between the maternal diet and tumorigenic potential in offspring.

5.3 PATERNAL NUTRITION AND OFFSPRING HEALTH

In addition to the increasingly well-documented contribution of maternal nutrition to determining the risk for chronic disease in offspring, we and others are beginning to question whether paternal exposures may also be of importance in this regard. We reason that because spermatogenesis is a continuously occurring process from puberty to death, as opposed to oogenesis, which has already halted by birth, the integrity of the male germline may be even more sensitive to environmental exposures that that of the female. Continuous sperm cell production requires a continuous and adequate supply of methyl donor and other nutrients to support DNA synthesis and maintenance of epigenetic marks and therefore fluctuations in dietary intake of these nutrients has the potential to directly impact on the fidelity of the code being passed on to offspring.

Landmark epidemiological studies of the Overkalix parish in Northern Sweden do support the notion that a father's diet is associated with health outcomes in offspring. Highly detailed historical crop records were reviewed going back to the late 1800s, and included several instances of crops surpluses and failures. Food availability of parents and grandparents was then associated with disease risk on offspring. Offspring were found to have a lower risk of cardiovascular death (OR = 0.42; 95% CI = 0.18–0.99) if their fathers experienced poor food availability during their slow growth period (9–12 years old) [2,3]. Additionally, offspring were determined to have an increased risk of diabetes mortality if their paternal grandfather had a surfeit of food during his slow growth period (OR = 4.1; 95% CI = 1.33–12.93). A follow-up study [4] revealed sex-specific effects. If paternal grandfathers had a high availability of food, only their grandsons had an increased mortality risk (RR = 1.67, $p = 0.009$), and if paternal grandmothers had a high availability of food, only their granddaughters had an increased mortality risk (RR = 2.13, $p = 0.001$). Similarly, in a separate, contemporary cohort, the investigators found that paternal onset of smoking during the slow growth period was associated with increased BMI in their 9-year-old sons, and this effect was not seen in daughters or when paternal onset of smoking was at a later age.

Several rodent studies also support a role for paternal nutrition in determining offspring health. In mice, paternal fasting resulted in reduced serum glucose in offspring, especially in males [76]. Furthermore, male offspring of fasted fathers had significantly lower serum corticosterone levels, while IGF-1 was lower in male offspring and higher in female offspring compared with offspring of control fathers.

Diet-induced obesity in fathers resulted in female offspring with impaired glucose tolerance (IGT), insulin secretion, and β-cell replication in rats [77]. Additionally, the expression of many pancreatic islet genes was altered, and the gene that showed the highest fold difference in expression (1.76-fold increase), *Il13ra2* (interleukin 13 receptor α 2), was also shown to be hypomethylated. Paternal high-fat diet induced obesity has also been shown to impair mouse embryo development, embryo mitochondrial activity, and pregnancy health [78,79], and the use of in vitro fertilization suggests that these effects are attributed to alterations in the sperm itself, rather than changes in mating behavior, seminal fluid, or time of fertilization [78].

Similarly, paternal high-fat diet results in subfertility in both the F1 and F2 generations in mice, which is transmitted through the male line [80]. These investigators also report that high-fat feeding is also associated with a repression of the histone deacetylase SIRT6 and increased in H3K9 acetylation in sperm [81]. In humans, paternal BMI is positively correlated with birth weight, body size, and cortisol levels in male, but not female offspring [82]. Paternal obesity is also associated with hypomethylation of the imprinted gene *Igf2* in newborns [83], which is of clinical importance because *Igf2* hypomethylation is associated with an increased risk of developing certain cancers [84–86].

Not only does altering the caloric intake of fathers affect their offspring, but altering dietary composition has yielded interesting results as well. In mice, the offspring of fathers fed a low-protein (high sucrose) diet had decreased levels of cholesterol esters and increased hepatic expression of many genes involved in cholesterol and lipid biosynthesis [87]. In offspring livers, there were widespread differences in CpG methylation and several microRNAs were differentially expressed, but these differences did not correlate overall with differences in gene expression. However, some of the differentially methylated regions were associated with lipid- and cholesterol-related genes, and the most notable difference was a substantial increase in methylation at a likely enhancer for the *Ppara* gene, a key regulator for lipid metabolism. Therefore, methylation differences in a small number of loci may be responsible for widespread gene expression differences. Another potential mechanism for the observed effects in offspring is chromatin remodeling of the paternal sperm genome, as the *Maoa* promoter was found to be significantly depleted of the histone modification H3K27me3.

Studies on the impact of paternal micronutrient intake on offspring health are just beginning. Kim et al. [88] report that paternal folate deficiency resulted in reduced hepatic folate in rat offspring. Additionally, paternal folate deficiency was associated with genomic hypomethylation and decreased *Igf2* expression in the fetal brain, which was positively correlated with paternal liver and testes folate content. Fetuses of deficient fathers also had decreased body weight and length, which may be explained by decreased *Igf2*, a growth factor. Another study in rats found that, although there was no difference in fetal weight, there was a significant reduction in placental weight (by 10%) and folate content (by 35%) in the paternal folate-deficient group [89]. Folate receptor α expression was also higher in the placenta (by 130%); this is likely a compensatory effect of low folate status, since this receptor mediates uptake of folate into cells. Since it is the father that contributes little more than DNA toward the developing offspring, it is interesting that paternal and fetal or placental vitamin status are correlated. Although the mechanism for these effects is unknown, it is possible that altered paternal folate status resulted in altered methylation patterns of folate-regulating genes in sperm, which were transmitted to and retained in the offspring.

Excess alcohol intake leads to folate deficiency [90] and has also been shown to exert detrimental effects on offspring when consumed by fathers prior to conception. The mechanism for such effects may or may not be related to its interference with folate bioavailability, but nevertheless, studies in this area provide supporting evidence that paternal exposures affect offspring health. In mice and rats, males exposed to alcohol sired offspring with developmental retardation [91,92], increased susceptibility to infection [93] increased mortality [92], and cognitive/behavioral deficits [91,92,94–96]. In humans, children of alcoholic fathers have presented with similar characteristics, which resemble those of fetal alcohol spectrum disorders, in which children are exposed to alcohol in utero [97,98]. Although the mechanism for these effects is unclear, some studies have found epigenetic effects of paternal alcohol consumption on offspring, such as a significant hypomethylation at the *H19* imprinting control region, which was significantly correlated with reduced weight at postnatal days 35–42 in mice [99]. Similarly, a trend of demethylation with moderate alcohol use was found at the *H19* region in human sperm DNA, as well as a significant hypomethylation of another imprinted region, IG-DMR, in heavy drinkers compared with nondrinkers [100]. Hypomethylation at the *H19* region leads to decreased expression of *Igf2*, which is a key embryonic growth regulator, and loss of paternal *Igf2* has resulted in severe growth deficiency in mice [101]. Likewise, IG-DMR demethylation is associated with developmental abnormalities in humans [102–104]. Therefore, if these epigenetic alterations in sperm are passed on to offspring, they may help explain paternal alcohol induced developmental abnormalities.

Evidence for a role of paternal nutrition in determining offspring health can also be found in mutigenerational maternal studies where male offspring are exposed in utero and impacts are seen in their subsequent progeny. For example, Jimenez-Chillaron et al. [105] report that caloric undernutrition during pregnancy in mice resulted in offspring with reduced birth weight and IGT and that these phenotypes progressed to the F2 generation through the paternal line. IGT was associated with impaired β-cell function, which was explained, in part, by dysregulated expression of *Sur1* (sulfonylurea receptor 1), an intracellular energy sensor. In another study, male offspring of dams with streptozotocin-induced gestational diabetes mellitus (GDM) sired offspring (F2 generation) with increased birth weight and IGT [106], with IGT being more pronounced in males than in females. These offspring also exhibited decreased expression of the imprinted genes *Igf2* and *H19* in pancreatic islets, an effect that was also more pronounced in males. This was accompanied by increased methylation of these genes in islets as well as decreased expression of these genes in sperm of F1 males. These findings suggest a possible epigenetic mode of inheritance through the male germline. Dunn and Bale have also found that maternal high-fat diet–induced obesity results in increased body length and reduced insulin sensitivity in two generations of offspring that was transmitted through both the male and female lineages in mice [107]. Another study found that maternal protein restriction during pregnancy resulted in increased blood pressure in both the F1 and F2 generations, which was transmitted through the paternal line, but no effect was seen in the F3 generation [108]. Since the F1 generation's primordial germ cells may be affected by the in utero environment, it is important to investigate whether effects persist into the F3 generation to determine whether stable epigenetic programming has occurred. Dunn and Bale conducted a follow-up study [109] for this reason and found that the effects in the F3 generation were sex-specific in terms of both transmission and inheritance. The only significant difference found was that females from the paternal lineage had an increased body weight and length compared with the control group. An analysis of growth-related imprinted genes revealed a nonstatistically significant trend toward greater volatility of expression in paternally expressed genes from the paternal lineage compared with the maternal lineage. Although the exact mechanisms of the sex-dependent effects in this study are unknown, the fact that the maternal phenotype is transmitted to the F3 generation through the male line provides evidence that epigenetic programming in the male germline occurred. Another study found that maternal protein restriction during pregnancy resulted in increased blood pressure in both the F1 and F2 generations, which was transmitted through the paternal line, but no effect was seen in the F3 generation [108]. This suggests that not all in utero exposures result in stable epigenetic programming in the male germline.

In summary, there is increasing evidence that paternal nutrition affects offspring metabolism and health. Although both genetic and epigenetic mechanisms are likely involved, the studies that we have reviewed suggest an epigenetic mode of inheritance, as changes in offspring gene expression and epigenetic marks have been found, in some cases with corresponding changes in fathers' sperm.

5.4 CONCLUSIONS

A growing body of literature is beginning to reveal the importance of maternal, and to a lesser degree paternal, nutrition in determining the health of offspring. Although the impact of maternal adiposity and nutrition on adiposity and cardiovascular disease risk in offspring are the topic of a great deal of research, here we have focused on the impact of maternal one-carbon nutrients on offspring tumorigenesis. With regards to an individual's own risk for cancer, the strongest evidence for a protective effect of these nutrients in the maternal diet has been for sporadic cancers of the breast and colorectum. The long latency for these cancers, however, makes them challenging to study maternal risk factors in epidemiological and clinical settings. For this reason, human studies have focused mainly on those cancers that present in childhood. As described above these studies generally indicate a protective effect of high maternal folate intakes [24].

Offspring studies in rodents, however, have focused on these cancers for which there is most evidence for a protective effect of folate and related one-carbon nutrients in individuals. In this literature, there is less consistency on whether folate supplementation suppresses or promotes tumorigenesis in offspring. The data can, however, be reconciled after careful consideration of the characteristics of the models and designs used. Our interpretation of this literature is that when a model with a mild propensity for tumorigenesis is used, folate supplementation is protective as seen in work by us and others [29,30,33]. Meanwhile, because of the role of folate in nucleotide synthesis there is a potential for the dual role of folate to be expressed in highly tumorigenic models, especially when supplementation is continued through the offspring's life [32,34]. Unlike folate, because choline only participates in homocysteine remethylation and not thymidine synthesis, a tumor promoting effect of choline has not been documented despite its use in highly tumorigenic models [35,36].

While the evidence for an effect of paternal nutrition on offspring is less extensive, animal studies do suggest that paternal nutrition can impact on offspring health, especially on various aspects of metabolism [76,77]. Studies on tumor development have not yet been published; however, our own unpublished work suggests that neither B-vitamin supplementation nor depletion impacts on intestinal tumorigenesis in offspring (in preparation).

Much work remains to be done, before we can achieve a comprehensive understanding of how maternal and paternal nutrition impact on chronic disease in offspring. In addition, we are only just beginning to unravel the mechanisms involved; however, altered methylation of key genes in disease-related signaling or metabolic pathways appears to be centrally involved.

REFERENCES

1. Barker DJ, Winter PD, Osmond C et al., Weight in infancy and death from ischaemic heart disease. *Lancet*, 1989. **2**(8663): pp. 577–80.
2. Kaati G, Bygren LO and Edvinsson S, Cardiovascular and diabetes mortality determined by nutrition during parents' and grandparents' slow growth period. *Eur J Hum Genet*, 2002. **10**(11): pp. 682–8.
3. Kaati G, Bygren LO, Pembrey M et al., Transgenerational response to nutrition, early life circumstances and longevity. *Eur J Hum Genet*, 2007. **15**(7): pp. 784–90.
4. Pembrey ME, Bygren LO, Kaati G et al., Sex-specific, male-line transgenerational responses in humans. *Eur J Hum Genet*, 2006. **14**(2): pp. 159–66.
5. Baylin SB and Jones PA, A decade of exploring the cancer epigenome—biological and translational implications. *Nat Rev Cancer*, 2011. **11**(10): pp. 726–34.
6. Wen W, Shu XO, Potter JD et al., Parental medication use and risk of childhood acute lymphoblastic leukemia. *Cancer*, 2002. **95**(8): pp. 1786–94.
7. Thompson JR, Gerald PF, Willoughby MLN et al., Maternal folate supplementation in pregnancy and protection against acute lymphoblastic leukaemia in childhood: A case-control study. *Lancet*, 2001. **358**(9297): pp. 1935–40.
8. Ross JA, Blair CK, Olshan AF et al., Periconceptional vitamin use and leukemia risk in children with Down syndrome: A Children's Oncology Group study. *Cancer*, 2005. **104**(2): pp. 405–10.

9. Milne E, Royle JA, Miller M et al., Maternal folate and other vitamin supplementation during pregnancy and risk of acute lymphoblastic leukemia in the offspring. *Int J Cancer*, 2010. **126**(11): pp. 2690–9.
10. Preston-Martin S, Pogoda JM, Mueller BA et al., Maternal consumption of cured meats and vitamins in relation to pediatric brain tumors. *Cancer Epidemiol Biomarkers Prev*, 1996. **5**(8): pp. 599–605.
11. Preston-Martin S, Pogoda JM, Mueller BA et al., Prenatal vitamin supplementation and risk of childhood brain tumors. *Int J Cancer*, 1998. **11 Suppl.**: pp. 17–22.
12. Bunin GR, Kuijten RR, Buckley JD et al., Relation between maternal diet and subsequent primitive neuroectodermal brain tumors in young children. *N Engl J Med*, 1993. **329**(8): pp. 536–41.
13. Bunin GR, Kuijten RR, Boesel CP et al., Maternal diet and risk of astrocytic glioma in children: A report from the Childrens Cancer Group (United States and Canada). *Cancer Causes Control*, 1994. **5**(2): pp. 177–87.
14. Lubin F, Farbstein H, Chetrit A et al., The role of nutritional habits during gestation and child life in pediatric brain tumor etiology. *Int J Cancer*, 2000. **86**(1): pp. 139–43.
15. Sarasua S and Savitz DA, Cured and broiled meat consumption in relation to childhood cancer: Denver, Colorado (United States). *Cancer Causes Control*, 1994. **5**(2): pp. 141–8.
16. Orjuela MA, Titievsky L, Liu X et al., Fruit and vegetable intake during pregnancy and risk for development of sporadic retinoblastoma. *Cancer Epidemiol Biomarkers Prev*, 2005. **14**(6): pp. 1433–40.
17. Bunin GR, Meadows AT, Emanuel BS et al., Pre- and postconception factors associated with sporadic heritable and nonheritable retinoblastoma. *Cancer Res*, 1989. **49**(20): pp. 5730–5.
18. Michalek AM, Buck GM, Nasca PC et al., Gravid health status, medication use, and risk of neuroblastoma. *Am J Epidemiol*, 1996. **143**(10): pp. 996–1001.
19. Olshan AF, Smith JC, Bondy ML et al., Maternal vitamin use and reduced risk of neuroblastoma. *Epidemiology*, 2002. **13**(5): pp. 575–80.
20. Schuz J, Weihkopf T and Kaatsch P, Medication use during pregnancy and the risk of childhood cancer in the offspring. *Eur J Pediatr*, 2007. **166**(5): pp. 433–41.
21. Grupp SG, Greenberg ML, Ray JG et al., Pediatric cancer rates after universal folic acid flour fortification in Ontario. *J Clin Pharmacol*, 2011. **51**(1): pp. 60–5.
22. French AE, Grant R, Weitzman S et al., Folic acid food fortification is associated with a decline in neuroblastoma. *Clin Pharmacol Ther*, 2003. **74**(3): pp. 288–94.
23. Linabery AM, Johnson KJ and Ross JA, Childhood cancer incidence trends in association with US folic acid fortification (1986–2008). *Pediatrics*, 2012. **129**(6): pp. 1125–33.
24. Goh YI, Bollano E, Einarson TR et al., Prenatal multivitamin supplementation and rates of pediatric cancers: A meta-analysis. *Clin Pharmacol Ther*, 2007. **81**(5): pp. 685–91.
25. Skibola CF, Smith MT, Kane E et al., Polymorphisms in the methylenetetrahydrofolate reductase gene are associated with susceptibility to acute leukemia in adults [see comments]. *Proc Natl Acad Sci U S A*, 1999. **96**(22): pp. 12810–5.
26. Lightfoot TJ, Johnston WT, Painter D et al., Genetic variation in the folate metabolic pathway and risk of childhood leukemia. *Blood*, 2010. **115**(19): pp. 3923–9.
27. Banelli B, Di Vinci A, Gelvi I et al., DNA methylation in neuroblastic tumors. *Cancer Lett*, 2005. **228**(1–2): pp. 37–41.
28. Garcia-Manero G, Yang H, Kuang SQ et al., Epigenetics of acute lymphocytic leukemia. *Semin Hematol*, 2009. **46**(1): pp. 24–32.
29. Ciappio ED, Liu Z, Brooks RS et al., Maternal B vitamin supplementation from preconception through weaning suppresses intestinal tumorigenesis in Apc1638N mouse offspring. *Gut*, 2011. **60**(12): pp. 1695–702.
30. Sie KK, Medline A, van Weel J et al., Effect of maternal and postweaning folic acid supplementation on colorectal cancer risk in the offspring. *Gut*, 2011. **60**(12): pp. 1687–94.

31. McKay JA, Williams EA and Mathers JC, Gender-specific modulation of tumorigenesis by folic acid supply in the Apc mouse during early neonatal life. *Br J Nutr*, 2008. **99**(3): pp. 550–8.
32. Lawrance AK, Deng L and Rozen R, Methylenetetrahydrofolate reductase deficiency and low dietary folate reduce tumorigenesis in Apc min/+ mice. *Gut*, 2009. **58**(6): pp. 805–11.
33. Sie KKY, Chen J, Sohn K-J et al., Folic acid supplementation provided in utero and during lactation reduces the number of terminal end buds of the developing mammary glands in the offspring. *Cancer Lett*, 2009. **280**(1): pp. 72–7.
34. Ly A, Lee H, Chen J et al., Effect of maternal and postweaning folic acid supplementation on mammary tumor risk in the offspring. *Cancer Res*, 2011. **71**(3): pp. 988–97.
35. Kovacheva VP, Davison JM, Mellott TJ et al., Raising gestational choline intake alters gene expression in DMBA-evoked mammary tumors and prolongs survival. *FASEB J*, 2009. **23**(4): pp. 1054–63.
36. Cho K, Mabasa L, Bae S et al., Maternal high-methyl diet suppresses mammary carcinogenesis in female rat offspring. *Carcinogenesis*, 2012. **33**(5): pp. 1106–12.
37. Sakatani T, Kaneda A, Iacobuzio-Donahue CA et al., Loss of imprinting of Igf2 alters intestinal maturation and tumorigenesis in mice. *Science*, 2005. **307**(5717): pp. 1976–8.
38. Cheng YW, Idrees K, Shattock R et al., Loss of imprinting and marked gene elevation are 2 forms of aberrant IGF2 expression in colorectal cancer. *Int J Cancer*, 2010. **127**(3): pp. 568–77.
39. Heijmans BT, Tobi EW, Stein AD et al., Persistent epigenetic differences associated with prenatal exposure to famine in humans. *Proc Natl Acad Sci U S A*, 2008. **105**(44): pp. 17046–9.
40. Tobi EW, Lumey LH, Talens RP et al., DNA methylation differences after exposure to prenatal famine are common and timing- and sex-specific. *Hum Mol Genet*, 2009. **18**(21): pp. 4046–53.
41. Waterland RA, Lin JR, Smith CA et al., Post-weaning diet affects genomic imprinting at the insulin-like growth factor 2 (Igf2) locus. *Hum Mol Genet*, 2006. **15**(5): pp. 705–16.
42. Kovacheva VP, Mellott TJ, Davison JM et al., Gestational choline deficiency causes global and Igf2 gene DNA hypermethylation by up-regulation of Dnmtl expression. *J Biol Chem*, 2007. **282**(43): pp. 31777–88.
43. Ba Y, Yu H, Liu F et al., Relationship of folate, vitamin B12 and methylation of insulin-like growth factor-II in maternal and cord blood. *Eur J Clin Nutr*, 2011. **65**(4): pp. 480–5.
44. Steegers-Theunissen RP, Obermann-Borst SA, Kremer D et al., Periconceptional maternal folic acid use of 400 microg per day is related to increased methylation of the IGF2 gene in the very young child. *PLoS One*, 2009. **4**(11): p. e7845.
45. Haggarty P, Hoad G, Campbell DM et al., Folate in pregnancy and imprinted gene and repeat element methylation in the offspring. *Am J Clin Nutr*, 2013. **97**(1): pp. 94–9.
46. Hoyo C, Murtha AP, Schildkraut JM et al., Methylation variation at IGF2 differentially methylated regions and maternal folic acid use before and during pregnancy. *Epigenetics*, 2011. **6**(7): pp. 928–36.
47. Cooper WN, Khulan B, Owens S et al., DNA methylation profiling at imprinted loci after periconceptional micronutrient supplementation in humans: Results of a pilot randomized controlled trial. *FASEB J*, 2012. **26**(5): pp. 1782–90.
48. Wolff GL, Kodell RL, Moore SR et al., Maternal epigenetics and methyl supplements affect agouti gene expression in Avy/a mice. *FASEB J*, 1998. **12**(11): pp. 949–57.
49. Waterland RA and Jirtle RL, Transposable elements: Targets for early nutritional effects on epigenetic gene regulation. *Mol Cell Biol*, 2003. **23**(15): pp. 5293–300.
50. Cooney CA, Dave AA and Wolff GL, Maternal methyl supplements in mice affect epigenetic variation and DNA methylation of offspring. *J Nutr*, 2002. **132**(8 Suppl): pp. 2393S–400S.

51. Yen T, Gill A, Frigeri L et al., Obesity, diabetes, and neoplasia in yellow A(vy)/- mice: Ectopic expression of the agouti gene. *FASEB J*, 1994. **8**(8): pp. 479–88.
52. Wolff GL, Morrissey RL and Chen JJ, Amplified response to phenobarbital promotion of hepatotumorigenesis in obese yellow Avy/A (C3H × VY) F-1 hybrid mice. *Carcinogenesis*, 1986. **7**(11): pp. 1895–8.
53. Wolff GL, Roberts DW, Morrissey RL et al., Tumorigenic responses to lindane in mice: Potentiation by a dominant mutation. *Carcinogenesis*, 1987. **8**(12): pp. 1889–97.
54. Waterland RA, Dolinoy DC, Lin JR et al., Maternal methyl supplements increase offspring DNA methylation at Axin Fused. *Genesis*, 2006. **44**(9): pp. 401–6.
55. Klaus A and Birchmeier W, Wnt signalling and its impact on development and cancer. *Nat Rev Cancer*, 2008. **8**(5): pp. 387–98.
56. Kinzler KW and Vogelstein B, Lessons from hereditary colorectal cancer. *Cell*, 1996. **87**(2): pp. 159–70.
57. Fearon ER and Vogelstein B, A genetic model for colorectal tumorigenesis. *Cell*, 1990. **61**(5): pp. 759–67.
58. Arnold CN, Goel A, Niedzwiecki D et al., APC promoter hypermethylation contributes to the loss of APC expression in colorectal cancers with allelic loss on 5q. *Cancer Biol Ther*, 2004. **3**(10): pp. 960–4.
59. Ying Y and Tao Q, Epigenetic disruption of the WNT/beta-catenin signaling pathway in human cancers. *Epigenetics*, 2009. **4**(5): pp. 307–12.
60. Suzuki H, Watkins DN, Jair K-W et al., Epigenetic inactivation of SFRP genes allows constitutive WNT signaling in colorectal cancer. *Nat Genet*, 2004. **36**(4): pp. 417–22.
61. Liu Z, Choi SW, Crott JW et al., Mild depletion of dietary folate combined with other B vitamins alters multiple components of the Wnt pathway in mouse colon. *J Nutr*, 2007. **137**(12): pp. 2701–8.
62. Liu Z, Ciappio ED, Crott JW et al., Combined inadequacies of multiple B vitamins amplify colonic Wnt signaling and promote intestinal tumorigenesis in BAT-LacZxApc1638N mice. *FASEB J*, 2011. **25**(9): pp. 3136–45.
63. Umbricht CB, Evron E, Gabrielson E et al., Hypermethylation of 14-3-3 sigma (stratifin) is an early event in breast cancer. *Oncogene*, 2001. **20**(26): pp. 3348–53.
64. Oshiro MM, Futscher BW, Lisberg A et al., Epigenetic regulation of the cell type-specific gene 14-3-3sigma. *Neoplasia*, 2005. **7**(9): pp. 799–808.
65. Mehedint MG, Niculescu MD, Craciunescu CN et al., Choline deficiency alters global histone methylation and epigenetic marking at the Re1 site of the calbindin 1 gene. *FASEB J*, 2010. **24**(1): pp. 184–95.
66. Jiang X, Yan J, West AA et al., Maternal choline intake alters the epigenetic state of fetal cortisol-regulating genes in humans. *FASEB J*, 2012. **26**(8): pp. 3563–74.
67. Jacob RA, Gretz DM, Taylor PC et al., Moderate folate depletion increases plasma homocysteine and decreases lymphocyte DNA methylation in postmenopausal women. *J Nutr*, 1998. **128**(7): pp. 1204–12.
68. Boeke CE, Baccarelli A, Kleinman KP et al., Gestational intake of methyl donors and global LINE-1 DNA methylation in maternal and cord blood: Prospective results from a folate-replete population. *Epigenetics*, 2012. **7**(3): pp. 253–60.
69. Sie KK, Li J, Ly A et al., Effect of maternal and postweaning folic acid supplementation on global and gene-specific DNA methylation in the liver of the rat offspring. *Mol Nutr Food Res*, 2013. **57**(4): pp. 677–85.
70. McKay JA, Waltham KJ, Williams EA et al., Folate depletion during pregnancy and lactation reduces genomic DNA methylation in murine adult offspring. *Genes Nutr*, 2011. **6**(2): pp. 189–96.
71. Kashiwagi K, Nimura K, Ura K et al., DNA methyltransferase 3b preferentially associates with condensed chromatin. *Nucleic Acids Res*, 2011. **39**(3): pp. 874–88.

72. Yang S-M, Kim BJ, Norwood Toro L et al., H1 linker histone promotes epigenetic silencing by regulating both DNA methylation and histone H3 methylation. *Proc Natl Acad Sci U S A*, 2013. **110**(5): pp. 1708–13.
73. Razin A, CpG methylation, chromatin structure and gene silencing-a three-way connection. *EMBO J*, 1998. **17**(17): pp. 4905–8.
74. Dobosy JR, Fu VX, Desotelle JA et al., A methyl-deficient diet modifies histone methylation and alters Igf2 and H19 repression in the prostate. *Prostate*, 2008. **68**(11): pp. 1187–95.
75. Dolinoy DC, Weinhouse C, Jones T et al., Variable histone modifications at the A(vy) metastable epiallele. *Epigenetics*, 2010. **5**(7): Epub ahead of print.
76. Anderson LM, Riffle L, Wilson R et al., Preconceptional fasting of fathers alters serum glucose in offspring of mice. *Nutrition*, 2006. **22**(3): pp. 327–31.
77. Ng SF, Lin RC, Laybutt DR et al., Chronic high-fat diet in fathers programs beta-cell dysfunction in female rat offspring. *Nature*, 2010. **467**(7318): pp. 963–6.
78. Binder NK, Hannan NJ and Gardner DK, Paternal diet-induced obesity retards early mouse embryo development, mitochondrial activity and pregnancy health. *PLoS One*, 2012. **7**(12): p. e52304.
79. Mitchell M, Bakos HW and Lane M, Paternal diet-induced obesity impairs embryo development and implantation in the mouse. *Fertil Steril*, 2011. **95**(4): pp. 1349–53.
80. Fullston T, Palmer NO, Owens JA et al., Diet-induced paternal obesity in the absence of diabetes diminishes the reproductive health of two subsequent generations of mice. *Hum Reprod*, 2012. **27**(5): pp. 1391–400.
81. Palmer NO, Fullston T, Mitchell M et al., SIRT6 in mouse spermatogenesis is modulated by diet-induced obesity. *Reprod Fertil Dev*, 2011. **23**(7): pp. 929–39.
82. Chen YP, Xiao XM, Li J et al., Paternal body mass index (BMI) is associated with offspring intrauterine growth in a gender dependent manner. *PLoS One*, 2012. **7**(5): p. e36329.
83. Soubry A, Schildkraut JM, Murtha A et al., Paternal obesity is associated with IGF2 hypomethylation in newborns: Results from a Newborn Epigenetics Study (NEST) cohort. *BMC Med*, 2013. **11**: p. 29.
84. Sullivan MJ, Taniguchi T, Jhee A et al., Relaxation of IGF2 imprinting in Wilms tumours associated with specific changes in IGF2 methylation. *Oncogene*, 1999. **18**(52): pp. 7527–34.
85. Cui H, Onyango P, Brandenburg S et al., Loss of imprinting in colorectal cancer linked to hypomethylation of H19 and IGF2. *Cancer Res*, 2002. **62**(22): pp. 6442–6.
86. Murphy SK, Huang Z, Wen Y et al., Frequent IGF2/H19 domain epigenetic alterations and elevated IGF2 expression in epithelial ovarian cancer. *Mol Cancer Res*, 2006. **4**(4): pp. 283–92.
87. Carone BR, Fauquier L, Habib N et al., Paternally induced transgenerational environmental reprogramming of metabolic gene expression in mammals. *Cell*, 2010. **143**(7): pp. 1084–96.
88. Kim HW, Kim KN, Choi YJ et al., Effects of paternal folate deficiency on the expression of insulin-like growth factor-2 and global DNA methylation in the fetal brain. *Mol Nutr Food Res*, 2013. **57**(4): pp. 671–6.
89. Kim HW, Choi YJ, Kim KN et al., Effect of paternal folate deficiency on placental folate content and folate receptor alpha expression in rats. *Nutr Res Pract*, 2011. **5**(2): pp. 112–6.
90. Medici V and Halsted CH, Folate, alcohol, and liver disease. *Mol Nutr Food Res*, 2013. **57**(4): pp. 596–606.
91. Ledig M, Misslin R, Vogel E et al., Paternal alcohol exposure: Developmental and behavioral effects on the offspring of rats. *Neuropharmacology*, 1998. **37**(1): pp. 57–66.

92. Meek LR, Myren K, Sturm J et al., Acute paternal alcohol use affects offspring development and adult behavior. *Physiol Behav*, 2007. **91**(1): pp. 154–60.
93. Berk RS, Montgomery IN, Hazlett LD et al., Paternal alcohol consumption: Effects on ocular response and serum antibody response to Pseudomonas aeruginosa infection in offspring. *Alcohol Clin Exp Res*, 1989. **13**(6): pp. 795–8.
94. Abel EL and Tan SE, Effects of paternal alcohol consumption on pregnancy outcome in rats. *Neurotoxicol Teratol*, 1988. **10**(3): pp. 187–92.
95. Abel EL and Bilitzke P, Paternal alcohol exposure: Paradoxical effect in mice and rats. *Psychopharmacology (Berl)*, 1990. **100**(2): pp. 159–64.
96. Wozniak DF, Cicero TJ, Kettinger L III et al., Paternal alcohol consumption in the rat impairs spatial learning performance in male offspring. *Psychopharmacology (Berl)*, 1991. **105**(2): pp. 289–302.
97. Lemoine P, Harousseau H, Borteyru JP et al., Children of alcoholic parents—observed anomalies: Discussion of 127 cases. *Ther Drug Monit*, 2003. **25**(2): pp. 132–6.
98. Gabrielli WF Jr. and Mednick SA, Intellectual performance in children of alcoholics. *J Nerv Ment Dis*, 1983. **171**(7): pp. 444–7.
99. Knezovich JG and Ramsay M, The effect of preconception paternal alcohol exposure on epigenetic remodeling of the h19 and rasgrf1 imprinting control regions in mouse offspring. *Front Genet*, 2012. **3**: p. 10.
100. Ouko LA, Shantikumar K, Knezovich J et al., Effect of alcohol consumption on CpG methylation in the differentially methylated regions of H19 and IG-DMR in male gametes: Implications for fetal alcohol spectrum disorders. *Alcohol Clin Exp Res*, 2009. **33**(9): pp. 1615–27.
101. DeChiara TM, Robertson EJ and Efstratiadis A, Parental imprinting of the mouse insulin-like growth factor II gene. *Cell*, 1991. **64**(4): pp. 849–59.
102. Delaval K and Feil R, Epigenetic regulation of mammalian genomic imprinting. *Curr Opin Genet Dev*, 2004. **14**(2): pp. 188–95.
103. Kagami M, Sekita Y, Nishimura G et al., Deletions and epimutations affecting the human 14q32.2 imprinted region in individuals with paternal and maternal upd(14)-like phenotypes. *Nat Genet*, 2008. **40**(2): pp. 237–42.
104. Sutton VR and Shaffer LG, Search for imprinted regions on chromosome 14: Comparison of maternal and paternal UPD cases with cases of chromosome 14 deletion. *Am J Med Genet*, 2000. **93**(5): pp. 381–7.
105. Jimenez-Chillaron JC, Isganaitis E, Charalambous M et al., Intergenerational transmission of glucose intolerance and obesity by in utero undernutrition in mice. *Diabetes*, 2009. **58**(2): pp. 460–8.
106. Ding GL, Wang FF, Shu J et al., Transgenerational glucose intolerance with Igf2/H19 epigenetic alterations in mouse islet induced by intrauterine hyperglycemia. *Diabetes*, 2012. **61**(5): pp. 1133–42.
107. Dunn GA and Bale TL, Maternal high-fat diet promotes body length increases and insulin insensitivity in second-generation mice. *Endocrinology*, 2009. **150**(11): pp. 4999–5009.
108. Harrison M and Langley-Evans SC, Intergenerational programming of impaired nephrogenesis and hypertension in rats following maternal protein restriction during pregnancy. *Br J Nutr*, 2009. **101**(7): pp. 1020–30.
109. Dunn GA and Bale TL, Maternal high-fat diet effects on third-generation female body size via the paternal lineage. *Endocrinology*, 2011. **152**(6): pp. 2228–36.

6 Alcohol and DNA Methylation

Silvia Udali, Simonetta Friso, and Sang-Woon Choi

CONTENTS

6.1 INTRODUCTION

Alcohol is an environmental toxicant that is associated with several major human diseases because it has harmful effects on many different tissues and organs. Alcohol effects cellular toxicity by several mechanisms, mainly through acetaldehyde, the first metabolite produced during ethanol degradation, and through the formation of reactive oxygen species (ROS). Acetaldehyde interferes with DNA synthesis and repair mechanisms and is well recognized as playing a role in cancer of upper and lower gastrointestinal tract. Meanwhile, the hazardous effects of alcohol on the liver are more likely to be mediated by oxidative stress (Poschl and Seitz, 2004; Dey and Cederbaum, 2006; Seitz and Stickel, 2007). Several studies have also demonstrated that alcohol impairs one-carbon metabolism leading to an aberrant methyl group transfer, and it is believed that this molecular event may play a role in the development of cancer and other alcohol-related diseases. In this chapter, we focused on the effects of alcohol on one-carbon metabolism and its influence on DNA methylation, which could be one of the mechanisms involved in the epigenetic effects of alcohol (Christensen and Marsit, 2011; Kim et al., 2012).

DNA methylation is one of the well-characterized epigenetic mechanisms, a phenomenon that affects gene expression and thereby modifies the phenotype without altering the base sequence of DNA (Bird, 2007; Goldberg et al., 2007). This epigenetic mechanism is heritable and potentially reversible in response to cues from the environment (Christensen and Marsit, 2011). DNA methylation consists in the covalent binding of a methyl group to the 5′-carbon of cytosine occurring at CpG dinucleotide sequences in the mammalian genome (Feinberg, 2007). This reaction is catalyzed by DNA methyltransferases (DNMTs), a class of enzymes that include both maintenance DNMT (DNMT1) that preserves the DNA methylation levels during mitotic processes, and *de novo* DNMTs (DNMT3A and 3B) that regulate the acquisition of novel cytosine methylation sites (Boland and Christman, 2009; Jia and Cheng, 2009). The main function of DNA methylation is to modulate the expression of genes by modifying the accessibility of transcriptional machinery to DNA. DNA methylation in the promoter region is classically associated with transcriptional inhibition, while demethylation seems to be necessary for the gene activation (Luczak and Jagodzinski, 2006). DNA methylation has crucial physiological roles in the cell acting in the stabilization of chromosomes (Jones and Takai, 2001), genomic imprinting (Falls et al., 1999; Tilghman, 1999; Kiefer, 2007), X-chromosome inactivation (Riggs and Pfeifer, 1992), mammalian embryogenesis (Kafri et al., 1992), and fetal development (Monk et al., 1987). It is also important in the transcriptional inhibition of repeat elements and transposons. DNA methylation has been studied in relation to several pathologic conditions, mostly cancer (Jones, 1986; Jones and Laird, 1999; Ehrlich, 2002; Feinberg and Tycko, 2004; Feinberg, 2007) but also other chronic diseases including cardiovascular diseases (Udali et al., 2013), neurocognitive diseases, endocrine diseases, and immune diseases.

The methyl groups for DNA methylation are derived from one-carbon metabolism, a network of methyl-transfer reactions in which folate is the principal carrier of methyl units in its reduced and active form, i.e., tetrahydrofolate (THF) (Choi and Mason, 2002) (Figure 6.1). This metabolic pathway is involved in both nucleotide synthesis and biological methylation of cellular biomolecules such as DNA, RNA, proteins, and lipids. Several enzymes play essential roles in the methyl-groups transfer reactions. Serine hydroxymethyltransferase (SHMT) catalyzes the transfer of a methyl group from serine to THF to form 5,10-methyleneTHF, which is then converted to 5-methylTHF by 5,10 methylenetetrahydrofolatereductase (MTHFR). Vitamin B_6 is a coenzyme for SHMT and vitamin B_2 is a coenzyme for MTHFR. Methionine synthase (MS), a vitamin B_{12}–dependent enzyme, catalyzes the transfer of a methyl group from 5-methylTHF to homocysteine (Hcy), leading to the formation of methionine and THF. Methionine adenosyltransferase (MAT) then catalyzes the addition of ATP to methionine for the generation of *S*-adenosylmethionine (SAM), the universal methyl donor for biological methylation reactions including DNA methylation. After the transfer of a methyl unit (one-carbon), SAM is converted into *S*-adenosylhomocysteine (SAH), which is then hydrolyzed to adenosine and Hcy by SAH hydrolase (SAHH) (Choi and Mason, 2002). All of these enzyme reactions are potentially affected by alcohol.

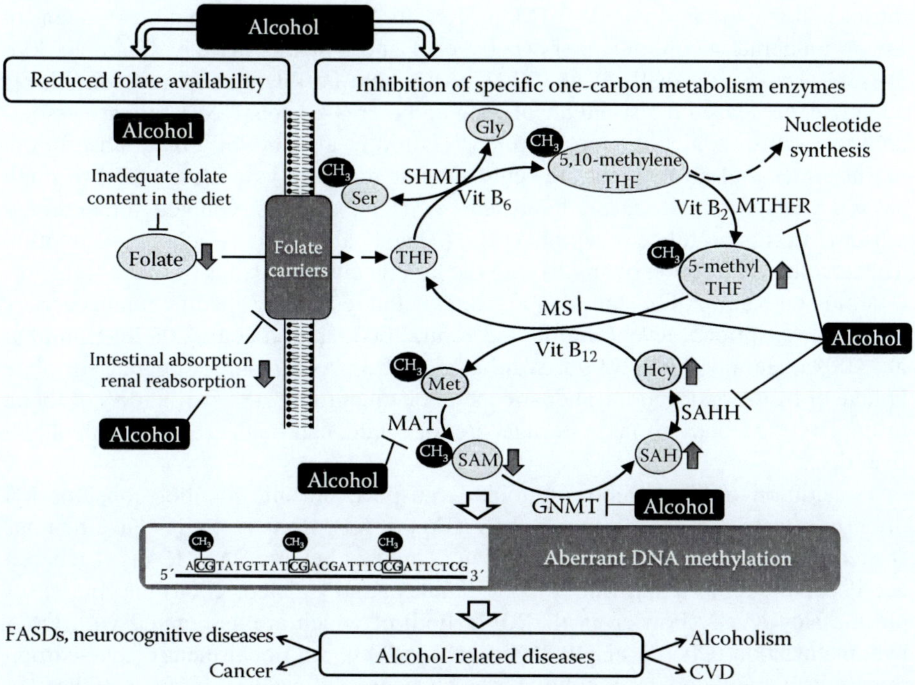

FIGURE 6.1 Chronic alcohol consumption impairs folate-dependent one-carbon metabolism by reducing folate availability and inhibiting specific enzymes, such as MTHFR, MS, MAT, SAHH, and GNMT. In this figure, one-carbon metabolism is schematically represented with a particular focus on the reactions that transfer one-carbon unit (CH_3) to *S*-adenosylmethionine (SAM), the universal methyl donor for biological methylation. Chronic alcohol consumption results in the increase of 5-methylTHF, Hcy, and SAH levels and the decrease of folate availability and SAM levels, all of which could be related with the aberrant DNA methylation observed in alcohol associated human diseases, such as FASDs, cancer, and alcoholism.

In this chapter, we focus on the effects of alcohol on one-carbon metabolism, a means by which alcohol interferes with DNA methylation, and subsequent effects on the alcohol-associated diseases, such as fetal alcohol spectrum disorders (FASDs), cancer, and alcoholism.

6.2 EFFECTS OF ALCOHOL ON ONE-CARBON METABOLISM

Chronic alcohol intake interferes with folate-dependent one-carbon metabolism by reducing the folate availability and directly inhibiting enzymes in one-carbon metabolism (Mason and Choi, 2005). Folate is an essential water soluble B vitamin, and its availability is compromised in alcoholic subjects due to decreased folate content in the diet, intestinal malabsorption, and increased renal excretion (Figure 6.1). Inadequate folate intake is common among heavy abusers of alcohol (at least 100 g ethanol per day) (Manari et al., 2003), but it is not so frequent in low to moderate alcohol drinkers (between 1 and 2 drinks per day, less than 30 g ethanol per day) (Planells et al., 2003). Although the folate content in the diet is adequate, decreased intestinal absorption and increased renal excretion are associated with chronic alcohol consumption (Hamid et al., 2009b). Folate absorption, mainly intestinal absorption and renal tubular reabsorption, is mediated through several transporters: reduced folate carrier (RFC), proton-coupled folate transporter (PCFT) and folate receptors (FRs) (Zhao et al., 2009). Alcohol consumption affects folate absorption by modifying the kinetic characteristics of folate transport (Hamid and Kaur, 2005; Hamid et al., 2007a) and repressing specific folate carriers at the transcriptional level (Hamid and Kaur, 2007; Hamid et al., 2007b; Hamid et al., 2009a; Hamid et al., 2009b; Wani et al., 2013). As a result of decreased folate uptake or from abnormal intrahepatic metabolism of folate, the availability of folate in the liver, the body's major storage site for folate, can be decreased in alcoholic liver disease.

In addition to the effects on folate transport, chronic alcohol consumption directly affects one-carbon metabolism (Figure 6.1). Rodent studies have demonstrated that chronic alcohol consumption decreases hepatic SAM levels (Lieber et al., 1990; Tsukamoto and Lu, 2001; Villanueva and Halsted, 2004) and increases plasma Hcy levels (Fowler et al., 2012), both of which are associated with DNA hypomethylation (Choi et al., 1999; Lu et al., 2000). An impairment of one-carbon metabolism mediated by alcohol has also been observed in human studies, in which a reduced blood folate and an elevated plasma Hcy were reported in chronic alcoholic patients (Cravo et al., 1996). Biochemically alcohol impairs one-carbon metabolism through inhibition of MS activity (Barak et al., 1987; Kenyon et al., 1998; Barak et al., 2002; Halsted et al., 2002), leading to hyperhomocysteinemia and sequestration of folate in the form of 5-methylTHF (the methylfolate trap). This inhibition of MS also reduces the synthesis of methionine and subsequently reduces the synthesis of SAM (Lu and Mato, 2005; Lu et al., 2006), which may affect biological methylation reactions in which SAM is the universal methyl donor. In a minipig model alcohol has also been described to inhibit other one-carbon metabolism enzymes, such as MTHFR, glycine *N*-methyltransferase (GNMT), and SAHH in the liver, which further deranges one-carbon metabolism and increases alcoholic liver injury (Villanueva and Halsted, 2004). In a clinical study using biopsy samples from patients hospitalized for alcoholic hepatitis, the hepatic mRNA levels of most enzymes involved in methionine metabolism and glutathione (GSH) synthesis such as MAT1A, MS, betaine homocysteine methyltransferase (BHMT), cystathionine β synthase (CBS), and GNMT were decreased, whereas albumin expression was unchanged. Additionally, hepatic SAM levels were reduced by 50%, and methionine, GSH, and cysteine levels were reduced by 70%–80% (Lee et al., 2004).

In summary, alcohol limits folate availability and inhibits specific methyl-transfer reactions, inducing an overall impairment of one-carbon metabolism. Since this metabolic pathway is involved both in biological methylation and nucleic acids synthesis, alcohol consumption interferes with many cellular processes.

6.3 EFFECTS OF ALCOHOL ON DNA METHYLTRANSFERASES

It has been suggested that alcohol reduces DNA methylation by inhibiting the activity of DNMTs (Shukla et al., 2008). In a recent in vitro embryonic model system, murine embryonic fibroblasts exposed to 200 mM ethanol in media demonstrated a 5% reduction in global DNA methylation. In addition, ethanol induced proteosomal degradation of all three DNMTs (DNMT1, DNMT3a, and DNMT3b) and the methyl CpG-binding proteins (Mukhopadhyay et al., 2013). In an animal study, combined prenatal alcohol exposure (dam treated with 4.5 g/kg body weight) and postnatal alcohol exposure (pup treated with 3.0 g/kg body weight) resulted in a significant increase in DNMT activity in the hippocampus. Developmental alcohol exposure also caused a change in gene expression of DNMT1, DNMT3a, and methyl CpG binding protein 2 (MeCP2) (Perkins et al., 2013). Collectively, alcohol alters DNA methylation by modifying the activity of DNMTs and proteins associated with DNA methylation.

6.4 ALCOHOL AFFECTS FETAL GROWTH AND NEURONAL DEVELOPMENT THROUGH DNA METHYLATION

Prenatal alcohol exposure can affect fetal development, leading to FASDs. FASDs are a wide range of physical and behavioral abnormalities that includes fetal alcohol syndrome, partial fetal alcohol syndrome, alcohol-related birth defects, and alcohol-related neurodevelopmental disorders. Fetal alcohol syndrome, the most severe form, is characterized by prenatal and/or postnatal growth retardation, cognitive and behavioral disabilities, and a peculiar pattern of facial anomalies such as short palpebral fissures, smooth philtrum, and a thin and vermillion border of the upper lip. Alcohol influences brain development through different pathways during all stages of embryogenesis, and the same processes that lead to facial dysmorphism are likely the same as those involved in brain damage (Riley et al., 2011). Severity of FASDs depends on alcohol dosage, frequency of exposure (chronic or acute), and timing, i.e., developmental stage (preconception, preimplantation, and gastrulation) (Kobor and Weinberg, 2011).

The mechanisms underlying FASDs are not yet fully understood, but it is expected that epigenetic phenomena, in particular DNA methylation, might be involved. DNA methylation plays an essential role in embryogenesis and is influenced by environmental factors such as alcohol exposure. Garro and colleagues were one of the first to investigate the effects of prenatal alcohol exposure on DNA methylation in mice. They measured the DNA methylation levels of fetuses from ethanol-fed dams (acute ethanol administration, 3 g/kg body weight, from gestational days, GDs, 9–11) and found lower DNA methylation levels and methylase activities in fetuses from alcohol-fed dams compared with isocalorically fed controls (Garro et al., 1991). This early evidence opened the door to later studies that aimed to clarify the effects of prenatal alcohol exposure on DNA methylation, focusing on the timing of alcohol exposure. Here we summarize the current knowledge, separated by three groups of studies: (a) preconceptional and preimplantational alcohol exposure, (b) alcohol exposure during gastrulation with particular attention on the early neurulation and neuroteratogenic effects of alcohol, and (c) paternal alcohol exposure.

6.4.1 Preconceptional and Preimplantational Maternal Alcohol Exposure

Preconceptional and preimplantational alcohol exposure is difficult to study in human subjects because alcoholic behavior usually persists during the whole pregnancy. As a consequence, the effects of alcohol consumption during those periods are usually masked and modified by the alcohol intake during the whole pregnancy. However, animal models allow for a more feasible evaluation of a narrow window of alcohol exposure. Kaminen-Ahola and colleagues developed a mouse model of preconceptional (maternal voluntary consumption of 10% v/v ethanol for 10 weeks prior to fertilization) and preimplantational alcohol exposure (ethanol consumption from 0.5 to 8.5 GD) (Kaminen-Ahola et al., 2010). They used the agouti viable yellow (A^{vy}) locus as an epigenetic sensitive allele in mice and analyzed the methylation status by bisulfite sequencing. The activity of A^{vy} is variable and influenced by the locus specific methylation status: hypomethylation is associated with the full expression of agouti gene with yellow coat color and hypermethylation is associated with the transcriptional silencing with black coat color (pseudoagouti), while the coat is mottled with intermediate levels of methylation. Both preconceptional and preimplantational alcohol exposure produced a higher proportion of pseudoagouti A^{vy} mice relative to mice fed a control diet, as ethanol increased the transcriptional silencing at this particular locus. The transcriptional silencing correlated with hypermethylation at A^{vy} in tail genomic DNA, which demonstrated that ethanol may affect adult phenotype by altering epigenetic marks of the early embryo (Kaminen-Ahola et al., 2010).

Preimplantational alcohol exposure was also investigated using another animal model generated by mating C57BL/6 female mice with *Mus musculus castaneus* males (Haycock and Ramsay, 2009). Using two different mouse strains researchers could distinguish two different parental alleles and analyze the corresponding DNA methylation levels. The alcohol treatment (2.9 g/kg body weight) was performed from 1.5 to 2.5 GD and the DNA methylation levels of the *H19* imprinting control region were determined from embryos and placentae by bisulfite sequencing. The *Igf2/H19* domain, one of the well characterized imprinted gene clusters, contains the paternally expressed *Igf2* and the maternally expressed *H19* genes. This domain was selected because aberrant methylation in the imprinting control region influences the expression of *Igf2*, thereby affecting fetal growth. In placentae, DNA methylation was reduced in response to the alcohol treatment, but there was no change in embryos. The authors suggested a functional effect of alcohol-induced paternal allele demethylation in placentae, perhaps as a result of disrupted imprinting control in the *H19/Igf2* domain (Haycock and Ramsay, 2009). These results do not fully explain the effects of preconceptional and preimplantational alcohol exposure on fetal development, but they do give some evidence of the role of DNA methylation in the development of FASDs.

6.4.2 Alcohol Exposure during Gastrulation and Early Neurulation

Differentiating cells seem to be particularly vulnerable to the teratogenic effects of alcohol; therefore, previous studies analyzing the effect of alcohol on embryonic development have focused on the gastrulation period (Kobor and Weinberg, 2011). Stouder and colleagues analyzed the effects of low alcohol doses (0.5 g/kg body weight) during the gastrulation period (from 10 to 18 GD) on the offspring's germinal and somatic tissues in mice (Stouder et al., 2011). They analyzed the methylation levels of 5 imprinted genes: *H19*, gene trap locus 2 (*Gtl2*), paternally expressed gene 1 (*Peg1*), small nuclear ribonucleoprotein-associated protein N (*Snrpn*), and paternally expressed gene 3 (*Peg3*) by pyrosequencing. They observed demethylation of *H19* in the sperm of F1 offspring but not in somatic tissues (tail, liver, muscle, and hippocampus). However, *H19* demethylation in somatic tissues, particularly in whole brain, was observed in F2 offspring, which may have been inherited from the F1 offspring sperm (Stouder et al., 2011).

In an effort to clarify the neuroteratogenic effects of alcohol in gastrulation several studies have centered on the central nervous system (Goodlett et al., 2005). Marutha Ravindran and Ticku analyzed the effects of chronic alcohol consumption (75 mM ethanol in culture medium) on mouse fetal cortical neurons (Marutha Ravindran and Ticku, 2004). Long-term ethanol treatment is known to affect the expression of *N*-methyl-D-aspartate (*NMDA*) receptors and excitatory neurotransmitter receptors in the mammalian brain. In particular, it upregulates the *NMDA* receptor subtype 2B (*NR2B*) subunit. The authors evaluated the methylation status of the *NR2B* gene to clarify the molecular mechanism involved in the ethanol-induced transcriptional upregulation of this particular gene. They observed DNA demethylation following chronic ethanol treatment, which was reverted after 48h of ethanol withdrawal. It seems that alcohol-induced DNA hypomethylation, which is associated with the upregulation of specific genes in cortical neurons, could be a possible mechanism underlying the alcoholic injury to the central nervous system (Marutha Ravindran and Ticku, 2004).

To elucidate the effects of alcohol exposure during neurulation, Liu and colleagues performed genome-wide DNA methylation and gene expression analyses in mouse embryos cultured with or without ethanol in the medium (Liu et al., 2009). C57BL/6 mice embryos were removed from the uterus at GD 8.25 and subdivided into three groups: alcohol-treated with neural tube closed, alcohol-treated with neural tube opened, and controls. The alcohol treatment (88 mM ethanol for 44 h), which is comparable to the in vivo dosage of binge-like alcohol intake, was associated with a delayed growth and reduced overall growth of embryos accompanied by alterations in the development of the heart, caudal neural tube, brain vesicles, optic system, and limb buds. DNA methylation profiles were obtained by methylated DNA immunoprecipitation (MeDIP) followed by microarray analysis, and the data were merged with the gene expression levels obtained under the same treatment conditions. They observed significant changes in methylation levels associated with alcohol treatment and, moreover, they also found differences between neural tube closed and opened groups, indicating that alcohol induced aberrant changes in DNA methylation patterns with associated changes in gene expression together may contribute to the observed abnormal fetal development (Liu et al., 2009).

The same experimental model presented above was also used to analyze the effects of alcohol on global DNA methylation by measuring 5-methylcytosine (5mC) and DNA methylation binding domain 1 (MBD1), using immunocytochemistry analysis

(Zhou et al., 2011). This study also demonstrated that alcohol exposure (88 mM ethanol for 6 or 44 h) influences the embryo's development, especially neural tube formation, and reduces DNA methylation during neurulation (Zhou et al., 2011). The finding that chronic alcohol exposure interferes with the acquisition of methylation marks was supported by the same authors in another study, in which they observed a decrease in 5mC and methyl CpG binding protein 2 (MeCp2) in embryos obtained from dams exposed to a moderate high chronic drinking (4% v/v ethanol) from 7 to 16 GD. This phenomenon correlates with hippocampus developmental retardation, suggesting a role of DNA methylation in neurulation (Chen et al., 2013).

6.4.3 Paternal Alcohol Exposure

The effects of alcohol have extensively been studied regarding maternal exposure, but a growing body of epidemiological studies have also suggested an effect of paternal exposure in the development of FASDs (Abel, 2004). Human and animal studies have therefore been designed to analyze the effects of paternal alcohol exposure with particular interest in DNA methylation. One of the first pieces of evidence comes from a study that showed an association between paternal alcohol consumption and decreased fetal weight, and demonstrated for the first time that paternal chronic alcohol exposure (6 g/kg body weight for 9 weeks) reduces cytosine methyltransferase mRNA levels in rat sperm. Lower cytosine methyltransferase mRNA levels could lead to DNA hypomethylation, suggesting a mechanism by which paternal alcohol exposure could affect embryonic development (Bielawski et al., 2002).

Additional studies have focused on imprinted genes because the epigenetic effects of paternal alcohol exposure may compromise genomic imprinting. Knezovich and Ramsay analyzed the methylation status of two paternally imprinted loci, *H19* and Ras protein-specific guanine nucleotide-releasing factor 1 (*Rasgrf1*), by pyrosequencing both in the sperm of exposed mice (about 6 g/kg body weight for 5 weeks) and in offspring tail biopsies (Knezovich and Ramsay, 2012). Sperm DNA did not show any methylation differences between alcohol-exposed and control mice, while somatic offspring DNA showed an alcohol-related hypomethylation of the developmentally significant locus *H19*. An alcohol-associated reduction in postnatal growth was also found. These results suggest an involvement of DNA methylation in underlying the effects of paternal alcohol exposure on offspring, although the data on sperm DNA also suggest the involvement of other epigenetic mechanisms such as noncoding RNA or chromatin remodeling (Knezovich and Ramsay, 2012).

The epigenetic effects of alcohol on imprinted genes in sperm DNA were investigated in a human study by analyzing differentially methylated regions (DMRs) at specific genomic regions: the *H19* locus and intergenic differentially methylated region (IG-DMR). Subjects were split into three categories, i.e., nondrinkers, moderate drinkers, and heavy drinkers, based on their self-reported drinking frequencies and drinks consumed per session. The authors described a correlation between demethylation of paternally methylated DMRs and alcohol consumption, although the results were significant only for the IG-DMR (Ouko et al., 2009).

6.5 ALCOHOL AFFECTS CARCINOGENESIS THROUGH DNA METHYLATION

Cancer epigenetics is the most studied area in epigenetics (Sandoval and Esteller, 2012). In a wide variety of cancers aberrant DNA methylation patterns such as global DNA hypomethylation and gene-specific hypermethylation have been found. It appears that global DNA hypomethylation may contribute to a genomic instability, while promoter hypermethylation may silence tumor suppressors (Ehrlich, 2006; Jones and Baylin, 2007). Interestingly, evidence indicates that alcohol alters both global and gene-specific DNA methylation, thereby accelerating carcinogenesis. Here we summarized the etiologic role of DNA methylation in alcohol associated cancers such as cancers of liver, colorectum, head, and neck, and breast.

6.5.1 Liver Cancer

The metabolism of alcohol in the liver results in an increase in oxidative stress, which is a major risk factor for development of liver diseases, i.e., steatosis, steatohepatitis, cirrhosis, and hepatocellular carcinoma (HCC). Moreover, relative to other organs one-carbon metabolism is predominant in the liver, so it is of particular interest to study the effects of alcohol on DNA methylation in this tissue.

HCC is the most frequent primary liver cancer (El-Serag, 2011) and the main etiological factors are chronic viral hepatitis B and C infections, nonalcoholic fatty liver disease, chronic alcohol intake, and aflatoxins exposure. The molecular mechanism of HCC pathogenesis are still poorly understood, but it is generally recognized that both genetic and epigenetic mechanisms may contribute to the carcinogenic process (Ozen et al., 2013). DNA methylation in particular has been deeply investigated in the development of HCC. Similar to the other cancers, global hypomethylation and aberrant gene-specific DNA methylation have been reported in HCC cells (Sceusi et al., 2011; Pogribny and Rusyn, 2014). Several recent studies analyzed HCC genome-wide methylation profiles by means of high throughput technology (Shin et al., 2010; Shen et al., 2012; Song et al., 2013).

Recent studies have attempted to analyze the epigenetic signatures of different risk factors for HCC (Herceg and Paliwal, 2011) but most of these studies were focused on viral infections, while reports on alcohol-related DNA methylation in HCC are few and inconclusive. Hernandez-Vargas and colleagues performed a large scale DNA promoter methylation profile assay using the Illumina bead array method to identify specific DNA methylation signatures associated with major risk factors and HCC progression (Hernandez-Vargas et al., 2010). They were able to identify three genes, deiodinase, iodothyroninetype III (*DIO3*), and signal transducer and activator of transcription 5A (*STAT5A*), which were specifically hypermethylated in association with alcohol consumption (Hernandez-Vargas et al., 2010). Similarly, Lambert and colleagues identified DNA methylation changes associated with alcohol abuse by pyrosequencing a selected panel of genes. They observed a significantly higher level of methylation of the O^6-methylguanine DNMT gene (*MGMT*) in alcohol-related HCC compared with non-alcohol-related HCC (Lambert et al., 2011).

Neuman and colleagues further demonstrated the importance of different etiological factors in determining a unique DNA methylation profile in HCC. They investigated the genome-wide methylation profile of HCC tumors and observed substantial differences between the four etiological groups: HBV, HCV, alcoholic, and cryptogenic (Neumann et al., 2012). Their data support the hypothesis that alcohol may exclusively affect the methylation status of specific genes, which may shed light on the mechanism underlying alcohol-mediated hepatocarcinogenesis.

6.5.2 Colorectal Cancer

Chronic alcohol consumption, together with older age and diet, is an important risk factor for colorectal cancer (Seitz et al., 2005; Arasaradnam et al., 2008). Moreover, aberrant DNA methylation, both global hypomethylation and gene-specific hypermethylation, has been widely reported in this cancer (Feinberg et al., 1988; Esteller et al., 2001; Bai et al., 2004; Ogino et al., 2008), suggesting that alcohol-mediated impairment of methyl transfer function may be involved in colon cancer development. Global DNA hypomethylation and *p16*-specific hypermethylation have been associated with ethanol consumption and reduced dietary folate intake in mice (Sauer et al., 2010), indicating that altered DNA methylation by reduced methyl transfer capacity may mediate the alcohol associated colonic carcinogenesis. In humans, Schernhammer and colleagues demonstrated that high alcohol consumption (≥15 g of alcohol per day) conferred a higher risk of LINE-1 hypomethylated colon cancer (Schernhammer et al., 2009).

Aberrant DNA methylation induced by chronic alcohol consumption has also been investigated using a gene-specific approach of analyzing the promoter methylation of six genes related to colorectal cancer: adenomatous polyposis coli (*APC*), *p14ARF*, *p16INK4A*, human mutLhomolog 1 (*hMLH1*), *MGMT*, and Ras association domain family 1 A (*RASSF1A*). Each of these gene's promoter was hypermethylated in colorectal cancer patients who consumed a diet low in folate (<215 g/day) and high in alcohol (≥5 g/day) when compared with those who reported high folate (≥215 g/day) and low alcohol intake (0–4 g/day), although the difference did not reach statistical significance (van Engeland et al., 2003).

6.5.3 Head and Neck Cancer

Epigenetic changes have been implicated in the pathogenesis of head and neck squamous cell carcinoma. Head and neck squamous cell carcinoma tissues are globally hypomethylated compared with the normal mucosal specimen. The degree of global hypomethylation is associated with alcohol use in univariate analysis ($p = 0.02$) (Smith et al., 2007). In a case and control study, history of alcohol use is associated with gene-specific methylation. Promoter hypermethylation of *hMLH1*, *MGMT*, and *p16* genes was commonly detected in cases of alcohol use, with the hypermethylation of the *MGMT* promoter significantly correlating with the history of alcohol use (Puri et al., 2005).

6.5.4 Breast Cancer

DNA methylation has been associated with breast cancer in a growing number of studies. One study that measured 1413 autosomal CpG loci associated with 773 cancer-related genes using a DNA methylation array in primary breast tumors demonstrated a significant independent association (controlling for age, dietary folate and other variables) between alcohol intake and tumor DNA methylation profile (Christensen et al., 2010). It appears that the exposure to alcohol as a breast cancer risk factor appears to be associated with gene-specific tumor methylation as well as overall methylation patterns.

From incident early-stage breast cancer (I–IIIa) subjects, a significant difference in leukocyte global DNA methylation was found when compared with controls that were free of cancer ($p < 0.001$). Interestingly, higher alcohol consumption was associated with lower global methylation both in controls ($p = 0.031$) and in cases and controls combined ($p = 0.036$) (Choi et al., 2009). This observation is further supported in a large pooled study of leukocyte DNA methylation of repeat elements, in which higher alcohol intake was inversely correlated with lower Alu methylation (r = −0.214) in healthy subjects. However, no association was found between alcohol drinking and LINE-1 methylation (Zhu et al., 2012). Among repetitive sequences, Alu and LINE-1 are the most plentiful families representing ~30% of the human genome. Because of their high representation throughout the genome, Alu and LINE-1 have been used as surrogate markers for estimating global DNA methylation levels.

6.6 ALCOHOLISM, ALCOHOL ABUSE, AND DNA METHYLATION

Recent findings highlighted the role of epigenetic modifications induced by alcoholism including those affecting DNA methylation (Starkman et al., 2012). Bönsch and colleagues, using a nonradioactive elongation assay, observed a significant increase in global DNA methylation in peripheral blood mononuclear cells (PBMCs) taken from alcoholics (more specifically patients with a history of alcohol consumption ranging from a duration of 6–42 years and a mean ethanol intake of 230 g/day) relative to controls (Bönsch et al., 2004). The same authors subsequently confirmed this increase of global PBMCs DNA methylation and also reported a significant decrease of mRNA expression of two DNMTs, DNMT-3a and DNMT-3b. The increase of global DNA methylation in patients with alcoholism was significantly associated with lowered DNMT-3b mRNA expression ($p = 0.014$). These results suggest that the transcriptional repression of DNMT-3b might be regulated through a negative feedback mechanism through increased global DNA methylation (Bönsch et al., 2006). Studies by Bönsch and colleagues (Bönsch et al., 2004; Bönsch et al., 2005; Bleich et al., 2006; Bönsch et al., 2006) specified that patients were taking no vitamin supplements and their nutritional assessment revealed no abnormalities. Not all the studies were, however, controlled for vitamins levels/intake (Hillemacher et al., 2009; Muschler et al., 2010; Manzardo et al., 2012).

Furthermore, there is some evidence indicating that alcohol may alter the methylation status of certain genes associated with brain function. In a human study that investigated whether the pattern of PBMCs DNA methylation can be altered in alcoholic subjects (with a history of alcohol consumption ranging from 6 to 42 years and with a mean ethanol intake of 237 g/day), a significant increase in DNA methylation was found in the promoter of α-synuclein gene, which encodes for a functional microtubule-associated protein (Bönsch et al., 2005). Meanwhile, no significant difference in the promoter DNA methylation within a control gene (*presenilin-1*) in alcoholics and controls was found, indicating that alcohol consumption may specifically affect the DNA methylation status of α-synuclein, which has been linked to alcohol craving, a common, although not uniform, feature of alcohol addiction (Bönsch et al., 2005).

Hillemacher and associates showed significantly higher methylation of the dopamine transporter (*DAT*) in patients undergoing alcohol withdrawal (Hillemacher et al., 2009). The dopaminergic neurotransmission is known to play an important role in the development and maintenance of alcohol dependence. Results from that study indicated that alcohol dependence might be associated with promoter DNA methylation of *DAT* that can influence the dopaminergic neurotransmission in alcoholism (Hillemacher et al., 2009).

It is also known that elevated plasma Hcy concentrations can influence both genomic and gene-specific DNA methylation in PBMCs. As described earlier, chronic alcohol consumption increases plasma Hcy levels by altering one-carbon metabolism. In a human study using alcohol-dependent patients with hyperhomocysteinemia, a significantly increased promoter DNA methylation of the Hcy-induced endoplasmic reticulum protein (*HERP*) gene ($p < 0.02$) was found in patients with alcohol dependence compared with healthy controls, which was also significantly associated with their elevated Hcy levels ($p < 0.007$). Furthermore, *HERP* mRNA expression was lower in patients with alcohol dependence compared with healthy controls ($p < 0.001$). These findings may be useful in the understanding of epigenetic mechanisms underlying alcohol dependence (Bleich et al., 2006). *HERP* is usually upregulated in response to endoplasmic reticulum stressing conditions, and the downregulation found in alcoholic patients may suggest an impairment in the ER stress response leading to a higher rate of vascular incidents, seizures, and other neurological damages. A correlation between promoter hypermethylation and transcriptional repression was also found for neuronal growth factor (*NGF*) (Heberlein et al., 2013) and for proopiomelanocortin (POMC) gene in PBMCs (Muschler et al., 2010). The epigenetic regulation of the *NGF* gene was observed during alcohol withdrawal, while the promoter hypermethylation and transcriptional repression of POMC was found in patients suffering from alcohol dependence compared with healthy controls.

Recently, Manzardo and colleagues analyzed genome-wide DNA methylation profiles in the frontal cortex of alcoholic patients compared with controls in an effort to provide insights into the epigenetic regulation in the human brain. The results of this study provided a detailed examination of DNA methylation profiles in the brain of alcoholic subjects. In particular, an increase in the degree of methylation in the promoter of several histone genes was found, suggesting the possibility of epigenetic disturbances in histone expression in alcoholism (Manzardo et al., 2012). This study demonstrated that alcohol is able to induce tissue-specific epigenetic changes, specifically in the brain, in addition to the "systemic" differences found in the blood.

The reports discussed here support the hypothesis that DNA methylation may play a role in the genesis and maintenance of alcohol-seeking behavior (Hillemacher, 2011). An important issue that should be considered are the different methods used to measure DNA methylation, both global as well as gene-specific, that may account for inconsistencies from different studies (Lim and Song, 2012).

6.7 CONCLUSIONS

Epigenetic effects of alcohol are matter of research for their crucial roles in alcohol-related diseases, such as FASDs, cancer, and alcoholism. In the present chapter, the effects of alcohol on DNA methylation have been illustrated in details referring to both animal and human studies. Also very interesting are alcohol-induced histone modifications, i.e., acetylation, methylation, and phosphorilation at histone tails sites, which, however, lie outside the specific object of the present chapter and deserve a distinct dissertation (for a review, see Shukla et al., 2008).

Many studies have demonstrated that DNA methylation is an important mechanism by which alcohol can cause major health problems. Alcohol is known to alter one-carbon metabolism that regulates the availability of methyl groups for DNA methylation. It is also known that both aberrant DNA methylation and altered one-carbon metabolism are highly associated with the development of alcohol-associated diseases such as cancer. However, most of human studies have suggested an associative, but not a causal, relationship. Among the studies presented here results are sometimes contradictory, but the comparison between different studies is difficult because (1) DNA methylation is tissue- and gene-specific; (2) DNA methylation is easily influenced by other environmental factors including diet, age, and genotypes; and (3) the type of DNA methylation markers (global or gene-specific) and methods to measure them can yield different results.

In the future, it needs to be clarified whether altered DNA methylation induced by alcohol consumption, either through altered one carbon metabolism or inhibition of DNMTs, is directly involved in the development of alcohol associated diseases such as FASD, cancer, and alcoholism. Furthermore, responsible genes and the location of altered DNA methylation in the genes needs to be investigated to find out better preventive and treatment modality for the alcohol associated diseases.

REFERENCES

Abel, E. "Paternal contribution to fetal alcohol syndrome." *Addict Biol* 9 no. 2 (2004): 127–33; discussion 135–6.

Arasaradnam, R. P., Commane, D. M., Bradburn, D. and Mathers, J. C. "A review of dietary factors and its influence on DNA methylation in colorectal carcinogenesis." *Epigenetics* 3 no. 4 (2008): 193–8.

Bai, A. H., Tong, J. H., To, K. F. et al. "Promoter hypermethylation of tumor-related genes in the progression of colorectal neoplasia." *Int J Cancer* 112 no. 5 (2004): 846–53.

Barak, A. J., Beckenhauer, H. C. and Tuma, D. J. "Methionine synthase: A possible prime site of the ethanolic lesion in liver." *Alcohol* 26 no. 2 (2002): 65–7.

Barak, A. J., Beckenhauer, H. C., Tuma, D. J. and Badakhsh, S. "Effects of prolonged ethanol feeding on methionine metabolism in rat liver." *Biochem Cell Biol* 65 no. 3 (1987): 230–3.

Bielawski, D. M., Zaher, F. M., Svinarich, D. M. and Abel, E. L. "Paternal alcohol exposure affects sperm cytosine methyltransferase messenger RNA levels." *Alcohol Clin Exp Res* 26 no. 3 (2002): 347–51.

Bird, A. "Perceptions of epigenetics." *Nature* 447 no. 7143 (2007): 396–8.

Bleich, S., Lenz, B., Ziegenbein, M. et al. "Epigenetic DNA hypermethylation of the HERP gene promoter induces down-regulation of its mRNA expression in patients with alcohol dependence." *Alcohol Clin Exp Res* 30 no. 4 (2006): 587–91.

Boland, M. J. and Christman, J. K. "Mammalian DNA methyltransferases." In *Nutrients and Epigenetics.* (eds. Choi, S.W. and Friso, S.) Boca Raton, FL: CRC Press, Taylor & Francis Group, 2009.

Bönsch, D., Lenz, B., Fiszer, R., Frieling, H., Kornhuber, J. and Bleich, S. "Lowered DNA methyltransferase (DNMT-3b) mRNA expression is associated with genomic DNA hypermethylation in patients with chronic alcoholism." *J Neural Transm* 113 no. 9 (2006): 1299–304.

Bönsch, D., Lenz, B., Kornhuber, J. and Bleich, S. "DNA hypermethylation of the alpha synuclein promoter in patients with alcoholism." *Neuroreport* 16 no. 2 (2005): 167–70.

Bönsch, D., Lenz, B., Reulbach, U., Kornhuber, J. and Bleich, S. "Homocysteine associated genomic DNA hypermethylation in patients with chronic alcoholism." *J Neural Transm* 111 no. 12 (2004): 1611–6.

Chen, Y., Ozturk, N. C. and Zhou, F. C. "DNA methylation program in developing hippocampus and its alteration by alcohol." *PLoS One* 8 no. 3 (2013): e60503.

Choi, J. Y., James, S. R., Link, P. A. et al. "Association between global DNA hypomethylation in leukocytes and risk of breast cancer." *Carcinogenesis* 30 no. 11 (2009): 1889–97.

Choi, S. W. and Mason, J. B. "Folate status: Effects on pathways of colorectal carcinogenesis." *J Nutr* 132 no. 8 Suppl (2002): 2413S–8S.

Choi, S. W., Stickel, F., Baik, H. W., Kim, Y. I., Seitz, H. K. and Mason, J. B. "Chronic alcohol consumption induces genomic but not p53-specific DNA hypomethylation in rat colon." *J Nutr* 129 no. 11 (1999): 1945–50.

Christensen, B. C., Kelsey, K. T., Zheng, S. et al. "Breast cancer DNA methylation profiles are associated with tumor size and alcohol and folate intake." *PLoS Genet* 6 no. 7 (2010): e1001043.

Christensen, B. C. and Marsit, C. J. "Epigenomics in environmental health." *Front Genet* 2 (2011): 84.

Cravo, M. L., Gloria, L. M., Selhub, J. et al. "Hyperhomocysteinemia in chronic alcoholism: Correlation with folate, vitamin B-12, and vitamin B-6 status." *Am J Clin Nutr* 63 no. 2 (1996): 220–4.

Dey, A. and Cederbaum, A. I. "Alcohol and oxidative liver injury." *Hepatology* 43 no. 2 Suppl 1 (2006): S63–74.

Ehrlich, M. "DNA methylation in cancer: Too much, but also too little." *Oncogene* 21 no. 35 (2002): 5400–13.

Ehrlich, M. "Cancer-linked DNA hypomethylation and its relationship to hypermethylation." *Curr Top Microbiol Immunol* 310 (2006): 251–74.

El-Serag, H. B. "Hepatocellular carcinoma." *N Engl J Med* 365 no. 12 (2011): 1118–27.

Esteller, M., Corn, P. G., Baylin, S. B. and Herman, J. G. "A gene hypermethylation profile of human cancer." *Cancer Res* 61 no. 8 (2001): 3225–9.

Falls, J. G., Pulford, D. J., Wylie, A. A. and Jirtle, R. L. "Genomic imprinting: Implications for human disease." *Am J Pathol* 154 no. 3 (1999): 635–47.

Feinberg, A. P. "Phenotypic plasticity and the epigenetics of human disease." *Nature* 447 no. 7143 (2007): 433–40.

Feinberg, A. P., Gehrke, C. W., Kuo, K. C. and Ehrlich, M. "Reduced genomic 5-methylcytosine content in human colonic neoplasia." *Cancer Res* 48 no. 5 (1988): 1159–61.

Feinberg, A. P. and Tycko, B. "The history of cancer epigenetics." *Nat Rev Cancer* 4 no. 2 (2004): 143–53.

Fowler, A. K., Hewetson, A., Agrawal, R. G. et al. "Alcohol-induced one-carbon metabolism impairment promotes dysfunction of DNA base excision repair in adult brain." *J Biol Chem* 287 no. 52 (2012): 43533–42.

Garro, A. J., Mcbeth, D. L., Lima, V. and Lieber, C. S. "Ethanol consumption inhibits fetal DNA methylation in mice: Implications for the fetal alcohol syndrome." *Alcohol Clin Exp Res* 15 no. 3 (1991): 395–8.

Goldberg, A. D., Allis, C. D. and Bernstein, E. "Epigenetics: A landscape takes shape." *Cell* 128 no. 4 (2007): 635–8.

Goodlett, C. R., Horn, K. H. and Zhou, F. C. "Alcohol teratogenesis: Mechanisms of damage and strategies for intervention." *Exp Biol Med (Maywood)* 230 no. 6 (2005): 394–406.

Halsted, C. H., Villanueva, J. A., Devlin, A. M. and Chandler, C. J. "Metabolic interactions of alcohol and folate." *J Nutr* 132 no. 8 Suppl (2002): 2367S–72S.

Hamid, A. and Kaur, J. "Kinetic characteristics of folate binding to rat renal brush border membrane in chronic alcoholism." *Mol Cell Biochem* 280 no. 1–2 (2005): 219–25.

Hamid, A. and Kaur, J. "Decreased expression of transporters reduces folate uptake across renal absorptive surfaces in experimental alcoholism." *J Membr Biol* 220 no. 1–3 (2007): 69–77.

Hamid, A., Kaur, J. and Mahmood, A. "Evaluation of the kinetic properties of the folate transport system in intestinal absorptive epithelium during experimental ethanol ingestion." *Mol Cell Biochem* 304 no. 1–2 (2007a): 265–71.

Hamid, A., Kiran, M., Rana, S. and Kaur, J. "Low folate transport across intestinal basolateral surface is associated with down-regulation of reduced folate carrier in in vivo model of folate malabsorption." *IUBMB Life* 61 no. 3 (2009a): 236–43.

Hamid, A., Wani, N. A. and Kaur, J. "New perspectives on folate transport in relation to alcoholism-induced folate malabsorption—Association with epigenome stability and cancer development." *FEBS J* 276 no. 8 (2009b): 2175–91.

Hamid, A., Wani, N. A., Rana, S., Vaiphei, K., Mahmood, A. and Kaur, J. "Down-regulation of reduced folate carrier may result in folate malabsorption across intestinal brush border membrane during experimental alcoholism." *FEBS J* 274 no. 24 (2007b): 6317–28.

Haycock, P. C. and Ramsay, M. "Exposure of mouse embryos to ethanol during preimplantation development: Effect on DNA methylation in the h19 imprinting control region." *Biol Reprod* 81 no. 4 (2009): 618–27.

Heberlein, A., Muschler, M., Frieling, H. et al. "Epigenetic down regulation of nerve growth factor during alcohol withdrawal." *Addict Biol* 18 no. 3 (2013): 508–10.

Herceg, Z. and Paliwal, A. "Epigenetic mechanisms in hepatocellular carcinoma: How environmental factors influence the epigenome." *Mutat Res* 727 no. 3 (2011): 55–61.

Hernandez-Vargas, H., Lambert, M. P., Le Calvez-Kelm, F. et al. "Hepatocellular carcinoma displays distinct DNA methylation signatures with potential as clinical predictors." *PLoS One* 5 no. 3 (2010): e9749.

Hillemacher, T. "Biological mechanisms in alcohol dependence—New perspectives." *Alcohol* 46 no. 3 (2011): 224–30.

Hillemacher, T., Frieling, H., Hartl, T., Wilhelm, J., Kornhuber, J. and Bleich, S. "Promoter specific methylation of the dopamine transporter gene is altered in alcohol dependence and associated with craving." *J Psychiatr Res* 43 no. 4 (2009): 388–92.

Jia, D. and Cheng, X. "Methylation on the nucleosome." In *Nutrients and Epigenetics*. (eds. Choi, S.W. and Friso, S.) Boca Raton, FL: CRC Press, Taylor & Francis Group, 2009.

Jones, P. A. "DNA methylation and cancer." *Cancer Res* 46 no. 2 (1986): 461–6.

Jones, P. A. and Baylin, S. B. "The epigenomics of cancer." *Cell* 128 no. 4 (2007): 683–92.

Jones, P. A. and Laird, P. W. "Cancer epigenetics comes of age." *Nat Genet* 21 no. 2 (1999): 163–7.

Jones, P. A. and Takai, D. "The role of DNA methylation in mammalian epigenetics." *Science* 293 no. 5532 (2001): 1068–70.

Kafri, T., Ariel, M., Brandeis, M. et al. "Developmental pattern of gene-specific DNA methylation in the mouse embryo and germ line." *Genes Dev* 6 no. 5 (1992): 705–14.

Kaminen-Ahola, N., Ahola, A., Maga, M. et al. "Maternal ethanol consumption alters the epigenotype and the phenotype of offspring in a mouse model." *PLoS Genet* 6 no. 1 (2010): e1000811.

Kenyon, S. H., Nicolaou, A. and Gibbons, W. A. "The effect of ethanol and its metabolites upon methionine synthase activity in vitro." *Alcohol* 15 no. 4 (1998): 305–9.

Kiefer, J. C. "Epigenetics in development." *Dev Dyn* 236 no. 4 (2007): 1144–56.

Kim, M., Bae, M., Na, H. and Yang, M. "Environmental toxicants–induced epigenetic alterations and their reversers." *J Environ Sci Health C Environ Carcinog Ecotoxicol Rev* 30 no. 4 (2012): 323–67.

Knezovich, J. G. and Ramsay, M. "The effect of preconception paternal alcohol exposure on epigenetic remodeling of the h19 and rasgrf1 imprinting control regions in mouse offspring." *Front Genet* 3 (2012): 10.

Kobor, M. S. and Weinberg, J. "Focus on: Epigenetics and fetal alcohol spectrum disorders." *Alcohol Res Health* 34 no. 1 (2011): 29–37.

Lambert, M. P., Paliwal, A., Vaissiere, T. et al. "Aberrant DNA methylation distinguishes hepatocellular carcinoma associated with HBV and HCV infection and alcohol intake." *J Hepatol* 54 no. 4 (2011): 705–15.

Lee, T. D., Sadda, M. R., Mendler, M. H. et al. "Abnormal hepatic methionine and glutathione metabolism in patients with alcoholic hepatitis." *Alcohol Clin Exp Res* 28 no. 1 (2004): 173–81.

Lieber, C. S., Casini, A., Decarli, L. M. et al. "S-adenosyl-L-methionine attenuates alcohol-induced liver injury in the baboon." *Hepatology* 11 no. 2 (1990): 165–72.

Lim, U. and Song, M. A. "Dietary and lifestyle factors of DNA methylation." *Methods Mol Biol* 863 (2012): 359–76.

Liu, Y., Balaraman, Y., Wang, G., Nephew, K. P. and Zhou, F. C. "Alcohol exposure alters DNA methylation profiles in mouse embryos at early neurulation." *Epigenetics* 4 no. 7 (2009): 500–11.

Lu, S. C., Huang, Z. Z., Yang, H., Mato, J. M., Avila, M. A. and Tsukamoto, H. "Changes in methionine adenosyltransferase and S-adenosylmethionine homeostasis in alcoholic rat liver." *Am J Physiol Gastrointest Liver Physiol* 279 no. 1 (2000): G178–85.

Lu, S. C., Martinez-Chantar, M. L. and Mato, J. M. "Methionine adenosyltransferase and S-adenosylmethionine in alcoholic liver disease." *J Gastroenterol Hepatol* 21 Suppl 3 (2006): S61–4.

Lu, S. C. and Mato, J. M. "Role of methionine adenosyltransferase and S-adenosylmethionine in alcohol-associated liver cancer." *Alcohol* 35 no. 3 (2005): 227–34.

Luczak, M. W. and Jagodzinski, P. P. "The role of DNA methylation in cancer development." *Folia Histochem Cytobiol* 44 no. 3 (2006): 143–54.

Manari, A. P., Preedy, V. R. and Peters, T. J. "Nutritional intake of hazardous drinkers and dependent alcoholics in the UK." *Addict Biol* 8 no. 2 (2003): 201–10.

Manzardo, A. M., Henkhaus, R. S. and Butler, M. G. "Global DNA promoter methylation in frontal cortex of alcoholics and controls." *Gene* 498 no. 1 (2012): 5–12.

Marutha Ravindran, C. R. and Ticku, M. K. "Changes in methylation pattern of NMDA receptor NR2B gene in cortical neurons after chronic ethanol treatment in mice." *Brain Res Mol Brain Res* 121 no. 1–2 (2004): 19–27.

Mason, J. B. and Choi, S. W. "Effects of alcohol on folate metabolism: Implications for carcinogenesis." *Alcohol* 35 no. 3 (2005): 235–41.

Monk, M., Boubelik, M. and Lehnert, S. "Temporal and regional changes in DNA methylation in the embryonic, extraembryonic and germ cell lineages during mouse embryo development." *Development* 99 no. 3 (1987): 371–82.

Mukhopadhyay, P., Rezzoug, F., Kaikaus, J., Greene, R. M. and Pisano, M. M. "Alcohol modulates expression of DNA methyltranferases and methyl CpG-/CpG domain-binding proteins in murine embryonic fibroblasts." *Reprod Toxicol* 37 (2013): 40–8.

Muschler, M. A., Hillemacher, T., Kraus, C., Kornhuber, J., Bleich, S. and Frieling, H. "DNA methylation of the POMC gene promoter is associated with craving in alcohol dependence." *J Neural Transm* 117 no. 4 (2010): 513–9.

Neumann, O., Kesselmeier, M., Geffers, R. et al. "Methylome analysis and integrative profiling of human HCCs identify novel protumorigenic factors." *Hepatology* 56 no. 5 (2012): 1817–27.

Ogino, S., Nosho, K., Kirkner, G. J. et al. "A cohort study of tumoral LINE-1 hypomethylation and prognosis in colon cancer." *J Natl Cancer Inst* 100 no. 23 (2008): 1734–8.

Ouko, L. A., Shantikumar, K., Knezovich, J., Haycock, P., Schnugh, D. J. and Ramsay, M. "Effect of alcohol consumption on CpG methylation in the differentially methylated regions of H19 and IG-DMR in male gametes: Implications for fetal alcohol spectrum disorders." *Alcohol Clin Exp Res* 33 no. 9 (2009): 1615–27.

Ozen, C., Yildiz, G., Dagcan, A. T. et al. "Genetics and epigenetics of liver cancer." *N Biotechnol* 30 no. 4 (2013): 381–4.

Perkins, A., Lehmann, C., Lawrence, R. C. and Kelly, S. J. "Alcohol exposure during development: Impact on the epigenome." *Int J Dev Neurosci* 31 no. 6 (2013): 391–7.

Planells, E., Sanchez, C., Montellano, M. A., Mataix, J. and Llopis, J. "Vitamins B6 and B12 and folate status in an adult Mediterranean population." *Eur J Clin Nutr* 57 no. 6 (2003): 777–85.

Pogribny, I. P. and Rusyn, I. "Role of epigenetic aberrations in the development and progression of human hepatocellular carcinoma." *Cancer Lett* 342 no. 2 (2014): 223–30.

Poschl, G. and Seitz, H. K. "Alcohol and cancer." *Alcohol* 39 no. 3 (2004): 155–65.

Puri, S. K., Si, L., Fan, C. Y. and Hanna, E. "Aberrant promoter hypermethylation of multiple genes in head and neck squamous cell carcinoma." *Am J Otolaryngol* 26 no. 1 (2005): 12–7.

Riggs, A. D. and Pfeifer, G. P. "X-chromosome inactivation and cell memory." *Trends Genet* 8 no. 5 (1992): 169–74.

Riley, E. P., Infante, M. A. and Warren, K. R. "Fetal alcohol spectrum disorders: An overview." *Neuropsychol Rev* 21 no. 2 (2011): 73–80.

Sandoval, J. and Esteller, M. "Cancer epigenomics: Beyond genomics." *Curr Opin Genet Dev* 22 no. 1 (2012): 50–5.

Sauer, J., Jang, H., Zimmerly, E. M. et al. "Ageing, chronic alcohol consumption and folate are determinants of genomic DNA methylation, p16 promoter methylation and the expression of p16 in the mouse colon." *Br J Nutr* 104 no. 1 (2010): 24–30.

Sceusi, E. L., Loose, D. S. and Wray, C. J. "Clinical implications of DNA methylation in hepatocellular carcinoma." *HPB (Oxford)* 13 no. 6 (2011): 369–76.

Schernhammer, E. S., Giovannucci, E., Kawasaki, T., Rosner, B., Fuchs, C. S. and Ogino, S. "Dietary folate, alcohol and B vitamins in relation to LINE-1 hypomethylation in colon cancer." *Gut* 59 no. 6 (2009): 794–9.

Seitz, H. K., Maurer, B. and Stickel, F. "Alcohol consumption and cancer of the gastrointestinal tract." *Dig Dis* 23 no. 3–4 (2005): 297–303.

Seitz, H. K. and Stickel, F. "Molecular mechanisms of alcohol-mediated carcinogenesis." *Nat Rev Cancer* 7 no. 8 (2007): 599–612.

Shen, J., Wang, S., Zhang, Y. J. et al. "Genome-wide DNA methylation profiles in hepatocellular carcinoma." *Hepatology* 55 no. 6 (2012): 1799–808.

Shin, S. H., Kim, B. H., Jang, J. J., Suh, K. S. and Kang, G. H. "Identification of novel methylation markers in hepatocellular carcinoma using a methylation array." *J Korean Med Sci* 25 no. 8 (2010): 1152–9.

Shukla, S. D., Velazquez, J., French, S. W., Lu, S. C., Ticku, M. K. and Zakhari, S. "Emerging role of epigenetics in the actions of alcohol." *Alcohol Clin Exp Res* 32 no. 9 (2008): 1525–34.

Smith, I. M., Mydlarz, W. K., Mithani, S. K. and Califano, J. A. "DNA global hypomethylation in squamous cell head and neck cancer associated with smoking, alcohol consumption and stage." *Int J Cancer* 121 no. 8 (2007): 1724–8.

Song, M. A., Tiirikainen, M., Kwee, S., Okimoto, G., Yu, H. and Wong, L. L. "Elucidating the landscape of aberrant DNA methylation in hepatocellular carcinoma." *PLoS One* 8 no. 2 (2013): e55761.

Starkman, B. G., Sakharkar, A. J. and Pandey, S. C. "Epigenetics-beyond the genome in alcoholism." *Alcohol Res* 34 no. 3 (2012): 293–305.

Stouder, C., Somm, E. and Paoloni-Giacobino, A. "Prenatal exposure to ethanol: A specific effect on the H19 gene in sperm." *Reprod Toxicol* 31 no. 4 (2011): 507–12.

Tilghman, S. M. "The sins of the fathers and mothers: Genomic imprinting in mammalian development." *Cell* 96 no. 2 (1999): 185–93.

Tsukamoto, H. and Lu, S. C. "Current concepts in the pathogenesis of alcoholic liver injury." *FASEB J* 15 no. 8 (2001): 1335–49.

Udali, S., Guarini, P., Moruzzi, S., Choi, S. W. and Friso, S. "Cardiovascular epigenetics: From DNA methylation to microRNAs." *Mol Aspects Med* 34 no. 4 (2013): 883–901.

Van Engeland, M., Weijenberg, M. P., Roemen, G. M. et al. "Effects of dietary folate and alcohol intake on promoter methylation in sporadic colorectal cancer: The Netherlands cohort study on diet and cancer." *Cancer Res* 63 no. 12 (2003): 3133–7.

Villanueva, J. A. and Halsted, C. H. "Hepatic transmethylation reactions in micropigs with alcoholic liver disease." *Hepatology* 39 no. 5 (2004): 1303–10.

Wani, N. A., Thakur, S., Najar, R. A., Nada, R., Khanduja, K. L. and Kaur, J. "Mechanistic insights of intestinal absorption and renal conservation of folate in chronic alcoholism." *Alcohol* 47 no. 2 (2013): 121–30.

Zhao, R., Matherly, L. H. and Goldman, I. D. "Membrane transporters and folate homeostasis: Intestinal absorption and transport into systemic compartments and tissues." *Expert Rev Mol Med* 11 (2009): e4.

Zhou, F. C., Chen, Y. and Love, A. "Cellular DNA methylation program during neurulation and its alteration by alcohol exposure." *Birth Defects Res A Clin Mol Teratol* 91 no. 8 (2011): 703–15.

Zhu, Z. Z., Hou, L., Bollati, V. et al. "Predictors of global methylation levels in blood DNA of healthy subjects: A combined analysis." *Int J Epidemiol* 41 no. 1 (2012): 126–39.

Section III

Other Essential Nutrients

7 Ascorbate as a Modulator of the Epigenome

Nilay Y. Thakar and Ernst J. Wolvetang

CONTENTS

7.1 INTRODUCTION

Vitamin C (ascorbate) is not produced in the human body, and we therefore rely on dietary intake of this important molecule. Hence, it is not surprising that we have evolved several mechanisms for efficient uptake and recycling of ascorbate (reviewed elsewhere).

Well before the discovery of its molecular identity the importance of vitamin C intake (in the form of fresh fruit) to prevent scurvy was recognized but it was not until 1953 that Van Robertson and Schwartz demonstrated the importance of vitamin C in cell biology and human health [1,2], particularly in regulation of collagen synthesis. It indeed plays an essential role in bone formation, wound healing, and the maintenance of healthy gums [3] Subsequently, several reports have implicated vitamin C as a potential drug for treatment and prevention of human ailments, ranging from the common cold and selected upper respiratory tract infections to diabetes, cataracts, glaucoma, macular degeneration, atherosclerosis, stroke, heart diseases, and cancer [3–8] and even attributed it an antiaging role [9,10]. These effects are largely ascribed to its antioxidant role or its function as a cofactor in several enzymatic reactions or its ability to affect hypoxia-inducible factor stability and genome-wide DNA methylation [5,6].

Vitamin C can at least qualitatively be considered one of the most important antioxidants given its ability to directly scavenge superoxide, hydrogen peroxide, the hydroxyl radical, hypochlorous acid, aqueous peroxyl radicals, and singlet oxygen [11,12]. It serves a key role in the metabolism of tyrosine, folic acid, conversion of tryptophan to the neurotransmitter serotonin and the synthesis of amino acids, carnitine, and catecholamines that regulate nervous system function [3,13]. It also helps in the conversion of cholesterol to bile acids and lower blood cholesterol, increases the absorption of iron in the gut, aids production of thyroid hormones, and breaks down histamine, the inflammatory component of several allergic reactions [3,13]. In agreement with these roles, deficiency of ascorbate can lead to anemia, scurvy, infections, bleeding gums, muscle degeneration, poor wound healing, atherosclerotic plaques, capillary hemorrhaging, and neurotic disturbances [3]. In vitro studies [3,12–14] have suggested ascorbate strengthens and protects the immune system by stimulating the activity of antibodies and immune system cells such as phagocytes and neutrophils, protecting against damage by the free radicals released by the body in its fight against the infection, and as an antiviral agent by elevating the body's interferon level [15]. It should be noted, however, that in particular, in the presence of divalent metal ions, ascorbate can also act in a prooxidant fashion [16–18].

In this review, we have chosen to focus on novel emerging roles of vitamin C in the regulation of the epigenome, given the increasing appreciation of the importance of this regulatory layer in governing gene expression and human health.

7.2 EPIGENOME REGULATORY LAYERS

Epigenetic mechanisms are defined as biologic processes that regulate mitotically or meiotically heritable changes in gene expression without altering the DNA sequence [19]. It is a well-known fact that DNA and histones are the basic components of a chromosome, in which the DNA helix is wrapped around core histones to form the simple "beads on a string" structure, which is then folded into higher-order chromatin [20]. Chromatin also contains various proteins that are required for its assembly and packaging, and for DNA replication, DNA and histone modification and transcription, and DNA repair and recombination. Chromatin is not uniform with respect to gene distribution and transcriptional activity. It is organized into domains, such as euchromatin and heterochromatin, which have different chromosomal architecture, transcriptional activity and replication timing [19–22]. Epigenetic marking by the covalent modification of DNA and of the core histones creates molecular landmarks that differentiate among active, inactive, and poised chromatin. Chromatin modification and chromatin remodeling constitute the core epigenetic mechanisms by which tissue-specific gene expression patterns and global gene silencing are established and maintained [20,21,23].

Epigenetic regulation can be divided into three major processes; DNA cytosine methylation, histone modifications such as acetylation and methylation of histone tails, and small non-coding RNA controlled pretranscriptional and posttranscriptional regulation of gene expression [19,22,24,25]. These epigenetic mechanisms are absolutely critical for regulation of gene expression, DNA replication and many other biological processes including maintenance of a stable cell identity [22]. Increasing evidence indicates that DNA- and histone-remodeling proteins interact with one another and with noncoding RNAs to form large complexes that regulate higher-order chromatin structures and the accessibility of chromatin to various factors [20,21,23].

DNA methylation is the process wherein DNA can be covalently modified through methylation of the carbon at the 5th position on the pyrimidine ring of the cytosine residue. In mammals, DNA methylation occurs primarily at symmetrical CpG dinucleotides. *De novo* cytosine methylation, catalyzed by two de novo DNA methyltransferases, Dnmt3a and Dnmt3b, adds methyl groups onto unmethylated DNA, which is important for the establishment of DNA methylation patterns. DNA methylation is primarily involved in establishing parental-specific imprinting during gametogenesis as well as silencing of retrotransposons and genes on the inactivated X-chromosome [19–21].

The posttranslational modification of histones involves acetylation and methylation of conserved lysine and arginine residues on the amino terminal tail, and these are dynamically regulated by chromatin modifying enzymes with opposing activities. Generally, lysine acetylation is mediated by histone acetyltransferases (HATs) and marks transcriptionally active regions. In contrast, histone deacetylases (HDACs) catalyze lysine deacetylation and the resulting hypoacetylated histones are usually associated with transcriptionally inactive chromatin structures [19–22].

Histone methylation is carried out by histone lysine or arginine methyltransferases, which generally deposit a methyl mark on the side chains of lysines and arginines of histones [26,27]. Unlike acetylation and phosphorylation, however, histone methylation does not alter the charge of the histone protein [27]. Adding to the complexity of this process lysines may be monomethylated, dimethylated, or trimethylated, whereas arginines may be monomethylated, symmetrically or asymmetrically dimethylated [19,21,22,28]. Histone methylation of lysine 4 on histone H3 (H3-K4) or H3-R17 is identified as an activating methylation mark since it is generally present within transcriptionally competent promoter regions [26]. Conversely, methylation of either lysine 9 or lysine 27 on histone H3 (H3-K9 or H3-K27) is a hallmark of silent chromatin regions and is predominantly associated with heterochromatic regions and/or inactive promoters [19,21,22,26].

Challenging earlier notions of histone methylation being irreversible, Bannister et al. [29] proposed several lysine and arginine demethylation mechanisms, which were subsequently verified experimentally. In 2004, lysine-specific demethylase 1 (LSD1) was identified. When LSD1 is complexed with the Co-REST repressor complex it can demethylate H3K4me1/2, but when LSD1 is complexed with the androgen receptor, it demethylates H3K9 [30]. This has the effect of switching the activity of LSD1 from a repressor function to that of a coactivator [31].

In 2006, lysine demethylases capable of demethylating trimethylated lysines were discovered that employed a distinct catalytic mechanism from that used by LSD1, using Fe(II) and α-ketoglutarate as cofactors, and a radical attack mechanism [32,33]. The first enzyme identified as a trimethyl lysine demethylase was JMJD2 (JmjC domain–containing histone demethylation protein 2) that demethylates H3K9me3 and H3K36me3 [32]. The enzymatic activity of JMJD2 resides within the *JmjC-jumonji* domain. Many histone lysine demethylases, except for LSD1, are all known to possess a catalytic jumonji domain [34].

Similar to the lysine methyltransferases, the demethylases possess a high level of substrate specificity with respect to their target lysine. They are also sensitive to the degree of lysine methylation; for instance, some of the enzymes are only capable of demethylating monomethyl and dimethyl substrates, whereas others can demethylate all three states of the methylated lysine [27]. The jumonji protein JMJD6 (JmjC domain–containing histone demethylation protein 6) was shown to demethylate arginines on histones H3-R2 and H4-R3 [35].

Some of the specificity and dynamics of these covalent epigenetic marks appear to be guided by noncoding RNAs, the third arm of epigenetic regulation. Small noncoding RNAs, miRNAs, lincRNA have indeed been shown to not only guide chromatin-modifying complexes to specific genomic loci but also to regulate the stability and translation of mRNAs and alter the formation of chromatin structures of targeted genomic sequences and ultimately regulate gene expression [36–39].

Taken together, the existence of opposing chromatin-modifying enzymes and cross-talk between the different epigenetic systems provides the molecular basis for the dynamic yet precise regulation of gene expression and escape from this regulatory constraint generally leads to loss of homeostasis and disease.

7.3 VITAMIN C ALTERS hPSC EPIGENOME

Although epigenetic regulation clearly plays a critical role in both development and homeostasis (or loss thereof), these processes, and the possible influence of vitamin C on the epigenome, have been difficult to study in large animal model systems and humans. Human embryonic stem cells (hESCs), cells derived from the inner cell mass of a day 5 preimplantation embryo (blastocyst), are therefore an appropriate and more accessible model system if one wants to understand the role of vitamin C in human health. HESCs display indefinite self-renewal and possess the ability to differentiate into the various cell types of the human body, the core characteristics that epitomize human pluripotent stem cells (hPSCs) [40–46]. Conversely, fully differentiated somatic cells can be reprogrammed into human induced pluripotent stem cells (hiPSCs) [40]. Both differentiation and reprogramming are accompanied by profound epigenetic changes that either consolidate or relax gene regulatory networks to impart cellular identity and homeostasis.

If ascorbate influences the regulation of the human epigenome then one would expect that it should affect hESC growth and differentiation as well as the cell reprogramming process. Because vitamin C is known to be a potent water soluble antioxidant it is commonly used as a supplement for serum-free hPSCs cell culture media (knockout serum replacement [KOSR] medium, mTeSR1 medium, StemPro medium), but is usually not present in serum-containing media. Allegruci et al. [47], found that switching hESCs from serum-containing (ascorbate-free) to serum-free (ascorbate-containing) conditions altered the hESC epigenome, accounting for an extra 90–120 CpG island methylation differences throughout the genome over only a few population doublings.

To further elucidate the role of ascorbate in hESC biology our laboratory next examined genome-wide DNA methylation profiles of hESCs cultured in KOSR medium *with and without* vitamin C and discovered that in hESCs physiological levels of ascorbate not only cause genome-wide demethylation but, rather surprisingly, is responsible for the consistent and specific DNA demethylation of 1847 genes with a clear bias toward demethylation of CpGs near the shores of CpG islands (Figure 7.1). A substantial subset of these DNA demethylated genes displayed concomitant gene expression changes and the position of the demethylated CpGs relative to the transcription start site was correlated with such changes.

We further showed that in response to physiological levels of ascorbate, 50% of demethylated genes that show a concomitant alteration in gene expression were identified by Irizarry et al. [48], as tissue differential DNA methylation regions (T-DMR). These T-DMRs were previously shown to significantly overlap with both differentially methylated regions (r-DMRs) in reprogrammed cells and cancer-specific DMRs (c-DMRs) and undergo shore CpG methylation changes in response to ascorbate. Hypomethylated R-DMRs in iPS cells have also been shown to colocalize with hypermethylated C-DMRs in cancer and bivalent chromatin marks, and hypermethylated R-DMRs with hypomethylated C-DMRs, and the absence of bivalent chromatin marks.

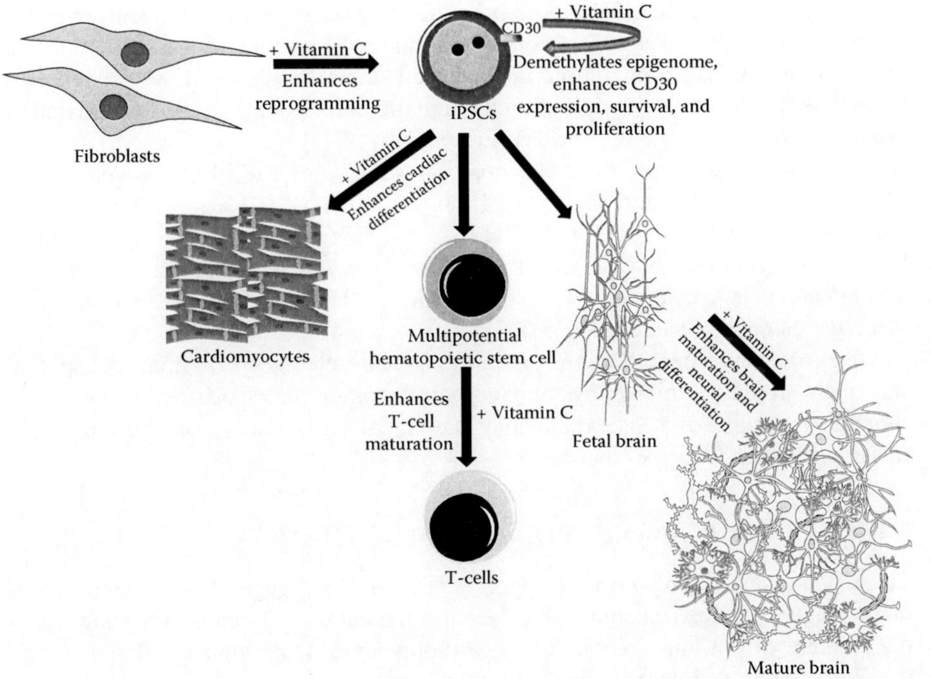

FIGURE 7.1 The impact on cellular identity via vitamin C–dependent modification of the epigenome. Vitamin C significantly enhances reprogramming efficiency and alters embryonic stem cell (ESC) and induced pluripotent stem cell (iPSC) epigenome with a significant role in governance of various biological processes. Vitamin C also markedly enhances cardiac differentiation chiefly by augmenting cardiac progenitor cell proliferation and collagen synthesis, enhances neural differentiation, and enhances T-cell maturation.

These data suggest that ascorbate can significantly influence epigenetic mechanisms involved in cell reprogramming to iPS cells and cancer. The first involves a loss of DNA methylation in iPS and a chromatin-dependent gain of DNA methylation in cancer and the other involves a gain of methylation in iPS and a chromatin-independent loss of DNA methylation in cancer [48,49].

We previously showed that CD30, a biomarker for Hodgkin lymphoma, anaplastic large cell lymphoma, and embryonal carcinoma cells, was one of the 1847 genes that were epigenetically activated by ascorbate-induced loss of DNA methylation of a CpG island, in this case, in the CD30 promoter [43]. Importantly, CD30 expression leads to inhibition of apoptosis, enhanced single-cell growth, and transcriptome changes that are associated with altered cell signaling, lipid metabolism, and tissue development (Figure 7.1) [43] and is considered to be a driver of Hodgkin lymphoma and anaplastic large cell lymphoma pathogenesis.

Despite the fact that this research uncovered unexpected and profound effects of ascorbate on the epigenetic makeup of somatic, hESCs, and iPSCs and emphasized the importance of cell culture conditions when examining hPSC behavior in vitro, the precise molecular mechanisms that underlie the specificity of the ascorbate-induced DNA methylation changes and its bias for CpG island boundaries remained largely unclear. The first clues as to how this may occur became evident when we examined the effect of ascorbate on histone methylation and demonstrated that in hESCs ascorbate causes increased histone acetylation, decreased H3K9 methylation as well as increased expression of JmjC domain–containing histone demethylation proteins (JMJD) 1A, 1B, 1C, and 2B (Figure 7.1) [42,43].

We hypothesized that ascorbate not only modulates the hPSC epigenome by altering the expression and activity of JMJD1 and JMJD2 affecting DNA topology [50,51] but may also directly affect DNA methylation through ascorbate-stimulated DNA demethylases such as FTO (fat mass and obesity associated), a 2-oxoglutarate-dependent nucleic acid demethylase [52] and/or TET1 (ten-eleven translocation methylcytosine dioxygenase 1) [53].

Given the reproducible and profound epigenetic changes in human pluripotent stem cells brought about by physiological concentrations of ascorbate it is not unreasonable to assume that ascorbate may have analogous effects on the epigenetic makeup of other cell types as well.

7.4 VITAMIN C AND CARDIAC DIFFERENTIATION

Cardiogenesis is a highly organized process that is tightly regulated by key developmental signals and extracellular microenvironmental cues. Because for many years the efficiency of in vitro generation of cardiomyocytes from mouse (m) and human (h) iPSCs remained relatively low [54], many laboratories explored various protocols aimed at improving this [54]. Several reports not only demonstrated an important role for vitamin C in enhancing cardiac differentiation of embryonic stem cells [54–60] but also offered new insights into the underlying molecular mechanisms (Figure 7.1). Takahashi et al. [59], showed that treatment of ES cells with 10^{-4} M ascorbic acid (physiological range) markedly increased the rate of differentiation into beating cardiac myocytes and induced the expression of important cardiac genes, such as

GATA4, α-MHC, and β-MHC in ES cells. Importantly, this effect was not mimicked by the other antioxidants such as *N*-acetylcysteine, Tiron, or vitamin E, supporting the notion that ascorbate at physiological levels enhanced cardiac differentiation independently of its antioxidant activity.

This same conclusion was reached by Cao et al. [54] who examined cardiac differentiation of mouse iPSCs [54,58]. They showed that only ascorbic acid consistently and robustly enhanced cardiac differentiation but also showed that alternative antioxidants (including Vb1, GMEE, and NAC) repeatedly failed to mimic the cardiomyocyte-promoting role of ascorbate in iPSCs. Significantly, ascorbate addition at physiological levels led to a 7.3-fold (miPSCs) and 30.2-fold (human iPSCs) increase in the yield of iPS-derived cardiomyocytes [54]. Furthermore, they showed that ascorbate exerted its enhancing effect on cardiac differentiation between day 2 and day 6 of differentiation, during which the specification of cardiac progenitor cells (CPCs) occurs. In support of a specific effect of ascorbate on CPCs, ascorbate increased the expression of cardiovascular markers but not pan-mesodermal markers. The authors attributed this effect to a specific increase in the proliferation of CPCs via the MEK-ERK1/2 pathway, which increased collagen synthesis. Interestingly, following maturation of the CPCs into more mature cardiomyocytes, ascorbate-induced cardiomyocytes showed better sarcomeric organization and enhanced responses of action potentials and calcium transients to β-adrenergic and muscarinic stimulations [54], suggesting that early ascorbate treatment not only increased cardiomyocyte numbers but also their quality.

Martinez et al. [55] further substantiated these findings by demonstrating that ascorbic acid supplementation greatly improved cardiomyoblast survival and increased vascularisation in H9C2 cardiomyoblasts-derived myocardial grafts in vivo. Kim et al. [56] next demonstrated that even post-engraftment ascorbate has positive effects on cardiomyocytes by inhibiting cellular aging.

At a cellular level, aging is typically characterized by the shortening of telomeres, accumulation of DNA mutations, and development of mitochondrial dysfunction and increased ROS production [61]. Not surprisingly, cardiomyocytes indeed show increased reactive oxidative species formation, and mitochondrial damage [62] with age. Kim et al. [56] showed that in cells were treated with 100 μM of vitamin C for 48 h, a concentration that is close to the recommended vitamin C dietary allowance, it conferred an antiaging cardioprotective effect in terms of expression of telomere-related genes, mitochondrial membrane potential, and overall cardiomyocyte function (Figure 7.1).

It remains at present unclear as to whether these antiaging effects of ascorbate are solely due to the antioxidant properties of ascorbate or are, perhaps in part, also related to its epigenetic modulatory roles. In any case, these data suggest that deficiency of ascorbate during cardiac development as well as in the aging heart may be a potential risk factor.

7.5 EPIGENETIC REGULATION OF CELLULAR ASCORBATE HOMEOSTASIS

If one accepts that ascorbate is a, perhaps currently insufficiently recognized, modulator of the epigenome then it would make sense that the cell would regulate the cellular concentration of ascorbate through an epigenetic mechanism. Recent evidence suggests that this is indeed the case and that modulation of DNA demethylation by ascorbate is part of an autoregulatory mechanism that controls cellular ascorbate concentration [63].

Since most mammalian cells and all human cells are unable to synthesize vitamin C (ascorbic acid), they are dependent upon uptake of the vitamin from their surroundings. This is primarily mediated by one of two sodium- and energy-dependent vitamin C transporters, termed SVCT (sodium–vitamin C cotransporter) 1 (SLC23A1) and SVCT2 (SLC23A2) [63]. SVCT2 (sodium–vitamin C cotransporter 2) is the major transporter mediating vitamin C uptake in most organs and its expression is critical for life, since its targeted deletion in the mouse fails to produce viable pups [64], further highlighting the importance of vitamin C.

A recent study mapped discrete elements within the proximal CpG-poor promoter of the SVCT2 gene, which is responsible for transcription exon 1a and identified two E boxes for USF (upstream stimulating factor) binding and one Y box for NF-Y (nuclear factor Y) binding [63]. The analysis of the CpG site located at the upstream USF-binding site in the promoter showed a strong correlation between expression and demethylation and the specific methylation of this CpG site impaired both the binding of USF and the formation of the functional NF-Y–USF complex as well as promoter activity, suggesting that it is important for cell-specific transcription of human ascorbate transporter SVCT2 exon 1a [63].

These data therefore strongly argue for an epigenetically controlled autoregulatory mechanism that directs cellular ascorbate concentrations through CpG demethylation–mediated promoter regulation of ascorbate transporter genes.

7.6 VITAMIN C AND CELL REPROGRAMMING

While hESCs are isolated from an early blastocyst, iPSCs offer an additional source of pluripotent stem cells for application in regenerative medicine. Mouse iPSCs were first generated in 2006 [65], followed by human iPSCs (hiPSCs) in 2007. The delivery and transient ectopic expression of just four genes, *OCT4*, *SOX2*, *KLF4*, and *MYC* as demonstrated by Yamanaka [65], or the six-gene approach consisting of *OCT4*, *SOX2*, *KLF4*, *C-MYC*, *LIN28*, and *NANOG* as demonstrated by Thomson have been shown to enable the reprogramming of adult somatic cells to an embryonic stem cell–like state. Somatic cell reprogramming not only constituted a paradigm shift in the stem cell field, since it allows the generation of patient-specific iPSC lines, provides a virtually limitless source of autologous, pluripotent cells and bypasses most of the ethical issues associated with using hESCs, but also challenged the dogma that somatic cells are epigenetically locked into a permanent cellular identity.

Since cell reprogramming requires extensive remodeling of the cellular epigenomic makeup and given the profound DNA demethylating effects of vitamin C and its potential role in altering histone methylation, it was recognized early that ascorbate had the potential to affect cell reprogramming from fibroblasts into iPSCs. Indeed, Esteban et al. [66] reported that vitamin C at physiological levels enhanced reprogramming of mouse and human fibroblasts into iPSCs (Figure 7.1) [66]. They conclusively demonstrated that vitamin C enhances the kinetics of mice and human iPSC generation, particularly in conversion of pre-iPSCs into fully reprogrammed iPSCs. Generally, the efficiency of alkaline phosphatase-positive (AP^+) colony formation with the four Yamanaka factors (*SOX2*, *KLF4*, *OCT4*, *C-MYC*; SKOM) in mouse fibroblasts is about 1% of the starting population, but only around 1 in 10 of those colonies is sufficiently reprogrammed to be chimera competent [63]. In human fibroblasts, only about 0.01% of cells transduced with SKOM form AP^+ iPSC colonies [40,41]. With the addition of ascorbate at physiological levels, this efficiency increased 8- to 10-fold for mice iPSCs and 3- to 6-fold for human iPSCs [66]. The authors suggested that this effect of vitamin C was at least in part due to its antioxidant effect and its ability to alleviate cell senescence by reducing p53 expression.

The observation that ascorbate also accelerated transcriptome changes during reprogramming that allow for the conversion of pre-iPSCs to iPSCs [66] suggested additional epigenetic effects of ascorbate perhaps related to the DNA-demethylating and histone-modifying effects of ascorbate. Strong support for this notion was presented by Stadtfeld et al. [64], who showed that reprogramming in the presence of ascorbic acid attenuates hypermethylation of Dlk1-Dio3, thus facilitating the generation of all-iPSC mice from mature B cells, which had until then failed to support the development of exclusively iPSC-derived postnatal animals.

Importantly, prevention of Dlk1–Dio3 silencing by ascorbic acid was not due to any potential effect of ascorbate on reprogramming factor expression or stoichiometry, since relative protein levels of OKSM factors between ascorbic acid–treated and untreated reprogrammable MEFs were quite similar. Furthermore, all of the MEF-derived iPSC lines generated by exogenous expression of only three factors (OKS) in the absence of ascorbic acid showed hypermethylation of the IG-DMR and Meg3 DMR [64]. Moreover, iPSCs generated with or without ascorbic acid displayed complete DNA demethylation and acquisition of the activating histone H3 lysine 4 trimethylation (H3K4me3) chromatin mark at the Pou5f1 promoter while methylation levels of other imprinted genes such as paternally imprinted H19 and Rasgrf1 genes and the maternally imprinted Mest and Peg3 genes remained unchanged. These results indicate that ascorbic acid affects neither epigenetic remodeling of key pluripotency genes nor global imprinted-gene methylation. These results were further corroborated by the observation that the efficiency and kinetics of reprogramming of B-cells was unaffected by ascorbate [64] and that ascorbic acid exposure did not cause major global epigenetic aberrations, at least at the level of genome-wide H3K4me3 patterns in ESCs [64].

Surprisingly, the expression status of maternal Dlk1–Dio3 transcripts was the only apparent difference between clones derived in the presence and absence of ascorbic acid in global gene expression profiling suggesting a remarkable specificity for ascorbic acid. However, it is possible that ascorbic acid influences other chromatin marks in iPSCs that do not adversely affect iPSC developmental potential as measured by their ability to generate adult all-iPSC mice. Given its strong effect on reprogramming speed and efficiency and demonstrated ability to alter hESC epigenome [42], it is plausible that ascorbic acid drives widespread but transient alterations in chromatin that are resolved upon entry into a pluripotent state and hence are undetectable in established iPSCs.

Of note, exposure of established Meg3^{off} iPSCs (iPSC lines that showed aberrant silencing of Dlk1–Dio3 transcripts) to ascorbic acid for fifteen passages was insufficient to reverse silencing due to sustained hypermethylation of both the IG-DMR and Meg3 DMR, which shows that ascorbic acid treatment prevents but cannot reverse aberrant DNA methylation of the Dlk1–Dio3 locus. Their observations suggest that the rapid loss of activating histone marks in the absence of ascorbic acid facilitates the recruitment of Dnmt3a to the IG-DMR, allowing Dnmt3a to catalyze DNA hypermethylation and leading to stable epigenetic silencing of maternal Dlk1–Dio3 transcripts [64]. Remarkably, Stadtfeld et al. demonstrated that simply adding ascorbic acid to cells undergoing reprogramming allowed the derivation of Meg3^{on} iPSCs from mature B cells that supported the development of adult all-iPSC mice.

This was the first demonstration that full developmental potential can be reinstated in iPSCs produced from a terminally differentiated adult cell of defined origin. This report [64] adds to the recent accumulation of reports [42,43,45,66–68] that draw attention to the critical role that environmental conditions play during cellular reprogramming and the profound effects it has on the epigenetic and biological properties of hESC and hiPSCs.

7.7 HISTONE MODIFICATION BY VITAMIN C IN CANCER AND HYPOXIA

The control over gene expression in eukaryotes relies in part on the methylation status of histone proteins. The methylation marks within the genome can be "erased" by protein lysine demethylases that include flavin-dependent monoamine oxidase LSD1 [30] and α-ketoglutarate-Fe^{2+}-dependent dioxygenases containing jumonji domains [33,51].

Vitamin C is a cofactor in reactions driven by dioxygenases including collagen prolyl hydroxylases, hypoxia-inducible factor (HIF) prolyl hydroxylases [6], and several histone demethylases [69], and it is tempting to speculate this might be the mechanism via which ascorbate increases reprogramming efficiency and kinetics and alters the hESC and hiPSC epigenome (Figure 7.2).

The jumonji family of histone demethylases are likely candidates that mediate this ascorbate-mediated epigenetic regulation. Vitamin C acts as a critical cofactor for enzymatic reactions that involve iron and α-ketoglutarate-dependent dioxygenases such as members of the jumonji family (Figure 7.2). For example, jumonji domain–containing protein 1A (JMJD1A) is an iron- and 2-oxoglutarate-dependent dioxygenase as well as a hypoxic response gene and its expression is upregulated by HIF-1α [70]. Abnormal expression of histone demethylase JMJD1A has been associated with various cancers, particularly renal carcinoma [70,71].

They are typically comprised of N-terminal jumonji domain and C-terminal prolyl hydroxylase domain (PHD) and tudor domains, such as is the case for JMJD2A [72]. Structural studies reveal that while protein modules such as PHD finger protein (PHFs) are reader modules detecting the methylation status of histones by recognizing lysine in methylated [73–76] and unmethylated states [77,78], the jumonji domain, meanwhile, predominantly recognizes the backbone of the histone peptides (unusually for a sequence-specific enzyme), allowing the enzyme to demethylate both

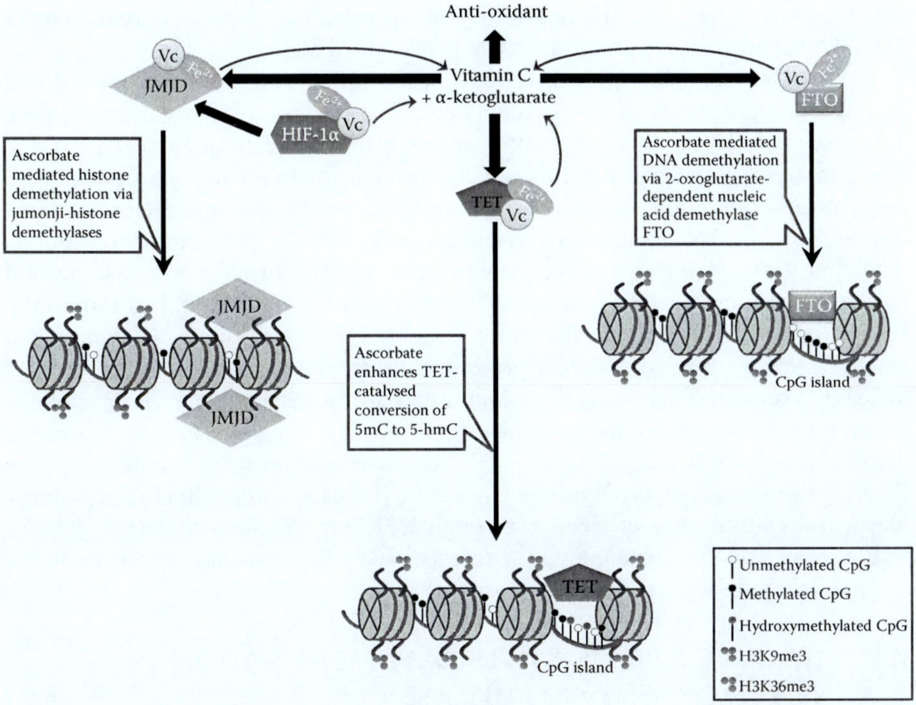

FIGURE 7.2 Various epigenetic regulatory roles of vitamin C. Vitamin C, possibly by recycling of the cofactor Fe^{2+}, acts as a critical cofactor for enzymatic reactions that involve iron and α-ketoglutarate-dependent dioxygenases such as members of the JMJD (histone demethylating jumonji) family proteins or DNA demethylases such as FTO (fat mass and obesity-associated) and/or TET1 (ten-eleven translocation methylcytosine dioxygenase). Vitamin C's role as a potent antioxidant is of course well established, however, its antiaging role could either be due to its anti-oxidative properties or via epigenetic regulatory mechanisms.

H3K9me3 and H3K9me2 as well as H3K36me3 and H3K36me2 (Figure 7.2) [79–81]. Meanwhile, the JMJD2A tudor domain binds two different histone sequences (H3K4me3 and H4K20me3) via very different approaches [82]. The jumonji domain often associates with at least one additional recognizable protein domain within the same polypeptide [33,83]. The JMJD2A jumonji domain alone is capable of demethylating H3K9me3, H3K9me2, H3K36me3, and H3K36me2 [72], albeit with a very low turnover rate [79], and the Jhdm1a/1b (sharing 56% homology) have both been shown to demethylate H3K6me2/3 [84]. The functional connection between the methyl mark reader and eraser in JMJD2A is, however, not clear.

Nickel is a known carcinogen [85], which increases JMJD1A expression [70] but interestingly has been suggested to inhibit JMJD1A enzymatic activity by replacing the ferrous iron in the catalytic centres in vitro [86]. While previous reports have indicated that nickel depleted intracellular ascorbate, stabilized HIF-1α protein, and increased the expression of hypoxia-response genes [87,88], Guo et al. [70] showed that ascorbate suppressed HIF-1α stability and reversed Ni^{2+}-induced upregulation of JMJD1A supporting its purported role in cancer prevention [85].

Previously published epidemiological data showed that ascorbate indeed had an important preventive effect on renal cell carcinoma, particularly in males [89]. Two nonmutually exclusive regulatory mechanisms controlling JMJD1A expression can be envisaged. Either this occurs through the hypoxia signal pathway, a notion supported by the fact that PHD-catalyzed hydroxylation of HIF-α requires oxygen, iron, α-ketoglutarate, and ascorbate (Figure 7.2) [90], and/or that this is mediated through reactive oxygen species (ROS) [91], an idea that is supported by the fact that ascorbate act as a radical scavenger [92], and ROS can inhibit the PHD enzymatic activity [93]. Therefore, under hypoxic conditions, metal ions such as Ni^{2+} and Co^{2+} increase HIF-1α protein stability and JMJD1A gene expression plausibly promoting carcinogenesis and ascorbate supplementation at physiological levels can both reduce the hypoxia effect and JMJD1A expression by inhibiting divalent metal ion toxicity (Figure 7.2). Treatment with vitamin C enhanced proliferation as well, as reported previously before [66,84,94]. Jhdm1b regulates fibroblast proliferation and senescence through H3K36me2/3 demethylation [84,95]. These studies further substantiate the role ascorbate plays in cancer prevention via epigenetic regulation.

7.8 VITAMIN C AND DNA HYDROXYL-METHYLATION IN DEVELOPMENT AND DISEASE

In addition to the jumonji family, there are several other histone demelthylases as well as DNA methylases that depend on ascorbate for enzymatic activity. FTO was shown to be able to demethylate N^6-methyladenosine in RNA, which launched the subfield of RNA epigenetics (Figure 7.2) [96].

Similarly, TET enzymes are 2-oxoglutarate (2OG)- and Fe(II)-dependent enzymes that catalyze the conversion of 5mC to 5-hydroxymethylcytosine (hmC) and, hence, play an important role in epigenetic regulation via DNA methylation (Figure 7.2) [53]. TET proteins are frequently mutated in haematopoietic tumors and are involved in cell lineage specification in mouse embryonic stem cells and homeostasis and differentiation of hematopoietic stem cells, making them critical for normal myelopoiesis [97,98]. Meanwhile TET3 has been shown to be a contributory factor in erasure of global DNA methylation during the zygote stage of embryonic development [99–101].

Ascorbate has been recently shown to directly enhance the catalytic activity of Tet dioxygenases to enhance the oxidation of 5-methylcytosine and thus increase 5-hmC generation [102–104] (Figure 7.2). In fact ascorbate can uniquely interact with the C-terminal catalytic domain of Tet enzymes, presumably by promoting their correct folding and/or recycling of the cofactor Fe^{2+} [104] (Figure 7.2). By acting as a critical cofactor for Tet methylcytosine dioxygenases, the cellular concentration of ascorbate is potentially directly linked to controlling DNA hydroxyl methylation through TET enzymes [102–104]. Of note, the effects of vitamin C are reversible and highly specific since several other antioxidants tested such as *N*-acetylcysteine, vitamin E, and lipoic acid did not increase global 5-hmC or cause any noticeable epigenetic alterations [66,68,102]. 5-hmC occurs preferentially in promoters and gene bodies where it was shown to be positively correlated with gene expression such as in the frontal lobe tissue of human brain [101].

Interestingly, differentially hydroxymethylated regions (DhMRs) that discriminate between fetal and adult brain samples are strongly associated with CpG island shores (Figure 7.2), suggesting that hydoxy-methylation of brain developmental genes may be affected by ascorbate availability (Figure 7.1). Several of these genes were previously associated with autism. Since it is believed that DNA hydroxymethylation may serve as an intermediate step that is a precursor to, or regulator of DNA methylation, it is intriguing that CpG shore hydroxyl methylation changes mediated by ascorbate-dependent TET enzymes appear to correlate with DNA methylation dynamics outside of CpG island regions during early brain development, aging, and cancer differences [48,105].

7.9 EPIGENETIC EFFECTS OF ASCORBATE IN THE IMMUNE SYSTEM

The important role vitamin C in immune system regulation has been well established. For example, ascorbate plays a central role in the ability of neutrophils to kill bacteria. In the presence of bacteria there is a 30-fold increase in ascorbate levels in neutrophils, [106] plausibly to protect these cells from the protein damage inflicted by ROS/lipid-derived 2-alkenals [107,108].

Ascorbate also augments NO-mediated generation of ROS in polymorphonuclear leukocytes [109], prolongs neutrophil NOS expression, NOS catalysis, and oxidative burst [110], and regulates neutrophil apoptosis during resolution of inflammation, possibly via HIF suppression [111].

However, the precise mechanistic basis of this effect has yet to be fully understood. Some of the plausible mechanisms via which ascorbic acid may promote immune response include modulation of phosphatase activity [112,113], posttranslational activation of AP-1 transcription factors [114], and epigenetic regulation of gene expression [42,43]. A recent study by Manning et al. [115] alludes to an important role for ascorbic acid in regulating both development and function of T cells, extending the role of vitamin C from immune system regulation to immune system development (Figure 7.1). They used in vitro T-cell maturation as a model for studying the role of ascorbic acid in the developing immune system and demonstrated that ascorbate added at physiological levels was able to, both in vitro and in vivo, modulates expression of genes encoding accessory molecules involved in signal transduction through TCRαβ, which subsequently resulted in positive selection of functional, adult-type T cells. They also speculated that ascorbate may execute this via its ability to regulate JmjC domain–containing enzymes, which implores further investigation in this area.

7.10 CONCLUSION

Humans have lost the ability to synthesize ascorbic acid due to a defect in L-gulono-1,4-lactone oxidase, an enzyme that catalyzes the conversion of L-gluconolactone into ascorbic acid [7], which makes dietary supplementation of vitamin C essential.

Vitamin C, apart from preventing scurvy, is a water-soluble antioxidant that reduces oxidative stress in vivo, aids cancer chemoprevention [116,117], cancer treatment [118], sepsis [119], and neurodegenerative diseases [120–123]. Several recent studies [121–123] discuss the critical roles of vitamin C cellular uptake, recycling, and neuroprotective effects of vitamin C in the neonatal brain by prevention of oxidative stress. Previously, vitamin C supplementation has also been suggested to induce vasodilation and decrease blood pressure [124,125].

The role of ascorbate in immune system regulation further emphasizes the need for appropriate dietary supplementation of vitamin C especially in individuals with low vitamin C status, e.g., in hospitalized, elderly patients who are at enhanced risk for being infected with bacteria [7].

It is becoming increasingly clear that in addition to its role as an antioxidant vitamin C can have profound effects on the epigenetic makeup of cells by acting as a cofactor for histone and DNA modifying enzymes. These roles were initially most clearly seen in pluripotent stem cell populations and during cell reprogramming, where careful examination of the epigenome is common place, but now clearly extend to development of the brain and immune system as well.

In light of the increasing amount of evidence of the important nexus between metabolism and epigenetic regulation of gene expression in health and disease, we anticipate that with the advent of high throughput analysis tools of histone modifications and DNA methylation and hydroxyl methylation an increasing proportion of the epigenome will be found to be affected by vitamin C.

REFERENCES

1. Van Robertson WB, Schwartz B. Ascorbic acid and the formation of collagen. *The Journal of Biological Chemistry.* 1953;201:689–696.
2. Gould BS. Collagen formation and fibrogenesis with special reference to the role of ascorbic acid. *International Review of Cytology.* 1963;15:301–361.
3. Iqbal K, Khan A, Khattak M. Biological significance of ascorbic acid (vitamin C) in human health—A review. *Pakistan Journal of Nutrition.* 2004;3:5–13.
4. Wilson CWM, Loh HS. Vitamin C and colds. *The Lancet.* 1973;301:1058–1059.
5. Piyathilake CJ, Bell WC, Johanning GL et al. The accumulation of ascorbic acid by squamous cell carcinomas of the lung and larynx is associated with global methylation of DNA. *Cancer.* 2000;89:171–176.
6. Knowles HJ, Raval RR, Harris AL et al. Effect of ascorbate on the activity of hypoxia-inducible factor in cancer cells. *Cancer Research.* 2003;63:1764–1768.
7. Traber MG, Stevens JF. Vitamins C and E: Beneficial effects from a mechanistic perspective. *Free Radical Biology and Medicine.* 2011;51:1000–1013.
8. Welsh JL, Wagner BA, van't Erve TJ et al. Pharmacological ascorbate with gemcitabine for the control of metastatic and node-positive pancreatic cancer (PACMAN): Results from a phase I clinical trial. *Cancer Chemotherapy and Pharmacology.* 2013;71:765–775.
9. Massip L, Garand C, Paquet ER et al. Vitamin C restores healthy aging in a mouse model for Werner syndrome. *FASEB Journal: Official Publication of the Federation of American Societies for Experimental Biology.* 2010;24:158–172.
10. Harman D. Aging: A theory based on free radical and radiation chemistry. *Journal of Gerontology.* 1956;11:298–300.
11. Levine M, Rumsey SC, Daruwala R et al. Criteria and recommendations for vitamin C intake. *JAMA.* 1999;281:1415–1423.

12. Young IS, Woodside JV. Antioxidants in health and disease. *Journal of Clinical Pathology.* 2001;54:176–186.
13. Levine M. New concepts in the biology and biochemistry of ascorbic acid. *New England Journal of Medicine.* 1986;314:892–902.
14. Bullen JJ, Rogers HJ, Spalding PB et al. Natural resistance, iron, and infection: A challenge for clinical medicine. *Journal of Medical Microbiology.* 2006;55:251–258.
15. Geber WF, Lefkowitz SS, Hung CY. Effect of ascorbic acid, sodium salicylate, and caffeine on the serum interferon level in response to viral infection. *Pharmacology.* 1975;13:228–233.
16. Abuja PM. Ascorbate prevents prooxidant effects of urate in oxidation of human low density lipoprotein. *FEBS Letters.* 1999;446:305–308.
17. Girotti AW, Thomas JP, Jordan JE. Prooxidant and antioxidant effects of ascorbate on photosensitized peroxidation of lipids in erythrocyte membranes. *Photochemistry and Photobiology.* 1985;41:267–276.
18. Chen Q, Espey MG, Sun AY et al. Pharmacologic doses of ascorbate act as a prooxidant and decrease growth of aggressive tumor xenografts in mice. *Proceedings of the National Academy of Sciences of the United States of America.* 2008;105:11105–11109.
19. Wu H, Sun YE. Epigenetic regulation of stem cell differentiation. *Pediatric Research.* 2006;59:21R–25R.
20. Li E. Chromatin modification and epigenetic reprogramming in mammalian development. *Nature Reviews Genetics.* 2002;3:662–673.
21. Jaenisch R, Bird A. Epigenetic regulation of gene expression: How the genome integrates intrinsic and environmental signals. *Nature Genetics.* 2003;33 Suppl:245–254.
22. Wutz A. Epigenetic regulation of stem cells: The role of chromatin in cell differentiation. *Advances in Experimental Medicine and Biology.* 2013;786:307–328.
23. Futscher BW, Oshiro MM, Wozniak RJ et al. Role for DNA methylation in the control of cell type specific maspin expression. *Nature Genetics.* 2002;31:175–179.
24. Wang T, Nandakumar V, Jiang XX et al. The control of hematopoietic stem cell maintenance, self-renewal, and differentiation by Mysm1-mediated epigenetic regulation. *Blood.* 2013;122:2812–2822.
25. Juliandi B, Abematsu M, Nakashima K. Epigenetic regulation in neural stem cell differentiation. *Development, Growth, and Differentiation.* 2010;52:493–504.
26. Kouzarides T. Histone methylation in transcriptional control. *Current Opinion in Genetics and Development.* 2002;12:198–209.
27. Bannister AJ, Kouzarides T. Regulation of chromatin by histone modifications. *Cell Research.* 2011;21:381–395.
28. Bedford MT, Clarke SG. Protein arginine methylation in mammals: Who, what, and why. *Molecular Cell.* 2009;33:1–13.
29. Bannister AJ, Schneider R, Kouzarides T. Histone methylation: Dynamic or static? *Cell.* 2002;109:801–806.
30. Shi Y, Lan F, Matson C et al. Histone demethylation mediated by the nuclear amine oxidase homolog LSD1. *Cell.* 2004;119:941–953.
31. Klose RJ, Zhang Y. Regulation of histone methylation by demethylimination and demethylation. *Nature Reviews Molecular Cell Biology.* 2007;8:307–318.
32. Whetstine JR, Nottke A, Lan F et al. Reversal of histone lysine trimethylation by the JMJD2 family of histone demethylases. *Cell.* 2006;125:467–481.
33. Tsukada Y, Fang J, Erdjument-Bromage H et al. Histone demethylation by a family of JmjC domain-containing proteins. *Nature.* 2006;439:811–816.
34. Mosammaparast N, Shi Y. Reversal of histone methylation: Biochemical and molecular mechanisms of histone demethylases. *Annual Review of Biochemistry.* 2010;79:155–179.
35. Chang B, Chen Y, Zhao Y et al. JMJD6 is a histone arginine demethylase. *Science.* 2007;318:444–447.

36. Carpenter S, Aiello D, Atianand MK et al. A long noncoding RNA mediates both activation and repression of immune response genes. *Science.* 2013;341:789–792.
37. Bartel DP. MicroRNAs: Genomics, biogenesis, mechanism, and function. *Cell.* 2004;116:281–297.
38. Almeida R, Allshire RC. RNA silencing and genome regulation. *Trends in Cell Biology.* 2005;15:251–258.
39. Khalil AM, Guttman M, Huarte M et al. Many human large intergenic noncoding RNAs associate with chromatin-modifying complexes and affect gene expression. *Proceedings of the National Academy of Sciences of the United States of America.* 2009;106:11667–11672.
40. Takahashi K, Tanabe K, Ohnuki M et al. Induction of pluripotent stem cells from adult human fibroblasts by defined factors. *Cell.* 2007;131:861–872.
41. Yu J, Vodyanik MA, Smuga-Otto K et al. Induced pluripotent stem cell lines derived from human somatic cells. *Science.* 2007;318:1917–1920.
42. Chung T-L, Brena RM, Kolle G et al. Vitamin C promotes widespread yet specific DNA demethylation of the epigenome in human embryonic stem cells. *Stem Cells.* 2010;28:1848–1855.
43. Chung T-L, Turner JP, Thaker NY et al. Ascorbate promotes epigenetic activation of CD30 in human embryonic stem cells. *Stem Cells.* 2010;28:1782–1793.
44. Ovchinnikov DA, Turner JP, Titmarsh DM et al. Generation of a human embryonic stem cell line stably expressing high levels of the fluorescent protein mCherry. *World J Stem Cells.* 2012;4:71–79.
45. Chung T-L, Thakar NY, Wolvetang EJ. Genetic and epigenetic instability of human pluripotent stem cells. *The Open Stem Cell Journal.* 2011;3:52–61.
46. Briggs JA, Sun J, Shepherd J et al. Integration-free induced pluripotent stem cells model genetic and neural developmental features of Down syndrome etiology. *Stem Cells.* 2012;31(3):467–478.
47. Allegrucci C, Wu Y-Z, Thurston A et al. Restriction landmark genome scanning identifies culture-induced DNA methylation instability in the human embryonic stem cell epigenome. *Human Molecular Genetics.* 2007;16:1253–1268.
48. Irizarry RA, Ladd-Acosta C, Wen B et al. The human colon cancer methylome shows similar hypo- and hypermethylation at conserved tissue-specific CpG island shores. *Nature Genetics.* 2009;41:178–186.
49. Doi A, Park IH, Wen B et al. Differential methylation of tissue- and cancer-specific CpG island shores distinguishes human induced pluripotent stem cells, embryonic stem cells and fibroblasts. *Nature Genetics.* 2009;41:1350–1353.
50. Klose RJ, Yamane K, Bae Y et al. The transcriptional repressor JHDM3A demethylates trimethyl histone H3 lysine 9 and lysine 36. *Nature.* 2006;442:312–316.
51. Yamane K, Toumazou C, Tsukada Y et al. JHDM2A, a JmjC-containing H3K9 demethylase, facilitates transcription activation by androgen receptor. *Cell.* 2006;125:483–495.
52. Gerken T, Girard CA, Tung YC et al. The obesity-associated FTO gene encodes a 2-oxoglutarate-dependent nucleic acid demethylase. *Science.* 2007;318:1469–1472.
53. Tahiliani M, Koh KP, Shen Y et al. Conversion of 5-methylcytosine to 5-hydroxymethylcytosine in mammalian DNA by MLL partner TET1. *Science.* 2009;324:930–935.
54. Cao N, Liu Z, Chen Z et al. Ascorbic acid enhances the cardiac differentiation of induced pluripotent stem cells through promoting the proliferation of cardiac progenitor cells. *Cell Research.* 2012;22:219–236.
55. Martinez EC, Wang J, Gan SU et al. Ascorbic acid improves embryonic cardiomyoblast cell survival and promotes vascularization in potential myocardial grafts in vivo. *Tissue Engineering Part A.* 2010;16:1349–1361.
56. Kim Y, Ku S-Y, Huh Y et al. Anti-aging effects of vitamin C on human pluripotent stem cell-derived cardiomyocytes. *Age.* 2013;35:1545–1557.

57. Shima A, Pham J, Blanco E et al. IGF-I and vitamin C promote myogenic differentiation of mouse and human skeletal muscle cells at low temperatures. *Experimental Cell Research.* 2011;317:356–366.
58. Liang Y, Xiao-Gang Z, Xin-Wei W. Vitamin C promotes differentiation of mouse induced pluripotent stem cells into cardiomyocyte in vitro. *Heart.* 2011;97:A8–A9.
59. Takahashi T, Lord B, Schulze PC et al. Ascorbic acid enhances differentiation of embryonic stem cells into cardiac myocytes. *Circulation.* 2003;107:1912–1916.
60. Chan SS-K, Chen J-H, Hwang S-M et al. Salvianolic acid B–vitamin C synergy in cardiac differentiation from embryonic stem cells. *Biochemical and Biophysical Research Communications.* 2009;387:723–728.
61. Finkel T, Holbrook NJ. Oxidants, oxidative stress and the biology of ageing. *Nature.* 2000;408:239–247.
62. Terman A, Brunk UT. The aging myocardium: Roles of mitochondrial damage and lysosomal degradation. *Heart, Lung, and Circulation.* 2005;14:107–114.
63. Silva J, Barrandon O, Nichols J et al. Promotion of reprogramming to ground state pluripotency by signal inhibition. *PLoS Biology.* 2008;6:e253.
64. Stadtfeld M, Apostolou E, Ferrari F et al. Ascorbic acid prevents loss of Dlk1-Dio3 imprinting and facilitates generation of all-iPS cell mice from terminally differentiated B cells. *Nature Genetics.* 2012;44:398–405, S391–S392.
65. Takahashi K, Yamanaka S. Induction of pluripotent stem cells from mouse embryonic and adult fibroblast cultures by defined factors. *Cell.* 2006;126:663–676.
66. Esteban MA, Wang T, Qin B et al. Vitamin C enhances the generation of mouse and human induced pluripotent stem cells. *Cell Stem Cell.* 2010;6:71–79.
67. Esteban MA, Pei D. Vitamin C improves the quality of somatic cell reprogramming. *Nature Genetics.* 2012;44:366–367.
68. Wang T, Chen K, Zeng X et al. The histone demethylases Jhdm1a/1b enhance somatic cell reprogramming in a vitamin-C-dependent manner. *Cell Stem Cell.* 2011;9:575–587.
69. Shi Y. Histone lysine demethylases: Emerging roles in development, physiology and disease. *Nature Reviews Genetics.* 2007;8:829–833.
70. Guo X, Lu J, Wang Y et al. Ascorbate antagonizes nickel ion to regulate JMJD1A expression in kidney cancer cells. *Acta Biochimica et Biophysica Sinica.* 2012;44:330–338.
71. Suikki HE, Kujala PM, Tammela TL et al. Genetic alterations and changes in expression of histone demethylases in prostate cancer. *The Prostate.* 2010;70:889–898.
72. Horton JR, Upadhyay AK, Qi HH et al. Enzymatic and structural insights for substrate specificity of a family of jumonji histone lysine demethylases. *Nature Structural and Molecular Biology.* 2010;17:38–43.
73. Pena PV, Davrazou F, Shi X et al. Molecular mechanism of histone H3K4me3 recognition by plant homeodomain of ING2. *Nature.* 2006;442:100–103.
74. Li H, Ilin S, Wang W et al. Molecular basis for site-specific read-out of histone H3K4me3 by the BPTF PHD finger of NURF. *Nature.* 2006;442:91–95.
75. Shi X, Hong T, Walter KL et al. ING2 PHD domain links histone H3 lysine 4 methylation to active gene repression. *Nature.* 2006;442:96–99.
76. Wysocka J, Swigut T, Xiao H et al. A PHD finger of NURF couples histone H3 lysine 4 trimethylation with chromatin remodelling. *Nature.* 2006;442:86–90.
77. Ooi SK, Qiu C, Bernstein E et al. DNMT3L connects unmethylated lysine 4 of histone H3 to de novo methylation of DNA. *Nature.* 2007;448:714–717.
78. Lan F, Collins RE, De Cegli R et al. Recognition of unmethylated histone H3 lysine 4 links BHC80 to LSD1-mediated gene repression. *Nature.* 2007;448:718–722.
79. Couture JF, Collazo E, Ortiz-Tello PA et al. Specificity and mechanism of JMJD2A, a trimethyllysine-specific histone demethylase. *Nature Structural and Molecular Biology.* 2007;14:689–695.

80. Ng SS, Kavanagh KL, McDonough MA et al. Crystal structures of histone demethylase JMJD2A reveal basis for substrate specificity. *Nature.* 2007;448:87–91.
81. Chen Z, Zang J, Kappler J et al. Structural basis of the recognition of a methylated histone tail by JMJD2A. *Proceedings of the National Academy of Sciences of the United States of America.* 2007;104:10818–10823.
82. Huang Y, Fang J, Bedford MT et al. Recognition of histone H3 lysine-4 methylation by the double tudor domain of JMJD2A. *Science.* 2006;312:748–751.
83. Klose RJ, Kallin EM, Zhang Y. JmjC-domain-containing proteins and histone demethylation. *Nature Reviews Genetics.* 2006;7:715–727.
84. He J, Kallin EM, Tsukada Y et al. The H3K36 demethylase Jhdm1b/Kdm2b regulates cell proliferation and senescence through p15(Ink4b). *Nature Structural and Molecular Biology.* 2008;15:1169–1175.
85. Chen C, Tang P, Yue J et al. Effect of siRNA targeting HIF-1alpha combined L-ascorbate on biological behavior of hypoxic MiaPaCa2 cells. *Technology in Cancer Research & Treatment.* 2009;8:235–240.
86. Chen H, Giri NC, Zhang R et al. Nickel ions inhibit histone demethylase JMJD1A and DNA repair enzyme ABH2 by replacing the ferrous iron in the catalytic centers. *The Journal of Biological Chemistry.* 2010;285:7374–7383.
87. Salnikow K, Kasprzak KS. Ascorbate depletion: A critical step in nickel carcinogenesis? *Environmental Health Perspectives.* 2005;113:577–584.
88. Kaczmarek M, Timofeeva OA, Karaczyn A et al. The role of ascorbate in the modulation of HIF-1alpha protein and HIF-dependent transcription by chromium(VI) and nickel(II). *Free Radical Biology and Medicine.* 2007;42:1246–1257.
89. Lee JE, Giovannucci E, Smith-Warner SA et al. Intakes of fruits, vegetables, vitamins A, C, and E, and carotenoids and risk of renal cell cancer. *Cancer Epidemiology, Biomarkers and Prevention: A Publication of the American Association for Cancer Research, Cosponsored by the American Society of Preventive Oncology.* 2006;15:2445–2452.
90. Fong GH, Takeda K. Role and regulation of prolyl hydroxylase domain proteins. *Cell Death and Differentiation.* 2008;15:635–641.
91. Huang X, Zhuang Z, Frenkel K et al. The role of nickel and nickel-mediated reactive oxygen species in the mechanism of nickel carcinogenesis. *Environmental Health Perspectives.* 1994;102 Suppl 3:281–284.
92. Niki E. Action of ascorbic acid as a scavenger of active and stable oxygen radicals. *The American Journal of Clinical Nutrition.* 1991;54:1119S–1124S.
93. Gunaratnam L, Bonventre JV. HIF in kidney disease and development. *Journal of the American Society of Nephrology: JASN.* 2009;20:1877–1887.
94. Hata R, Senoo H. L-Ascorbic acid 2-phosphate stimulates collagen accumulation, cell proliferation, and formation of a three-dimensional tissuelike substance by skin fibroblasts. *Journal of Cellular Physiology.* 1989;138:8–16.
95. Tzatsos A, Pfau R, Kampranis SC et al. Ndy1/KDM2B immortalizes mouse embryonic fibroblasts by repressing the Ink4a/Arf locus. *Proceedings of the National Academy of Sciences.* 2009;106:2641–2646.
96. Jia G, Fu Y, Zhao X et al. N^6-Methyladenosine in nuclear RNA is a major substrate of the obesity-associated FTO. *Nature Chemical Biology.* 2011;7:885–887.
97. Koh KP, Yabuuchi A, Rao S et al. Tet1 and Tet2 regulate 5-hydroxymethylcytosine production and cell lineage specification in mouse embryonic stem cells. *Cell Stem Cell.* 2011;8:200–213.
98. Ko M, Bandukwala HS, An J et al. Ten-eleven-translocation 2 (TET2) negatively regulates homeostasis and differentiation of hematopoietic stem cells in mice. *Proceedings of the National Academy of Sciences.* 2011;108:14566–14571.
99. Inoue A, Zhang Y. Replication-dependent loss of 5-hydroxymethylcytosine in mouse preimplantation embryos. *Science.* 2011;334:194.

100. Gu TP, Guo F, Yang H et al. The role of Tet3 DNA dioxygenase in epigenetic reprogramming by oocytes. *Nature.* 2011;477:606–610.
101. Wang T, Pan Q, Lin L et al. Genome-wide DNA hydroxymethylation changes are associated with neurodevelopmental genes in the developing human cerebellum. *Human Molecular Genetics.* 2012;21(26):5500–5510.
102. Blaschke K, Ebata KT, Karimi MM et al. Vitamin C induces Tet-dependent DNA demethylation and a blastocyst-like state in ES cells. *Nature.* 2013;500:222–226.
103. Minor EA, Court BL, Young JI et al. Ascorbate induces ten-eleven translocation (Tet) methylcytosine dioxygenase-mediated generation of 5-hydroxymethylcytosine. *Journal of Biological Chemistry.* 2013;288:13669–13674.
104. Yin R, Mao SQ, Zhao B et al. Ascorbic acid enhances Tet-mediated 5-methylcytosine oxidation and promotes DNA demethylation in mammals. *Journal of the American Chemical Society.* 2013;135:10396–10403.
105. Numata S, Ye T, Hyde TM et al. DNA methylation signatures in development and aging of the human prefrontal cortex. *The American Journal of Human Genetics.* 2012;90:260–272.
106. Wang Y, Russo TA, Kwon O et al. Ascorbate recycling in human neutrophils: Induction by bacteria. *Proceedings of the National Academy of Sciences of the United States of America.* 1997;94:13816–13819.
107. Chavez J, Chung W-G, Miranda CL et al. Site-specific protein adducts of 4-hydroxy-2(E)-nonenal in human THP-1 monocytic cells: Protein carbonylation is diminished by ascorbic acid. *Chemical Research in Toxicology.* 2009;23:37–47.
108. Miranda CL, Reed RL, Kuiper HC et al. Ascorbic acid promotes detoxification and elimination of 4-hydroxy-2(E)-nonenal in human monocytic THP-1 cells. *Chemical Research in Toxicology.* 2009;22:863–874.
109. Sharma P, Raghavan SA, Dikshit M. Role of ascorbate in the regulation of nitric oxide generation by polymorphonuclear leukocytes. *Biochemical and Biophysical Research Communications.* 2003;309:12–17.
110. Chatterjee M, Saluja R, Kumar V et al. Ascorbate sustains neutrophil NOS expression, catalysis, and oxidative burst. *Free Radical Biology and Medicine.* 2008;45:1084–1093.
111. Mateo J, Garcia-Lecea M, Cadenas S et al. Regulation of hypoxia-inducible factor-1alpha by nitric oxide through mitochondria-dependent and -independent pathways. *The Biochemical Journal.* 2003;376:537–544.
112. Yu JS. Activation of protein phosphatase 2A by the Fe^{2+}/ascorbate system. *Journal of Biochemistry.* 1998;124:225–230.
113. Sommer D, Fakata KL, Swanson SA et al. Modulation of the phosphatase activity of calcineurin by oxidants and antioxidants in vitro. *European Journal of Biochemistry/FEBS.* 2000;267:2312–2322.
114. Arkan MC, Leonarduzzi G, Biasi F et al. Physiological amounts of ascorbate potentiate phorbol ester-induced nuclear-binding of AP-1 transcription factor in cells of macrophagic lineage. *Free Radical Biology and Medicine.* 2001;31:374–382.
115. Manning J, Mitchell B, Appadurai DA et al. Vitamin C promotes maturation of T-cells. *Antioxidants & Redox Signaling.* 2013;19(17):2054–2067.
116. Gaziano J, Glynn RJ, Christen WG et al. Vitamins E and C in the prevention of prostate and total cancer in men: The Physicians' Health Study: II. Randomized controlled trial. *JAMA.* 2009;301:52–62.
117. Gann PH. Randomized trials of antioxidant supplementation for cancer prevention: First bias, now chance—Next, cause. *JAMA.* 2009;301:102–103.
118. Padayatty SJ, Sun AY, Chen Q et al. Vitamin C: Intravenous use by complementary and alternative medicine practitioners and adverse effects. *PLoS One.* 2010;5:e11414.
119. Wilson JX. Mechanism of action of vitamin C in sepsis: Ascorbate modulates redox signaling in endothelium. *BioFactors.* 2009;35:5–13.

120. Bowman GL, Dodge H, Frei B et al. Ascorbic acid and rates of cognitive decline in Alzheimer's disease. *Journal of Alzheimer's Disease.* 2009;16:93–98.
121. Tveden-Nyborg P, Lykkesfeldt J. Does vitamin C deficiency result in impaired brain development in infants? *Redox Report.* 2009;14:2–6.
122. Harrison FE, Meredith ME, Dawes SM et al. Low ascorbic acid and increased oxidative stress in gulo(–/–) mice during development. *Brain Research.* 2010;1349:143–152.
123. Harrison FE, May JM. Vitamin C function in the brain: Vital role of the ascorbate transporter SVCT2. *Free Radical Biology and Medicine.* 2009;46:719–730.
124. Duffy SJ, Gokce N, Holbrook M et al. Treatment of hypertension with ascorbic acid. *Lancet.* 1999;354:2048–2049.
125. Gokce N, Keaney JF, Jr., Frei B et al. Long-term ascorbic acid administration reverses endothelial vasomotor dysfunction in patients with coronary artery disease. *Circulation.* 1999;99:3234–3240.

8 Epigenetic Regulation of Cellular Responses toward Vitamin D

Prashant K. Singh and Moray J. Campbell

CONTENTS

8.1 INTRODUCTION

The bioactive metabolite of vitamin D, 1α,25-dihydroxyvitamin D3 ($1\alpha,25(OH)_2D_3$), regulates cellular growth and differentiation and thereby influences many physiological processes including bone formation, immune functions and skin formation. The genomic actions of $1\alpha,25(OH)_2D_3$ are mediated through the transcription factor, vitamin D receptor (VDR), a member of the nuclear receptor (NR) superfamily. The VDR regulates gene expression by binding as part of large and dynamic transcriptional regulatory complexes to genomic sequences on target genes. These complexes contain various proteins that govern epigenetic modifications. For example, upon ligand activation of the VDR, recruitment of coactivators leads to histone acetylation, and thereby induces the reorganization of local chromatin regions at the target gene promoters and induces gene activation. Conversely, unliganded VDR recruits corepressors (e.g., NCOR1, NCOR2) leading to histone deacetylation and transcriptional silencing of the targets genes. These transient histone states also act as platforms to allow effector complexes to regulate other aspects of chromatin architecture; histone deacetylation can lead to more stable DNA CpG methylation. These functions of VDR are distorted in cancer. For example, many cancers show resistance to antiproliferative effect of vitamin D compounds, and this reflects both the altered epigenetic state of cancer genomes, altered expression of key transcriptional process and distorted VDR complex interactions with the target genes. Distortions to histone modification processes appear to be driven by altered corepressor recruitment and can lead to stable gene silencing. These processes associated with stable gene silencing may shed light on why the VDR and other NRs are often expressed in cancer cells that are phenotypically resistant to the antiproliferative actions of the receptor.

8.2 VITAMIN D RECEPTOR (NR1I1/VDR)

8.2.1 The VDR Is a Member of the NR Superfamily

The NR superfamily is one of the largest families of transcription factors in humans and the actions of the 48 different members contribute to the regulation of a large part of the genome and contribute to a wide variety of cell functions. NRs in general are transcription factors that commonly function within multicomponent transcriptional complexes to regulate target gene expression. Indeed, a great deal of general understanding of gene transcription has arisen from the study of NR-mediated transactivation and transrepression. The primary structure of NRs have five or six contiguous regions of variable degrees of homology running from the N-terminus to the C-terminus and nominated as A/B, C, D, E, and F. This representation of nuclear receptors arose from early comparisons made of the human and chicken estrogen receptor amino acid sequences that identified three regions of high homology (A, C, and E) interspersed with regions of lesser homology (B, D, and F).[1] The C region is entirely conserved between human and chicken estrogen receptor amino acid sequences, and this region was later recognized as the most highly conserved across the nuclear receptor family. The region encodes a DNA-binding domain (DBD) that is composed of two zinc fingers. This DNA-binding motif is characteristic of nuclear receptors and sets them apart from other DNA-binding proteins.[2]

Following on from these discoveries, in 1987, the cloning of the avian vitamin D receptor was described using newly available receptor-specific monoclonal antibodies to recover the appropriate cDNA.[3] The identification that glucocorticoid, estrogen, thyroid, and vitamin D receptors were structurally related prompted researchers to seek additional family members. Thus, in just 3 years, the field went from being unaware of how cells perceived diverse lipophilic signaling hormones to having identified four receptors that displayed cross species conservation. Most remarkable was the realization that chemically distinct hormones and other small lipophilic molecules were demonstrably signaling via highly related receptors providing the first evidence of a new receptor family. Subsequently, it emerged that the NR superfamily is the largest superfamily of human transcription factors.

NRs became broadly divided into classes based on their binding to DNA as homodimers, heterodimers, or as monomers. The receptors of the classical steroids, the glucocorticoid, mineralocorticoid, progesterone, androgen, and estrogen receptors (GR, MR, PR, AR, and ER) were found to bind DNA as homodimers and formed the "group I" receptors. The group II and III receptors bind DNA predominantly as heterodimers. The group II receptors were distinct from group III receptors in as much as they had defined ligands whereas group III receptors were orphan receptors. This distinction is becoming less clear as a number of receptors once considered as orphan receptors have now been assigned ligands or putative ligands. The group IV receptors bind DNA as monomers and until recently were almost exclusively orphan receptors, although ligand identification for orphan receptors is a developing field.

Set against this grouping according to function are other classifications in terms of phylogenetics. In this manner the family is divided in 6 subclasses. In this classification the VDR falls in group 1, the Thyroid hormone receptor like nuclear receptors. As a result the official NR name for the VDR is NR1I1, and it specifically binds the seco-steroid hormone, 1α,25-dihydroxyvitamin D_3 (1α,25$(OH)_2D_3$), with high affinity. Interestingly, the group 1 subfamily also includes LXRs and FXR and more distantly the PPARs. The receptors within this subfamily preferentially form homodimers or heterodimeric complexes with RXR acting as a common central partner for VDR, PPARs, LXRs, and FXR. Thus, the receptors in the group appear to be all responsive to either bile acid or xenobiotics and therefore widely integrated with bile acid homeostasis and detoxification. In keeping with this capacity the bile acid, lithocholic acid (LCA), has recently been shown to be a potent ligand for the VDR albeit with lower micromolar affinity.[4]

A highly conserved VDR has been found in animals with a calcified skeleton and is undetectable in nonchordate species, although it is present in certain noncalcified chordates such as the lamprey (reviewed by Krasowski[5]). Within prokaryotes there appears to be the capacity to undertake UV-catalyzed metabolism of cholesterol compounds that may suggest that the evolution of vitamin D biochemistry is very ancient. This also suggests that the VDR system has adapted to regulate calcium function, and retains other functions that are calcium independent. Some of these functions, in common with other NRs, include the capacity to be regulated by ligand-independent events and may reflect a pre-ligand–activated evolutionary state.[6] It is therefore reasonable to speculate that the VDR originated during early chordate evolution as a transcription factor, probably involved in detoxification, and subsequently developed a role in calcium homeostasis in response to UV-catalyzed synthesis of 1α,25$(OH)_2D_3$.

1α,25$(OH)_2D_3$ and its precursor 25$(OH)D_3$, in common with most other NR ligands, are highly hydrophobic and are therefore transported in the aqueous blood stream associated with a specific binding protein (DBP).[7,8] At the cell membrane, they are free to diffuse across the lipid membrane, although the identification of LRP2/Megalin as an active transport protein for 25$(OH)D_3$ suggests that transport into the cell of vitamin D_3 metabolites maybe more tightly regulated than merely by passive diffusion alone.[9] Once in target cells, 1α,25$(OH)_2D_3$ associates with the VDR.

8.2.2 VDR Signaling Occurs through Endocrine and Autocrine Mechanisms

A central physiological VDR function is the regulation of serum calcium levels and a downstream consequence is the control of bone formation and maintenance. Awareness of the implications of insufficient VDR endocrine signaling, namely in the form of rickets, precedes isolation of the receptor by approximately 300 years.

Rickets is the primary disease associated with 1α,25$(OH)_2D_3$ deficiency and was first described in the seventeenth century by Daniel Whistler in the Netherlands, and subsequently, by other physicians in London and elsewhere. The molecular etiology for rickets remained unresolved until the beginning of the twentieth century when the central deficiency of an active vitamin D hormone was revealed. Sir Edward Mellanby in 1919 discovered that the dietary deficiency that caused rickets could be ameliorated by fish oil extracts and that the active ingredient was identified as vitamin D_2 (ergocalciferol). Around the same time, Huldschinsky, Hess, and Unger found rickets could be cured by exposure to UV radiation or vegetable oil.

The increased understanding of rickets and its prevention by dietary and environmental factors (UV) were entirely coincident with the rapid rise of the syndrome in the urbanized and industrialized living conditions of the cities of northern Europe where poor access to sunlight and a restricted diet made the effects of underactive VDR all too common. Further studies revealed that vitamin D_2 and vitamin D_3 are actually not vitamins but are seco-steroids derived from ergosterol and 7-dehydrocholesterol, respectively. Indeed, such was the vigor and significance of the increase in understanding the role of 1α,25$(OH)_2D_3$ that its chemical identification and syntheses by Adolf Windaus was sufficient to receive the Nobel Prize in Chemistry (1928) (reviewed Bouillon et al.[10]). Continued work in the 1960s by Norman and colleagues, among others, led to the identification of 25(OH)D and analyses of 1α,25$(OH)_2D_3$ metabolism followed by the identification of vitamin D binding proteins in target cells.[11–13] In the subsequent decades, remarkable strides have been made in describing the diverse biology that the VDR participates in. Researchers accommodated this diversity of biological actions by separating functions into the so-called classical (the regulation of serum calcium levels) and nonclassical (everything else).

Systemic monitoring and regulation of serum calcium levels are fundamentally important processes owing to the vital function that calcium plays in a wide range of cellular functions. As a result, the VDR plays a well-established endocrine role in the regulation of calcium homeostasis by regulating calcium absorption in the gut and kidney and bone mineralization. The status of active VDR ligand availability is dependent upon cutaneous synthesis initiated by solar radiation and also on dietary intake—a reduction of either one or both sources leads to 1α,25$(OH)_2D_3$ insufficiency. The contribution from the UV-initiated cutaneous conversion of 7-dehydrocholesterol to vitamin D_3 is the greater, contributing over 90% toward final 1α,25$(OH)_2D_3$ synthesis in a vitamin D sufficient individual. The importance of the relationships between solar exposure and the ability to capture UV-mediated energy is underscored by the inverse correlation between human skin pigmentation and latitude. That is, the individual capacity to generate vitamin D_3 in response to solar UV exposure is intimately associated with forebear environmental adaptation. The correct and sufficient level of solar exposure and serum vitamin D_3 are matters of considerable debate and in 2010 was the subject of an Institute of Medicine report, which recommended daily vitamin D_3 intake at the levels of 800 IU/day for older adults who may be at risk of deficiency. However, this is not without controversy as parallel reassessment of the $1\alpha,25(OH)_2D_3$ impact on the prevention of osteoporosis has suggested that the correct level may be as high as 2000–3000 IU/day, which may reflect more accurately "ancestral" serum levels.[14]

The central relationship between UV exposure and calcium homeostasis has underpinned the development of the endocrine view of $1\alpha,25(OH)_2D_3$ signaling, with spatially distinct sites within the body of incremental vitamin D activation. Thus, vitamin D_3 produced in the skin is converted in the liver to 25-hydroxyvitamin D_3 (25(OH)D) and circulating levels of this metabolite serve as a useful index of vitamin D status. A further hydroxylation occurs in the kidney at the carbon 1 position by 25-hydroxyvitamin D-1α-hydroxylase (encoded by *CYP27B1*) produces the biologically active hormone $1\alpha,25(OH)_2D_3$. A second mitochondrial cytochrome P450 enzyme, the 24-hydroxylase (encoded by *CYP24A1*), can utilize both 25(OH)D and $1\alpha,25(OH)_2D_3$ as substrates and is the first step in the inactivation pathway for these metabolites.

Another role of $1\alpha,25(OH)_2D_3$ in the kidney is regulation of its own production, mediated by the VDR via a negative feedback mechanism. The presence of $1\alpha,25(OH)_2D_3$ inhibits renal expression of 1α-hydroxylase and enhances the production of 24-hydroxylase, thus catabolizing $1\alpha,25(OH)_2D_3$ into less active metabolites and allowing their excretion. Thus, elevated levels of $1\alpha,25(OH)_2D_3$ appear to block its synthesis and induce its own inactivation[15] in a classical negative feedback loop.

Expression of 1α-hydroxylase has also been identified in keratinocytes and a wide range of other cell types, suggesting an autocrine/paracrine role for the local synthesis and signaling of $1\alpha,25(OH)_2D_3$.[16–21] Thus, not only the liver and kidney endocrine loop, but in multiple target tissues, 25(OH)D may enter into an intracellular VDR signaling axis that coordinates the local autocrine synthesis, metabolism, and signal transduction of $1\alpha,25(OH)_2D_3$. The components of this axis are regulated dynamically, similarly to the kidney, with *CYP27B1* being repressed by $1\alpha,25(OH)_2D_3$ and correspondingly *CYP24A1* positively regulated by $1\alpha,25(OH)_2D_3$. The biological significance of these autocrine actions suggests that $1\alpha,25(OH)_2D_3$ availability is regulated locally and governs many of the more diverse physiological functions, which include modulation of immune responses and the control of cell differentiation.

8.2.3 VDR Actions in Health and Disease

Key insights into diverse biological actions of VDR have been gained through the use of transgenic and knockout mice.[22–26] VDR disruption results in a profound phenotype in these models, which is principally observed post weaning and is associated with the alteration of duodenal calcium absorption and bone mineralization, resulting in hypocalcemia, secondary hyperparathyroidism, osteomalacia, rickets, impaired bone formation, and elevated serum levels of $1\alpha,25(OH)_2D_3$. In parallel, a range of more subtle effects are seen more clearly when the animals are rescued with dietary calcium supplementation. These include uterine hypoplasia, impaired ovarian folliculogenesis, reproductive dysfunction, cardiac hypertrophy, and enhanced thrombogenicity.[27–39]

One site that typifies the integrated actions of the VDR is the mammary gland. VDR actions in this gland have been studied in a comprehensive series of experiments by Welsh and colleagues.[34,40] In mammals, the requirement for calcium increases during pregnancy and lactation to meet the increased needs of the fetal skeleton for mineralization. Analysis of the mammary gland in *VDR*-deficient mice supports important antimitotic and prodifferentiative VDR roles. The glands in the knockout mice display enhanced growth compared with their wild-type littermates and are heavier, with an elevated number of terminal end buds, greater ductal outgrowth, and enhanced secondary branch points. This accelerated growth is exacerbated further during the pregnancy-associated proliferative burst and the postlactation apoptosis associated with involution is delayed. Similarly, in cells derived from these mice and in ex vivo organ cultures, the response to exogenous estrogen and progesterone appears enhanced. Thus, the mammary gland represents an intriguing tissue where endocrine (calcemic) and autocrine (antimitotic, prodifferentiative, proapoptotic) effects of the VDR appear to converge.

Another epithelial system, the skin, also reveals key aspects of VDR function. WNT signaling is one of the major processes regulating postmorphogenic hair follicle development. Interestingly, the development of dermal cysts and increase in sebaceous glands is observed in the *VDR* and *Hairless*$^{-/-}$ mice are also similar to mice expressing a keratinocyte-specific disruption to β-catenin.[41,42] These findings have raised the possibility that one function of the *VDR* may be to coregulate aspects of Wnt signaling, a concept that is supported further by the physical association of VDR in a complex with β-catenin and other WNT components.[43] Another unexpected finding of the *Vdr*$^{-/-}$ animals was the uterine hypoplasia and impaired ovarian function in the females that leads to dramatically reduced fertility. Similar to the hair phenotype, this was not restored by the rescue diet of high calcium.[44] Estradiol supplementation, however, of the female mice restored uterine function and fertility and suggests the fault lies with an inability to generate estrogen.

Also reflecting the cooperative and integrated nature of VDR function, a number of workers have identified mechanisms by which more dominant signaling process are able either to ablate or attenuate VDR signaling. For example, Munoz and colleagues have dissected the interrelationships among the VDR, E-cadherin, and the WNT signaling pathway in colon cancer cell lines and primary tumors. In these studies, the induction of *CDH1* (encodes E-cadherin) was seen in subpopulations of SW480 colon cancer cells, which express the VDR and respond to $1\alpha,25(OH)_2D_3$. The VDR thereby limits the transcriptional effects of β-catenin by physically and directly binding it in the nucleus, and by upregulating E-cadherin to sequestrate β-catenin in the cytoplasm. In malignancy, these actions are corrupted through downregulation of *VDR* mRNA, which appears to be a direct consequence of binding by the transcriptional repressor SNAIL—a key regulator of the epithelial–mesenchyme transition, which is overexpressed in colon cancer.[45–47]

The VDR also plays a functional role in various myeloid and lymphoid cells. Recently, unbiased interrogation of all transcriptional networks that are involved in hematopoietic differentiation have identified a significant role for the VDR to cooperate functionally in a module of transcription factors that includes CEBPA and PU-1 to govern granulocyte and monocyte differentiation.[48] Similarly, many cells of the immune system express VDR and respond to $1\alpha,25(OH)_2D_3$, notably antigen-presenting cells (APCs), including macrophages and dendritic cells (DCs). Thus, $1\alpha,25(OH)_2D_3$ may participate in the physiological control of immune responses, and be involved in maintaining tolerance to self-antigens, as suggested by the enlarged lymph nodes in VDR-deficient mice containing a higher frequency of mature DCs. Somewhat counterbalancing these immunomodulatory and tolerogenic actions, VDR signaling also exerts antimicrobial actions by upregulating a number of target genes encoding antimicrobial peptides, for example, LL-37 (also known as cathelicidin) in humans and nonhuman primates only, but not mice.[49] These actions appear to be integrated with enhanced wound healing and tissue remodeling activities, and interplay with angiogenesis and growth factor–mediated cell proliferation.

An important and emergent area, both in terms of physiology and therapeutic exploitation, is the role the VDR appears to play in maintaining genomic integrity and facilitating DNA repair. There appears to be close cooperation between VDR actions and the p53 tumor suppressor pathway. The maintenance of genomic fidelity against a backdrop of self-renewal is central to the normal development and adult function of many tissues including the mammary and prostate glands, and the colon. For example, in the mammary gland p53 family members play a role in gland development and maintenance. $p63^{-/-}$ animals have an absence of mammary and other epithelial structures, associated with a failure of lineage commitment (reviewed by Barbieri and Pietenpol[50]), whereas $p53^{-/-}$ animals have delayed mammary gland involution, reflecting the $Vdr^{-/-}$ animals, and wider tumor susceptibility (reviewed by Blackburn and Jerry[51]).

The overlap between p53 and VDR appears to extend beyond cellular phenotypes. The *VDR* is a common transcriptional target of both p53 and p63[52,53] and VDR and p53 share a cohort of direct target genes associated with cell cycle arrest, signal transduction and programmed cell death including *CDKN1A*, *GADD45A*, *IGFBP3*, and *TGFB1/2*.[54–61] At the transcriptional level, both VDR heterodimers and p53 tetramers associate, for example, with chromatin remodeling factors CBP/p300 and the SWI/SNF to initiate transactivation.[62,63] By contrast, in the gene-repressive state VDR and p53 appear to associate with distinct repressor proteins, for example, p53 with SnoN[64] and VDR with NCOR1, suggesting the possible association with distinct sets of histone deacetylases. Furthermore, *CDKN1A* promoter-dissection studies revealed adjacent p53 and VDR binding sites, suggesting composite responsive regions.[61] Together, these findings suggest that $1\alpha,25(OH)_2D_3$-replete environments enhance p53 signaling to regulate mitosis negatively.

Similarly, the role of $1\alpha,25(OH)_2D_3$ in the skin is also suggestive of its genome-protective effects. UV light from sun exposure has several effects in the skin; UVA light induces DNA damage through increasing the level of reactive oxygen species (ROS), but importantly UVB light is critical for synthesis of 25(OH)-D and induces the expression of VDR in the skin. In addition, antimicrobial and anti-inflammatory

VDR target genes are induced by UV radiation. Suppression of the adaptive inflammatory response is thought to be protective for several reasons. Inflamed tissues contain more ROS, which in turn can damage DNA and prevent proper function of DNA repair machinery. In addition, the induction of cytokines and growth factors associated with inflammation act to increase the proliferative potential of the cells. NF-κB is a key mediator of inflammation and the VDR attenuates this process by negatively regulating NF-κB signaling.[65] This control by VDR is underscored by studies showing *Vdr*$^{-/-}$ mice are more sensitive to chemicals that induce inflammation than their wild-type counterparts.[66] The normally protective effect of inflammation that occurs under other conditions is lost through VDR mediated suppression but is compensated for by the induction of a cohort of antimicrobial and antifungal genes.[67–69] The induction of antimicrobial genes not only prevents infection in damaged tissue but can be cytotoxic for cells with increased levels of anion phospholipids within their membranes, a common feature of transformed cells.[70] Finally, and most recently, network strategies have been used in different strains of mice with altered sensitivity toward skin cancer. Remarkably, in such unbiased screens, the Vdr emerges as a key nodal control point in determining sensitivity toward skin tumors as it regulates both turnover of self-renewal and inflammatory infiltrate.[71]

The key question, and central to exploiting any therapeutic potential of this receptor, is why should the VDR exert such pleiotropic actions? One possible explanation for this pleiotropism is that it represents an adaptation of the skin to UV exposure, coupling the paramount importance of initiating $1\alpha,25(OH)_2D_3$ synthesis with protection of cell and tissue integrity. Thus, VDR actions are able to maximize UV-initiated synthesis of $1\alpha,25(OH)_2D_3$ while controlling the extent of local inflammation that can result from sun exposure. To compensate for the potential loss of protection associated with immunosuppression, the VDR mediates a range of antimicrobial actions. Equally, local genomic protection is ensured through the upregulation of target genes that induce G_0/G_1 arrest, cooperation with p53, and the induction of cell differentiation. It remains a tantalizing possibility that the functional convergence between p53 family and VDR signaling, arose in the dermis as an evolutionary adaptation to counterbalance the conflicting physiological requirements of vitamin D synthesis.

8.3 TRANSCRIPTIONAL REGULATION BY THE VDR

8.3.1 The Choreography of Transactivation by the VDR

The VDR is distributed in various cellular compartments including the cell membrane and cytoplasm, and there is evidence of cell membrane–associated VDR, which may mediate rapid nongenomic responses.[72] Similarly, there is evidence for trafficking of cytoplasmic VDR into the nucleus upon ligand activation in tandem with RXRs,[73] each in association with specific importins.[74] However, the VDR appears to be predominantly located in the nucleus.

The choreography of NR complex binding to chromatin has been investigated in the context of the VDR[61,75–80] and other NRs.[81–84] Collectively, these approaches have revealed that the exchange of these factors is highly dynamic and involves cyclical rounds of promoter-specific complex assembly, gene transactivation, complex disassembly, and proteosome-mediated receptor degradation coincident with corepressor complex binding and silencing of transcription. More recent data have emerged to support chromatin looping within the same VDR target gene loci.[85] Collectively, these events give rise to the characteristic periodicity of NR transcriptional activation and pulsatile mRNA and protein accumulation. These actions appear to be highly specific to individual genes and the specific VDR binding sites located within a given genes. Therefore, the periodicity of VDR induced mRNA accumulation of target genes is not shared, but rather tends toward patterns that are specific for individual target genes and suggests that promoter-specific complexes combine to determine the precise periodicity.[61,76]

In the absence of ligand, the VDR exist in an *apo* state, associated with corepressors (e.g., NCOR1, NCOR2/SMRT) as part of large complexes (~2.0 MDa)[56,86,87] and at specific binding site sequences. These complexes in turn actively recruit a range of enzymes that exert posttranslational modification of histone tails, for example, histone deacetylases (HDACs) and methyltransferases, and thereby maintain a locally condensed chromatin structure around response element sequences.[75,87–92] This is in contrast to the classical steroid hormone receptors, AR and ERα, which have more pronounced expression in the cytoplasm, in the absence of ligand, and shuttle into the nucleus upon ligand activation.

In the nucleus, in the presence of ligand, the VDR commonly forms a heterodimer with RXR.[93] In this *holo* state, the VDR ligand-binding pocket contains hydrophobic residues that are important in the binding of $1\alpha,25(OH)_2D_3$. Ligand binding induces a conformational change, which in turn releases corepressors and recruits coactivator proteins. Some of these coactivators have histone acetyltransferase (HAT) activity, they acetylate histone tails and create local chromatin relaxation for gene transcription. Important VDR coactivators with HAT activity include SRC/p160 family members, SRC-3/ACTR, SRC-1/NCoA-1, GRIP1/TIF2/NCoA-2. SRC/p160 family member form a larger coactivator complex together with CBP/p300, p/CIP, and PCAF.[94]

8.3.2 VDR Binding Sites through the Genome

To identify VDR binding sites through the genome various investigators have undertaken ChIP-seq studies in different human cell types including immortalized lymphoblastoids,[95] hepatic stellate cells,[96] and cancer cell lines representing monocytic leukemia[97] and colon cancer.[98] These studies share the identification of approximately the same number of binding sites, on the order of 2000 different binding sites following treatment with $1\alpha,25(OH)_2D_3$ in these different cell types. This number appears to be lower than for some other NRs, and significantly lower than major transcription factors, for example, MYC. Furthermore, the intersection of these data sets reveals that fewer than 20% of the VDR binding sites are in common between the different cell types.

Within each cell type, the VDR binding sites are normally distributed around the TSS of target genes and the distribution includes sites that appear many kilobases away. This reflects the distributions found for other transcription factors, as established very clearly within the ENCODE project.[99] This clearly implies that VDR binding sites are both upstream and downstream of the TSS region of the target genes. The importance of the sites that are extremely distal (>100 kb) is an area of active investigation, but it appears that there is good evidence to suggest that distal sites are playing a role in controlling expression most likely through looping mechanisms.

Finally, an interesting commonality between the data sets is that the consensus binding site for the VDR, the so-called direct repeat (DR) spaced by three nucleotides (DR-3), appears to be the minority genomic element that directly binds the receptor; perhaps only 30% of genomic VDR binding sites contain a DR-3. Following ligand treatment there is, however, increased enrichment for VDR binding to DR-3 elements. Nonetheless, a range of other genomic elements were enriched in VDR binding peaks suggesting that the VDR cooperates closely with other factors to associate with the genome, both in the absence and presence of ligand. Together, these findings suggest there is considerable tailoring, or guiding of both the type and position of binding sites for the VDR depending on cell phenotype and disease state. The diversity of VDR expression sites, being expressed in virtually all cells of the human, and the disparate phenotypic effects, from regulating calcium transport to sensing redox potential and DNA damage beg the question of cell specificity of actions. That is, what governs the spatial–temporal regulation of VDR-dependent transcriptomes in different cell types? It is unclear why a ubiquitous receptor binds to such divergent genomic sites in different cell types, but this is observed for a number of NRs and other transcription factors and is also seen in different disease states, for example, in cancer progression, and suggests that the VDR displays considerable plasticity of genomic binding and transcriptional control. The specificity of VDR signaling may arise due to integration with other perhaps more dominant transcription factors. Again, for other NRs (e.g., AR, ERα), the so-called pioneer factors appear to be highly influential in determining choice and magnitude of transcriptional actions.[100] It remains to be established to what extent the VDR interacts similarly with other transcription factors.

8.3.3 Regulation of Cell Cycle Arrest by the VDR as a Model of Integrated Transcriptional Control

Much progress in understanding the actions of the VDR has emerged from investigating its potential to be exploited as a chemotherapeutic agent in cancer cells. Defining the mechanisms by which the VDR exerts desirable anticancer effects has been an area of significant investigation. In 1981, Colston and colleagues were first to demonstrate that $1\alpha,25(OH)_2D_3$ inhibited human melanoma cell proliferation significantly in vitro at nanomolar concentrations.[101] Parallel studies in the same year also found $1\alpha,25(OH)_2D_3$ could induce differentiation in cultured mouse and human myeloid leukemia cells.[102,103] Following these studies, antiproliferative effects have been demonstrated in a wide variety of cancer cell lines, including those from prostate, breast, and colon.[104–111] To identify critical target genes that mediate these actions, comprehensive genome-wide transcriptomic screens have analyzed the antiproliferative VDR transcriptome and revealed broad consensus on certain targets but has also highlighted variability.[54,104,112,113] This heterogeneity may in part reflect experimental conditions, cell-line differences, and genuine tissue-specific differences of cofactor expression that alter the amplitude and periodicity of VDR transcriptional actions.

A common antiproliferative VDR function is associated with arrest at G_0/G_1 of the cell cycle, coupled with upregulation of a number of cell cycle inhibitors including p21$^{(waf1/cip1)}$ and p27$^{(kip1)}$. Promoter characterization studies have demonstrated a series of VDR binding sites in the promoter/enhancer region of *CDKN1A*.[61,114] By contrast, the regulation of the related CDKI p27$^{(kip1)}$ is mechanistically enigmatic, reflecting both transcriptional and translational regulation, which include enhanced mRNA translation, and attenuating degradative mechanisms.[115–118] The upregulation of p21$^{(waf1/cip1)}$ and p27$^{(kip1)}$ principally mediate G_1 cell cycle arrest, but $1\alpha,25(OH)_2D_3$ has been shown to mediate a G_2/M cell cycle arrest in a number of cancer cell lines via direct induction of *GADD45*α.[56,113,119] These studies highlight the difficulty of establishing strict transcriptional effects of the VDR, as a range of posttranscriptional effects act in concert to regulate target gene levels. Concomitant with changes in the cell cycle, there is some evidence that $1\alpha,25(OH)_2D_3$ not only induces differentiation, most clearly evidenced in myeloid cell lines, but is also supported by other cell types and most likely reflects the intimate links that exist between the regulation of the G_1 transition, the expression of CDKIs including p21$^{(waf1/cip1)}$ and the induction of cellular differentiation.[120]

Historically, hematological malignancies combined an ease of interrogation with robust classification of cellular differentiation capacity that were envied by investigators of solid tumors. It is therefore no coincidence that these cell systems yielded many important insights, including understanding of chromosomal translocations, genomic instability, and the role of committed adult stem cells. Indeed, the capacity to readily differentiate in response to external and internal signals has fascinated leukemia researchers as they have sought to understand why leukemia cells appear to fail at certain stages of differentiation. It is within this context that investigators[121,122] considered a role for the VDR and the related retinoic acid receptor (RAR) to reactivate dormant differentiation programs in so-called differentiation therapies. Despite these efforts, clinical exploitation of the VDR has largely proved to be equivocal. The one exception to this translational failure has been the exploitation of RAR signaling in patients with acute promyelocytic leukemia. Again, understanding the basic signaling behind this application proved significant to the developing understanding of epigenetic regulation of transcription and the promise of histone deacetylase (HDAC) inhibitors.[123]

Novershtern et al.[48] measured the transcriptome profiles of a large number of hematopoietic stem cells, multiple progenitor states, and terminally differentiated cell types. They found distinct regulatory circuits in both stem cells and differentiated cells, which implicated dozens of new regulators in hematopoiesis. They identified 80 distinct modules of tightly coexpressed genes in the hematopoietic system. One of these modules is expressed in granulocytes and monocytes and includes genes encoding enzymes and cytokine receptors that are essential for inflammatory responses. Major players in this module are VDR together with the pioneer factors CEBPA and SPI1. This indicates that VDR works together with this small set of transcription factors to regulate granulocyte and monocyte differentiation. It is reasonable to anticipate that such modules exist in multiple cell types.

8.4 EPIGENETIC REGULATION OF VDR SIGNALING

8.4.1 A Primer on Epigenetic Events Involved in Transcription

Epigenetic events play a central role for transcriptional complexes, and the various components in these multimeric complexes sequentially initiate, sustain, and finally terminate transcription.[124] In this manner, transcription can work as a type of biological ratchet, with differing histone modifications associated with the various states by generating chromatin states that are either receptive or resistant to transcription (reviewed by Taverna et al.[125]). For example, different histone modifications can control the rate and magnitude of transcription (reviewed by Goldberg et al.[126]). These events are intertwined with levels of CpG methylation.[127–129] Thus, the histone modifications and other epigenetic events including DNA methylation processes combine during transcription to generate highly flexible chromatin states that are either transcriptionally receptive and resistant.[130] That is, the specific transcriptional potential of a gene is flexibly controlled by the combination of epigenetic events. These events are varied in space across the genomic loci, and in time through the course of the transcriptional cycle.

The diversity of histone modifications and their association with different DNA functions formed the basis for the histone code hypothesis. This concept, first proposed in 1993, held that these modifications were governed in a coordinated manner and formed a code that mirrored the underlying DNA code to convey heritable information on transcription and expression.[131] Given the rapid expansion of the understanding in the number of histone modifications, their genomic distribution and their combinatorial manner, it is actually only relatively recently that the true diversity of the range of binding options available, and their functional outcomes, has become apparent.[126] The strongest evidence that histone modifications at the level of meta-chromatin architecture form a stable and heritable "histone code" is perhaps seen with X chromosome inactivation (reviewed by Turner[132]). The extent to which similar processes operate to govern the activity of microchromatin contexts at gene promoter regions is an area of debate.[133,134] The regulation of transcription and the patterns of mRNA expression have been related to the expression of these histone modifications through a wide range of correlative and functional studies.

In this manner, the status of histone modifications appears to control the transcriptional action of gene loci. For example, histone H3 lysine 4 trimethylation (H3K4me3) is found in the promoter regions of actively transcribed genes. This mark is mutually exclusive with H3K9me, which instead is associated with transcriptionally silent promoter regions. Acetylation of H3K9 is found along with methylation of H3K4 at active promoter regions. An individual lysine in a histone tail can be either methylated or acetylated at any one moment but not both. Therefore, methylation and acetylation of H3K9 are mutually exclusive. Similarly, H3K27 can be either acetylated or methylated, with acetylation associated with active gene transcription and methylation associated with gene silencing. H3K36 is typically found as acetylated in promoters of active genes, but methylated in the bodies of those genes while K36 is unmodified at silent genes regardless of position in the promoter or gene body. At enhancer regions of active genes, H3K4 monomethylation is found along with acetylation of K9 and K27, while enhancers of silent genes retain H3K4me, but do not have acetylation at K9 and K27.

Two further points are particularly important; first, the steady state level of each modification represents a highly dynamic balance between the antagonistic effects of key enzymes (with turnover likely to vary from one part of the genome to another and between cell types). Second, many, if not all, of the enzymes are dependent upon, or influenced by either metabolites or components present in the intracellular or extracellular environment. Thus, the nucleosome, through the array of histone modifications it carries and the enzymes that put them in place, is a finely tuned sensor of the metabolic state of the cell and the composition of its environment. Consequently it provides a platform through which external environmental and internal variables can influence genomic function. In this manner, epigenetic states are a key modulator of transcriptional capacity and are regulated directly by cell context, cell cycle status, cell–cell interactions, and cell lineage commitment (reviewed by Thorne et al.[135]).

Finally, it is important to note that histone modifications, DNA methylation, and miRNA function are all intertwined. Thus, transient histone states also act as platforms to allow effector complexes to regulate DNA CpG methylation. For example, gene-repressive H3K9me2 associates with CpG methylation and heterochromatin. At high density regions of CpG methylation, spanning hundreds of base pairs, these marks act as triggers to recruit heterochromatin binding protein 1 (HP1).[130] The recruitment of HP1 through interaction with the methyl-CpG binding protein MBD1 leads to recruitment of both KMT1A (SUV39H1)[136] and DNA methyltransferases (DNMTs)[137] and thereby the entire region acquires H3K9 and H3K27 methylation and loses H3K4 methylation (reviewed by Cheng and Blumenthal[138]). It has also emerged that in actively regulated regions, dynamic changes in DNA methylation appear to occur. For example, these have been measured in response to NR actions.[127–129] In parallel, increased transcriptional corepressor binding promotes direct association with the DNA methylation-dependent transcriptional repressor ZBTB33/KAISO[139] and targets DNA methylation.

miRNAs control gene expression posttranscriptionally by translational inhibition when base pairing between miRNA and target sequence is imperfect, while perfect or near-perfect complementarity can induce the degradation of the target mRNA.[140] For example, expression of miRNAs can be regulated by DNA methylation in promoter regions of miRNAs.[141] Similarly, miRNAs can directly target chromatin remodeling enzymes[142] and DNA methyltransferases (DNMTs)[143] and in turn modulate the downstream effects. Although the biogenesis of miRNA has been well described, the regulation of miRNA expression remains less well explored. Several miRNAs have well-defined promoters, but for many, the regulatory elements are not as clear. They are often present in introns of protein coding genes and the promoters for the genes are considered to be promoters for the intragenic miRNAs.[140,144] Interestingly, there is evidence for differential regulation between the host mRNA and miRNA and may suggest an emergent role for alternative regulation processes.

miRNAs can also directly or indirectly target key effectors of the epigenetic machinery. For example, miR-29 family members (29a, 29b, and 29c) are downregulated in several cancers and they are predicted to target DNA methyltransferase *DNMT3A*, *DNMT3B*, and *DNMT1*.[145,146] Recently, miR-34b has been shown to target *DNMT1*, *HDAC1*, *HDAC2*, and *HDAC4*.[147] Interestingly, miR-34b was also epigenetically silenced by DNA methylation and ectopic expression of miR-34b resulted in partial demethylation of 5′-upstream sequence of the miR-34b gene and also showed enrichment of trimethylated histone H3 lysine 4 (H3K4me3), mark for an active chromatin.[147]

Various groups, including that of Studzinski, have worked consistently exploring mechanisms of resistance to VDR signaling and methods of exploitation and recently demonstrated, elegantly, a role for VDR to downregulate miR-181a, which when left unchecked degrades $p27^{(kip1)}$.[148] Thus, indirectly, VDR activation elevates expression of $p27^{(kip1)}$, initiates cell cycle arrest, and commits cells toward differentiation. Transcriptional control of miRNAs and their biological effects are a field of rapid expansion, and members of the NR superfamily are implicated in their regulation.[149,150] A role for the VDR to govern the expression of this regulatory miRNA and, importantly, place its role in well-understood map of differentiation is highly novel.

Similar integration of miRNA and mRNA was revealed to control the regulation of *CDKN1A* (encodes $p21^{(waf1/cip1)}$). In nonmalignant prostate epithelial cells, rapid and dynamic patterns of *CDKN1A* mRNA accumulation occurred.[144] These patterns of mRNA accumulation reflected receptor and corepressor exchanges occurring in a unique manner at each of three VDREs. These results support the concept that in nonmalignant systems VDR transcriptional responses can be rapid and functional and faithfully lead to changes in protein. The transition through activated epigenetic states appears to allow the integration of transcriptional signals, for example, strong cooperation with p53 at 1 hour and suggests an epigenetic basis for the observed cooperation between VDR and p53 pathways.[52,53,61,151,152] Further, these studies in RWPE-1 cells revealed that the magnitude of epigenetic modulation was refined by the cell cycle status. ChIP approaches in FACS-sorted cells revealed that G_1 phase cells were characterized by enhanced VDR-induced activating histone modifications (for example, H3K9ac), with S and G_2/M phases being largely repressive, for example, with H3K27me3 enrichment on the *CDKN1A* promoter. Thus, events that were not significant when considering bulk populations emerged with significant clarity when considering each specific cell cycle phase and underscore the fact that bulk culture findings represent an average event of potentially very different populations. While other NRs have been demonstrated to display cell cycle–specific phases of activation,[153,154] this study revealed an underlying role for differential regulation of histone modifications through the cell cycle to govern these actions. VDR-dependent coregulation of miR-106b was also revealed to modulate the precise timing of *CDKN1A* accumulation and also the expression of $p21^{(waf1/cip1)}$. Together, these data demonstrate that VDR induced regulation of $p21^{(waf1/cip1)}$ is determined by interplay of histone modifications and miRNA expression that combine in a feed-forward loop and determine the final extent of the cell cycle arrest.

Together, these data suggest the magnitude of *CDKN1A* activation, at least at early time points, is influenced significantly by the stage of the cell cycle that changes the epigenetic state of key VDR binding regions that allow regulation of mRNA and potentially targeting miRNA. In this regard, the epigenome dictates how effectively the VDR can control expression of target genes. It is not unreasonable to anticipate the changes in the epigenome that occur through the cell cycle, through lineage commitment, through cell–cell interactions, and other biological processes will determine how effectively the VDR function. Indeed, this paradigm of the epigenomic determining VDR binding and transcriptional capacity would also contribute to an explanation of the extreme diversity observed for VDR binding site distribution observed in the four published ChIP-Seq studies.

8.4.2 VDR Regulation by Corepressors and Coactivator Complexes

Independent of binding, the temporal regulation of VDR-dependent transcriptomes appears to differ between cell types and disease states. A high level of specificity over the timing and choice of VDR cofactor interactions may also provide a mechanistic basis for signaling specificity. In this manner, the expression and choice of interacting cofactors may determine the choice and timing of gene regulation. A further emerging theme concerns the hierarchical organization; the extent to which VDR is upstream or downstream of other regulatory events.

There is good evidence that specific histone modifications determine the assembly of transcription factors on the promoter, and control individual promoter transcriptional responsiveness.[155–157] For example, the VDR may recognize basal histone modifications on target gene promoters; functional studies of the SANT motif contained in the corepressor NCOR2/SMRT supports this latter idea[158,159] and again support the cyclical nature of gene regulation being profoundly influenced by the epigenome. It remains an enticing prospect that the variability in gene regulatory patterns reflects the organization of different subsets of gene targets, perhaps organized by phenotype. Thus, VDR regulated networks may be differentially governed by the chromatin context at response elements and include near and long-term chromatin associations.

Interactions with corepressor illustrate how transcriptional choices can be filtered. Of the principal corepressors, it remains to be established to what extent specificity and redundancy occurs. The expression, localization, and isoforms of NCOR1 and NCOR2/SMRT corepressors strongly influences the spatiotemporal equilibrium between repressing and activating transcription complexes and transcriptional outputs[160] (reviewed by Rosenfeld et al.[159]). These dynamic exchanges between corepressors and coactivators and their epigenetic impact have been illustrated clearly with several members of the NR superfamily including the AR, VDR, and ERα[61,82,85,144,161] and also with AP-1 members and NF-κB factors (reviewed by Thorne et al.[135]).

The principal corepressors NCOR1 and NCOR2/SMRT have been investigated in vivo, and these studies revealed specificities with the regards to tissue and function with respect to interactions with RAR and PPARs. NCOR1 and NCOR2/SMRT expression, localization, and isoforms have emerged as critical to determine the spatiotemporal equilibrium between the antagonistic actions of the ligand-bound and unbound NR complexes and thus determine the dynamics of target gene promoter responsiveness. Knockout of these proteins is embryonically lethal, most likely due to the interaction of NCOR1 and NCOR2/SMRT with a significant number of transcription factors including MYC and FOXO family members other than NRs. More recently, stem cell components from *Ncor1*$^{-/-}$ and *Ncor2/Smrt*$^{-/-}$ mice and other studies using conditional approaches have revealed interaction networks with NRs and other transcription factors.[91,162–164] In brain development, NCOR2/SMRT represses Rar-induced expression of jumonji domain–containing histone demethylases that are able to demethylate H3K27me3 and induce genes associated with neuronal differentiation.[162] Similarly, a knockin approach was used to circumvent the embryonic lethality and allow the specific disruption of the interaction of NCOR2/SMRT with NRs. These models revealed dramatically enhanced differentiation rates, notably in adipocyte differentiation mediated by PPAR-γ.[165] These findings suggest that corepressors significantly regulate transcriptional actions, over both choice and periodicity of target sequence regulation.

VDR transcriptional actions in nonmalignant human systems support the concept corepressors play specific levels of interactions. For example, NCOR1 and NCOR2/SMRT display both parallel loss and gain of enrichment at specific VDRE on the *CDKN1A* promoter in response to ligand activation. That is, both corepressor loss and recruitment occurred simultaneously on the same promoter at different regions. The altered NCOR1 enrichment patterns associated with both loss and gain and activating histone modifications and collectively appeared to contribute to the oscillation of mRNA.[144] This ligand-induced corepressor associations have now also been established on genome-wide scale for the related glucocorticoid receptor where ligand activation induces receptor binding to NCOR1 and NCOR2/SMRT at specific response elements and drives transrepression.[166]

Beyond NCOR1 and NCOR2/SMRT the list continues to grow of novel corepressor proteins that the VDR interacts with (Table 8.1). Compared with the relatively massive size of the corepressors NCOR1 and NCOR2/SMRT, a number of smaller molecules have emerged as showing corepressor function. TRIP15/COPS2/Alien has been demonstrated to interact with the VDR and act as a corepressor and may not require the same interactions with HDACs that NCOR1 does.[167] Intriguingly, this protein contributes to the lid subcomplex of the 26S proteasome and thereby potentially links VDR function with the regulation of protein stability.[168] Similarly, SLIRP[169] has also emerged as a repressive factor for the VDR, although, to date, very little is known about the specificity, in terms of tissue and target gene.

Other repressors appear to demonstrate more explicit phenotypic specificity. *Hairless* blocks VDR-mediated differentiation of keratinocytes, whereas addition of $1\alpha,25(OH)_2D_3$ displaces *Hairless* from the promoter of target genes and recruits coactivators to promote differentiation.[170–172] Similarly, downstream regulatory element antagonist modulator (DREAM) usually binds to direct repeat response elements in the promoters of target genes to enhance transcription of VDR and RAR target genes, in a calcium-dependent manner, and suggests that specificity arises from the interactions of VDR with further tissue-specific cofactors.[173]

As discussed previously, ligand binding to VDR induces a conformational change to expose docking surfaces for transcriptional coactivator complexes.[174,175] A large number of interacting coactivator proteins have been described, which can

TABLE 8.1
List of Important VDR Corepressors and Coactivators and Their Associated Mouse Phenotype

Name	Function	MGI Mutant Phenotypes
	Corepressors	
NCOR1	Part of a complex that promotes histone deacetylation and the formation of repressive chromatin structures	Mice homozygous for a targeted mutation in this gene exhibit embryonic lethality with erythrocytic, thymocytic, and central nervous system development abnormalities.
NCOR2/SMRT	Part of a complex that promotes histone deacetylation and the formation of repressive chromatin structures	Mice homozygous for a null allele die before E16.5 of heart defects and exhibit neural defects
TRIP15/COPS2/Alien	Ubiquitination, essential regulator of the ubiquitin (Ubl) conjugation pathway	Mice homozygous for disruptions in this gene die as embryos very soon after implantation
SLIRP	Repress the SRA-mediated nuclear receptor coactivation	NA
Hairless	Histone deacetylases	Mutant homozygotes exhibit hair loss, usually wrinkled skin with epidermal cysts. Females do not nurse their pups well.
DREAM	Functions as a calcium-regulated transcriptional repressor, and interacts with presenilins; DREAM can function as an activator of transcription on certain promoters.	Mice homozygous for a null allele exhibit abnormal spatial learning, hyperactivity, hypophagia, increased sensitivity to shock, and enhanced long term potentiation.
	Coactivators	
TCF4	Activates transcription by binding to the E box (5′-CANNTG-3′)	Homozygotes for a null allele show a partial block in early thymopoiesis, increased double-negative T cell count, and increased sensitivity to anti-CD3-induced apoptosis. Homozygotes for another null allele show neonatal or postnatal lethality, reduced pro-B cell number, and abnormal pontine nuclei.
DRIP/TRAP/ARC	Component of the mediator complex, a transcriptional coactivator complex. The mediator complex is recruited by transcriptional activators or nuclear receptors to induce gene expression, possibly by interacting with RNA polymerase II and promoting the formation of a transcriptional preinitiation complex.	Mice homozygous for a conditional allele exhibited in the heart exhibit increased susceptibility to obesity and worsened glucose intolerance when fed a high-fat diet.

(*continued*)

TABLE 8.1 (Continued)
List of Important VDR Corepressors and Coactivators and Their Associated Mouse Phenotype

Name	Function	MGI Mutant Phenotypes
	Coactivators	
NCoA-62	A member of the *SNW* gene family, coactivator that enhances transcription from some Pol II promoters.	NA
	P160/SRC Family Members	
NCoA-1/SRC-1	A member of the p160/steroid receptor coactivator (SRC) family and like other family members has histone acetyltransferase (HAT) activity. Binds nuclear receptors directly and stimulates the transcriptional activities in a hormone-dependent fashion.	Homozygotes for a null allele show osteopenia, increased serum sex hormone levels, altered bone remodeling, and skeletal responses to sex hormones, and obesity. Homozygotes for another null allele show thyroid and steroid hormone resistance, delayed Purkinje cell development, and behavioral deficits.
NCoA-2/GRIP1/TIF2	Transcriptional coactivator for steroid receptors and nuclear receptors. Coactivator of the steroid binding domain (AF-2) but not of the modulating N-terminal domain (AF-1).	Homozygous null mice exhibit a transient postnatal growth deficiency and hypofertility. Male hypofertility is due to defects in spermiogenesis and an age-dependent testicular degeneration preceded by defective lipid metabolism in Sertoli cells. Female hypofertility is due to a placental hypoplasia.
NCoA-3/pCIP/RAC3/ACTR/AIB1	A nuclear receptor coactivator that interacts with nuclear hormone receptors to enhance their transcriptional activator functions. The encoded protein has histone acetyltransferase activity and recruits p300/CBP-associated factor and CREB binding protein as part of a multisubunit coactivation complex.	Nullizygous mice exhibit growth defects and reduced serum IGF-1 levels, and may show impaired proliferative responses to various factors, delayed mammary gland growth and puberty, reproductive dysfunction, susceptibility to endotoxin shock, altered lymphopoiesis, and protection against obesity.

be divided into multiple families including the p160 family, the non-p160 members, and members of the large "bridging" DRIP/TRAP/ARC complex, which links the receptor complex to the cointegrators CBP/p300 and basal transcriptional machinery.[176,177] Although the DRIPs were first thought to be VDR specific, they were also identified in the regulation of the other NRs including the GR and TR.[178,179] Indeed, the exact specificity/selectivity of many of the coregulatory factors remains to be established fully, although there is some suggestion that certain coactivators are specific for the VDR, for example, NCoA-62.[180] Binding of DRIP205 and other coactivators to VDR target gene promoters are 1α,25$(OH)_2D_3$-dependent. Following the DRIP complex binding the coactivator binding leads to the coordinate activation of an antagonistic battery of chromatin remodeling enzymes, including histone acetyltransferases, and thereby induce the reorganization of local chromatin regions at the target gene promoter.

In essence, expression and function of these key proteins alters the equilibrium of key histone modifications and thereby allowing the gene regulatory actions of a given transcription factor to become more or less pronounced. For example, NCOA3/SRC3 is situated within a common area of chromosomal amplification in breast cancer on chromosome 20q. Initially, a cDNAs was isolated from this region that contained a putative target gene that was termed AIB1 (for "amplified in breast cancer 1"). Subsequently, this gene was found to be a member of the SRC coactivator family and was amplified and overexpressed in breast and ovarian cancer cell lines as well as in breast cancer biopsies.[181] NCOA3/SRC3/AIB1 interacts with ERs in a ligand-dependent fashion and enhances the regulation of target genes. Specifically, the protein has intrinsic HAT activity and also acts to recruit other CBP/p300 in an allosteric manner.[182] Therefore, increased expression increases the ability of the ERα to transactivate a given gene target. Subsequently, this protein was identified as NCOA3 and shown to be a potent histone acetyl transferase able to enhance the function of multiple NRs.[183–185]

8.4.3 Epigenetic Mechanisms of Resistance to the VDR Regulation

In cancer cells, there is often a lack of an antiproliferative response toward the VDR, and this is reflected by a suppression of the transcriptional responsiveness of antiproliferative target genes (e.g., *CDKN1A*, *GADD45A*, *IGFBPs*, *BRCA1*[56,106,186,187]). Paradoxically, VDR transactivation of other targets is sustained or even enhanced, as measured by induction of the highly 1α,25$(OH)_2D_3$-inducible *CYP24* gene.[188,189] Together, these data suggest that the lack of functional VDR alone cannot explain resistance and instead the VDR transcriptome is skewed to disfavor antiproliferative target genes. It has been proposed that this apparent 1α,25$(OH)_2D_3$-insensitivity is the result of epigenetic events that selectively suppress the ability of the VDR to transactivate target genes;[190] that is, the cancer epigenome disfavors VDR regulation of genes that exert mitotic restraint.

The epigenetic basis for such transcriptional discrepancies has been investigated intensively in prostate cancer. There is compelling evidence that histone and DNA methylation processes disrupt NR transcriptional actions, both alone and together. Despite these important roles for corepressors to regulate transcription and potentially trigger DNA methylation, they remain somewhat overlooked compared with their coactivator cousins. Ambiguity remains over how and to what extent these actions are distorted in cancer (reviewed by Battaglia et al.[191]). The sheer diversity of transcription factors and corepressors interactions contributes significantly to this uncertainty. This is compounded by the fact that there are functionally different corepressor isoforms[160,192] and that corepressor actions appear specific to each phase of the cell cycle.[144,193,194]

Increased NCOR1 and NCOR2/SMRT expression and localization occurs in prostate and other cancers,[56,86,195–198] for example, to suppress VDR and PPARs responsiveness.[56,193,199] In contrast, putative loss of function mutations in NCOR1 have been identified contributing to breast cancer.[200] The recruitment to the AR of corepressors contributes to efficacy of androgen deprivation therapy and consequent loss of expression of NCOR1 and NCOR2/SMRT is also associated with recurrent prostate cancer.[201,202]

Thus, corepressor expression and interactions appear to change during the course of disease progression. These changes most likely reflect the conflicting selection pressures on corepressor expression, targeting, and activity. Set against these uncertainties, dissecting corepressor actions may address the key question of separating epigenetic processes that drive cancer initiation and progression, from those that are merely a consequence of altered genomic structure. This understanding has importance for defining clinical targeting strategies, given that altered corepressor expression generates critical targets for agents that target the regulation of H3K9 acetylation and methylation states.

Expression profiling in solid tumors has revealed elevated NCOR1 and NCOR2/SMRT expression, for example, in breast, bladder, and prostate cancers.[56,86,195–198] In both prostate cell lines and primary cultures, elevated NCOR2/SMRT mRNA levels were proportional to a loss of sensitivity toward the antiproliferative actions of $1\alpha,25(OH)_2D_3$. Targeting the HDAC capacity of the corepressor complex with the HDAC inhibitor TSA, in combination with $1\alpha,25(OH)_2D_3$, synergistically augmented the antiproliferative effects. Microarray studies identified that $1\alpha,25(OH)_2D_3$ plus TSA uniquely upregulated a cohort of genes that were silenced in $1\alpha,25(OH)_2D_3$ resistant prostate cancer cell lines and primary cultures. Knockdown of NCOR2/SMRT in lieu of HDAC inhibitor treatment also led to restored VDR regulation of specific genes, resulting in restored antiproliferative responses in prostate cell lines.[56,186]

The consequences of enhanced and sustained corepressor recruitment would result in a critical loss of H3K9ac at, and around, the VDRE binding regions and may in turn allow KMT enzymes to modify this lysine and sustain H3K9me2 levels. Supportively, in other noncancer systems, increased targeting of KMT1A/SUV39H1 to the *CDKN1A* promoter sustained H3K9me2.[203,204] Supportive of this mechanism, multiple KMTs are commonly overexpressed, including *KHMT1C/EHMT2* and *KMT1B/SUV39H2* in prostate cancer cell lines. Cotreatment of $1\alpha,25(OH)_2D_3$ with the KMT inhibitor, chaetocin, reversed the gene transrepression in PC-3 cells, supporting a role for these enzymes to alter the patterns of *CDKN1A* regulation.[205]

Supporting a role for the H3K9me2 levels to attract the machinery that drives DNA CpG methylation, basal and regulated DNA methylation patterns on the *CDKN1A* promoter differed between RWPE-1 and PC-3 cells. Significantly, both the basal and $1\alpha,25(OH)_2D_3$-regulated CpG methylation differed in a position-specific manner. Basal differences were evident, but more surprisingly, $1\alpha,25(OH)_2D_3$ treatment resulted in clear changes in the site-specific methylation in both models, and notably, in PC-3 cells, key sites remained either highly methylated or displayed increased methylation. Again, these findings also reflected previous studies in RWPE-1 cells that identified VDRE2 as a critical response element for the activation of *CDKN1A*.[144] In parallel, Carlberg and colleagues revealed that VDRE2 was a key responsive element involved in chromatin looping and gene activation.[85]

Taken together, these findings supported the concept that inappropriate NCOR1 recruitment coupled with elevated levels of key KMTs, including *KMT1B/SUV39H2*, can sustain H3K9me2 levels that in turn attract the DNA methylation machinery, for example, through HP-1, and sustain transcriptional silencing by inducing DNA methylation[130,139] (reviewed by Cheng and Blumenthal[138]). A parallel inference may also be that this mechanism could contribute to the transrepression by the VDR of targets such as c-MYC,[206] and therefore, it is tempting to speculate that gene regulation behavior reflects how sustained the interactions are with NCOR1 and NCOR2/SMRT.

The consequences of distortions to this process of altered histone modifications leading to distorted CpG methylation could therefore be the targeted methylation of genes where NCOR1 and other corepressors are recruited by the VDR or other nuclear resident transcription factors. Therefore, the VDR and other NRs may provide a route for the silencing of critical transcriptional programs by selective corepressor recruitment and thereby allow cancer cells to escape mitotic restraint. Given the role for corepressors to sequestrate and direct histone deacetylases and methyltransferases, these findings have important implications for the effective targeting of epigenetic therapies during prostate cancer progression with current and next generation epigenetic drugs.

The distributions of these histone modifications and DNA methylation patterns in cell-line models are being organized by research consortia, for example, ENCODE.[207] Again, these genome-wide data sets also appear to support the idea that these histone marks are strongly associated with features of genomic architecture, TSS, and enhancer regions where regulatory transcription factors can bind. Given that certain cell lines have been analyzed by multiple approaches to dissect epigenetic events, there is the possibility to generate highly integrative understanding of how different epigenomic aspects relate to transcription factor binding and mRNA expression. Comparison of repressive histone modifications in the enhancer or promoter regions of a gene locus and altered DNA methylation levels can be analyzed in terms of significant associations with specific transcription factors, gene networks, and drug sensitivities.[208–214] Importantly, epigenetic lesions are individually highly targetable with clinically available small molecular weight inhibitors targeted to specific histone deacetylation events, and more recently, this has been extended to include histone methylation events,[215] coupled with agents that target CpG methylation (reviewed by Graham et al.[216]). Targeting efficiency extends beyond HDACs, and new and potentially more selective epigenetic drugs are emerging.[217–219] Thus, comprehensive understanding of the key corepressors in malignancy, delineating the key transcription factors interactions and the critical targets that are thereby dysregulated, may have considerable prognostic utility, specifically through the capacity to stratify patients for specific tailored epigenetic therapies.

8.5 SUMMARY

It is now over 30 years since the initial reports demonstrated the anticancer actions of $1\alpha,25(OH)_2D_3$.[101–103] Following these studies, antiproliferative effects were demonstrated in a wide variety of cancer cell lines. As the anticancer effects of the ligand emerged, large-scale epidemiological studies found inverse associations between circulating $25(OH)D_3$ and cancer risk and advanced disease, for example, of prostate cancer.[220–228] However, while in vitro, in vivo, and epidemiological data support links among replete VDR signaling, growth restraint, and broad anticancer activities, clinical exploitation of this receptor has been limited. A significant impediment to translation remains the inability to predict accurately which patients will respond to either chemoprevention or chemotherapy strategies centered on vitamin D compounds. The impact of the epigenome, both in physiology and pathophysiology, has emerged as significant in determining how the VDR functions in different cell types and disease states. In turn, the VDR is a central part of a transcriptional complex that can modify the local epigenetic state at the sites of binding. For example, in cancer, the differential recruitment of corepressors may address these apparent resistance of the VDR.[56,86,193,195–199] Given that the epigenetic states, and the enzymes that control them, are emerging as exciting therapeutic targets, there appear to be opportunities to exploit VDR signaling in combination therapies to induce cancer differentiation, or modify immune cell states, or to target other pleiotropic actions of the VDR.

REFERENCES

1. Krust, A. et al. The chicken oestrogen receptor sequence: Homology with v-erbA and the human oestrogen and glucocorticoid receptors. *EMBO J* **5**, 891–897 (1986).
2. Mangelsdorf, D.J. et al. The nuclear receptor superfamily: The second decade. *Cell* **83**, 835–839 (1995).
3. McDonnell, D.P., Mangelsdorf, D.J., Pike, J.W., Haussler, M.R. & O'Malley, B.W. Molecular cloning of complementary DNA encoding the avian receptor for vitamin D. *Science* **235**, 1214–1217 (1987).
4. Makishima, M. et al. Vitamin D receptor as an intestinal bile acid sensor. *Science* **296**, 1313–1316 (2002).
5. Krasowski, M.D., Yasuda, K., Hagey, L.R. & Schuetz, E.G. Evolutionary selection across the nuclear hormone receptor superfamily with a focus on the NR1I subfamily (vitamin D, pregnane X, and constitutive androstane receptors). *Nucl Recept* **3**, 2 (2005).
6. Ellison, T.I., Eckert, R.L. & MacDonald, P.N. Evidence for 1,25-dihydroxyvitamin D3-independent transactivation by the vitamin D receptor: Uncoupling the receptor and ligand in keratinocytes. *J Biol Chem* **282**, 10953–10962 (2007).
7. Imawari, M., Kida, K. & Goodman, D.S. The transport of vitamin D and its 25-hydroxy metabolite in human plasma. Isolation and partial characterization of vitamin D and 25-hydroxyvitamin D binding protein. *J Clin Invest* **58**, 514–523 (1976).
8. Bouillon, R., Van Assche, F.A., Van Baelen, H., Heyns, W. & De Moor, P. Influence of the vitamin D-binding protein on the serum concentration of 1,25-dihydroxyvitamin D3. Significance of the free 1,25-dihydroxyvitamin D3 concentration. *J Clin Invest* **67**, 589–596 (1981).
9. Nykjaer, A. et al. An endocytic pathway essential for renal uptake and activation of the steroid 25-(OH) vitamin D3. *Cell* **96**, 507–515 (1999).
10. Bouillon, R. et al. Vitamin D and human health: Lessons from vitamin D receptor null mice. *Endocr Rev* **29**, 726–776 (2008).
11. Adams, T.H. & Norman, A.W. Studies on the mechanism of action of calciferol. I. Basic parameters of vitamin D-mediated calcium transport. *J Biol Chem* **245**, 4421–4431 (1970).
12. Adams, T.H., Wong, R.G. & Norman, A.W. Studies on the mechanism of action of calciferol. II. Effects of the polyene antibiotic, filipin, on vitamin D-mediated calcium transport. *J Biol Chem* **245**, 4432–4442 (1970).

13. Myrtle, J.F. & Norman, A.W. Vitamin D: A cholecalciferol metabolite highly active in promoting intestinal calcium transport. *Science* **171**, 79–82 (1971).
14. wson-Hughes, B. et al. Estimates of optimal vitamin D status. *Osteoporos Int* **16**, 713–716 (2005).
15. Takeyama, K. et al. 25-Hydroxyvitamin D3 1alpha-hydroxylase and vitamin D synthesis. *Science* **277**, 1827–1830 (1997).
16. Zehnder, D. et al. Extrarenal expression of 25-hydroxyvitamin d(3)-1 alpha-hydroxylase. *J Clin Endocrinol Metab* **86**, 888–894 (2001).
17. Schwartz, G.G., Whitlatch, L.W., Chen, T.C., Lokeshwar, B.L. & Holick, M.F. Human prostate cells synthesize 1,25-dihydroxyvitamin D3 from 25-hydroxyvitamin D3. *Cancer Epidemiol Biomarkers Prev* **7**, 391–395 (1998).
18. Schwartz, G.G. et al. Pancreatic cancer cells express 25-hydroxyvitamin D-1 alpha-hydroxylase and their proliferation is inhibited by the prohormone 25-hydroxyvitamin D3. *Carcinogenesis* **25**, 1015–1026 (2004).
19. Diaz, L. et al. Identification of a 25-hydroxyvitamin D3 1alpha-hydroxylase gene transcription product in cultures of human syncytiotrophoblast cells. *J Clin Endocrinol Metab* **85**, 2543–2549 (2000).
20. Friedrich, M. et al. Analysis of 25-hydroxyvitamin D3-1alpha-hydroxylase in cervical tissue. *Anticancer Res* **22**, 183–186 (2002).
21. Townsend, K. et al. Autocrine metabolism of vitamin D in normal and malignant breast tissue. *Clin Cancer Res* **11**, 3579–3586 (2005).
22. Li, Y.C., Pirro, A.E. & Demay, M.B. Analysis of vitamin D-dependent calcium-binding protein messenger ribonucleic acid expression in mice lacking the vitamin D receptor. *Endocrinology* **139**, 847–851 (1998).
23. Amling, M. et al. Rescue of the skeletal phenotype of vitamin D receptor-ablated mice in the setting of normal mineral ion homeostasis: Formal histomorphometric and biomechanical analyses. *Endocrinology* **140**, 4982–4987 (1999).
24. Kato, S. et al. In vivo function of VDR in gene expression-VDR knock-out mice. *J Steroid Biochem Mol Biol* **69**, 247–251 (1999).
25. Panda, D.K. et al. Targeted ablation of the 25-hydroxyvitamin D 1alpha-hydroxylase enzyme: Evidence for skeletal, reproductive, and immune dysfunction. *Proc Natl Acad Sci U S A* **98**, 7498–7503 (2001).
26. Liu, W. et al. Intestinal epithelial vitamin D receptor signaling inhibits experimental colitis. *J Clin Invest* **123**, 3983–3996 (2013).
27. Cui, M., Li, Q., Johnson, R. & Fleet, J.C. Villin promoter-mediated transgenic expression of TRPV6 increases intestinal calcium absorption in wild-type and VDR knockout mice. *J Bone Miner Res* **27**, 2097–2107 (2012).
28. Kriebitzsch, C. et al. 1,25-dihydroxyvitamin D3 influences cellular homocysteine levels in murine preosteoblastic MC3T3-E1 cells by direct regulation of cystathionine beta-synthase. *J Bone Miner Res* **26**, 2991–3000 (2011).
29. Blomberg Jensen, M. et al. Vitamin D receptor and vitamin D metabolizing enzymes are expressed in the human male reproductive tract. *Hum Reprod* **25**, 1303–1311 (2010).
30. Narvaez, C.J., Matthews, D., Broun, E., Chan, M. & Welsh, J. Lean phenotype and resistance to diet-induced obesity in VDR knockout mice correlates with induction of uncoupling protein-1 in white adipose tissue. *Endocrinology* **150**, 651–661 (2008).
31. Zhang, X., Rahemtulla, F.G., MacDougall, M.J. & Thomas, H.F. Vitamin D receptor deficiency affects dentin maturation in mice. *Arch Oral Biol* **52**, 1172–1179 (2007).
32. Froicu, M., Zhu, Y. & Cantorna, M.T. Vitamin D receptor is required to control gastrointestinal immunity in IL-10 knockout mice. *Immunology* **117**, 310–318 (2006).
33. Dong, X. et al. Regulation of relB in dendritic cells by means of modulated association of vitamin D receptor and histone deacetylase 3 with the promoter. *Proc Natl Acad Sci U S A* **102**, 16007–16012 (2005).

34. Zinser, G.M. & Welsh, J. Accelerated mammary gland development during pregnancy and delayed postlactational involution in vitamin D3 receptor null mice. *Mol Endocrinol* **18**, 2208–2223 (2004).
35. Wittke, A., Weaver, V., Mahon, B.D., August, A. & Cantorna, M.T. Vitamin D receptor-deficient mice fail to develop experimental allergic asthma. *J Immunol* **173**, 3432–3436 (2004).
36. Kalueff, A.V., Lou, Y.R., Laaksi, I. & Tuohimaa, P. Increased grooming behavior in mice lacking vitamin D receptors. *Physiol Behav* **82**, 405–409 (2004).
37. Bolt, M.J. et al. Critical role of vitamin D in sulfate homeostasis: Regulation of the sodium-sulfate cotransporter by 1,25-dihydroxyvitamin D3. *Am J Physiol Endocrinol Metab* **287**, E744–E749 (2004).
38. Li, X., Zheng, W. & Li, Y.C. Altered gene expression profile in the kidney of vitamin D receptor knockout mice. *J Cell Biochem* **89**, 709–719 (2003).
39. O'Kelly, J. et al. Normal myelopoiesis but abnormal T lymphocyte responses in vitamin D receptor knockout mice. *J Clin Invest* **109**, 1091–1099 (2002).
40. Zinser, G., Packman, K. & Welsh, J. Vitamin D(3) receptor ablation alters mammary gland morphogenesis. *Development* **129**, 3067–3076 (2002).
41. Beaudoin, G.M. III, Sisk, J.M., Coulombe, P.A. & Thompson, C.C. Hairless triggers reactivation of hair growth by promoting Wnt signaling. *Proc Natl Acad Sci U S A* **102**, 14653–14658 (2005).
42. Thompson, C.C., Sisk, J.M. & Beaudoin, G.M. III. Hairless and Wnt signaling: Allies in epithelial stem cell differentiation. *Cell Cycle* **5**, 1913–1917 (2006).
43. Palmer, H.G., Anjos-Afonso, F., Carmeliet, G., Takeda, H. & Watt, F.M. The vitamin D receptor is a Wnt effector that controls hair follicle differentiation and specifies tumor type in adult epidermis. *PLoS One* **3**, e1483 (2008).
44. Yoshizawa, T. et al. Mice lacking the vitamin D receptor exhibit impaired bone formation, uterine hypoplasia and growth retardation after weaning. *Nat Genet* **16**, 391–396 (1997).
45. Pendas-Franco, N. et al. DICKKOPF-4 is induced by TCF/beta-catenin and upregulated in human colon cancer, promotes tumour cell invasion and angiogenesis and is repressed by 1alpha,25-dihydroxyvitamin D3. *Oncogene* **27**, 4467–4477 (2008).
46. Palmer, H.G. et al. The transcription factor SNAIL represses vitamin D receptor expression and responsiveness in human colon cancer. *Nat Med* **10**, 917–919 (2004).
47. Palmer, H.G. et al. Vitamin D(3) promotes the differentiation of colon carcinoma cells by the induction of E-cadherin and the inhibition of beta-catenin signaling. *J Cell Biol* **154**, 369–387 (2001).
48. Novershtern, N. et al. Densely interconnected transcriptional circuits control cell states in human hematopoiesis. *Cell* **144**, 296–309 (2011).
49. Adams, J.S. et al. Vitamin D-directed rheostatic regulation of monocyte antibacterial responses. *J Immunol* **182**, 4289–4295 (2009).
50. Barbieri, C.E. & Pietenpol, J.A. p63 and epithelial biology. *Exp Cell Res* **312**, 695–706 (2006).
51. Blackburn, A.C. & Jerry, D.J. Knockout and transgenic mice of Trp53: What have we learned about p53 in breast cancer? *Breast Cancer Res* **4**, 101–111 (2002).
52. Kommagani, R., Caserta, T.M. & Kadakia, M.P. Identification of vitamin D receptor as a target of p63. *Oncogene* **25**, 3745–3751 (2006).
53. Maruyama, R. et al. Comparative genome analysis identifies the vitamin D receptor gene as a direct target of p53-mediated transcriptional activation. *Cancer Res* **66**, 4574–4583 (2006).
54. Wang, T.T. et al. Large-scale in silico and microarray-based identification of direct 1,25-dihydroxyvitamin D3 target genes. *Mol Endocrinol* **19**, 2685–2695 (2005).

55. Lu, J. et al. Transcriptional profiling of keratinocytes reveals a vitamin D-regulated epidermal differentiation network. *J Invest Dermatol* **124**, 778–785 (2005).
56. Khanim, F.L. et al. Altered SMRT levels disrupt vitamin D3 receptor signalling in prostate cancer cells. *Oncogene* **23**, 6712–6725 (2004).
57. Yang, L., Yang, J., Venkateswarlu, S., Ko, T. & Brattain, M.G. Autocrine TGFbeta signaling mediates vitamin D3 analog-induced growth inhibition in breast cells. *J Cell Physiol* **188**, 383–393 (2001).
58. Wu, Y., Craig, T.A., Lutz, W.H. & Kumar, R. Identification of 1 alpha,25-dihydroxyvitamin D3 response elements in the human transforming growth factor beta 2 gene. *Biochemistry* **38**, 2654–2660 (1999).
59. Matilainen, M., Malinen, M., Saavalainen, K. & Carlberg, C. Regulation of multiple insulin-like growth factor binding protein genes by 1alpha,25-dihydroxyvitamin D3. *Nucleic Acids Res* **33**, 5521–5532 (2005).
60. Vousden, K.H. & Lu, X. Live or let die: the cell's response to p53. *Nat Rev Cancer* **2**, 594–604 (2002).
61. Saramaki, A., Banwell, C.M., Campbell, M.J. & Carlberg, C. Regulation of the human p21(waf1/cip1) gene promoter via multiple binding sites for p53 and the vitamin D3 receptor. *Nucleic Acids Res* **34**, 543–554 (2006).
62. Lee, D. et al. SWI/SNF complex interacts with tumor suppressor p53 and is necessary for the activation of p53-mediated transcription. *J Biol Chem* **277**, 22330–22337 (2002).
63. Kitagawa, H. et al. The chromatin-remodeling complex WINAC targets a nuclear receptor to promoters and is impaired in Williams syndrome. *Cell* **113**, 905–917 (2003).
64. Wilkinson, D.S. et al. A direct intersection between p53 and transforming growth factor beta pathways targets chromatin modification and transcription repression of the alpha-fetoprotein gene. *Mol Cell Biol* **25**, 1200–1212 (2005).
65. Szeto, F.L. et al. Involvement of the vitamin D receptor in the regulation of NF-[kappa] B activity in fibroblasts. *J Steroid Biochem Mol Biol* **103**, 563–566 (2007).
66. Froicu, M. & Cantorna, M. Vitamin D and the vitamin D receptor are critical for control of the innate immune response to colonic injury. *BMC Immunol* **8**, 5 (2007).
67. Gombart, A.F., Borregaard, N. & Koeffler, H.P. Human cathelicidin antimicrobial peptide (CAMP) gene is a direct target of the vitamin D receptor and is strongly up-regulated in myeloid cells by 1,25-dihydroxyvitamin D3. *FASEB J* **19**, 1067–1077 (2005).
68. Wang, T.-T. et al. Cutting edge: 1,25-dihydroxyvitamin D3 is a direct inducer of antimicrobial peptide gene expression. *J Immunol* **173**, 2909–2912 (2004).
69. Mallbris, L., Edström, D.W., Sundblad, L., Granath, F. & Stahle, M. UVB Upregulates the antimicrobial protein hCAP18 mRNA in human skin. *J Invest Dermatol* **125**, 1072–1074 (2005).
70. Zasloff, M. Sunlight, vitamin D, and the innate immune defenses of the human skin. *J Invest Dermatol* **125**, xvi–xvii (2005).
71. Quigley, D.A. et al. Genetic architecture of mouse skin inflammation and tumour susceptibility. *Nature* **458**, 505–508 (2009).
72. Vertino, A.M. et al. Nongenotropic, anti-apoptotic signaling of 1alpha,25(OH)2-vitamin D3 and analogs through the ligand binding domain of the vitamin D receptor in osteoblasts and osteocytes. Mediation by Src, phosphatidylinositol 3-, and JNK kinases. *J Biol Chem* **280**, 14130–14137 (2005).
73. Prufer, K., Racz, A., Lin, G.C. & Barsony, J. Dimerization with retinoid X receptors promotes nuclear localization and subnuclear targeting of vitamin D receptors. *J Biol Chem* **275**, 41114–41123 (2000).
74. Yasmin, R., Williams, R.M., Xu, M. & Noy, N. Nuclear import of the retinoid X receptor, the vitamin D receptor, and their mutual heterodimer. *J Biol Chem* **280**, 40152–40160 (2005).

75. Malinen, M. et al. Distinct HDACs regulate the transcriptional response of human cyclin-dependent kinase inhibitor genes to trichostatin A and 1alpha,25-dihydroxyvitamin D3. *Nucleic Acids Res* **36**, 121–132 (2007).
76. Vaisanen, S., Dunlop, T.W., Sinkkonen, L., Frank, C. & Carlberg, C. Spatio-temporal activation of chromatin on the human CYP24 gene promoter in the presence of 1alpha,25-dihydroxyvitamin D3. *J Mol Biol* **350**, 65–77 (2005).
77. Zella, L.A., Kim, S., Shevde, N.K. & Pike, J.W. Enhancers located within two introns of the vitamin D receptor gene mediate transcriptional autoregulation by 1,25-dihydroxyvitamin D3. *Mol Endocrinol* **20**, 1231–1247 (2006).
78. Kim, S., Shevde, N.K. & Pike, J.W. 1,25-Dihydroxyvitamin D3 stimulates cyclic vitamin D receptor/retinoid X receptor DNA-binding, co-activator recruitment, and histone acetylation in intact osteoblasts. *J Bone Miner Res* **20**, 305–317 (2005).
79. Seo, Y.K. et al. Xenobiotic- and vitamin D-responsive induction of the steroid/bile acid-sulfotransferase Sult2A1 in young and old mice: The role of a gene enhancer in the liver chromatin. *Gene* **386**, 218–223 (2007).
80. Meyer, M.B., Watanuki, M., Kim, S., Shevde, N.K. & Pike, J.W. The human TRPV6 distal promoter contains multiple vitamin D receptor binding sites that mediate activation by 1,25-dihydroxyvitamin D3 in intestinal cells. *Mol Endocrinol* **20**, 1447–1461 (2006).
81. Reid, G. et al. Cyclic, proteasome-mediated turnover of unliganded and liganded ERalpha on responsive promoters is an integral feature of estrogen signaling. *Mol Cell* **11**, 695–707 (2003).
82. Metivier, R. et al. Estrogen receptor-alpha directs ordered, cyclical, and combinatorial recruitment of cofactors on a natural target promoter. *Cell* **115**, 751–763 (2003).
83. Yang, X. et al. Nuclear receptor expression links the circadian clock to metabolism. *Cell* **126**, 801–810 (2006).
84. Carroll, J.S. et al. Genome-wide analysis of estrogen receptor binding sites. *Nat Genet* **38**, 1289–1297 (2006).
85. Saramaki, A. et al. Cyclical chromatin looping and transcription factor association on the regulatory regions of the p21 (CDKN1A) gene in response to 1alpha,25-dihydroxyvitamin D3. *J Biol Chem* **284**, 8073–8082 (2009).
86. Banwell, C.M. et al. Altered nuclear receptor corepressor expression attenuates vitamin D receptor signaling in breast cancer cells. *Clin Cancer Res* **12**, 2004–2013 (2006).
87. Li, J. et al. Both corepressor proteins SMRT and N-CoR exist in large protein complexes containing HDAC3. *EMBO J* **19**, 4342–4350 (2000).
88. Yoon, H.G. et al. Purification and functional characterization of the human N-CoR complex: The roles of HDAC3, TBL1 and TBLR1. *EMBO J* **22**, 1336–1346 (2003).
89. Alenghat, T., Yu, J. & Lazar, M.A. The N-CoR complex enables chromatin remodeler SNF2H to enhance repression by thyroid hormone receptor. *EMBO J* **25**, 3966–3974 (2006).
90. Picard, F. et al. Sirt1 promotes fat mobilization in white adipocytes by repressing PPAR-gamma. *Nature* **429**, 771–776 (2004).
91. Yu, C. et al. The nuclear receptor corepressors NCoR and SMRT decrease PPARgamma transcriptional activity and repress 3T3-L1 adipogenesis. *J Biol Chem* **280**, 13600–13605 (2005).
92. Takada, I. et al. A histone lysine methyltransferase activated by non-canonical Wnt signalling suppresses PPAR-gamma transactivation. *Nat Cell Biol* **9**, 1273–1285 (2007).
93. Quack, M. & Carlberg, C. The impact of functional vitamin D(3) receptor conformations on DNA-dependent vitamin D(3) signaling. *Mol Pharmacol* **57**, 375–384 (2000).
94. Perissi, V., Jepsen, K., Glass, C.K. & Rosenfeld, M.G. Deconstructing repression: Evolving models of co-repressor action. *Nat Rev Genet* **11**, 109–123 (2010).
95. Ramagopalan, S.V. et al. A ChIP-seq defined genome-wide map of vitamin D receptor binding: Associations with disease and evolution. *Genome Res* **20**, 1352–1360 (2010).

96. Ding, N. et al. A vitamin D receptor/SMAD genomic circuit gates hepatic fibrotic response. *Cell* **153**, 601–613 (2013).
97. Heikkinen, S. et al. Nuclear hormone 1α,25-dihydroxyvitamin D_3 elicts a genome-wide shift in the locations of VDR chromatin accupancy. *Nucleic Acids Res* **39**, 9181–9193 (2011).
98. Meyer, M.B., Goetsch, P.D. & Pike, J.W. VDR/RXR and TCF4/β-catenin cistromes in colonic cells of colorectal tumor origin: Impact on c-FOS and c-MYC gene expression. *Mol Endocrinol (Baltimore, Md)* **26**, 37–51 (2012).
99. ENCODE-Project-Consortium et al. An integrated encyclopedia of DNA elements in the human genome. *Nature* **489**, 57–74 (2012).
100. Eeckhoute, J., Carroll, J.S., Geistlinger, T.R., Torres-Arzayus, M.I. & Brown, M. A cell-type-specific transcriptional network required for estrogen regulation of cyclin D1 and cell cycle progression in breast cancer. *Genes Dev* **20**, 2513–2526 (2006).
101. Colston, K., Colston, M.J. & Feldman, D. 1,25-dihydroxyvitamin D3 and malignant melanoma: The presence of receptors and inhibition of cell growth in culture. *Endocrinology* **108**, 1083–1086 (1981).
102. Miyaura, C. et al. 1 alpha,25-Dihydroxyvitamin D3 induces differentiation of human myeloid leukemia cells. *Biochem Biophys Res Commun* **102**, 937–943 (1981).
103. Abe, E. et al. Differentiation of mouse myeloid leukemia cells induced by 1 alpha,25-dihydroxyvitamin D3. *Proc Natl Acad Sci U S A* **78**, 4990–4994 (1981).
104. Palmer, H.G. et al. Genetic signatures of differentiation induced by 1alpha,25-dihydroxyvitamin D3 in human colon cancer cells. *Cancer Res* **63**, 7799–7806 (2003).
105. Koike, M. et al. 19-nor-hexafluoride analogue of vitamin D3: A novel class of potent inhibitors of proliferation of human breast cell lines. *Cancer Res* **57**, 4545–4550 (1997).
106. Campbell, M.J., Elstner, E., Holden, S., Uskokovic, M. & Koeffler, H.P. Inhibition of proliferation of prostate cancer cells by a 19-nor-hexafluoride vitamin D3 analogue involves the induction of p21waf1, p27kip1 and E-cadherin. *J Mol Endocrinol* **19**, 15–27 (1997).
107. Elstner, E. et al. Novel 20-epi-vitamin D3 analog combined with 9-cis-retinoic acid markedly inhibits colony growth of prostate cancer cells. *Prostate* **40**, 141–149 (1999).
108. Peehl, D.M. et al. Antiproliferative effects of 1,25-dihydroxyvitamin D3 on primary cultures of human prostatic cells. *Cancer Res* **54**, 805–810 (1994).
109. Welsh, J. et al. Impact of the vitamin D3 receptor on growth-regulatory pathways in mammary gland and breast cancer. *J Steroid Biochem Mol Biol* **83**, 85–92 (2002).
110. Colston, K.W., Berger, U. & Coombes, R.C. Possible role for vitamin D in controlling breast cancer cell proliferation. *Lancet* **1**, 188–191 (1989).
111. Colston, K., Colston, M.J., Fieldsteel, A.H. & Feldman, D. 1,25-dihydroxyvitamin D3 receptors in human epithelial cancer cell lines. *Cancer Res* **42**, 856–859 (1982).
112. Eelen, G. et al. Microarray analysis of 1alpha,25-dihydroxyvitamin D3-treated MC3T3-E1 cells. *J Steroid Biochem Mol Biol* **89–90**, 405–407 (2004).
113. Akutsu, N. et al. Regulation of gene expression by 1alpha,25-dihydroxyvitamin D3 and Its analog EB1089 under growth-inhibitory conditions in squamous carcinoma cells. *Mol Endocrinol* **15**, 1127–1139 (2001).
114. Liu, M., Lee, M.H., Cohen, M., Bommakanti, M. & Freedman, L.P. Transcriptional activation of the Cdk inhibitor p21 by vitamin D3 leads to the induced differentiation of the myelomonocytic cell line U937. *Genes Dev* **10**, 142–153 (1996).
115. Wang, Q.M., Jones, J.B. & Studzinski, G.P. Cyclin-dependent kinase inhibitor p27 as a mediator of the G1-S phase block induced by 1,25-dihydroxyvitamin D3 in HL60 cells. *Cancer Res* **56**, 264–267 (1996).
116. Li, P. et al. p27(Kip1) stabilization and G(1) arrest by 1,25-dihydroxyvitamin D(3) in ovarian cancer cells mediated through down-regulation of cyclin E/cyclin-dependent kinase 2 and Skp1-Cullin-F-box protein/Skp2 ubiquitin ligase. *J Biol Chem* **279**, 25260–25267 (2004).

117. Huang, Y.C., Chen, J.Y. & Hung, W.C. Vitamin D(3) receptor/Sp1 complex is required for the induction of p27(Kip1) expression by vitamin D(3). *Oncogene* **23**, 4856–4861 (2004).
118. Hengst, L. & Reed, S.I. Translational control of p27Kip1 accumulation during the cell cycle. *Science* **271**, 1861–1864 (1996).
119. Jiang, F., Li, P., Fornace, A.J. Jr., Nicosia, S.V. & Bai, W. G2/M arrest by 1,25-dihydroxyvitamin D3 in ovarian cancer cells mediated through the induction of GADD45 via an exonic enhancer. *J Biol Chem* **278**, 48030–48040 (2003).
120. Lubbert, M. et al. Stable methylation patterns of MYC and other genes regulated during terminal myeloid differentiation. *Leukemia* **5**, 533–539 (1991).
121. Koeffler, H.P. Induction of differentiation of human acute myelogenous leukemia cells: therapeutic implications. *Blood* **62**, 709–721 (1983).
122. Studzinski, G.P., Bhandal, A.K. & Brelvi, Z.S. Potentiation by 1-alpha,25-dihydroxyvitamin D3 of cytotoxicity to HL-60 cells produced by cytarabine and hydroxyurea. *J Natl Cancer Inst* **76**, 641–648 (1986).
123. Lin, R.J. et al. Role of the histone deacetylase complex in acute promyelocytic leukaemia. *Nature* **391**, 811–814 (1998).
124. Dobrzynski, M. & Bruggeman, F.J. Elongation dynamics shape bursty transcription and translation. *Proc Natl Acad Sci U S A* **106**, 2583–2588 (2009).
125. Taverna, S.D., Li, H., Ruthenburg, A.J., Allis, C.D. & Patel, D.J. How chromatin-binding modules interpret histone modifications: Lessons from professional pocket pickers. *Nat Struct Mol Biol* **14**, 1025–1040 (2007).
126. Goldberg, A.D., Allis, C.D. & Bernstein, E. Epigenetics: A landscape takes shape. *Cell* **128**, 635–638 (2007).
127. Le May, N. et al. NER factors are recruited to active promoters and facilitate chromatin modification for transcription in the absence of exogenous genotoxic attack. *Mol Cell* **38**, 54–66 (2010).
128. Metivier, R. et al. Cyclical DNA methylation of a transcriptionally active promoter. *Nature* **452**, 45–50 (2008).
129. Kangaspeska, S. et al. Transient cyclical methylation of promoter DNA. *Nature* **452**, 112–115 (2008).
130. Mohn, F. & Schubeler, D. Genetics and epigenetics: Stability and plasticity during cellular differentiation. *Trends Genet* **25**, 129–136 (2009).
131. Turner, B.M. Decoding the nucleosome. *Cell* **75**, 5–8 (1993).
132. Turner, B.M. Histone acetylation as an epigenetic determinant of long-term transcriptional competence. *Cell Mol Life Sci* **54**, 21–31 (1998).
133. Jenuwein, T. & Allis, C.D. Translating the histone code. *Science* **293**, 1074–1080 (2001).
134. Turner, B.M. Cellular memory and the histone code. *Cell* **111**, 285–291 (2002).
135. Thorne, J.L., Campbell, M.J. & Turner, B.M. Transcription factors, chromatin and cancer. *Int J Biochem Cell Biol* **41**, 164–175 (2009).
136. Fujita, N. et al. Methyl-CpG binding domain 1 (MBD1) interacts with the Suv39h1-HP1 heterochromatic complex for DNA methylation-based transcriptional repression. *J Biol Chem* **278**, 24132–24138 (2003).
137. Esteve, P.O. et al. Direct interaction between DNMT1 and G9a coordinates DNA and histone methylation during replication. *Genes Dev* **20**, 3089–3103 (2006).
138. Cheng, X. & Blumenthal, R.M. Coordinated chromatin control: Structural and functional linkage of DNA and histone methylation. *Biochemistry* **49**, 2999–3008 (2010).
139. Yoon, H.G., Chan, D.W., Reynolds, A.B., Qin, J. & Wong, J. N-CoR mediates DNA methylation-dependent repression through a methyl CpG binding protein Kaiso. *Mol Cell* **12**, 723–734 (2003).
140. Filipowicz, W., Bhattacharyya, S.N. & Sonenberg, N. Mechanisms of post-transcriptional regulation by microRNAs: Are the answers in sight? *Nat Rev Genet* **9**, 102–114 (2008).

141. Saito, Y. & Jones, P.A. Epigenetic activation of tumor suppressor microRNAs in human cancer cells. *Cell Cycle* **5**, 2220–2222 (2006).
142. Chen, J.F. et al. The role of microRNA-1 and microRNA-133 in skeletal muscle proliferation and differentiation. *Nat Genet* **38**, 228–233 (2006).
143. Duursma, A.M., Kedde, M., Schrier, M., le Sage, C. & Agami, R. miR-148 targets human DNMT3b protein coding region. *RNA* **14**, 872–877 (2008).
144. Thorne, J.L. et al. Epigenetic control of a VDR-governed feed-forward loop that regulates p21(waf1/cip1) expression and function in non-malignant prostate cells. *Nucleic Acids Res* **39**, 2045–2056 (2011).
145. Garzon, R. et al. MicroRNA-29b induces global DNA hypomethylation and tumor suppressor gene reexpression in acute myeloid leukemia by targeting directly DNMT3A and 3B and indirectly DNMT1. *Blood* **113**, 6411–6418 (2009).
146. Fabbri, M. et al. MicroRNA-29 family reverts aberrant methylation in lung cancer by targeting DNA methyltransferases 3A and 3B. *Proc Natl Acad Sci U S A* **104**, 15805–15810 (2007).
147. Majid, S. et al. miRNA-34b inhibits prostate cancer through demethylation, active chromatin modifications, and AKT pathways. *Clin Cancer Res* **19**, 73–84 (2013).
148. Wang, X., Gocek, E., Liu, C.G. & Studzinski, G.P. MicroRNAs181 regulate the expression of p27(Kip1) in human myeloid leukemia cells induced to differentiate by 1,25-dihydroxyvitamin D(3). *Cell Cycle* **8**, 736–741 (2009).
149. Song, G. & Wang, L. Transcriptional mechanism for the paired miR-433 and miR-127 genes by nuclear receptors SHP and ERRgamma. *Nucleic Acids Res* **36**, 5727–5735 (2008).
150. Shah, Y.M. et al. Peroxisome proliferator-activated receptor alpha regulates a microRNA-mediated signaling cascade responsible for hepatocellular proliferation. *Mol Cell Biol* **27**, 4238–4247 (2007).
151. Ellison, T.I., Smith, M.K., Gilliam, A.C. & MacDonald, P.N. Inactivation of the vitamin D receptor enhances susceptibility of murine skin to UV-induced tumorigenesis. *J Invest Dermatol* **128**, 2508–2517 (2008).
152. Lambert, J.R. et al. Prostate derived factor in human prostate cancer cells: Gene induction by vitamin D via a p53-dependent mechanism and inhibition of prostate cancer cell growth. *J Cell Physiol* **208**, 566–574 (2006).
153. Jin, F. & Fondell, J.D. A novel androgen receptor-binding element modulates Cdc6 transcription in prostate cancer cells during cell-cycle progression. *Nucleic Acids Res* **37**, 4826–4838 (2009).
154. Okada, M. et al. Switching of chromatin-remodelling complexes for oestrogen receptor-alpha. *EMBO Rep* **9**, 563–568 (2008).
155. Shogren-Knaak, M. et al. Histone H4-K16 acetylation controls chromatin structure and protein interactions. *Science* **311**, 844–847 (2006).
156. Shi, X. et al. ING2 PHD domain links histone H3 lysine 4 methylation to active gene repression. *Nature* **442**, 96–99 (2006).
157. Varambally, S. et al. The polycomb group protein EZH2 is involved in progression of prostate cancer. *Nature* **419**, 624–629 (2002).
158. Yu, J., Li, Y., Ishizuka, T., Guenther, M.G. & Lazar, M.A. A SANT motif in the SMRT corepressor interprets the histone code and promotes histone deacetylation. *EMBO J* **22**, 3403–3410 (2003).
159. Rosenfeld, M.G., Lunyak, V.V. & Glass, C.K. Sensors and signals: A coactivator/corepressor/epigenetic code for integrating signal-dependent programs of transcriptional response. *Genes Dev* **20**, 1405–1428 (2006).
160. Goodson, M.L., Jonas, B.A. & Privalsky, M.L. Alternative mRNA splicing of SMRT creates functional diversity by generating corepressor isoforms with different affinities for different nuclear receptors. *J Biol Chem* **280**, 7493–7503 (2005).

161. Kang, Z., Janne, O.A. & Palvimo, J.J. Coregulator recruitment and histone modifications in transcriptional regulation by the androgen receptor. *Mol Endocrinol* **18**, 2633–2648 (2004).
162. Jepsen, K. et al. SMRT-mediated repression of an H3K27 demethylase in progression from neural stem cell to neuron. *Nature* **450**, 415–419 (2007).
163. Alenghat, T. et al. Nuclear receptor corepressor and histone deacetylase 3 govern circadian metabolic physiology. *Nature* **456**, 997–1000 (2008).
164. Astapova, I. et al. The nuclear corepressor, NCoR, regulates thyroid hormone action in vivo. *Proc Natl Acad Sci U S A* **105**, 19544–19549 (2008).
165. Nofsinger, R.R. et al. SMRT repression of nuclear receptors controls the adipogenic set point and metabolic homeostasis. *Proc Natl Acad Sci U S A* **105**, 20021–20026 (2008).
166. Surjit, M. et al. Widespread negative response elements mediate direct repression by agonist-liganded glucocorticoid receptor. *Cell* **145**, 224–241 (2011).
167. Polly, P. et al. VDR-Alien: A novel, DNA-selective vitamin D(3) receptor-corepressor partnership. *FASEB J* **14**, 1455–1463 (2000).
168. Lykke-Andersen, K. et al. Disruption of the COP9 signalosome Csn2 subunit in mice causes deficient cell proliferation, accumulation of p53 and cyclin E, and early embryonic death. *Mol Cell Biol* **23**, 6790–6797 (2003).
169. Hatchell, E.C. et al. SLIRP, a small SRA binding protein, is a nuclear receptor corepressor. *Mol Cell* **22**, 657–668 (2006).
170. Hsieh, J.C. et al. Physical and functional interaction between the vitamin D receptor and hairless corepressor, two proteins required for hair cycling. *J Biol Chem* **278**, 38665–38674 (2003).
171. Miller, J. et al. Atrichia caused by mutations in the vitamin D receptor gene is a phenocopy of generalized atrichia caused by mutations in the hairless gene. *J Invest Dermatol* **117**, 612–617 (2001).
172. Xie, Z., Chang, S., Oda, Y. & Bikle, D.D. Hairless suppresses vitamin D receptor transactivation in human keratinocytes. *Endocrinology* **147**, 314–323 (2005).
173. Scsucova, S. et al. The repressor DREAM acts as a transcriptional activator on vitamin D and retinoic acid response elements. *Nucleic Acids Res* **33**, 2269–2279 (2005).
174. Renaud, J.P. et al. Crystal structure of the RAR-gamma ligand-binding domain bound to all-trans retinoic acid. *Nature* **378**, 681–689 (1995).
175. Nakabayashi, M. et al. Crystal structures of rat vitamin D receptor bound to adamantyl vitamin D analogs: Structural basis for vitamin D receptor antagonism and partial agonism. *J Med Chem* **51**, 5320–5329 (2008).
176. Oda, Y. et al. Two distinct coactivators, DRIP/mediator and SRC/p160, are differentially involved in vitamin D receptor transactivation during keratinocyte differentiation. *Mol Endocrinol* **17**, 2329–2339 (2003).
177. Rachez, C. et al. The DRIP complex and SRC-1/p160 coactivators share similar nuclear receptor binding determinants but constitute functionally distinct complexes. *Mol Cell Biol* **20**, 2718–2726 (2000).
178. Ito, M. & Roeder, R.G. The TRAP/SMCC/Mediator complex and thyroid hormone receptor function. *Trends Endocrinol Metab* **12**, 127–134 (2001).
179. Ding, X.F. et al. Nuclear receptor-binding sites of coactivators glucocorticoid receptor interacting protein 1 (GRIP1) and steroid receptor coactivator 1 (SRC-1): Multiple motifs with different binding specificities. *Mol Endocrinol* **12**, 302–313 (1998).
180. Zhang, C. et al. Nuclear coactivator-62 kDa/Ski-interacting protein is a nuclear matrix-associated coactivator that may couple vitamin D receptor-mediated transcription and RNA splicing. *J Biol Chem* **278**, 35325–35336 (2003).
181. Anzick, S.L. et al. AIB1, a steroid receptor coactivator amplified in breast and ovarian cancer. *Science* **277**, 965–968 (1997).

182. Demarest, S.J. et al. Mutual synergistic folding in recruitment of CBP/p300 by p160 nuclear receptor coactivators. *Nature* **415**, 549–553 (2002).
183. Zhao, C. et al. Elevated expression levels of NCOA3, TOP1, and TFAP2C in breast tumors as predictors of poor prognosis. *Cancer* **98**, 18–23 (2003).
184. Doi, M., Hirayama, J. & Sassone-Corsi, P. Circadian regulator CLOCK is a histone acetyltransferase. *Cell* **125**, 497–508 (2006).
185. Esteyries, S. et al. NCOA3, a new fusion partner for MOZ/MYST3 in M5 acute myeloid leukemia. *Leukemia* **22**, 663–665 (2008).
186. Rashid, S.F. et al. Synergistic growth inhibition of prostate cancer cells by 1 alpha,25 dihydroxyvitamin D(3) and its 19-nor-hexafluoride analogs in combination with either sodium butyrate or trichostatin A. *Oncogene* **20**, 1860–1872 (2001).
187. Campbell, M.J., Gombart, A.F., Kwok, S.H., Park, S. & Koeffler, H.P. The antiproliferative effects of 1alpha,25(OH)2D3 on breast and prostate cancer cells are associated with induction of BRCA1 gene expression. *Oncogene* **19**, 5091–5097 (2000).
188. Miller, G.J., Stapleton, G.E., Hedlund, T.E. & Moffat, K.A. Vitamin D receptor expression, 24-hydroxylase activity, and inhibition of growth by 1alpha,25-dihydroxyvitamin D3 in seven human prostatic carcinoma cell lines. *Clin Cancer Res* **1**, 997–1003 (1995).
189. Rashid, S.F., Mountford, J.C., Gombart, A.F. & Campbell, M.J. 1alpha,25-dihydroxyvitamin D(3) displays divergent growth effects in both normal and malignant cells. *Steroids* **66**, 433–440 (2001).
190. Campbell, M.J. & Adorini, L. The vitamin D receptor as a therapeutic target. *Expert Opin Ther Targets* **10**, 735–748 (2006).
191. Battaglia, S., Maguire, O. & Campbell, M.J. Transcription factor co-repressors in cancer biology: Roles and targeting. *Int J Cancer* **126**, 2511–2519 (2010).
192. Chen, J.D., Umesono, K. & Evans, R.M. SMRT isoforms mediate repression and antirepression of nuclear receptor heterodimers. *Proc Natl Acad Sci U S A* **93**, 7567–7571 (1996).
193. Battaglia, S. et al. Elevated NCOR1 disrupts PPARalpha/gamma signaling in prostate cancer and forms a targetable epigenetic lesion. *Carcinogenesis* **31**, 1650–1660 (2010).
194. Altintas, D.M., Vlaeminck, V., Angelov, D., Dimitrov, S. & Samarut, J. Cell cycle regulated expression of NCoR might control cyclic expression of androgen responsive genes in an immortalized prostate cell line. *Mol Cell Endocrinol* **332**, 149–162 (2011).
195. Girault, I. et al. Expression analysis of estrogen receptor alpha coregulators in breast carcinoma: Evidence that NCOR1 expression is predictive of the response to tamoxifen. *Clin Cancer Res* **9**, 1259–1266 (2003).
196. Abedin, S.A. et al. Elevated NCOR1 disrupts a network of dietary-sensing nuclear receptors in bladder cancer cells. *Carcinogenesis* **30**, 449–456 (2009).
197. Kim, J.Y., Son, Y.L. & Lee, Y.C. Involvement of SMRT corepressor in transcriptional repression by the vitamin D receptor. *Mol Endocrinol* **23**, 251–264 (2009).
198. Zhang, Z. et al. NCOR1 mRNA is an independent prognostic factor for breast cancer. *Cancer Lett* **237**, 123–129 (2006).
199. Chang, T.H. & Szabo, E. Enhanced growth inhibition by combination differentiation therapy with ligands of peroxisome proliferator-activated receptor-gamma and inhibitors of histone deacetylase in adenocarcinoma of the lung. *Clin Cancer Res* **8**, 1206–1212 (2002).
200. Stephens, P.J. et al. The landscape of cancer genes and mutational processes in breast cancer. *Nature* **486**, 400–404 (2012).
201. Liao, G. et al. Regulation of androgen receptor activity by the nuclear receptor corepressor SMRT. *J Biol Chem* **278**, 5052–5061 (2003).
202. Hodgson, M.C. et al. The androgen receptor recruits nuclear receptor Corepressor (N-CoR) in the presence of mifepristone via its N and C termini revealing a novel molecular mechanism for androgen receptor antagonists. *J Biol Chem* **280**, 6511–6519 (2005).

203. Cherrier, T. et al. p21(WAF1) gene promoter is epigenetically silenced by CTIP2 and SUV39H1. *Oncogene* **28**, 3380–3389 (2009).
204. Pollard, P.J. et al. Regulation of Jumonji-domain-containing histone demethylases by hypoxia-inducible factor (HIF)–1alpha. *Biochem J* **416**, 387–394 (2008).
205. Doig, C.L. et al. Recruitment of NCOR1 to VDR target genes is enhanced in prostate cancer cells and associates with altered DNA methylation patterns. *Carcinogenesis* **34**, 248–256 (2013).
206. Toropainen, S., Vaisanen, S., Heikkinen, S. & Carlberg, C. The down-regulation of the human MYC gene by the nuclear hormone 1alpha,25-dihydroxyvitamin D3 is associated with cycling of corepressors and histone deacetylases. *J Mol Biol* **400**, 284–294 (2010).
207. Birney, E. et al. Identification and analysis of functional elements in 1% of the human genome by the ENCODE pilot project. *Nature* **447**, 799–816 (2007).
208. Rosenbloom, K.R. et al. ENCODE data in the UCSC genome browser: Year 5 update. *Nucleic Acids Res* **41**, D56–D63 (2013).
209. Skipper, M., Dhand, R. & Campbell, P. Presenting ENCODE. *Nature* **489**, 45 (2012).
210. Sastre, L. Clinical implications of the ENCODE project. *Clin Transl Oncol* **14**, 801–802 (2012).
211. Paik, Y.K. & Hancock, W.S. Uniting ENCODE with genome-wide proteomics. *Nat Biotechnol* **30**, 1065–1067 (2012).
212. Su, G., Burant, C.F., Beecher, C.W., Athey, B.D. & Meng, F. Integrated metabolome and transcriptome analysis of the NCI60 dataset. *BMC Bioinformatics* **12 Suppl 1**, S36 (2011).
213. Cavill, R. et al. Consensus-phenotype integration of transcriptomic and metabolomic data implies a role for metabolism in the chemosensitivity of tumour cells. *PLoS Comput Biol* **7**, e1001113 (2011).
214. Garraway, L.A. et al. Integrative genomic analyses identify MITF as a lineage survival oncogene amplified in malignant melanoma. *Nature* **436**, 117–122 (2005).
215. Schulte, J.H. et al. Lysine-specific demethylase 1 is strongly expressed in poorly differentiated neuroblastoma: Implications for therapy. *Cancer Res* **69**, 2065–2071 (2009).
216. Graham, J.S., Kaye, S.B. & Brown, R. The promises and pitfalls of epigenetic therapies in solid tumours. *Eur J Cancer* **45**, 1129–1136 (2009).
217. Bandopadhayay, P. et al. BET bromodomain inhibition of MYC-amplified medulloblastoma. *Clin Cancer Res* **20**, 912–925 (2014).
218. Loven, J. et al. Selective inhibition of tumor oncogenes by disruption of super-enhancers. *Cell* **153**, 320–334 (2013).
219. Lockwood, W.W., Zejnullahu, K., Bradner, J.E. & Varmus, H. Sensitivity of human lung adenocarcinoma cell lines to targeted inhibition of BET epigenetic signaling proteins. *Proc Natl Acad Sci U S A* **109**, 19408–19413 (2012).
220. Garland, C.F. & Garland, F.C. Do sunlight and vitamin D reduce the likelihood of colon cancer? *Int J Epidemiol* **9**, 227–231 (1980).
221. Garland, F.C., Garland, C.F., Gorham, E.D. & Young, J.F. Geographic variation in breast cancer mortality in the United States: A hypothesis involving exposure to solar radiation. *Prev Med* **19**, 614–622 (1990).
222. Luscombe, C.J. et al. Prostate cancer risk: Associations with ultraviolet radiation, tyrosinase and melanocortin-1 receptor genotypes. *Br J Cancer* **85**, 1504–1509 (2001).
223. Giovannucci, E. The epidemiology of vitamin D and cancer incidence and mortality: A review (United States). *Cancer Causes Control* **16**, 83–95 (2005).
224. Schwartz, G.G. & Hulka, B.S. Is vitamin D deficiency a risk factor for prostate cancer? (Hypothesis). *Anticancer Res* **10**, 1307–1311 (1990).

225. John, E.M., Schwartz, G.G., Koo, J., Van Den, B.D. & Ingles, S.A. Sun exposure, vitamin D receptor gene polymorphisms, and risk of advanced prostate cancer. *Cancer Res* **65**, 5470–5479 (2005).
226. Chen, T.C., Wang, L., Whitlatch, L.W., Flanagan, J.N. & Holick, M.F. Prostatic 25-hydroxyvitamin D-1alpha-hydroxylase and its implication in prostate cancer. *J Cell Biochem* **88**, 315–322 (2003).
227. Hsu, J.Y., Feldman, D., McNeal, J.E. & Peehl, D.M. Reduced 1alpha-hydroxylase activity in human prostate cancer cells correlates with decreased susceptibility to 25-hydroxyvitamin D3-induced growth inhibition. *Cancer Res* **61**, 2852–2856 (2001).
228. Ahonen, M.H., Tenkanen, L., Teppo, L., Hakama, M. & Tuohimaa, P. Prostate cancer risk and prediagnostic serum 25-hydroxyvitamin D levels (Finland). *Cancer Causes Control* **11**, 847–852 (2000).

9 Role of Iron in Epigenetic Regulation of Gene Expression

Laura Cartularo and Max Costa

CONTENTS

9.1 INTRODUCTION: WHY IS IRON ESSENTIAL?

Iron is the fourth most abundant element in the earth's crust and is involved in many different important biological and chemical processes. All aerobic organisms require iron as a catalytic cofactor for various reactions, and thus, it is an essential dietary nutrient. Iron is involved in heme synthesis, oxygen transport via hemoglobin, mitochondrial oxidative phosphorylation, changes in chromatin structure, xenobiotic metabolism, and oxidative stress defense (Li et al., 2006; Vergara and Thiele, 2008). When complexed within a heterocyclic ring of protoporphyrins, iron forms heme. Heme can bind to several different proteins, forming hemeproteins that have diverse essential functions, including diatomic gas transport, electron transfer, and transcription regulation (Li et al., 2011). Heme iron is coordinated by α- and β-globin helices in hemoglobin, the oxygen transporting metalloprotein present in all red blood cells.

Iron can exist in several different oxidation states because it is a transition metal. Iron can easily gain and lose electrons and participate in one-electron redox reactions between its ferric (+3) and ferrous (+2) states (De Domenico et al., 2011; Vergara and Thiele, 2008). Iron's ability to occupy different oxidation states contributes to both its biological importance and to its toxicity. Free iron in its ferrous form can donate electrons to O_2 and H_2O_2, generating superoxide anion and hydroxyl radicals, respectively. These reactive oxygen species can oxidize proteins, lipids, and nucleic acids in Fenton-type reactions. The combination of high levels of free iron and oxygen is particularly damaging to biological molecules. Additionally, iron can replace other transition metals in their protein binding sites, generating aberrant proteins and inactive enzymes (Mole, 2010; Ward and Kaplan, 2012).

Because of its involvement in many different biological functions, and its potential toxicity, iron levels must be carefully regulated in mammals. In humans, normal serum iron concentrations typically vary between 12 and 25 μM (Doyard et al., 2012). Altered iron levels can lead to anemia or iron overload. Anemia, due to low plasma iron, causes reduced oxygen transport in blood and leads to nonspecific clinical symptoms such as weakness, fatigue, and shortness of breath. Anemia is one of the leading causes of morbidity and mortality, according to the World Health Organization (Halwachs-Baumann, 2012). Iron levels up to 50 μM are indicative of iron overload. Iron overload is a hallmark of hereditary hemochromatosis and can also occur due to blood transfusion. Symptoms of hemochromatosis can include weakness, lethargy, joint pain, and impotence. Over time, iron overload can lead to osteoporosis, cirrhosis, hepatocellular cancer, cardiomyopathy, dysrhythmia, and diabetes mellitus (Crownover and Covey, 2013). Osteoporosis is commonly observed in both hemochromatosis patients and postmenopausal woman as blood iron levels increase in women after menopause (Doyard et al., 2012; Jian et al., 2009).

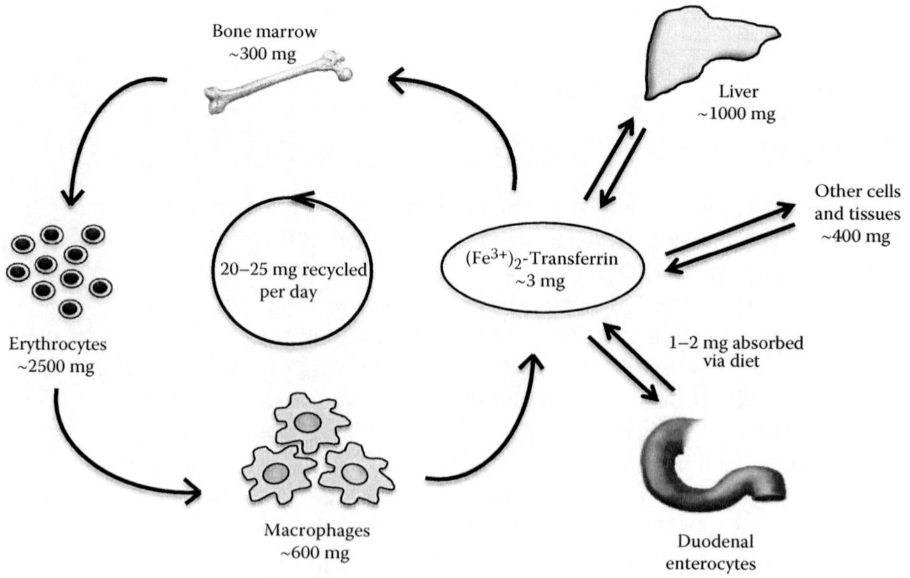

FIGURE 9.1 Iron stores in the major iron-containing cells and tissues.

The average human infant is born with approximately 270 mg of iron in their body (Fuqua et al., 2012) and 3–4 g are present in the average adult, half of which is contained in hemoglobin. About 1 g of total body iron is contained in body stores, predominately in the liver (Peyssonnaux et al., 2008). About 80% of iron demand is required for the daily production of hemoglobin to fill the nearly 200 billion new erythrocytes produced each day in the bone marrow (Hentze et al., 2004; Peyssonnaux et al., 2008). The majority of this iron is recycled from old erythrocytes as they are turned over, or by release of stores in the liver, macrophages, or other tissues. Due to iron loss via hemorrhage and sloughing of skin and mucosal surfaces, about 1–2 mg per day of iron must be absorbed via the diet (Hentze et al., 2010; Mole, 2010) (Figure 9.1).

9.2 IRON ABSORPTION, TRANSPORT, AND STORAGE

Dietary iron can come from animal sources (heme iron) or nonanimal sources (non-heme iron). Sources of non-heme iron include fortified cereals, soybeans, lentils and other legumes, spinach, and other plant foods. Iron is present in the diet mainly in its ferric form, or as heme (Mole, 2010). Heme iron, from animal sources, is ingested in the form of hemoglobin or myoglobin. Heme iron is typically more bioavailable and more readily absorbed than non-heme iron. In enterocytes of the brush border of the duodenum, the upper portion of the small intestine, dietary ferric [Fe(III)] iron is reduced to ferrous [Fe(II)] iron by the membrane-associated apical ferric reductase duodenal cytochrome B (DcytB) (Figure 9.2). Reduced ferrous iron is exported into the cytoplasm by the enterocyte transporter divalent metal transporter 1 (*DMT1*). If the ingested iron is not exported across the enterocyte membrane into the plasma, it is not absorbed, as these enterocytes are rapidly shed from the intestinal villi (Kaminsky and Zhang, 2003; Mole, 2010). Only 1–2 mg of dietary iron per day are absorbed by duodenal enterocytes, whereas about 20–25 mg are recycled from senescent erythrocytes by macrophages (Hentze et al., 2010).

When needed, iron is exported across the basolateral enterocyte membrane into plasma by ferroportin (FPN). In the plasma, iron is preferentially transported bound to the glycoprotein transferrin and taken up by transferrin receptors on the surface of various cell types. Heme iron is absorbed independently, by mechanisms that are not well understood (Fuqua et al., 2012; Hentze et al., 2010). Export of iron by ferroportin requires a multicopper ferroxidase, which oxidizes Fe(II) to Fe(III). Iron must be oxidized back to its ferric state, by either ceruloplasmin (Cp) or hephaestin, before it can be exported (Peyssonnaux et al., 2008; Ward and Kaplan, 2012).

Ferroportin is most highly expressed in cells that have a role in iron transport, and its expression is regulated both transcriptionally and posttranscriptionally. In macrophages, ferroportin expression is increased in response to iron or heme levels. Ferroportin transcription is regulated by Nrf2, Bach1, or MTF1 transcription factors (De Domenico et al., 2011) and can be modulated by iron deficiency, hypoxia, transition metals, heme, and inflammatory cytokines (Ward and Kaplan, 2012). Both erythropoiesis and erythrophagocytosis can upregulate ferroportin transcription. Ferroportin mRNA is regulated posttranscriptionally by an iron response element (IRE) in its 5′-untranslated region. Increased iron upregulates ferroportin translation

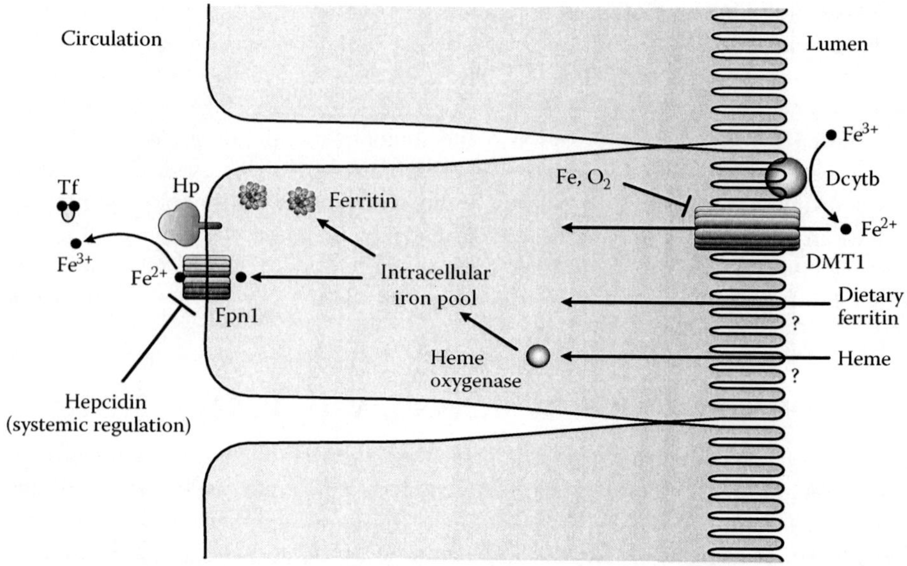

FIGURE 9.2 Passage of iron across the duodenal enterocyte. Tf, transferrin; Hp, hephaestin; Fpn1, ferroportin 1; Dcytb, duodenal cytochrome B; *DMT1*, divalent metal transporter 1. (Reprinted from *Journal of Trace Elements in Medicine and Biology*, Vol. 26, Brie K. Fuqua et al., Intestinal iron absorption, 115–119, Copyright 2012, with permission from Elsevier.)

due to release of iron regulatory proteins (IRPs) from the IRE. The increase of iron export by ferroportin leads to a decrease in cytosolic iron. Ferroportin protein stability is modulated by hepcidin, the hormone regulator of iron homeostasis. Hepcidin binds ferroportin, triggers its internalization, ubiquitination, and degradation. Mutations in ferroportin can lead to iron linked disorders, including iron overload syndromes (De Domenico et al., 2011).

Iron is transported and stored in its less toxic ferric (+3) state, bound to protective proteins such as transferrin (Mole, 2010). Transferrin, which contains two high-affinity Fe(III) binding sites, plays many key roles in iron absorption and storage. Transferrin maintains iron in a soluble form, facilitates transport of iron from plasma into cells, and limits the generation of toxic radicals. Plasma transferrin is normally about 30% saturated with iron. Saturation levels below 16% indicate iron deficiency, while levels above 45% indicate iron overload. Above 60% saturation, free iron accumulates in the circulation and can cause cellular and tissue damage (Hentze et al., 2010). Transferrin–iron complexes bind to transferrin receptors on the surface of various cell types, leading to endocytosis. After binding, the transferrin receptor *TfR1* internalizes to form endosomes, where acidification releases iron. The released Fe(III) is reduced by ferrireductase to Fe(II), which is then transported to the cytoplasm by *DMT1*. Iron that is not immediately required for synthesis of biological molecules is stored bound to ferritin (Mole, 2010). A small portion of cellular iron remains available for biological processes in the form of a chelatable iron pool, commonly referred to as the labile iron pool (LIP).

The majority of circulating iron comes from macrophage recycling of erythrocyte heme. Under normal physiological conditions, less than 10% of daily iron needs are met by intestinal absorption of dietary iron (Hentze et al., 2010; Ward and Kaplan, 2012). Each day, approximately 20–25 mg of iron is recycled by macrophages from senescent erythrocytes in the spleen and liver (Peyssonnaux et al., 2008). Phagocytic macrophages engulf aged or damaged erythrocytes and utilize the endoplasmic reticulum enzyme hemeoxygenase to catabolize heme and release iron (Hentze et al., 2010). The released iron may be stored or exported into circulation for use by various cell types (Ward and Kaplan, 2012).

9.3 REGULATION OF IRON HOMEOSTASIS

Iron levels must be tightly regulated within cells and biological fluids (Ward and Kaplan, 2012). Within cells, the free iron level is maintained within a narrow range of about 0.8 μM to minimize the production of reactive oxygen species (Chen et al., 2010a). Several mechanisms have evolved to strictly regulate iron levels in biological systems. The master regulator of iron homeostasis is hepcidin (Ganz, 2003), an antimicrobial peptide produced and expressed in the liver (Peyssonnaux et al., 2008). Hepatocytic expression of hepcidin is regulated by systemic iron levels, hepatic iron stores, erythropoiesis, hypoxia, and inflammatory states (Hentze et al., 2010). Binding of hepcidin to ferroportin causes cytosolic tyrosine kinase Jak2 to bind ferroportin. Two Jak2 monomers are autophosphorylated and then phosphorylate ferroportin on either tyrosine 302 or 303, leading to internalization of the ferroportin–hepcidin complex via endocytosis. Internalized ferroportin is monoubuquitinated, leading to its degradation, preventing further uptake of iron (De Domenico et al., 2007, 2009).

Hepcidin controls iron transport and iron regulatory genes to ensure a proper balance between absorption of dietary iron and recycling of iron by macrophages (Bayele and Srai, 2009). By controlling iron absorption, recycling, and transport, hepcidin regulates systemic iron levels, erythropoiesis, and inflammation. Hepcidin inhibits ferroportin-mediated iron absorption in the intestine as well as the release of iron into serum from macrophages and hepatocytes, acting as a negative regulator of iron uptake (Mole, 2010; Ward and Kaplan, 2012). During iron deficiency and hypoxia the liver produces less hepcidin, and intestinal iron uptake is enhanced. During infection and inflammation, hepcidin expression is induced, leading to the sequestration of iron in cells and decreased release of iron into plasma, causing "anemia of chronic disease" (Haase, 2013; Peyssonnaux et al., 2008).

Iron levels are regulated posttranslationally by iron responsive proteins (IRPs). Two iron responsive proteins, IRP1 and IRP2, monitor cytosolic iron concentrations (Rouault, 2006). IRP1 and 2 control expression of many iron sensitive mRNAs in response to intracellular iron levels, including transferrin receptor 1, ferritin, and *DMT1* (Haase, 2013; Li et al., 2006). IRP1 binds to IREs when oxygen tension is high (Peyssonnaux et al., 2008). IRP1 exists in two structural forms, depending on the presence of an iron sulfur (4Fe–4S) cluster between its two domains. When iron levels are high, the 4Fe–4S cluster forms part of the IRP1 active site that converts citrate to isocitrate in the tricarboxylic acid cycle of aerobic respiration. In this orientation, the enzyme is in its aconitase form and lacks RNA binding capability (Hanson and Leibold, 1998; Li et al., 2006; Mole, 2010). When iron levels are low, IRP1 loses its iron sulfur cluster and takes on a more open structure and is able to bind the IRE stem loop of regulated transcripts with high affinity. Binding of IRPs to IREs in the 3′-untranslated region (UTR) stabilizes the mRNA, increasing expression of genes such as *TfR1* to restore cytosolic iron levels. Binding to IREs in the 5′-UTR of mRNAs inhibits translation initiation to downregulate expression of genes such as ferritin, mitochondrial aconitase, and ferroportin (Mole, 2010). IRP2 is regulated by iron-dependent proteasomal degradation and does not convert to an aconitase form (Haase, 2013; Thompson and Bruick, 2012). High iron levels limit the ability of IRPs to interact with IREs in mRNAs of iron related genes. This ensures proper regulation of cytoplasmic iron pool levels to avoid toxicity (Doyard et al., 2012).

Humans with iron overload diseases or inherited iron-related anemias often have alterations in hepcidin levels or resistance of ferroportin to its effects. Most cases of hereditary hemochromatosis result from hepcidin deficiency, causing increased iron uptake (Ward and Kaplan, 2012), while increased hepcidin expression results in iron-associated anemias (Doyard et al., 2012). Hepcidin expression is quite varied among different individuals and may depend on age, sex, disease state, or even time of day. Expression levels also depend on various regulatory factors. Diet and environment are important epigenetic regulators of hepcidin because of their potential to modify histones and chromatin. Infectious disease exposure or a diet rich in iron may lead to increased hepcidin expression. Alternatively, sterile living environments, diets low in iron, or chronic hypoxic conditions may cause suppressed hepcidin expression (Bayele and Srai, 2009).

Iron absorption and storage must be tightly regulated in eukaryotes. When iron absorption exceeds transferrin binding capacity, it is removed from plasma and deposited into hepatocytes in the liver (De Domenico et al., 2011). There is no regulated excretory pathway for iron; thus, systemic iron homeostasis is controlled through regulation of dietary absorption (Ward and Kaplan, 2012). When iron stores are in excess, iron absorption is partially suppressed, and in times of iron deficiency, iron absorption is increased (Peyssonnaux et al., 2008). Iron absorption can also be enhanced by hypoxia or anemia as well as erythropoiesis (De Domenico et al., 2011). Iron can be lost due to epithelial cell loss, menstruation, hemorrhage, or childbearing. Dysregulation of iron loss can lead to iron overload, tissue damage, and organ dysfunction (Ward and Kaplan, 2012). Iron losses or insufficient dietary iron uptake and absorption can lead to iron deficiency, which causes anemia (Hentze et al., 2010). Thus, iron levels must be carefully regulated to ensure adequate oxygen transport and proper functioning of the various iron-containing and iron-dependent enzymes in the body.

9.4 ROLE OF IRON IN DIOXYGENASES

Several classes of enzymes require iron as a necessary cofactor for their catalytic activity, including enzymes with various epigenetic roles. Fe(II)- and 2-oxoglutarate-dependent dioxygenases are a family of non-heme iron-containing enzymes that catalyze a wide array of oxidation reactions in cells. Dioxygenases contain a two histidine–one carboxylate facial triad at their core. This class of enzymes includes hypoxia-inducible factor (HIF) prolyl hydroxylases, factor inhibiting HIF (FIH), the ten-eleven translocation (Tet) family of enzymes, several histone demethylases, and the DNA repair enzymes ABH2 and ABH3, all of which catalyze various oxidation reactions. HIF prolyl hydroxylases and FIH are involved in regulation of hypoxia-inducible factors (HIFs), and will be further discussed in the following section. Tet proteins are involved in DNA demethylation, an important epigenetic modification that can dramatically alter gene expression. Histone demethylases can dynamically remove methyl groups from histone lysines. One class of histone demethylases, the jumonji C domain containing histone demethylases (JHDM), are a family of dioxygenases that utilize Fe(II) and 2-oxoglutarate to carry out their enzymatic activities. This class of dioxygenases will be discussed further in Section 9.6.2. ABH2 and ABH3 catalyze oxidative demethylation of aberrantly methylated DNA bases, including 1-methyladenine (1-meA), 3-methylcytosine (3-meC), and 3-methylthymine (3-meT) (Chen et al., 2010a; Li et al., 2006; Yi et al., 2009), whereas fat mass and obesity associated protein (FTO) catalyzes the removal of 6-methyladenine from RNA (Fu and He, 2012). The proteasomal degradation of iron regulatory protein IRP2 may also be mediated by Fe(II) dependent dioxygenases (Wang et al., 2004). These varied dioxygenases use Fe(II), 2-oxoglutarate, oxygen, and ascorbic acid as cofactors for their enzymatic activity.

9.5 THE ROLE OF HYPOXIA-INDUCIBLE FACTOR IN IRON METABOLISM

9.5.1 Background

Hypoxia-inducible factor (HIF), a master regulator of oxygen homeostasis, was first identified as a hypoxia inducible transcription factor that stimulates expression of the erythropoietin (*EPO*) gene (Wang and Semenza, 1993). *EPO* is a glycoprotein hormone that controls erythropoiesis, or red blood cell production, and is also important in iron homeostasis. When oxygen levels are low, *EPO* stimulates the bone marrow to promote the proliferation of red blood cells to increase oxygen transport (Mole, 2010). As reviewed below, under hypoxia, HIF associates with transcriptional coactivators p300 and CBP. These proteins function as histone acetyltransferases, important epigenetic enzymes that facilitate transcription via the opening up of chromatin in gene promoter regions.

HIF is a heterodimeric basic helix–loop–helix transcription factor comprised of an O_2-sensitive α subunit and a constitutive β subunit, which is also known as the aryl hydrocarbon nuclear translocator (ARNT) (Ke and Costa, 2006). There are three HIFα subunits; HIF1α, HIF2α, and HIF3α, and three HIFβ subunits; HIF1β, HIF2β, and HIF3β. HIF1α and HIF2α contain two transactivation domains required for recruiting cofactors that regulate transcription of many different genes (Mole, 2010). HIF2 regulates *EPO* synthesis and iron metabolism and can be stabilized by iron deficiency (Li et al., 2006).

In addition to EPO, HIF1α modulates a widespread network of over 100 oxygen-dependent genes. HIF heterodimers bind to DNA consensus sequences called hypoxia response elements (HREs) and recruit transcriptional cofactors to regulate transcription. Many genes under regulation by HIF1α are involved in restoring oxygen homeostasis and in reducing oxygen demand. In response to hypoxia, HIF stimulates angiogenesis, anaerobic glucose metabolism, mitochondrial biogenesis, and regulation of iron homeostasis (Ke and Costa, 2006; Li et al., 2006).

9.5.2 Regulation of HIF by Hydroxylation

HIF proteins are regulated by several Fe(II)- and oxygen-dependent enzymes, including HIF prolyl hydroxylases (PHD) and asparagine hydroxylase, or factor inhibiting HIF (FIH) (Kaczmarek et al., 2009). Under normal oxygen tension, or normoxia, prolyl-4-hydroxylase domain proteins 1–3 (PHD1–3) hydroxylate prolines 402 and 564 in the oxygen-dependent degradation domain of HIF1α or HIF2α (Chan et al., 2005). PHD2 appears to be the most important isoform in sensing oxygen (Kaczmarek et al., 2009; Mole, 2010). These dioxygenases use non-heme Fe(II), oxygen, and 2-oxoglutarate as cofactors to catalyze the splitting of molecular dioxygen, oxidative decarboxylation of 2-oxoglutarate to succinate and carbon dioxide, and the hydroxylation of HIF1α (Mole, 2010). Ascorbic acid is required for maintaining iron in its +2 oxidation state (Kaczmarek et al., 2009). Hydroxylation under normoxia causes HIFα subunits to associate with the Von Hippel-Lindau (VHL) tumor suppressor-E3 ligase complex, leading to their proteasomal degradation via polyubiquitination. Degradation of these subunits prevents the formation of transcriptionally active HIF heterodimers (Ke and Costa, 2006; Li et al., 2006). Hypoxia, due to anemia, ischemia, high altitude, or other causes, leads to HIF stabilization via inhibition of PHDs. Anemia causes reduced O_2 transfer to tissues, leading to tissue hypoxia. As anemia is corrected and oxygen levels rise, HIF then is degraded via hydroxylation and VHL-mediated ubiquitination (Peyssonnaux et al., 2008).

In the presence of oxygen, HIF1α is also hydroxylated on asparagine 803 by asparagine hydroxylase, or factor inhibiting HIF (FIH), preventing its interaction with the transcriptional coactivators CREB binding protein (CPB) and p300. (Kaczmarek et al., 2009) (Figure 9.3). Both prolyl and asparaginyl hydroxylases are inhibited by hypoxia. During hypoxia, stabilized HIF1α protein accumulates and translocates to the nucleus, where it binds to HIF1β to form an active transcription factor that recruits transcriptional coactivators and histone acetyltransferases CBP and p300. The dimerized HIF transcription factor can then bind to HREs in target genes, leading to upregulation of transcription (Peyssonnaux et al., 2008; Ward and Kaplan, 2012).

Fe(II) loosely coordinated at the catalytic center of HIF prolyl and asparaginyl hydroxylases is an essential cofactor for these enzymes (Mole, 2010). HIF proteins can be dysregulated by inhibition of these dioxygenases due to iron deficiency or chelation, interference by other metal ions, nitric oxide, or oxidative stress. Low intracellular iron levels causes reduced PHD hydroxylation and upregulated HIF activity, leading to enhanced expression of hypoxia related genes (Peyssonnaux et al., 2008).

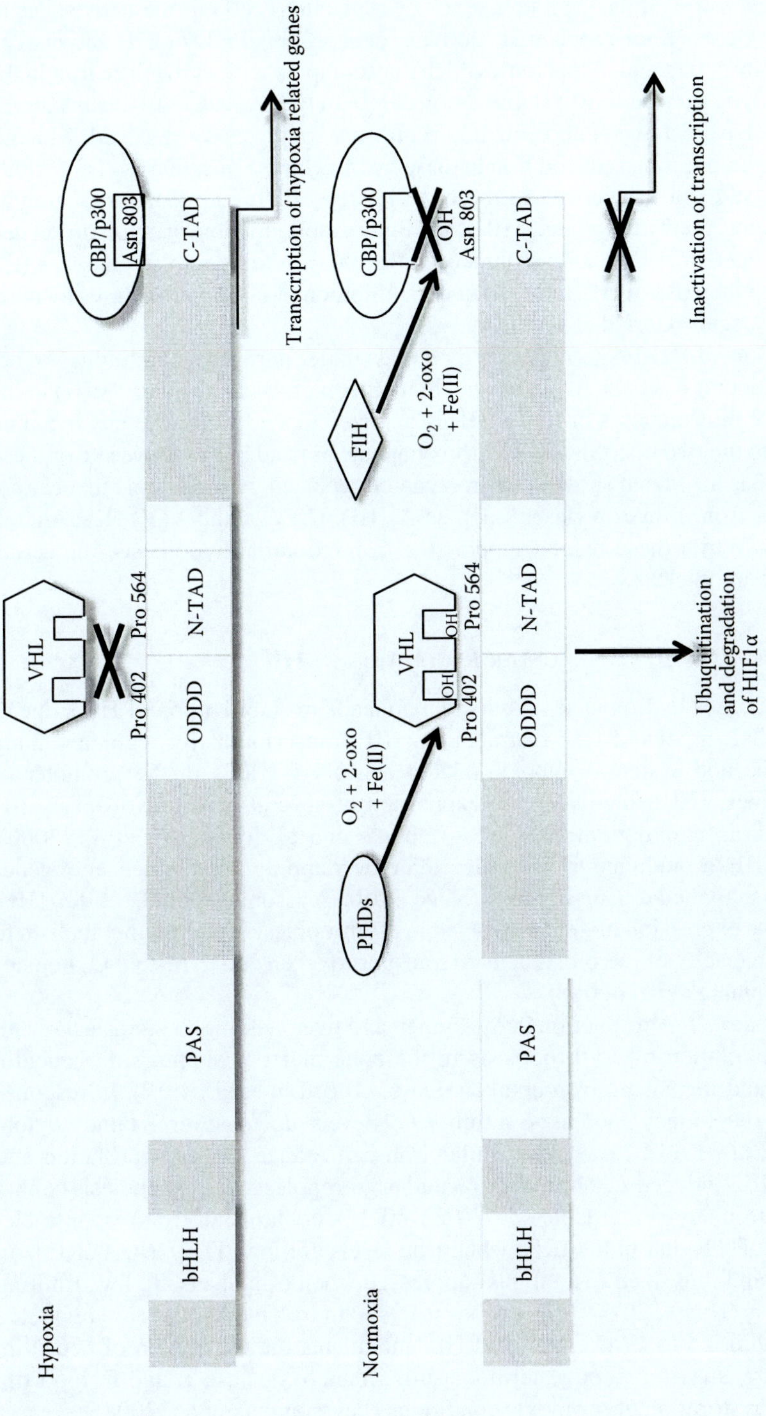

FIGURE 9.3 Regulation of HIF1α by hydroxylation. bHLH, basic helix–loop–helix; PAS, Per/Arnt/Sim domain; ODDD, oxygen-dependent degradation domain; N-TAD, N-terminal transactivation domain; VHL, Von Hippel-Lindau; CBP, CREB-binding protein; C-TAD, C-terminal transactivation domain; Asn, asparagine; PHD, prolyl dydroxylase domain containing proteins; Pro, proline; FIH, factor inhibiting HIF; OH, hydroxyl group.

In addition, other divalent metal ions can potentially decrease dioxygenase function by interfering with cellular iron uptake, or by replacing Fe(II) in the active sites of these hydroxylases (Kaczmarek et al., 2009; Li et al., 2006). PHD or FIH activity can be inhibited by cobalt and nickel ions, which may compete with or displace iron in the enzyme catalytic site, leading to stabilization of HIF1α (Chen et al., 2010a; Kaczmarek et al., 2009). Divalent nickel and cobalt ions may also oxidize ascorbic acid, which is essential for maintaining reduced iron in these hydroxylases (Kaczmarek et al., 2009). Nitric oxide and iron chelators can inhibit PHDs by binding or removing catalytic iron, stabilizing nonhydroxylated HIF1, leading to upregulation of hypoxia related genes like *EPO* (Mole, 2010; Peyssonnaux et al., 2008). Additionally, oxidative stress can lead to an increase in HIF1α levels due to inhibition of HIF hydroxylases by reactive oxygen species (Gerald et al., 2004).

Mechanisms of HIF1α stabilization by metals under normoxic conditions are not well understood. It has been hypothesized that other metals can displace Fe(II) in the active sites of dioxygenases or that Fe(II) is oxidized to a +3 state whereby it can no longer bind to the two histidine–one carboxylate facial triad in the active site of these enzymes. It has also been shown that iron can be replaced by metal ions that cannot substitute for iron in hydroxylases, such as As(III), Cr(VI), and V(V) (Kaczmarek et al., 2009). While these results support the iron oxidation hypothesis, the actual mechanisms are unclear.

9.5.3 Iron-Related Genes under Regulation by HIF

Hypoxia decreases hydroxylase activity, which leads to stabilization of HIFs due to decreased VHL-mediated HIF degradation. HIF transcriptionally regulates many tissue-specific and systemic hypoxia-related genes via HREs in the promoter of regulated genes. HIF1α activates the expression of several genes involved in iron metabolism, vascular regulation, glucose uptake and glycolysis (Li et al., 2006). HIF1α and HIF2α modulate transcription of an overlapping but distinct set of genes and are both activated during hypoxia (Mole, 2010; Peyssonnaux et al., 2008). HIFs upregulate the expression of genes involved in erythropoiesis and iron metabolism to increase the capacity of red blood cells to transport oxygen and to restore O_2 homeostasis (Peyssonnaux et al., 2008).

HIF2 induces *EPO* production in the kidney and liver, leading to increased serum *EPO* and stimulation of erythropoiesis in the bone marrow, suppressing hepcidin in the liver, and increasing iron uptake (Haase, 2013; Liu et al., 2012). In response to hypoxia, the kidney increases serum *EPO* levels up to several hundred fold (Goldberg et al., 1991). Low intracellular iron can reduce *HIF2A* translation and thus limit HIF2-induced *EPO* production and erythropoiesis. Erythropoiesis cannot occur when iron stores are depleted. *HIF2A* mRNA contains an iron response element in its 5′-UTR that binds IRPs when iron levels are low. The *HIF2A* IRE is in the 5′-UTR, and thus, its translation is suppressed when iron levels are low, limiting *EPO* synthesis (Haase, 2013). Many genes involved in iron metabolism are regulated by HIF. In addition to *EPO*, stabilized HIF modulates the expression of hepcidin, transferrin, transferrin receptor, ceruloplasmin, hemeoxygenase 1, and ferroportin, allowing the restoration of normoxic conditions (Peyssonnaux et al., 2008).

9.6 ROLE OF IRON IN EPIGENETIC REGULATION

Epigenetic modifications are heritable changes in gene expression caused by mechanisms other than alterations at the level of the DNA sequence. DNA methylation and posttranslational histone modifications are two such epigenetic changes that can lead to gene expression changes (Goldberg et al., 1991). Environmental factors such as diet can affect epigenetic machinery and lead to alterations in gene expression patterns. Epigenetic modifications can help organisms adapt to the environment by modulating gene expression levels in response to certain stimuli, but can also lead to disease such as cancer and neurological disorders (Chen et al., 2010b). Cancers and other diseases often display an aberrant epigenetic landscape. Several key proteins that are involved in regulating important epigenetic changes to histones and DNA use Fe(II) and oxygen as essential cofactors.

9.6.1 Methylation and Demethylation of Cytosines

DNA methylation is mediated by DNA methyltransferases (DNMTs). DNMTs catalyze the addition of a methyl group from the universal methyl donor *S*-adenosylmethionine (SAM) to the 5-carbon position of cytosines to form 5-methylcytosine (5meC) (Robertson, 2002). There are both maintenance and de novo DNMTs at work at different stages of development. Three DNA methyltransferases have been identified in mammalian cells, DNMT1, DNMT3A, and DNMT3B. DNMT1 is described as a maintenance DNA methyltransferase, as it preferentially methylates hemimethylated DNA. Alternatively, DNMT3A and 3B function as de novo DNA methyltransferases that can establish new patterns of DNA methylation (Huidobro et al., 2013). 5meC, which is the most abundant epigenetic alteration in the human genome, controls pluripotency status, and cell fate via its role in gene silencing. This repressive mark recruits 5mCpG binding proteins to methylated cytosines and functions in the regulation of gene expression, genomic imprinting, and retrotransposon silencing (Fu and He, 2012). Cytosine methylation usually occurs at cytosine–guanine (CpG) dinucleotides in CpG islands, short interspersed cytosine and guanine-rich stretches of DNA that typically occur near sites of transcription initiation (Deaton and Bird, 2011). 5meC is a minor base in mammalian DNA, representing only 1% of all DNA and found almost exclusively in CpG dinucleotides (Ehrlich and Wang, 1981). The majority of 5meC is found in repetitive DNA elements (Chen et al., 2010b). DNA methylation in repetitive elements and gene promoters leads to silencing (Suva et al., 2013). DNA methylation patterns are often highly dysregulated in cancers (Chen et al., 2010b). Hypermethylation of CpG islands is frequently observed in cancers (Suva et al., 2013). Hypermethylation of tumor suppressor gene promoters leads to gene silencing, which, accompanied by a host of other genetic mutations or genomic instability, can be a major initiator or promoter of carcinogenesis. Hypermethylation of the Von Hippel-Lindau and *CDK2NA* tumor suppressor genes in several different tumor types highlights the importance of DNA methylation in the carcinogenesis pathway (Baylin and Jones, 2011). Gene-specific promoter hypermethylation is often accompanied by genome-wide hypomethylation in cancers and other diseases. Although gene-specific promoter hypermethylation relates to gene silencing, genome-wide hypomethylation often correlates with genomic instability and can be indicative of overall disease progression (Ehrlich, 2009; Li et al., 2012).

While methylation of cytosines can drastically alter cell fate and gene expression patterns, this epigenetic modification is not immutable. 5meC can be lost during DNA replication or via active demethylation pathways independent of DNA replication. One such demethylation pathway involves the Tet proteins. Tet1–3 are a family of non-heme Fe(II)- and 2-oxoglutarate-dependent dioxygenases (Fu and He, 2012). Tet1 is expressed mainly in embryonic stem cells, whereas Tet2 and Tet3 are more ubiquitous (Franchini et al., 2012). These proteins contain a conserved double-stranded β helix domain (DSBH) core, with a cysteine-rich N-terminus and a CpG binding protein (CXXC)–type zinc binding domain (Fu and He, 2012). In 2009, it was discovered that the oncogenic protein TET1, a fusion partner of the mixed-lineage leukemia (MLL) methyltransferase, converts 5mC to 5-hydroxymethylcytosine (5hmC) in cultured cells and *in vitro* (Tahiliani et al., 2009). Tet dioxygenases catalyze the conversion of 5mC to 5hmC using Fe(II) and 2-oxoglutarate as cofactors. Tet proteins can hydroxylate an inert carbon–hydrogen (C–H) bond in 5mC via activation of dioxygen. Hydroxylation is a two-electron oxidation. The cofactor 2-oxoglutarate provides the other two electrons required for the four-electron reduction. The cytosine hydroxylation reaction takes place in two steps. The first step involves the activation of dioxygen to form an Fe(IV)-oxo intermediate. The 5mC substrate is oxidized in the second step, which involves abstraction of hydrogen from the C–H bond by Fe(IV)-oxo to yield the hydroxylated product 5hmC. 5hmC is further oxidized to 5-formylcytosine (5fC) and 5-carboxycytosine (5caC) via a similar mechanism. 5fC and 5caC are recognized and excised by the base excision repair enzyme thymine DNA glycosylase (TDG). Oxidation of 5mC by Tet1 and Tet2 to 5hmC, followed by TDG mediated removal of 5fC and 5caC and base excision repair of the excised base to an unmethylated cytosine represent a mechanism of demethylation of DNA that requires Fe(II) as a necessary cofactor (Fu and He, 2012; Zhao and Chen, 2013) (Figure 9.4). Demethylation of cytosines in gene promoters can lead to the activation of previously silenced genes, which can have dramatic effects on gene expression or differentiation. In fact, Tet proteins have been found to be important in epigenetic reprogramming in early embryos and primordial germ cells (Zhao and Chen, 2013).

9.6.2 Methylation and Demethylation of Histones

Another important epigenetic change involves covalent modifications to the N-terminal tails of histones. Posttranslational modifications to core histones include methylation, acetylation, ubiquitination, and SUMOylation. These changes can have significant effects on chromatin structure and gene expression, depending on the type of functional group added and the location and number of modifications. Regulation of these histone alterations plays an important role in carcinogenesis and other disease development (Hickok et al., 2013).

Histone methylation is involved in gene expression regulation, maintenance of chromatin structure, X chromosome inactivation, and genomic imprinting (Klose et al., 2006). Histone lysines are methylated by SET domain–containing proteins and DOT-1–like proteins, while arginine residues are methylated by members of the

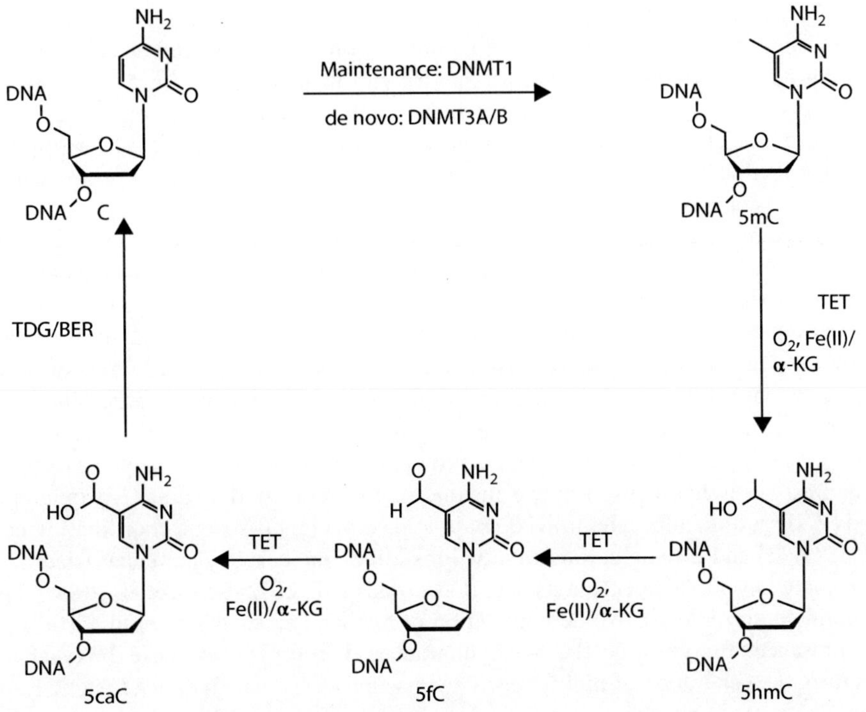

FIGURE 9.4 Overview of the demethylation pathway of 5-methylcytosine by TET dioxygenases. DNMT, DNA methyltransferase; TDG, thymine DNA glycosylase; BER, base excision repair; TET, ten-eleven translocate protein; α-KG, α-ketoglutarate; 5mC, 5 methylcytosine; 5hmC, 5 hydroxymethylcytosine; 5fC, 5 forrmylcytosine; 5caC, 5 carboxycytosine. (Reprinted from *Current Opinion in Chemical Biology*, Vol. 16, Ye Fu and Chuan He, Nucleic acid modifications with epigenetic significance, 516–524, Copyright 2012, with permission from Elsevier.)

protein arginine *N*-methyltransferase (PRMT) family. Histone methyltransferases transfer methyl groups from SAM to lysine or arginine residues on the tails of histone H3 or H4. Histone lysines can be monomethylated, dimethylated, or trimethylated while histone arginines can be monomethylated, symmetrically dimethylated, or asymmetrically dimethylated. Many different lysine and arginine methyltransferases have been characterized, most of which have specific target residues. For example, G9a dimethylates histone 3 lysine 9 (H3K9), EZH1 and EZH2 dimethylate and trimethylate histone 3 lysine 27 (H3K27), and SETD1A and SETD1B can monomethylate, dimethylate, or trimethylate histone 3 lysine 4 (H3K4) (Greer and Shi, 2012). Histone methylation can activate or suppress gene expression, depending on the specific residue modified and the number of methyl groups present. Methylation of histone H3, lysine 4 and 36 (H3K4 and H3K36) is associated with an open chromatin state and active transcription, whereas methylation of H3K9 and H3K27 is associated with condensation of chromatin and repression of transcription (Huidobro et al., 2013).

Specific histone methylation marks are recognized by various proteins with methyl binding domains, which can include chromatin-modifying enzymes (Greer and Shi, 2012). Methylated histones may recruit histone acetyltransferases (HATs) or histone deacetylases (HDACs), leading to either the unwinding or condensation of chromatin. Thus, histone methylation can either facilitate or repress transcription. Histone methylation is a dynamic process, regulated by the balance between histone methyltransferases and histone demethylases (Hickok et al., 2013).

Histone methylation marks are not permanent and can be dynamically removed by histone demethylases. There are two families of histone demethylases, the LSD family of lysine-specific flavin-dependent amine oxidases, and the jumonji C (JmjC) domain–containing histone demethylases (JHDMs) (Hickok et al., 2013). Flavin-dependent amine oxidases, which include lysine-specific demethylase 1 (LSD1), contain an amine oxidase catalytic center domain. These enzymes can demethylate lysines via a flavin adenine dinucleotide (FAD)–dependent oxidation reaction (Culhane and Cole, 2007). The JmjC family of histone demethylases, which is composed of 30 different proteins, can dynamically remove histone methylation marks using 2-oxoglutarate and Fe(II) as cofactors. The JmjC domain folds into eight β-sheets, coordinating non-heme Fe(II) and 2-oxoglutarate in a two histidine–one carboxylate facial triad in an enzymatically active pocket at its core (Chervona et al., 2012; Klose et al., 2006). This domain catalyzes oxidative demethylation of histone lysines (Chen et al., 2010b). In the presence of oxygen, Fe(II), 2-oxoglutarate, and ascorbic acid, these dioxygenases catalyze the generation of highly reactive oxygen species, which attack histone lysine methyl groups to produce unstable hydroxyl methyl intermediates that are spontaneously lost as formaldehyde. This leads to the removal of these methyl groups from the histone lysines (Figure 9.5). Loss of histone methylation marks can lead to repression or activation of transcription of specific genes, depending on the residues altered and the degree of methylation or demethylation.

Lysine (K)-specific demethylase 3A (KDM3A/JHDM2A/JMJD1A), one member of this enzyme family, acts as a transcriptional coactivator and has roles in a diverse array of cellular functions including hormone receptor signaling, spermatogenesis, and

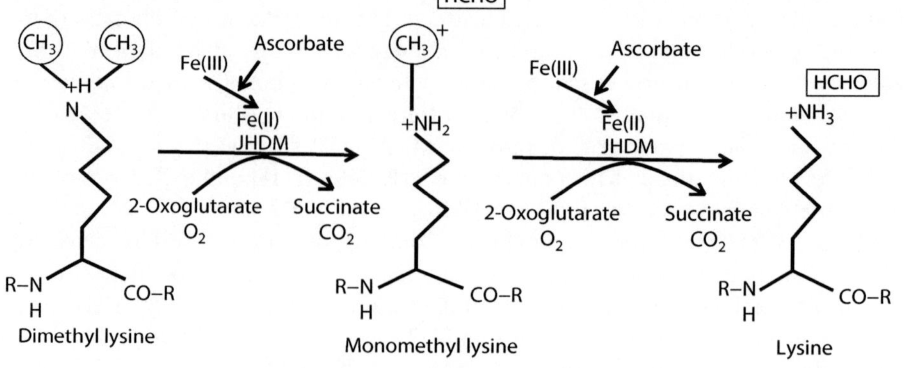

FIGURE 9.5 Model for Fe(II)- and 2-oxoglutarate–dependent dioxygenase demethylation of lysines by jumonji C domain–containing demethylases (JHDMs).

energy metabolism. JMJD1A can remove histone 3 (H3) lysine 9 (K9) monomethylation (H3K9me) and dimethylation (H3K9me2), but not trimethylation (H3K9me3). The JMJD2 dioxygenases can demethylate H3K9 dimethylation and trimethylation. H3K9me2 is a critical mark of gene silencing and establishment of DNA methylation patterns. Removal of this repressive mark activates transcription. Environmental carcinogens as well as hypoxia can increase global histone methylation, most likely by inactivating these demethylases. JmjC domain–containing demethylases are less active under hypoxic conditions as they require oxygen for optimum enzymatic activity. Carcinogenic metals may inactivate these demethylases by displacing the cofactor Fe(II). Inactivation of these demethylases may contribute to carcinogenesis, as this could lead to aberrant demethylation of histones and dysregulation of gene expression. Several studies have also shown that JMJD1, JMJ2B, and JMJD2C expression can be induced during hypoxia by HIF1α, which binds to the HRE in the promoter of these genes (Chen et al., 2010a,b; Chervona et al., 2012).

9.6.3 Histone Deacetylation

Another Fe(II)- and 2-oxoglutarate-dependent dioxygenase with a role in epigenetic regulation of gene expression is HIF1α asparagine hydroxylase, or factor inhibiting HIF (FIH). In the presence of oxygen, FIH hydroxylates asparagine 803 on HIF1α, preventing its association with transcriptional coactivators CREB-binding protein (CBP) and p300 (Li et al., 2006). CBP and p300 function as histone acetyltransferases, relaxing chromatin around gene promoters, facilitating transcription (Holmqvist and Mannervik, 2013). Acetylation neutralizes the positive charge of the histone lysine, relaxing the chromatin and facilitating loading of transcription machinery onto the DNA (Barneda-Zahonero and Parra, 2012). The CBP/p300 complex may also recruit RNA polymerase II and/or act as a scaffold for the various DNA–protein and protein–protein interactions at the transcription start site (Holmqvist and Mannervik, 2013). Hydroxylation by FIH under normoxia inhibits the transcriptional activity of HIF1α by preventing its binding with CBP and p300 (Li et al., 2006). FIH is inactivated under hypoxia, leading to activation of HIF transcription factors. Similarly, inhibition of FIH due to low iron levels or iron chelation can lead to upregulation of various hypoxia related HIF target genes (Cho et al., 2013).

9.6.4 Inhibition of Iron-Dependent Dioxygenases

Fe(II)- and 2-oxoglutarate-dependent dioxygenases can be inhibited by divalent metals, iron chelators, and nitric oxide (Hickok et al., 2013). Diminished dioxygenase activity could lead to aberrant DNA and histone methylation, which can cause dramatic alterations in gene expression. These epigenetic changes could lead to disease or tumorigenesis, depending on the genes and pathways affected.

Divalent metals such as nickel(II) cobalt(II) can inhibit these enzymes by displacing Fe(II) in the catalytic core. Dioxygenases are known to be major targets of nickel toxicity. Ni(II) has been shown to inhibit the demethylase activity of both JMJD1 and ABH2 in vitro. The affinity constant for Ni(II) in these dioxygenases is at least three orders of magnitude greater than that of Fe(II). Using X-ray absorption spectroscopy,

Chen et al. showed that Ni(II) displaces Fe(II) from its binding site in ABH2. Co(II) has also been observed to inhibit the H3K9/H3K36 trimethyldemethylase JMJD2A in vitro (Li et al., 2009). Additionally, chromate and arsenic could also deplete the reduced ascorbate required by these demethylases, leading to diminished activity (Chen et al., 2010b; Chervona et al., 2012).

Iron chelators such as desferrioxamine and 2-hydroxy-1-napthylaldehyde isonicotinoyl hydrazone can also inhibit dioxygenases by binding or reducing the iron pool. Inhibition of dioxygenases due to iron chelation or inhibitor drugs can lead to upregulation of HIF regulated genes due to inactivation of PHDs and FIH. This inhibition can also lead to increased DNA and histone methylation due to reduced activity of Tet and JmjC histone demethylases. Increased histone methylation can lead to dysregulation of gene expression (Hickok et al., 2013). Iron chelation is a common therapy for hemochromatosis, and it is being explored as a possible therapy for some cancers; thus, it is important that alterations in gene expression by these drugs be further studied to ensure patient safety (Saletta et al., 2010).

Nitric oxide (NO) can inhibit dioxygenases by binding or removing catalytic iron. Altered H3K9 methylation was observed in human breast carcinoma MDA-MB-231 cells treated with physiological concentrations of NO in both a time- and dose-dependent manner. The chelatable iron pool is the source of the non-heme iron cofactor for JmjC demethylases. This iron pool is also the source for dinitrosyl iron complex (DNIC) iron (Hickok et al., 2013). DNICs form in conditions of inflammation, cancer, and other disease states due to increased production of NO. DNICs and other NO complexes act to protect cells from the free radical activity of NO (Lewandowska et al., 2011). Results of this study suggest that the formation of DNIC due to inflammation inhibits dioxygenase activity by reducing the chelatable iron pool. NO can affect histone methylation either by direct inhibition of JmjC demethylases, by reducing the availability of non-heme iron, or by regulation of expression of histone lysine methyltransferases and demethylases (Hickok et al., 2013).

9.7 CONCLUSION

Iron is a very important transition metal, required for many crucial biological functions, including oxygen transport by hemoglobin, modification of chromatin, and xenobiotic metabolism. Although the recycling of senescent erythrocytes provides most of the iron required for these diverse functions, it remains an essential part of our diet. Intracellular and plasma levels must be carefully regulated to ensure proper functioning of numerous enzymes and to prevent the generation of reactive oxygen species. Iron is a necessary cofactor for many enzymes, several of which have key epigenetic roles. Demethylation of both DNA and histones requires Fe(II), underscoring the importance of this element in the epigenetic regulation of gene expression.

REFERENCES

Barneda-Zahonero, B., and Parra, M. (2012). Histone deacetylases and cancer. *Molecular Oncology* 6, 579–589.

Bayele, H.K., and Srai, S.K. (2009). Genetic variation in hepcidin expression and its implications for phenotypic differences in iron metabolism. *Haematologica 94*, 1185–1188.

Baylin, S.B., and Jones, P.A. (2011). A decade of exploring the cancer epigenome—Biological and translational implications. *Nature Reviews. Cancer 11*, 726–734.

Chan, D.A., Sutphin, P.D., Yen, S.E., and Giaccia, A.J. (2005). Coordinate regulation of the oxygen-dependent degradation domains of hypoxia-inducible factor 1 alpha. *Molecular and Cellular Biology 25*, 6415–6426.

Chen, H., Giri, N.C., Zhang, R. et al. (2010a). Nickel ions inhibit histone demethylase JMJD1A and DNA repair enzyme ABH2 by replacing the ferrous iron in the catalytic centers. *The Journal of Biological Chemistry 285*, 7374–7383.

Chen, H., Kluz, T., Zhang, R., and Costa, M. (2010b). Hypoxia and nickel inhibit histone demethylase JMJD1A and repress Spry2 expression in human bronchial epithelial BEAS-2B cells. *Carcinogenesis 31*, 2136–2144.

Chervona, Y., Arita, A., and Costa, M. (2012). Carcinogenic metals and the epigenome: Understanding the effect of nickel, arsenic, and chromium. *Metallomics: Integrated Biometal Science 4*, 619–627.

Cho, E.A., Song, H.K., Lee, S.H., Chung, B.H., Lim, H.M., and Lee, M.K. (2013). Differential in vitro and cellular effects of iron chelators for hypoxia inducible factor hydroxylases. *Journal of Cellular Biochemistry 114*, 864–873.

Crownover, B.K., and Covey, C.J. (2013). Hereditary hemochromatosis. *American Family Physician 87*, 183–190.

Culhane, J.C., and Cole, P.A. (2007). LSD1 and the chemistry of histone demethylation. *Current Opinion in Chemical Biology 11*, 561–568.

De Domenico, I., Lo, E., Ward, D.M., and Kaplan, J. (2009). Hepcidin-induced internalization of ferroportin requires binding and cooperative interaction with Jak2. *Proceedings of the National Academy of Sciences of the United States of America 106*, 3800–3805.

De Domenico, I., Ward, D.M., and Kaplan, J. (2011). Hepcidin and ferroportin: The new players in iron metabolism. *Seminars in Liver Disease 31*, 272–279.

De Domenico, I., Ward, D.M., Langelier, C. et al. (2007). The molecular mechanism of hepcidin-mediated ferroportin down-regulation. *Molecular Biology of the Cell 18*, 2569–2578.

Deaton, A.M., and Bird, A. (2011). CpG islands and the regulation of transcription. *Genes and Development 25*, 1010–1022.

Doyard, M., Fatih, N., Monnier, A. et al. (2012). Iron excess limits HHIPL-2 gene expression and decreases osteoblastic activity in human MG-63 cells. *Osteoporosis International: A Journal Established as Result of Cooperation between the European Foundation for Osteoporosis and the National Osteoporosis Foundation of the USA 23*, 2435–2445.

Ehrlich, M. (2009). DNA hypomethylation in cancer cells. *Epigenomics 1*, 239–259.

Ehrlich, M., and Wang, R.Y. (1981). 5-Methylcytosine in eukaryotic DNA. *Science 212*, 1350–1357.

Franchini, D.M., Schmitz, K.M., and Petersen-Mahrt, S.K. (2012). 5-Methylcytosine DNA demethylation: More than losing a methyl group. *Annual Review of Genetics 46*, 419–441.

Fu, Y., and He, C. (2012). Nucleic acid modifications with epigenetic significance. *Current Opinion in Chemical Biology 16*, 516–524.

Fuqua, B.K., Vulpe, C.D., and Anderson, G.J. (2012). Intestinal iron absorption. *Journal of Trace Elements in Medicine and Biology: Organ of the Society for Minerals and Trace Elements (GMS) 26*, 115–119.

Ganz, T. (2003). Hepcidin, a key regulator of iron metabolism and mediator of anemia of inflammation. *Blood 102*, 783–788.

Gerald, D., Berra, E., Frapart, Y.M. et al. (2004). JunD reduces tumor angiogenesis by protecting cells from oxidative stress. *Cell 118*, 781–794.

Goldberg, M.A., Gaut, C.C., and Bunn, H.F. (1991). Erythropoietin mRNA levels are governed by both the rate of gene transcription and posttranscriptional events. *Blood 77*, 271–277.

Greer, E.L., and Shi, Y. (2012). Histone methylation: A dynamic mark in health, disease and inheritance. *Nature Reviews. Genetics 13*, 343–357.

Haase, V.H. (2013). Regulation of erythropoiesis by hypoxia-inducible factors. *Blood Reviews 27*, 41–53.

Halwachs-Baumann, G. (2012). Diagnosis of anaemia: Old things rearranged. *Wiener Medizinische Wochenschrift 162*, 478–488.

Hanson, E.S., and Leibold, E.A. (1998). Regulation of iron regulatory protein 1 during hypoxia and hypoxia/reoxygenation. *The Journal of Biological Chemistry 273*, 7588–7593.

Hentze, M.W., Muckenthaler, M.U., and Andrews, N.C. (2004). Balancing acts: molecular control of mammalian iron metabolism. *Cell 117*, 285–297.

Hentze, M.W., Muckenthaler, M.U., Galy, B., and Camaschella, C. (2010). Two to tango: Regulation of Mammalian iron metabolism. *Cell 142*, 24–38.

Hickok, J.R., Vasudevan, D., Antholine, W.E., and Thomas, D.D. (2013). Nitric oxide modifies global histone methylation by inhibiting Jumonji C domain-containing demethylases. *The Journal of Biological Chemistry 288*, 16004–16015.

Holmqvist, P.H., and Mannervik, M. (2013). Genomic occupancy of the transcriptional co-activators p300 and CBP. *Transcription 4*, 18–23.

Huidobro, C., Fernandez, A.F., and Fraga, M.F. (2013). The role of genetics in the establishment and maintenance of the epigenome. *Cellular and Molecular Life Sciences: CMLS 70*, 1543–1573.

Jian, J., Pelle, E., and Huang, X. (2009). Iron and menopause: Does increased iron affect the health of postmenopausal women? *Antioxidants and Redox Signaling 11*, 2939–2943.

Kaczmarek, M., Cachau, R.E., Topol, I.A., Kasprzak, K.S., Ghio, A., and Salnikow, K. (2009). Metal ions-stimulated iron oxidation in hydroxylases facilitates stabilization of HIF-1 alpha protein. *Toxicological Sciences: An Official Journal of the Society of Toxicology 107*, 394–403.

Kaminsky, L.S., and Zhang, Q.Y. (2003). The small intestine as a xenobiotic-metabolizing organ. *Drug Metabolism and Disposition: The Biological Fate of Chemicals 31*, 1520–1525.

Ke, Q., and Costa, M. (2006). Hypoxia-inducible factor-1 (HIF-1). *Molecular Pharmacology 70*, 1469–1480.

Klose, R.J., Kallin, E.M., and Zhang, Y. (2006). JmjC-domain-containing proteins and histone demethylation. *Nature Reviews. Genetics 7*, 715–727.

Lewandowska, H., Kalinowska, M., Brzoska, K., Wojciuk, K., Wojciuk, G., and Kruszewski, M. (2011). Nitrosyl iron complexes—Synthesis, structure and biology. *Dalton Transactions: An International Journal of Inorganic Chemistry/RSoC 40*, 8273–8289.

Li, J., Harris, R.A., Cheung, S.W. et al. (2012). Genomic hypomethylation in the human germline associates with selective structural mutability in the human genome. *PLoS Genetics 8*, e1002692.

Li, Q., Chen, H., Huang, X., and Costa, M. (2006). Effects of 12 metal ions on iron regulatory protein 1 (IRP-1) and hypoxia-inducible factor-1 alpha (HIF-1alpha) and HIF-regulated genes. *Toxicology and Applied Pharmacology 213*, 245–255.

Li, T., Bonkovsky, H.L., and Guo, J.T. (2011). Structural analysis of heme proteins: Implications for design and prediction. *BMC Structural Biology 11*, 13.

Liu, Q., Davidoff, O., Niss, K., and Haase, V.H. (2012). Hypoxia-inducible factor regulates hepcidin via erythropoietin-induced erythropoiesis. *The Journal of Clinical Investigation 122*, 4635–4644.

Mole, D.R. (2010). Iron homeostasis and its interaction with prolyl hydroxylases. *Antioxidants and Redox Signaling 12*, 445–458.

Peyssonnaux, C., Nizet, V., and Johnson, R.S. (2008). Role of the hypoxia inducible factors HIF in iron metabolism. *Cell Cycle 7*, 28–32.

Robertson, K.D. (2002). DNA methylation and chromatin—Unraveling the tangled web. *Oncogene 21*, 5361–5379.

Rouault, T.A. (2006). The role of iron regulatory proteins in mammalian iron homeostasis and disease. *Nature Chemical Biology 2*, 406–414.

Saletta, F., Suryo Rahmanto, Y., Noulsri, E., and Richardson, D.R. (2010). Iron chelator-mediated alterations in gene expression: Identification of novel iron-regulated molecules that are molecular targets of hypoxia-inducible factor-1 alpha and p53. *Molecular Pharmacology 77*, 443–458.

Suva, M.L., Riggi, N., and Bernstein, B.E. (2013). Epigenetic reprogramming in cancer. *Science 339*, 1567–1570.

Tahiliani, M., Koh, K.P., Shen, Y. et al. (2009). Conversion of 5-methylcytosine to 5-hydroxymethylcytosine in mammalian DNA by MLL partner TET1. *Science 324*, 930–935.

Thompson, J.W., and Bruick, R.K. (2012). Protein degradation and iron homeostasis. *Biochimica et Biophysica Acta 1823*, 1484–1490.

Vergara, S.V., and Thiele, D.J. (2008). Post-transcriptional regulation of gene expression in response to iron deficiency: Co-ordinated metabolic reprogramming by yeast mRNA-binding proteins. *Biochemical Society Transactions 36*, 1088–1090.

Wang, G.L., and Semenza, G.L. (1993). General involvement of hypoxia-inducible factor 1 in transcriptional response to hypoxia. *Proceedings of the National Academy of Sciences of the United States of America 90*, 4304–4308.

Wang, J., Chen, G., Muckenthaler, M., Galy, B., Hentze, M.W., and Pantopoulos, K. (2004). Iron-mediated degradation of IRP2, an unexpected pathway involving a 2-oxoglutarate-dependent oxygenase activity. *Molecular and Cellular Biology 24*, 954–965.

Ward, D.M., and Kaplan, J. (2012). Ferroportin-mediated iron transport: Expression and regulation. *Biochimica et Biophysica Acta 1823*, 1426–1433.

Yi, C., Yang, C.G., and He, C. (2009). A non-heme iron-mediated chemical demethylation in DNA and RNA. *Accounts of Chemical Research 42*, 519–529.

Zhao, H., and Chen, T. (2013). Tet family of 5-methylcytosine dioxygenases in mammalian development. *Journal of Human Genetics 58*, 421–427.

10 Selenium and Epigenetic Effects on Histone Marks and DNA Methylation*

Wen-Hsing Cheng, Meltem Muftuoglu, and Ryan T. Y. Wu

CONTENTS

10.1 EARLY HISTORY: FROM SELENIUM TO SELENOPROTEINS

Selenium was named and identified by Jons Jacob Berzelius in 1817 during the study of sulfuric production [1]. In the first 125 years since the discovery, this element was considered toxic to farm animals or as a drug to shrink tumors. The proposed chronic selenium toxicity, known as "alkali disease" or "change hoof disease," observed in farm animals was first experimentally verified in 1934 by Kurt Franke, who observed that rats fed seleniferous plants grown on the lands rich in selenium

* Approved for publication as Book Chapter No. BC-12437 of the Mississippi Agricultural and Forestry Experimental Station, Mississippi State University.

in Northern Plains and Western states in the United States showed symptoms of selenium toxicity [2]. In medicine, bacteriologist August von Wassermann in 1911 discovered chemotherapeutic efficacy of selenite in the suppression of Ehrlich sarcoma in mice [3].

Meanwhile, beneficial effects of selenium in the diet were also noticed in the first half of the twentieth century. In 1941, W. E. Poley reported the first positive role of dietary selenium as chickens fed a diet supplemented with 2 ppm of selenium showed improved growth [4]. Furthermore, optimal dehydrogenase activity in the gut bacterium *Escherichia coli* requires minute amount of selenite [5]. In 1957, selenium was shown to prevent exudative diathesis in chickens [6,7] and liver necrosis in rats on a diet deficient in vitamin E and low in cysteine [8]. In 1958, selenium supplementation was demonstrated to prevent muscular dystrophy, also known as "white muscle disease" or "stiff lamb disease" [9,10]. Later on, selenium deficiency was confirmed as the cause of mulberry heart disease in pigs [11]. Although hints on selenium being an essential nutrient are provided as early as 1941 [4], nutrition scientists credit the discovery of selenium as a nutrient to Klaus Schwarz. In the seminal feeding study published in 1957 [8], liver necrosis in rats fed a basal diet was shown to be reversed by selenium supplementation. Selenium deficiency syndromes also exist in humans who live on the land where the soil contains low levels of this element. The first human selenium deficiency syndrome, Keshan cardiomyopathy, was not identified until 1979 by the Keshan Disease Research Group in China, followed by the discoveries of Kashin-Beck disease in 1988 [12,13] and myxoedematic cretinism in 1990 [14]. In the last two decades, genetic manipulations in mice and other species have revealed an array of metabolic and physiological functions of the selenocysteine-containing selenoproteins.

Since the recognition of selenium as a nutrient in 1957, a major research focus has been how functional roles of this element are performed. Although this question is now well addressed by knowing that selenium confers its physiological functions mainly through selenoproteins, decades of studies in various disciplines have enabled such conclusion to be made. In fact, Kurt Franke showed in 1934 based on his toxicity studies that selenium as a "toxicant" occurred in the protein fraction, although he did not propose or demonstrate the existence of selenium-containing proteins [15]. The first selenium-containing protein was experimentally discovered in 1972, when selenium was found to be the prosthetic group of glutathione peroxidase 1 (GPX1) based on the observation that the added ^{75}Se migrated to a protein fraction that was superimposed with GPX activity [16,17]. Unlike other peroxidases, GPX1 does not contain a heme group and GPX activity cannot be inhibited by azide or cyanide [18]. Interestingly, 1973 is a prolific year of discovering selenium-containing proteins. Two additional selenoproteins, "protein A" in the glycine reductase system and formate dehydrogenase, were identified in *Clostridium thermoaceticum* by employing ^{75}Se incorporation experiments [19,20]. The existence of selenocysteine in selenoproteins is first known in protein A in 1976 [21]. The mechanism of selenium incorporation into selenoproteins was elucidated by analyzing murine *Gpx1* sequence, from which selenocysteine was shown to be cotranslationally integrated into selenoproteins through UGA, a stop codon elsewhere [22]. The other two landmark discoveries on mammalian selenoprotein synthesis are the identification of the

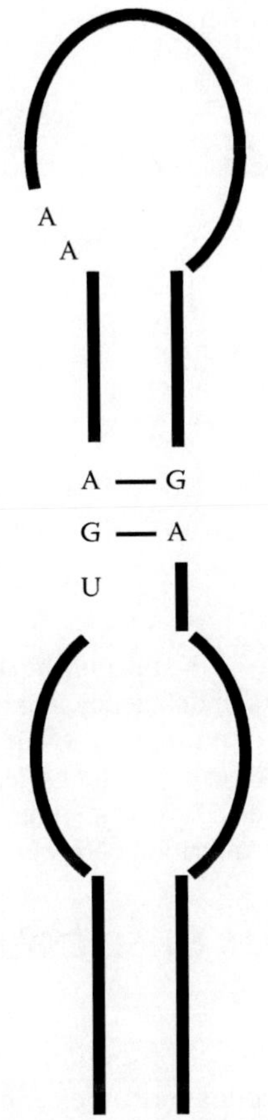

FIGURE 10.1 The stem-loop structure of mammalian SECIS element located in the 3′-UTR of all selenoprotein mRNAs. Since selenocysteine uses one of the UGA codons as the insertion signal, it is required to have SECIS (selenocysteine insertion sequence) element at the 3′-UTR to facilitate selenocysteine incorporation and to distinguish an in-frame from a stop UGA. There is a non–Watson-Crick base pair quartet with a conserved GA/AG tandem in the core of the hairpin structure of a SECIS element, which facilitates the binding of selenium binding protein 2 to insert a selenocysteine into selenoproteins.

selenocysteine tRNA and the selenocysteine insertion sequence (SECIS, Figure 10.1) that is located at the 3′-untranslated region (UTR) of a selenoprotein mRNA [23,24]. With these pieces of knowledge in mind, it is becoming possible to draw a broad picture of selenium incorporation into selenoproteins. In the process of selenoprotein translation, the SECIS stem loop structure serves as a platform to recruit proteins that interact with selenocysteine tRNA and stabilize the mRNA–protein–tRNA complex for selenocysteine incorporation by decoding the UGA codon. Detailed mechanism of selenoprotein incorporation can be found elsewhere [25].

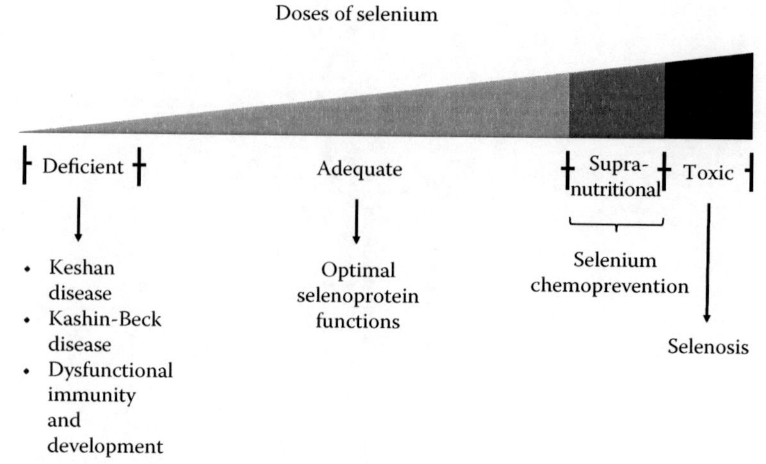

FIGURE 10.2 Current understanding on the relationship between the doses of selenium and the physiological impacts. Body selenium deficiency and in excess is highly correlated with the prevalence of selenium deficiency and toxicity syndromes, respectively. At nutritional levels, selenium in the body supports the expression of selenoproteins for optimal physiological functions. Selenium at supranutritional levels can induce mild stress, which is believed to offer some health benefits in a personal nutrition perspective.

10.2 PHYSIOLOGICAL ROLES OF SELENIUM AND SELENOPROTEINS

Like many other nutrients, the biological functions of selenium depend on the dosages (Figure 10.2). Selenium-deficient syndromes can be prevented by optimal selenoprotein functions at the nutritional levels of selenium. Selenium is believed to exert cancer chemoprevention functions at supranutritional levels, and can result in selenosis at toxic levels. Below the biological functions of selenoproteins at nutritional levels and selenium compounds at supranutritional level are discussed.

10.2.1 SELENOPROTEINS

The most important physiological role of selenium is believed to be exerted through selenoproteins. Body selenium deprivation suppresses the expression of selenoproteins to a various degree and in a tissue specific manner [26]. Computational sequence analyses of the entire human genome against SECIS element conclude that there is a total of 25 selenoproteins in humans [27]. The physiological and metabolic functions of selenoproteins have been unveiled largely by employing transgenic and knockout mouse models. Although some selenoproteins are dispensable, such as GPX1 [28], *Trsp*$^{-/-}$ mice with all the selenoproteins eliminated by null deletion of selenocysteine tRNA gene are embryonic lethal [29]. Thus, some of the selenoproteins have functional counterparts while others play distinct and essential roles. Selenoproteins are traditionally categorized as families of the glutathione peroxidase, the thioredoxin reductase, the deiodinase, and others that do not have established enzymatic activities or functions as a family. Table 10.1 provides an overview of the complete list of human selenoproteins and their major functions, if known. Functional and physiological roles of major selenoproteins are discussed below.

10.2.1.1 The Selenium-Dependent Glutathione Peroxidase Family

The selenium-dependent GPXs use glutathione as a reducing agent to decompose H_2O_2 and/or organic hydroperoxides [30]. Since 2003, it has been concluded that there are a total of five human selenium-dependent GPXs: GPX1-4 and GPX6 [27]. Although catalase is responsible for removing the majority of cellular H_2O_2, GPX1 is generally considered to be more physiological due to its relatively low K_M for H_2O_2. GPX1, previously known as GPX at the time of discovery, was first found to protect hemoglobin from oxidative breakdown in erythrocytes [31,32]. In February 1973, the

TABLE 10.1
Abbreviations and Functions of All Human Selenoproteins

Selenoproteins	Abbreviations/ Common Name	Enzymatic/Protein Functions
Selenium-Dependent Glutathione Peroxidase Family		
Cytosolic glutathione peroxidase	GPX1	Glutathione-dependent reduction of hydrogen peroxide and other peroxides
Gastrointestinal glutathione peroxidase	GPX2	
Plasma glutathione peroxidase	GPX3	
Phospholipid hydroperoxide glutathione peroxidase	GPX4	
Olfactory glutathine peroxidase	GPX6	Unknown
Selenoproteins Involved in Thioredoxin Signaling		
Thioredoxin reductase Type I	TrxR1	Recycling oxidized thioredoxin
Thioredoxin reductase Type II	TrxR2	
Thioredoxin reductase Type III	TrxR3	
Selenoproteins Involved in Thyroid Hormone Metabolism		
Deiodinase Type I	DIO1	T4 → T3 iodothyronine deiodination
Deiodinase Type II	DIO2	
Deiodinase III	DIO3	
Selenoproteins 15		
Selenoprotein 15	Sep 15	Proper protein folding
Selenoprotein M	Sel M	

(continued)

TABLE 10.1 (Continued)
Abbreviations and Functions of All Human Selenoproteins

Selenoproteins	Abbreviations/ Common Name	Enzymatic/Protein Functions
	Other Notable Selenoproteins	
Selenoprotein H	Sel H	Unknown
Selenoprotein I	Sel I	Phospholipid biosynthesis
Selenoprotein K	Sel K	Proper protein folding
Selenoprotein N	Sel N	Muscle
Selenoprotein O	Sel O	Unknown
Selenoprotein P	Sel P	Selenium storage and transportation
Selenoprotein R	Sel R, MsrB1	Methionine sulfoxide reductase
Selenoprotein S	Sel S	Endoplasmic reticulum stress regulation
Selenoprotein T	Sel T	Calcium mobilization
Selenoprotein V	Sel V	Unknown
Selenoprotein W	Sel W	Muscle
Selenoprotein synthetase	SPS2	Required for SectRNA$^{[Ser]Sec}$ biosynthesis

Hoekstra Laboratory reported the first case of selenium-dependent protein [16], which was accompanied by a publication from the Flohé Laboratory 3 months later [17]. The GPX family members display an expression pattern that is tissue- and organelle-specific. GPX1 is the most abundant selenoprotein and expresses in the body ubiquitously; however, the expression is the greatest in liver, kidney, blood, and placenta. The majority of GPX2 expresses in the gastrointestinal tract, followed by liver and mammary gland. The extracellular GPX3 resides mainly in plasma and is highly expressed in and secreted from epididymis and kidneys. Phospholipid hydroperoxide GPX (GPX4) expresses highly in liver and testis. GPX6 expression is rich in embryo and olfactory epithelium. In terms of cellular distribution, GPX1 locates mainly in cytosol while GPX4 has three isoforms resided in cytosol, mitochondria, and nucleus. Little is known about the cellular localization of GPX6. For a comprehensive review on glutathione peroxidases, see Brigelius-Flohé and Maiorino 2013 [33].

GPX1 plays a pivotal role in the mitigation of acute oxidative damage and accounts for 60% of total selenium in mouse liver [28]. While *Gpx1*$^{-/-}$ mice appear to develop normally without the development of age-related phenotypes except for cataract [34], they are hypersensitive to acute oxidative stress in a manner involving oxidation of NADH, NADPH, lipids and total proteins [35,36]. Considering the link between oxidative stress and genome instability, GPX1 in principle can prevent oxidative DNA damage and suppress tumorigenesis [37–39]. Indeed, genetic evidence indicates a positive association between GPX1 Pro198Leu polymorphism and colon cancer incidence [40] and the 198Leu GPX1 shows dramatic induction in cultured MCF7 breast cancer cells in the presence of selenium compounds at the nanomolar range [41]. Consistent with this notion, transactivation of GPX1 by p53, a tumor suppressor, results in a redox shift against oxidative stress [42]. Similarly, overexpression of GPX1 induces the expression of Gadd45 that reinforces cell cycle arrest and counteracts tumorigenesis [43]. GPX1 is also implicated in the protection against cardiovascular disease, neurodegeneration, and autoimmune disease [44]. Interestingly, both beneficial and adverse roles of GPX1 in glucose metabolism and diabetes have been reported. Mice overexpressing GPX1 develop insulin resistance and glucose intolerance [45], which is best explained by the scenario when ROS is overquenched to set a stage for reductive stress that hampers insulin signaling in the mice. Furthermore, *Gpx1*$^{-/-}$ mice on a high-fat diet are more resistant than wild type animals to develop insulin resistance and glucose intolerance [46]. However, β-cell specific overexpression of GPX1 in the leptin receptor defective *db/db* obesity mice can rescue the intrinsic β-cell dysfunction and diabetes in the 20-week-old mice [47,48]. Altogether, the physiological roles of GPX1 depend on the body health status, oxygen tone, and environmental factors.

Interestingly, GPX1 protein level is increased in the colon and ileum of *Gpx2*$^{-/-}$ mice [49], suggesting an adaptive response to oxidative stress through another GPX with similar enzymatic function. Although *Gpx1*$^{-/-}$ or *Gpx2*$^{-/-}$ mice are apparently normal, *Gpx1*$^{-/-}$*Gpx2*$^{-/-}$ double knockout mice display symptoms of inflammatory bowel disease and increased lipid peroxidation in the colon [50]. *Gpx3*$^{-/-}$ mice exhibit cerebral infarction, but this symptom can be alleviated by antioxidant treatment [51]. Different from GPX1–3, *Gpx4*$^{-/-}$ mice are embryonic lethal [52]. Various GPX4-inducible or conditional knockout mice have helped to get around this restraint. By applying these more advanced approaches, spermatocyte-specific or mitochondrial GPX4 knockout results in male infertility in mice [53,54]. Surprisingly, *Gpx4*$^{+/-}$ heterozygous mice show increased sensitivity to ROS-dependent apoptosis and live slightly longer than wild-type mice (1029 versus 963 days, $p < 0.05$) [55]. Results from forward genetics demonstrate that extra copies of GPX4 protect mice against liver apoptosis induced by *t*-butylhydroperoxide and diquat [56]. Furthermore, photoreceptor-specific *Gpx4*$^{-/-}$ mice display defective maturation of photoreceptor cells in conjunction with lipid peroxidation accumulation [57,58], and α-tocopherol supplementation is known to prevent neurodegeneration in neuron-specific *Gpx4*$^{-/-}$ mice [58]. These mouse models of GPX forward and reverse genetics have provided valuable physiological evidence and extended our understanding beyond the previously conceived roles of these selenoproteins merely as "antioxidant" enzymes.

10.2.1.2 Selenoproteins Involved in Thioredoxin (Trx) Signaling

In mammals, there are three thioredoxin reductases (TrxR) that play various cellular and physiological functions: the cytosolic and extracellular TrxR1 [59] and the mitochondrial TrxR2 and TrxR3 [60,61]. TrxR3 is also known as thioredoxin glutathione reductase. TrxRs are the members of the pyridine nucleotide–disulfide oxidoreductase family that recycles oxidized Trx in mammalian cells. The Trx system regulates the transactivation activity of NF-κB, AP-1, p53, and the glucocorticoid receptor, all of which contain cysteine residues in their DNA-binding domains [62]. *Txnrd1*$^{-/-}$ and *Txnrd2*$^{-/-}$ mice are embryonic lethal, express nonfunctional, oxidized ribonucleotide reductase and display impaired DNA synthesis [63], suggesting essential roles of TrxR1 and TrxR2 during embryonic development. To circumvent this limitation, many lines of conditional knockout mice have been generated and the roles in the pathogenesis of cancer and neurodegeneration have been shown. Liver-specific *Txnrd1*$^{-/-}$ mice are hypersensitive to diethylnitrosamine-induced hepatic carcinogenesis and show glutathione depletion and GPX2 upregulation [64], consistent with the observations that TrxR1 is overexpressed in malignant cells [65–67]. Therefore, TrxR1 is considered as a potential cancer therapeutic target [68]. Although neuron-specific *Txnrd1*$^{-/-}$ mice display symptoms of neurodegeneration [63], neuron-specific *Txnrd2*$^{-/-}$ and heart-specific *Txnrd1*$^{-/-}$ mice appear to develop normally [63,69]. Nonetheless, embryonic fibroblasts derived from *Txnrd2*$^{-/-}$ mice are hypersensitive to oxidative stress [70]. TrxR3 is specifically expressed in maturing spermatids and carries both a glutaredoxin domain and a thioredoxin motif [71]. *Txnrd3*$^{-/-}$ mice are not known to exist to date. Similar to GPX1 and GPX4, unexpected roles of TrxR1 exist. Liver-specific *Txnrd1*$^{-/-}$ mice are resistant to acetaminophen-induced acute toxicity [72]. Interestingly, mice deficient in selenoprotein M, a thioredoxin-like endoplasmic reticulum protein, are obese with elevated white muscle adipose tissue [73].

10.2.1.3 Selenoproteins Involved in Thyroid Hormone Metabolism

Iodothyronine deiodinases (DIOs) are critical enzymes that regulate thyroid hormone maturation through the conversions between thyroxine (T4), 3,5,3′-triiodothyronine (T3), and reverse triiodothyronine (rT3). There are three DIOs in this family. While DIO1 is mainly located in liver, kidney, and thyroid, DIO2 expresses highly in brain, pituitary, thyroid, skeletal muscle, and brown adipose tissue, and DIO3 is abundant in placenta and uterus during pregnancy, cerebral cortex, and skin [74]. T4, the inactive hormone secreted from thyroid gland, requires DIO1 and DIO2 to remove one iodine to form the active T3 in the body. Targeted disruption of DIO1 or DIO2 does not impact on serum T3 levels in the mice [75], but *Dio2*$^{-/-}$ mice retain higher T4 in the serum and show impaired hearing, thermogenesis, and neurocognition [76]. Recent results have indicated new roles of DIOs in metabolic and physiological regulation besides thyroid hormone maturation. Available lines of recent evidence by using *Dio3*$^{-/-}$ mice demonstrate physiological roles of DIOs in the control of thyroid hormone action during cerebellar development, heart function during the pathogenesis of restrictive cardiomyopathy, and impaired glucose-stimulated insulin secretion [77–79]. Furthermore, *Dio2*$^{-/-}$ mice are obese, show glucose intolerance and insulin resistance, and are susceptible to ventilator-induced lung injury. These results suggest that DIO1-3 carry both redundant and specific physiological functions [80–82].

Since thyroid hormones are involved in the signaling regulation of differentiation, proliferation and apoptosis, DIOs may also mediate carcinogenesis. DIO1 expression is decreased in papillary thyroid cancer or carcinoma, renal cell carcinoma, lung cancer, prostate cancer, and hepatic adenoma. DIO2 expression is elevated in follicular thyroid carcinoma, anaplastic or medullary thyroid cancer, astrocytoma, gliosarcoma, glioblastoma, and pituitary tumor [83]. Increased expression of DIO3 has been demonstrated to promote tumorigenesis, including liver hemangioma, gliomas, gliosarcoma, glioblastoma, and pituitary tumors [83,84]. Therefore, DIO1-3 may serve as potential targets for cancer diagnosis and therapy.

10.2.1.4 Selenoproteins Involved in Se Transportation and Storage

The role of selenoprotein P (Seppl), an extracellular selenoprotein containing 10 selenocysteines, in selenium transportation and store is becoming clear [85]. Mass spectrometric characterization of purified rat Seppl demonstrated four isoforms, the full-length and the C-terminal truncated ones [86]. The three shortened Seppl isoforms arise from alternatively using the second, third, and the seventh UGAs as stop signals, as UGA is a wobble codon for selenocysteine and translation termination. Seppl is mainly synthesized in the liver and secreted to extracellular fluids for usage by all organs [87]. In particular, Seppl is utilized by the testes and the brain through apolipoprotein E receptor-2 (apoER2) and by the kidneys through another Seppl receptor, Meglin [88–90]. Apparently, future studies are needed to understand the full list of Seppl receptors. Furthermore, *Sepp1*$^{-/-}$ rats show increased Se excretion through urine system [91]. In addition to the roles in selenium homeostasis, Seppl is also implicated in the development of spermatogenesis and the prevention of brain degeneration [92]. Interestingly, purified human Seppl has been shown to decompose phospholipid hydroperoxide but not hydrogen peroxide [93]. Another extracellular selenoprotein, GPX3, is not considered to play a major role in whole-body selenium homeostasis. This is because although GPX3 accounts for 97% of total plasma selenium together with Seppl, the majority of body GPX3 resides in the basement membranes of renal tubules [94].

10.2.1.5 Other Notable Selenoproteins

Sep15 and selenoprotein M are members of the Sep15 family that is believed to control the quality of protein folding in endoplasmic reticulum. Mice express relatively high levels of Sep15 in the liver, kidneys, testes, and prostate [74], whereas selenoprotein M is abundant in the brain. Sep15 cooperates with the chaperon protein, UDP-glucose:glycoprotein glucosyltransferase, to regulate protein folding [95] and may participate in disulfide bond modifications of unfolded and misfolded proteins [96]. Although the physiological function of Sep15 is still unknown, a recent report implicates Sep15 in the promotion of tumorigenesis and metastasis of colon cancer [97].

Selenoprotein H (SelH) is potentially a dual function protein involved in redox regulation and transactivation of phase II antioxidants [98], although the biological function of this nuclear selenoprotein is mainly unknown. SelH was initially identified as a thioredoxin reductase homologue with GPX-like activity [99]. The mRNA expression of SelH is abundant in mouse brain, thymus, testes, and uterus as well as in human cancers including the colorectal HCT116 and prostate LNCaP cells [99]. Overexpression of human SelH protects mouse HT22 neuronal cells from UVB irradiation through the suppression of superoxide production [100]. Moreover, SelH-overexpressed cells show elevated mitochondrial biogenesis and functions [101]. Analyses of SelH protein domains demonstrated an AT-hook domain, which can potentially bind DNA minor grooves and transactivate the target genes. This prediction has been verified using chromatin immunoprecipitation assays to demonstrate the appearance of GFP-tagged SelH on heat shock element and stress response element [98]. Upstream of SelH expression, metal transcription factor 1 has been shown to transactivate SelH by binding to its metal response element [102]. Therefore, SelH expression in principle can be upregulated by oxidative stress through heavy metals or other prooxidants, which in turn can upregulate an array of genes whose expression depends on SelH transactivation. Furthermore, GFP-tagged SelH is known to be localized to the nucleolus in rat 3T3 cells [99], suggesting a link between SelH and rDNA stability.

A thorough review of all selenoproteins is beyond the scope of this book chapter, but a couple of important new observations are discussed here. Mutations of selenoprotein N (SelN) result in a genetic disorder, rigid spine muscular dystrophy, through a possible mechanism by which loss of functional SelN promotes protein oxidation, calcium-handling abnormalities, and predisposition to oxidative stress [103,104]. Selenoprotein W (SelW) expresses mainly in the muscle, heart, spleen and brain of mammals and is proposed to play important roles in immune responses and thioredoxin-dependent pathways [105]. Selenoprotein R (SelR), also known as methionine sulfoxide reductase B (MsrB), is one of the enzymes in the Msr family. SelR expression is closely associated with the level of dietary selenium intake [26]. The Msr family prevents cells from the formation of protein–carbonyl adducts and has been implicated in delaying aging and age-related neurodegeneration [106–108]. Future studies on mechanistic and physiological investigation of all selenoproteins shall provide critical information toward the full understanding of selenium biology in optimal health.

10.2.2 Selenium Metabolites

Selenium is a nutrient from the soil. Amazingly, selenium content of the soil in the Western United States varies by 75,000-fold (0.02–1500 ppm [109]). This element can be incorporated into plants in either inorganic (such a selenide, selenite, and selenate) or organic forms (such as selenomethionine, selenocysteine, and methylseleninic acid). Early geographic studies indicate associations of selenium contents in the soil with the toxicity of domestic animals ranged on the land [110] and with cancer incidences in populations living on the land [111]. The best known function of selenium metabolites is the anticarcinogenic activities. The landmark Nutritional Prevention of Cancer (NPC) clinical trial strongly implicates selenium as an effective chemoprevention agent to reduce cancer risk of the colon, prostate, and lung [112]. Furthermore, a role for selenium in preventing cancer in patients with prostatic intraepithelial neoplasia has been inferred [113]. In cultured prostate cancer cells, selenite, and selenomethionine can induce an apoptotic response in a manner depending on the p53 protein and oxidative stress [114–116]. Consistent with these observations, animal studies have demonstrated that methylseleninic acid and selenomethionine can suppress tumorigenesis in the TRAMP mouse model of spontaneous prostate cancer or in nude mice inoculated with prostate cancer DU145 and PC-3 cells [117,118]. Similarly, we have found that dietary selenium at supranutritional levels (1 μg/kg diet in the form of sodium selenate) suppresses PC-3 cell tumorigenesis at the late stage in adult but not young nude mice [119]. Nonetheless, since the publication of the larger-scale Selenium and Vitamin E Cancer Prevention Clinical Trial (SELECT) in 2009, the selenium chemoprevention field has been fraught with confusion. The SELECT failed to demonstrate a role for selenomethionine in reducing prostate cancer risk, and was prematurely terminated in 2008 due to safety concerns and negative data for the formulations and doses given [120]. However, it is becoming clear in the last few years, based on careful and comprehensive reanalyses of the SELECT and other data [121], that the form of selenium and body selenium status are critical determinants for the efficacy of selenium chemoprevention. The seeming discrepancy between the two major selenium chemoprevention clinical trials, NPC and SELECT, conducted in the United States can be explained at least partially by forms of selenium compounds and body selenium status prior to entering the trials. While the NPC trial employed selenized yeast (selenomethionine, 65%–80%; other mainly unidentified selenium compounds, 20%–35%), the SELECT study used 100% selenomethionine. Clearly, selenium speciation other than selenomethionine in the selenized yeast confers selenium chemoprevention in the NPC study. Although the identity of the effective selenium species in chemoprevention clinically remains to be investigated, selenium compounds carrying a methyl group appear to offer superior chemoprevention efficacy [122]. In mice, methylseleninic acid is far more efficacious than selenomethionine in the suppression of chemically induced tumorigenesis [117]. With this in mind, a fair question to ask is whether one should take selenium supplements for cancer prevention. Because body selenium status is adequate in most people living in the United States and the safety zone for selenium above nutritional level is narrow, personal nutritional and health status must be taken into consideration regarding selenium supplementation. However, people living on selenium-poor lands such as part of China and Europe have relatively low body selenium status and should benefit from selenium supplementation. In conclusion, selenium is a very promising chemoprevention agent, but personal selenium status and nutritional genomics need to be placed into perspectives.

Metabolites of selenium compounds in excess can induce reactive oxygen species (ROS), leading to oxidative modifications and breaks on DNA. Furthermore, the cancer cell cycle could generate excessive amounts of single-stranded DNA and oxidatively modified DNA bases, resulting in a synergistic or synthetic effect on chemoprevention together with selenium supplementation. Previous studies focus on selenium-induced stress responses in cancer cells using toxic or high doses, results of which suggest that much of the selenium function against tumorigenesis is attributed to ROS-induced, p53-dependent apoptosis, or cell cycle arrest in cancer cells [123–125]. Recent results from our laboratory suggest that selenium compounds at doses ≤ LD_{50} can induce senescence and DNA damage responses, hallmarks of early tumorigenesis barriers, in normal but not in cancerous cells. This pathway depends on ROS, ataxia–telangiectasia mutated (ATM), p53, and the catalytic subunit of DNA-dependent protein kinase proteins [126–128]. Therefore, we propose that moderate amount of selenium supplementation can stifle tumorigenesis at the very early stage by the activation of early tumorigenesis barriers in individuals with normal or suboptimal selenium status. It is imperative to stress that personal body selenium status must be taken into consideration for future selenium chemoprevention clinical trials.

Meanwhile, body selenium status has to be assessed before an attempt to consume this nutrient for expecting optimal health. Furthermore, our unpublished results indicate that the selenium-induced, senescent noncancerous cells display features of senescence-associated heterochromatin, a phenomenon known to be mediated epigenetically. Detailed analyses on selenium metabolites and their physiological significance can be found elsewhere [129].

10.3 EPIGENETIC REGULATION BY SELENIUM

Like some other bioactive compounds, selenium has been proposed to function in cancer prevention and other pathophysiological responses through epigenetic regulation. DNA methylation and histone modifications are the major machineries for epigenetic regulation. As described below in detail, selenium and selenoproteins play important roles in tumor suppression and homeostasis, which can be directly and indirectly regulated by epigenetics.

10.3.1 Selenium and DNA Methylation

10.3.1.1 Effect of Selenium on DNA Methylation

Epigenetic regulation can control gene expression by direct methylation and demethylation of DNA bases, known as *cis*-epigenetics [130]. DNA methylation involves the addition of methyl groups by DNA methyltransferase (DNMT), resulting in the repression of gene expression. In mammals, five members of the DNMT protein family (DNMT1, DNMT2, DNMT3a, DNMT3b, and DNMT3L) have been discovered, of which only three are shown to possess methyltransferase activities (DNMT1, DNMT3a, DNMT3b). DNMT-1 is the most abundant form and is catalytically most active in dividing eukaryotic cells, while DNMT-3a and DNMT-3b are responsible for de novo methylation during embryonic development. Furthermore, DNMT1 is the key enzyme to mediate the transfer of a methyl group from *S*-adenosylmethionine (SAM) to cytosine residues at CpG dinucleotides (for a detailed review, see Li and Tollefsbol, 2010 [131]). The most frequently occurred epigenetic modification is CpG dinucleotide and/or CpG island methylation that exists in or near the promoter region of genes where transcription is initiated. In cancer cells, hypermethylation of CpG islands leads to abnormal silencing of the genes such as tumor suppressors, and global hypomethylation of genomic DNA promotes chromosome instability and translocation and gene disruption. Thus, dysregulation of the methylation processes may lead to negative impacts on health, including the development of cancer [132,133]. Several studies have demonstrated that bioactive food components such as folate, vitamin B_{12}, vitamin B_6, methionine, choline, and selenium can modulate both global and gene-specific DNA methylation, thus modulating gene expression and tumorigenesis. These components are involved in one-carbon metabolism and can directly affect the biochemical pathway of methylation processes and/or supply of methyl groups (reviewed by Davis and Uthus [134]). Several other bioactive food components, including genistein from soybeans, tea catechins, isothiocyanates, cadmium, and zinc, can also modulate DNA methylation by direct interactions with DNA methyltransferases (see detailed reviews by Choi and Friso [135] and Ho et al. [136]). For example, epigallocatechin-3-gallate, the major polyphenol compound isolated from green tea leaves, inhibits DNMT activity by binding to the protein, resulting in the reactivation of methylation silencing genes in cancer cells [137]. Furthermore, cadmium and zinc display their inhibitory effects by binding to the active center of DNMT [138]. Among the many bioactive dietary compounds, in this section, the effects of selenium on DNA methylation and cancer susceptibility are discussed.

Several in vitro studies have demonstrated that selenium at nutritional and supranutritional levels are effective in cancer prevention through impacting DNA methylation, thereby leading to altered expression profiles of certain key genes. For example, treatment of human colon carcinoma HCT116 cells with selenium metabolites such as phenylenebis-(methylene)-selenocyanate, sodium selenite, and benzyl selenocyanate results in inhibition of DNMT activity and the subsequent suppression of DNA methylation [139]. In addition, treatment of LNCaP prostate cancer cells with selenium compounds causes a decrease in the protein levels of DNMT1, resulting in demethylation in the promoter regions of π-glutathione *S*-transferase, the tumor suppressor adenomatous polyposis coli, and cellular stress response 1 genes [140]. Meanwhile, selenium-deficient rats are prone to carcinogen-induced aberrant crypts in rat colon [141,142]. Compared with a supranutritional selenium diet (2 ppm), rats fed a selenium-deficient diet show decreased aberrant crypts induced by dimethylhydrazine (25 mg/kg) [142,143]. In addition, selenium deficiency in Caco-2 cells decreases DNA methylation in the promoter region of the tumor suppressor gene, *p53* [143]. Considering that selenium at nutritional level is necessary to support selenoprotein expression and many selenoproteins display antioxidant functions, it is possible that oxidative stress induced by reduced expression of selenoprotein can induce DNA hypomethylation. Similarly, selenium in excess is known to induce the formation of ROS. Supranutritional selenium supplementation decreases total DNA methylation levels in the liver of rats [144]. However, the intestinal tumor model Apc/p21 mice fed a selenium-supplemented diet show no effect on methylation but marginal tumor inhibitory effect, whereas the combination of selenium and sulindac significantly inhibit intestinal tumorigenesis [145]. Altogether, it is likely that cancer predisposition induced by either selenium deficiency or in excess is associated with the suppression of DNMT activity and DNA methylation through induction of oxidative stress. This is consistent with the notion that global DNA hypomethylation is associated with carcinogenesis [146].

Body selenium status may also modulate DNA methylation through indirect interactions with folate metabolism pathways. It has been shown that dietary selenium can regulate aberrant crypt formation induced by folate deficiency in the rats. The mechanisms of actions are unknown, but dietary selenium can influence plasma homocysteine concentrations and the ratio between SAM and *S*-adenosylhomosysteine [142,147,148]. Plasma homocysteine and cysteine levels are significantly decreased in rats or mice fed a selenium-deficient diet as compared with those fed a selenium-adequate or a selenium-supranutritional diet. This is likely due to the upregulation of glutamate cysteine ligase, a rate limiting enzyme for the synthesis of glutathione. In addition to cancer, recent studies have begun to investigate the transgenerational effect of body selenium and folate status on global DNA methylation.

A recent example illustrates that postweanling supplementation with adequate levels of selenium and folate can alter global DNA methylation in the liver of offspring of female mice on high-fat diets deficient in these two nutrients during gestation and lactation. Further investigations focus on the different forms and doses of selenium and folate will shed light on the understanding of selenium and its interactions with other bioactive compounds in cancer-associated DNA methylation. Furthermore, the combinational effect of selenium with other bioactive compounds is proposed as a promising strategy for the prevention chronic diseases.

Furthermore, a cross-sectional study have assessed the methylation levels on age-related CpG islands in the rectal mucosa of healthy individuals and demonstrated that they are influenced by gender, the availability of folate, vitamin D, and selenium, and perhaps by factors related to systemic inflammation [149]. There is a weak inverse association between plasma selenium status and the methylation on the CpG islands of WIF1 gene in both genders. Based on most of the genes studied, this weak correlation was found predominantly in females. In contrast, selenium status is correlated positively with the methylation on LINE-1 promoter in females. This suggests the possibility of gender-dependent interactions among selenium, DNA methylation, and colorectal cancer risk. Further studies are needed to investigate the mechanisms of selenium-induced aberrant global DNA methylation and gene-specific methylation on CpG sites.

10.3.1.2 Effect of DNA Methylation on Selenoprotein Expression

Impressive lines of recent evidence have indicated that the expression of selenoproteins, in particular GPX1 and GPX3, can be suppressed by DNA methylation status in cancer cells. The epigenetic alterations of selenoproteins lead to a compromised antioxidant system, linking the expression of selenium-dependent GPX to chemoprevention. Several studies have reported that GPX3 transcription is silenced by promoter hypermethylation in prostate cancer cells, ovarian cancer cells, endometrial adenocarcinoma and gastric cancers, and in Barrett's disease [150–157]. Monoallelic hypermethylation and inactivation of GPX3 in benign precursor lesions, metaplasia, and dysplasia of the esophagus have been reported, while inactivation of both alleles occurs in invasive carcinoma [150]. Complete inactivation of the GPX3 gene by genetic loss of one allele and methylation-mediated silencing of the remaining allele has been reported in the etiology of prostate cancer [154]. GPX1 promoter methylation is also associated with GPX1 gene silencing in gastric cancer cell lines and breast cancer [158,159]. Taken together, the dysfunction of GPX1 and/or GPX3 in various cancer types is mediated by genetic and epigenetic alterations, suggesting impaired antioxidant mechanisms and the possible involvement in tumorigenesis and metastasis.

10.3.2 Selenium and Histone Modifications

10.3.2.1 Histone Modifications

In addition to DNA methylation, epigenetic regulation can control gene expression by proteins and RNAs that influence the accessibility of transcriptional machinery, known as *trans*-epigenetics [130]. Histone modifications can determine the accessibility of the associated DNA to transcriptional machinery and influence gene expression in various biological conditions including development and tumorigenesis. Histones are dynamically modified by various forms including methylation, acetylation, ubiquitination, and phosphorylation. Acetylation is one of the most studied histone modifications, and this process is enzymatically modulated by histone deacetylase (HDAC) and histone acetyltransferase (HAT). Recent studies have linked selenium to histone modifications. In rats developing colorectal carcinogenesis, dietary supplementation of green tea and selenium (1 ppm in the form of selenium-enriched milk protein) additively prevents the formation of aberrant crypt foci and induces histone H3 acetylation [160]. This event may be achieved through modulation of HDAC, as selenium derivation enhances the efficacy of suberoylanilide hydroxamic acid, an FDA-approved HDAC inhibitor, on the suppression of melanoma proliferation and lung cancer cells [161,162]. Similarly, methylseleninic acid has been shown to inhibit HDAC activity in B-cell lymphoma [163]. Nonetheless, although rats fed a high-fat, low-selenium, and low-folate diet results in a transgenerational effect on global DNA methylation in the liver, the same is not true on histone modifications [164]. Interestingly, it has been reported that glutamine transaminase K and L-amino acid oxidase can convert methylselenocysteine to α-keto acid metabolites, which in turn can inhibit HDAC activity in prostate cancer cells [165]. Similarly, selenite treatment inhibits HDAC activity, resulting in increased levels of acetylated lysine 9 on histone H3 [140]. The effect of selenoprotein on histone modifications is also reported. In a line of GPX1-overexpressing mice developing hyperinsulinemia, pancreatic islets in these mice show increased histone H3 and H4 acetylation in the promoter of *Pdx1* gene, a key transcription factor for β-cell differentiation, resulting in increased *Pdx1* expression [166]. However, what we have known on the epigenetic regulation of selenium and selenoproteins only represents the tip of the iceberg. It is unclear whether selenium compounds can impact on HAT activities and whether selenoproteins other than GPX1 can be influenced by histone modifications. It is also of strong future interest to study whether and how selenium affects other forms of histone modifications (methylation, phosphorylation, and ubiquitination), enzymes that modify or remove such marks, and their physiological significance.

10.3.2.2 DNA Damage Response

Upon DNA damage, the DNA damage response halts cell cycle from progressing into the next phase of the cell cycle to avoid the propagation with mutated DNA. One of the key DNA damage response kinase is ATM, which can be activated either by DNA breaks or oxidative stress [167,168]. Activated ATM phosphorylates various downstream substrates and initiates a cascade of signaling events. Interestingly, one of the ATM phosphorylation substrates is histone H2AX. Phosphorylated H2AX serves as a "mark" for sites of DNA damage and recruits DNA repair and checkpoint proteins to fix the lesion. DNA damage responses require the remodeling of local chromatin structure, thus offering an opportunity for selenium and selenoprotein to regulate gene expression epigenetically.

We have previously provided cellular evidence to suggest a new mechanism by which selenium can mitigate tumorigenesis through the activation of early tumorigenesis barriers [126]. In noncancerous but not in cancerous cells, selenium compounds (methylseleninic acids, methylselenocysteine, and selenite) can activate a senescence response through the ATM pathway activation, resulting in H2AX phosphorylation. This pathway depends on p53 [127], and requires DNA-PK$_{cs}$ that is positioned downstream of ATM to feed forward the selenium-induced ROS formation and maintain the senescence response [128]. Of note, this pathway of senescence induction is achievable by selenium doses as low as 0.1 μM in the form of selenite and methylseleninic acid. In cancer cells, we have shown that selenium compounds at lethal doses (10 μM selenite or methylseleninic acid) kill HCT116 colon cancer cells through the ATM-dependent DNA damage response in a manner depending on the mismatch repair protein MLH1 [169]. Furthermore, selenium-induced ATM pathway activation and H2AX phosphorylation can be sensitized in U-2OS osteosarcoma cells by underexpression of the Werner syndrome protein [170]. These results potentially provide a mechanistic basis for chemoprevention at upper range of nutritional or lower range of supranutritional levels of selenium, where selenium compounds may specifically induce senescence response at precancerous stage or tumor stem cells to stifle tumorigenesis. This can be tested using models for initiation, promotion and progression of carcinogenesis, clinical samples at precancerous stages such as polyps from the colon, and cancer cells derived from genome instability or premature aging syndromes.

10.4 CONCLUDING REMARKS

Significant progress has been made in recent years in advancing our understanding in how DNA methylation impacts on selenoprotein expression and is regulated by body selenium status. This is especially evident in the context of cancer. Surprisingly, little is known about the interactions between histone modifications and selenium status or selenoproteins. By interactions with the epigenetic processes, selenium shall profoundly modulate DNA damage response and repair, senescence, cell cycle arrest, cell proliferation, differentiation, and inflammation, among other processes, during the pathophysiology of chronic diseases. The reversible nature of epigenetic processes on DNA methylation and histone modifications serves as ideal targets for the actions of selenium compounds and selenoproteins in a timely manner for prevention or perhaps treatment of chronic diseases. This interesting approach exemplifies a new direction on the mechanistic study of selenium nutrigenomics. Considering the available evidence on epigenetic control by selenium and selenoproteins in cancer, it is promising and of great significance to elucidate the epigenetic marks in favor of or against other chronic diseases, and how these events are affected by selenium status and specific selenoproteins. Another line of interesting research is how and when noncoding RNA is expressed and affects selenoprotein expression.

Based on a cohort of selenium-adequate subjects (mean plasma selenium concentration as 87.6 ± 1.8 μg/L), body selenium status has been shown to be inversely associated with genomic DNA methylation [171]. This cross-sectional study is based on 287 Bangladesh adults with 86% of the participants considered selenium-adequate, with the cutoff value of plasma selenium concentration at 70 μg/L. Because global hypomethylation of genomic DNA promotes chromosome instability, it is possible that selenium supplementation at a critical, yet-to-be-defined range at the higher tertile of adequate level can promote mild oxidative stress for induction of DNA damage response that is at a tolerable level. It has been established that induction of DNA damage and senescence response are early barriers of tumorigenesis to hold carcinogenesis from entering the cancerous stage [172]. Considering that selenium compounds at doses $\leq LD_{50}$ can activate the ATM- and p53-dependent senescence response in noncancerous but not in cancerous cells [126,127], these in vitro data help to explain the mechanistic basis for the observed inverse association between body selenium status in selenium-adequate individuals and DNA methylation at the early stage tumorigenesis. Future studies are needed to study how selenium, selenium metabolites, and selenoproteins mediate DNA methylation and perhaps histone modifications on chemoprevention and other chronic diseases.

REFERENCES

1. Berzelius, J.J., and M. de Lettre, Berzelius a M Berthollet sur deux metaux nouveaux. *Ann Chim Phys*, 1818: pp. 199–202.
2. Franke, K.W., A new toxicant occurring naturally in certain samples of plant foodstuffs. I. Results obtained in preliminary feeding trials. *J Nutr*, 1934. 8: pp. 597–608.
3. von Wassermann, A.V., F. Keysser, and M. Wassermann, Beiträge zum Problem: Geschwülste von der Blubahn aus therapeutisch zu beeinflussen. *Dtsch Med Wschr*, 1911. 37: pp. 223–32.
4. Poley, W.E. et al., The effect of selenized grains on the rate of growth of chicks. *Poult Sci*, 1941. 20: pp. 171–9.
5. Pinsent, J., The need for selenite and molybdate in the formation of formic dehydrogenase by members of the coli-aerogenes group of bacteria. *Biochem J*, 1954. 57(1): pp. 10–6.
6. Patterson, E.L., R. Milstrey, and E.L. Stokstad, Effect of selenium in preventing exudative diathesis in chicks. *Proc Soc Exp Biol Med*, 1957. 95(4): pp. 617–20.
7. Schwarz, K. et al., Prevention of exudative diathesis in chicks by factor 3 and selenium. *Proc Soc Exp Biol Med*, 1957. 95(4): pp. 621–5.
8. Schwarz, K., and C.M. Foltz, Selenium as an integral part of factor 3 against dietary necrotic liver degeneration. *J Am Chem Soc*, 1957. 79: pp. 3292–3.
9. Muth, O.H. et al., Effects of selenium and vitamin E on white muscle disease. *Science*, 1958. 128(3331): p. 1090.
10. Hogue, D.E., Vitamin E, selenium and other factors related to nutritional muscular dystrophy in lambs. *Cornell Nutrition Conference for Feed Manufacturers Proceedings*, 1958: pp. 32–9.
11. Van Vleet, J.F., W. Carlton, and H.J. Olander, Hepatosis dietetica and mulberry heart disease associated with selenium deficiency in Indiana swine. *J Am Vet Med Assoc*, 1970. 157(9): pp. 1208–19.
12. Yang, F.Y. et al., Keshan disease—an endemic mitochondrial cardiomyopathy in China. *J Trace Elem Electrolytes Health Dis*, 1988. 2(3): pp. 157–63.
13. Moreno-Reyes, R. et al., Kashin-Beck osteoarthropathy in rural Tibet in relation to selenium and iodine status. *N Engl J Med*, 1998. 339(16): pp. 1112–20.
14. Vanderpas, J.B. et al., Iodine and selenium deficiency associated with cretinism in northern Zaire. *Am J Clin Nutr*, 1990. 52(6): pp. 1087–93.
15. Franke, K.W., A new toxicant occurring naturally in certain samples of plant foodstuffs. II. The occurrence of the toxicant in the protein fraction. *J Nutr*, 1934. 8: pp. 609–13.
16. Rotruck, J.T. et al., Selenium: Biochemical role as a component of glutathione peroxidase. *Science*, 1973. 179(4073): pp. 588–90.

17. Flohe, L., W.A. Gunzler, and H.H. Schock, Glutathione peroxidase: A selenoenzyme. *FEBS Lett*, 1973. 32(1): pp. 132–4.
18. Hoekstra, W.G., Biochemical function of selenium and its relation to vitamin E. *Fed Proc*, 1975. 34(11): pp. 2083–9.
19. Turner, D.C., and T.C. Stadtman, Purification of protein components of the clostridial glycine reductase system and characterization of protein A as a selenoprotein. *Arch Biochem Biophys*, 1973. 154(1): pp. 366–81.
20. Andreesen, J.R., and L.G. Ljungdahl, Formate dehydrogenase of Clostridium thermoaceticum: Incorporation of selenium-75, and the effects of selenite, molybdate, and tungstate on the enzyme. *J Bacteriol*, 1973. 116(2): pp. 867–73.
21. Cone, J.E. et al., Chemical characterization of the selenoprotein component of clostridial glycine reductase: Identification of selenocysteine as the organoselenium moiety. *Proc Natl Acad Sci U S A*, 1976. 73(8): pp. 2659–63.
22. Chambers, I. et al., The structure of the mouse glutathione peroxidase gene: The selenocysteine in the active site is encoded by the 'termination' codon, TGA. *EMBO J*, 1986. 5(6): pp. 1221–7.
23. Lee, B.J. et al., Identification of a selenocysteyl-tRNA(Ser) in mammalian cells that recognizes the nonsense codon, UGA. *J Biol Chem*, 1989. 264(17): pp. 9724–7.
24. Berry, M.J. et al., Recognition of UGA as a selenocysteine codon in type I deiodinase requires sequences in the 3′ untranslated region. *Nature*, 1991. 353(6341): pp. 273–6.
25. Allmang, C., L. Wurth, and A. Krol, The selenium to selenoprotein pathway in eukaryotes: More molecular partners than anticipated. *Biochim Biophys Acta*, 2009. 1790(11): pp. 1415–23.
26. Raines, A.M., and R.A. Sunde, Selenium toxicity but not deficient or super-nutritional selenium status vastly alters the transcriptome in rodents. *BMC Genomics*, 2011. 12: p. 26.
27. Kryukov, G.V. et al., Characterization of mammalian selenoproteomes. *Science*, 2003. 300(5624): pp. 1439–43.
28. Cheng, W.H. et al., Cellular glutathione peroxidase knockout mice express normal levels of selenium-dependent plasma and phospholipid hydroperoxide glutathione peroxidases in various tissues. *J Nutr*, 1997. 127(8): pp. 1445–50.
29. Bosl, M.R. et al., Early embryonic lethality caused by targeted disruption of the mouse selenocysteine tRNA gene (Trsp). *Proc Natl Acad Sci U S A*, 1997. 94(11): pp. 5531–4.
30. Cohen, G., and P. Hochstein, glutathione peroxidase: The primary agent for the elimination of hydrogen peroxide in erythrocytes. *Biochemistry*, 1963. 2: pp. 1420–8.
31. Mills, G.C., Hemoglobin catabolism. I. Glutathione peroxidase, an erythrocyte enzyme which protects hemoglobin from oxidative breakdown. *J Biol Chem*, 1957. 229(1): pp. 189–97.
32. Mills, G.C., The purification and properties of glutathione peroxidase of erythrocytes. *J Biol Chem*, 1959. 234(3): pp. 502–6.
33. Brigelius-Flohe, R., and M. Maiorino, Glutathione peroxidases. *Biochim Biophys Acta*, 2013. 1830(5): pp. 3289–303.
34. Ho, Y.S. et al., Mice deficient in cellular glutathione peroxidase develop normally and show no increased sensitivity to hyperoxia. *J Biol Chem*, 1997. 272(26): pp. 16644–51.
35. Cheng, W.H., B.A. Valentine, and X.G. Lei, High levels of dietary vitamin E do not replace cellular glutathione peroxidase in protecting mice from acute oxidative stress. *J Nutr*, 1999. 129(11): pp. 1951–7.
36. Cheng, W.H., G.F. Combs, Jr., and X.G. Lei, Knockout of cellular glutathione peroxidase affects selenium-dependent parameters similarly in mice fed adequate and excessive dietary selenium. *Biofactors*, 1998. 7(4): pp. 311–21.
37. Esposito, L.A. et al., Mitochondrial disease in mouse results in increased oxidative stress. *Proc Natl Acad Sci U S A*, 1999. 96(9): pp. 4820–5.

38. Sandstrom, B.E., and S.L. Marklund, Effects of variation in glutathione peroxidase activity on DNA damage and cell survival in human cells exposed to hydrogen peroxide and t-butyl hydroperoxide. *Biochem J*, 1990. 271(1): pp. 17–23.
39. Brigelius-Flohe, R., and A. Kipp, Glutathione peroxidases in different stages of carcinogenesis. *Biochim Biophys Acta*, 2009. 1790(11): pp. 1555–68.
40. Hu, Y. et al., Allelic loss of the gene for the GPX1 selenium-containing protein is a common event in cancer. *J Nutr*, 2005. 135(12 Suppl): pp. 3021S–4S.
41. Zhuo, P. et al., Molecular consequences of genetic variations in the glutathione peroxidase 1 selenoenzyme. *Cancer Res*, 2009. 69(20): pp. 8183–90.
42. Tan, M. et al., Transcriptional activation of the human glutathione peroxidase promoter by p53. *J Biol Chem*, 1999. 274(17): pp. 12061–6.
43. Nasr, M.A. et al., GPx-1 modulates Akt and P70S6K phosphorylation and Gadd45 levels in MCF-7 cells. *Free Radic Biol Med*, 2004. 37(2): pp. 187–95.
44. Lei, X.G., W.H. Cheng, and J.P. McClung, Metabolic regulation and function of glutathione peroxidase-1. *Annu Rev Nutr*, 2007. 27: pp. 41–61.
45. McClung, J.P. et al., Development of insulin resistance and obesity in mice overexpressing cellular glutathione peroxidase. *Proc Natl Acad Sci U S A*, 2004. 101(24): pp. 8852–7.
46. Loh, K. et al., Reactive oxygen species enhance insulin sensitivity. *Cell Metab*, 2009. 10(4): pp. 260–72.
47. Harmon, J.S. et al., beta-Cell-specific overexpression of glutathione peroxidase preserves intranuclear MafA and reverses diabetes in db/db mice. *Endocrinology*, 2009. 150(11): pp. 4855–62.
48. Guo, S. et al., Inactivation of specific beta cell transcription factors in type 2 diabetes. *J Clin Invest*, 2013. 123(8): pp. 3305–16.
49. Florian, S. et al., Loss of GPx2 increases apoptosis, mitosis, and GPx1 expression in the intestine of mice. *Free Radic Biol Med*, 2010. 49(11): pp. 1694–702.
50. Esworthy, R.S. et al., Mice with combined disruption of Gpx1 and Gpx2 genes have colitis. *Am J Physiol Gastrointest Liver Physiol*, 2001. 281(3): pp. G848–55.
51. Jin, R.C. et al., Glutathione peroxidase-3 deficiency promotes platelet-dependent thrombosis in vivo. *Circulation*, 2011. 123(18): pp. 1963–73.
52. Yant, L.J. et al., The selenoprotein GPX4 is essential for mouse development and protects from radiation and oxidative damage insults. *Free Radic Biol Med*, 2003. 34(4): pp. 496–502.
53. Imai, H. et al., Depletion of selenoprotein GPx4 in spermatocytes causes male infertility in mice. *J Biol Chem*, 2009. 284(47): pp. 32522–32.
54. Schneider, M. et al., Mitochondrial glutathione peroxidase 4 disruption causes male infertility. *FASEB J*, 2009. 23(9): pp. 3233–42.
55. Ran, Q. et al., Reduction in glutathione peroxidase 4 increases life span through increased sensitivity to apoptosis. *J Gerontol A Biol Sci Med Sci*, 2007. 62(9): pp. 932–42.
56. Ran, Q. et al., Transgenic mice overexpressing glutathione peroxidase 4 are protected against oxidative stress-induced apoptosis. *J Biol Chem*, 2004. 279(53): pp. 55137–46.
57. Ueta, T. et al., Glutathione peroxidase 4 is required for maturation of photoreceptor cells. *J Biol Chem*, 2012. 287(10): pp. 7675–82.
58. Seiler, A. et al., Glutathione peroxidase 4 senses and translates oxidative stress into 12/15-lipoxygenase dependent- and AIF-mediated cell death. *Cell Metab*, 2008. 8(3): pp. 237–48.
59. Tamura, T., and T.C. Stadtman, A new selenoprotein from human lung adenocarcinoma cells: Purification, properties, and thioredoxin reductase activity. *Proc Natl Acad Sci USA*, 1996. 93(3): pp. 1006–11.
60. Miranda-Vizuete, A. et al., Human mitochondrial thioredoxin reductase cDNA cloning, expression and genomic organization. *Eur J Biochem*, 1999. 261(2): pp. 405–12.

61. Sun, Q.A. et al., Redox regulation of cell signaling by selenocysteine in mammalian thioredoxin reductases. *J Biol Chem*, 1999. 274(35): pp. 24522–30.
62. Lu, J., and A. Holmgren, Selenoproteins. *J Biol Chem*, 2009. 284(2): pp. 723–7.
63. Jakupoglu, C. et al., Cytoplasmic thioredoxin reductase is essential for embryogenesis but dispensable for cardiac development. *Mol Cell Biol*, 2005. 25(5): pp. 1980–8.
64. Carlson, B.A. et al., Thioredoxin reductase 1 protects against chemically induced hepatocarcinogenesis via control of cellular redox homeostasis. *Carcinogenesis*, 2012. 33(9): pp. 1806–13.
65. Matsutani, Y. et al., Inverse correlation of thioredoxin expression with estrogen receptor- and p53-dependent tumor growth in breast cancer tissues. *Clin Cancer Res*, 2001. 7(11): pp. 3430–6.
66. Raffel, J. et al., Increased expression of thioredoxin-1 in human colorectal cancer is associated with decreased patient survival. *J Lab Clin Med*, 2003. 142(1): pp. 46–51.
67. Kakolyris, S. et al., Thioredoxin expression is associated with lymph node status and prognosis in early operable non-small cell lung cancer. *Clin Cancer Res*, 2001. 7(10): pp. 3087–91.
68. Klossowski, S. et al., Studies toward novel peptidomimetic inhibitors of thioredoxin-thioredoxin reductase system. *J Med Chem*, 2011. 55(1): pp. 55–67.
69. Soerensen, J. et al., The role of thioredoxin reductases in brain development. *PLoS One*, 2008. 3(3): p. e1813.
70. Conrad, M. et al., Essential role for mitochondrial thioredoxin reductase in hematopoiesis, heart development, and heart function. *Mol Cell Biol*, 2004. 24(21): pp. 9414–23.
71. Su, D. et al., Mammalian selenoprotein thioredoxin-glutathione reductase. Roles in disulfide bond formation and sperm maturation. *J Biol Chem*, 2005. 280(28): pp. 26491–8.
72. Patterson, A.D. et al., Disruption of thioredoxin reductase 1 protects mice from acute acetaminophen-induced hepatotoxicity through enhanced NRF2 activity. *Chem Res Toxicol*, 2013. 26(7): pp. 1088–96.
73. Pitts, M.W. et al., Deletion of selenoprotein M leads to obesity without cognitive deficits. *J Biol Chem*, 2013. 288(36): p. 26121–34.
74. Davis, C.D., P.A. Tsuji, and J.A. Milner, Selenoproteins and cancer prevention. *Annu Rev Nutr*, 2012. 32: pp. 73–95.
75. Galton, V.A. et al., Life without thyroxine to 3,5,3′-triiodothyronine conversion: Studies in mice devoid of the 5’-deiodinases. *Endocrinology*, 2009. 150(6): pp. 2957–63.
76. St Germain, D.L., V.A. Galton, and A. Hernandez, Minireview: Defining the roles of the iodothyronine deiodinases: Current concepts and challenges. *Endocrinology*, 2009. 150(3): pp. 1097–107.
77. Peeters, R.P. et al., Cerebellar abnormalities in mice lacking type 3 deiodinase and partial reversal of phenotype by deletion of thyroid hormone receptor alpha1. *Endocrinology*, 2013. 154(1): pp. 550–61.
78. Ueta, C.B. et al., Absence of myocardial thyroid hormone inactivating deiodinase results in restrictive cardiomyopathy in mice. *Mol Endocrinol*, 2012. 26(5): pp. 809–18.
79. Medina, M.C. et al., The thyroid hormone-inactivating type III deiodinase is expressed in mouse and human beta-cells and its targeted inactivation impairs insulin secretion. *Endocrinology*, 2011. 152(10): pp. 3717–27.
80. Castillo, M. et al., Disruption of thyroid hormone activation in type 2 deiodinase knockout mice causes obesity with glucose intolerance and liver steatosis only at thermoneutrality. *Diabetes*, 2011. 60(4): pp. 1082–9.
81. Marsili, A. et al., Mice with a targeted deletion of the type 2 deiodinase are insulin resistant and susceptible to diet induced obesity. *PLoS One*, 2011. 6(6): p. e20832.
82. Barca-Mayo, O. et al., Role of type 2 deiodinase in response to acute lung injury (ALI) in mice. *Proc Natl Acad Sci U S A*, 2011. 108(49): pp. E1321–9.

83. Piekielko-Witkowska, A., and A. Nauman, Iodothyronine deiodinases and cancer. *J Endocrinol Invest*, 2011. 34(9): pp. 716–28.
84. Casula, S., and A.C. Bianco, Thyroid hormone deiodinases and cancer. *Front Endocrinol (Lausanne)*, 2012. 3: p. 74.
85. Himeno, S., H.S. Chittum, and R.F. Burk, Isoforms of selenoprotein P in rat plasma. Evidence for a full-length form and another form that terminates at the second UGA in the open reading frame. *J Biol Chem*, 1996. 271(26): pp. 15769–75.
86. Burk, R.F., and K.E. Hill, Selenoprotein P: An extracellular protein with unique physical characteristics and a role in selenium homeostasis. *Annu Rev Nutr*, 2005. 25: pp. 215–35.
87. Burk, R.F., and K.E. Hill, Selenoprotein P. A selenium-rich extracellular glycoprotein. *J Nutr*, 1994. 124(10): pp. 1891–7.
88. Hill, K.E. et al., Deletion of selenoprotein P alters distribution of selenium in the mouse. *J Biol Chem*, 2003. 278(16): pp. 13640–6.
89. Schomburg, L. et al., Gene disruption discloses role of selenoprotein P in selenium delivery to target tissues. *Biochem J*, 2003. 370(Pt 2): pp. 397–402.
90. Olson, G.E. et al., Apolipoprotein E receptor-2 (ApoER2) mediates selenium uptake from selenoprotein P by the mouse testis. *J Biol Chem*, 2007. 282(16): pp. 12290–7.
91. Burk, R.F. et al., Deletion of selenoprotein P upregulates urinary selenium excretion and depresses whole-body selenium content. *Biochim Biophys Acta*, 2006. 1760(12): pp. 1789–93.
92. Burk, R.F., and K.E. Hill, Selenoprotein P-Expression, functions, and roles in mammals. *Biochim Biophys Acta*, 2009. 1790(11): pp. 1441–7.
93. Saito, Y. et al., Selenoprotein P in human plasma as an extracellular phospholipid hydroperoxide glutathione peroxidase. Isolation and enzymatic characterization of human selenoprotein p. *J Biol Chem*, 1999. 274(5): pp. 2866–71.
94. Olson, G.E. et al., Extracellular glutathione peroxidase (Gpx3) binds specifically to basement membranes of mouse renal cortex tubule cells. *Am J Physiol Renal Physiol*, 2010. 298(5): pp. F1244–53.
95. Korotkov, K.V. et al., Association between the 15-kDa selenoprotein and UDP-glucose:glycoprotein glucosyltransferase in the endoplasmic reticulum of mammalian cells. *J Biol Chem*, 2001. 276(18): pp. 15330–6.
96. Labunskyy, V.M., D.L. Hatfield, and V.N. Gladyshev, The Sep15 protein family: Roles in disulfide bond formation and quality control in the endoplasmic reticulum. *IUBMB Life*, 2007. 59(1): pp. 1–5.
97. Irons, R. et al., Deficiency in the 15-kDa selenoprotein inhibits tumorigenicity and metastasis of colon cancer cells. *Cancer Prev Res (Phila)*, 2010. 3(5): pp. 630–9.
98. Panee, J. et al., Selenoprotein H is a redox-sensing high mobility group family DNA-binding protein that up-regulates genes involved in glutathione synthesis and phase II detoxification. *J Biol Chem*, 2007. 282(33): pp. 23759–65.
99. Novoselov, S.V. et al., Selenoprotein H is a nucleolar thioredoxin-like protein with a unique expression pattern. *J Biol Chem*, 2007. 282(16): pp. 11960–8.
100. Ben Jilani, K.E. et al., Overexpression of selenoprotein H reduces Ht22 neuronal cell death after UVB irradiation by preventing superoxide formation. *Int J Biol Sci*, 2007. 3(4): pp. 198–204.
101. Mendelev, N. et al., Upregulation of human selenoprotein H in murine hippocampal neuronal cells promotes mitochondrial biogenesis and functional performance. *Mitochondrion*, 2011. 11(1): pp. 76–82.
102. Stoytcheva, Z.R. et al., Metal transcription factor-1 regulation via MREs in the transcribed regions of selenoprotein H and other metal-responsive genes. *Biochim Biophys Acta*, 2010. 1800(3): pp. 416–24.
103. Castets, P. et al., Selenoprotein N in skeletal muscle: From diseases to function. *J Mol Med (Berl)*, 2012. 90(10): pp. 1095–107.

104. Arbogast, S., and A. Ferreiro, Selenoproteins and protection against oxidative stress: Selenoprotein N as a novel player at the crossroads of redox signaling and calcium homeostasis. *Antioxid Redox Signal*, 2010. 12(7): pp. 893–904.
105. Whanger, P.D., Selenoprotein expression and function-Selenoprotein W. *Biochim Biophys Acta*, 2009. 1790(11): pp. 1448–52.
106. Novoselov, S.V. et al., Regulation of selenoproteins and methionine sulfoxide reductases A and B1 by age, calorie restriction, and dietary selenium in mice. *Antioxid Redox Signal*, 2010. 12(7): pp. 829–38.
107. Moskovitz, J., and D.B. Oien, Protein carbonyl and the methionine sulfoxide reductase system. *Antioxid Redox Signal*, 2010. 12(3): pp. 405–15.
108. Oien, D.B., and J. Moskovitz, Selenium and the methionine sulfoxide reductase system. *Molecules*, 2009. 14(7): pp. 2337–44.
109. Lakin, H.W., Vertical and lateral distribution of selenium in sedimentary rocks of Western United States. In *Selenium in Agriculture*, U.S. Department of Agriculture, Washington, DC, 1961. Agriculture Handbook No. 200: pp. 12–24.
110. Moxon, A.L., and O.E. Olson, Selenium content of plants and soils. *South Dakota Agric Exp Sta Ann Rep*, 1940. 53: p. 15.
111. Shamberger, R.J., and D.V. Frost, Possible protective effect of selenium against human cancer. *Can Med Assoc J*, 1969. 100(14): p. 682.
112. Clark, L.C. et al., Effects of selenium supplementation for cancer prevention in patients with carcinoma of the skin. A randomized controlled trial. Nutritional Prevention of Cancer Study Group. *JAMA*, 1996. 276(24): pp. 1957–63.
113. Marshall, J.R. et al., Design and progress of a trial of selenium to prevent prostate cancer among men with high-grade prostatic intraepithelial neoplasia. *Cancer Epidemiol Biomarkers Prev*, 2006. 15(8): pp. 1479–84.
114. Zhao, R. et al., Effects of selenite and genistein on G2/M cell cycle arrest and apoptosis in human prostate cancer cells. *Nutr Cancer*, 2009. 61(3): pp. 397–407.
115. Kulimova, E. et al., Growth inhibition and induction of apoptosis in acute myeloid leukemia cells by new indolinone derivatives targeting fibroblast growth factor, platelet-derived growth factor, and vascular endothelial growth factor receptors. *Mol Cancer Ther*, 2006. 5(12): pp. 3105–12.
116. Zhao, R. et al., Expression of p53 enhances selenite-induced superoxide production and apoptosis in human prostate cancer cells. *Cancer Res*, 2006. 66(4): pp. 2296–304.
117. Li, G.X. et al., Superior *in vivo* inhibitory efficacy of methylseleninic acid against human prostate cancer over selenomethionine or selenite. *Carcinogenesis*, 2008. 29(5): pp. 1005–12.
118. Wang, L. et al., Methyl-selenium compounds inhibit prostate carcinogenesis in the transgenic adenocarcinoma of mouse prostate model with survival benefit. *Cancer Prev Res (Phila)*, 2009. 2(5): pp. 484–95.
119. Holmstrom, A. et al., Nutritional and supranutritional levels of selenate differentially suppress prostate tumor growth in adult but not young nude mice. *J Nutr Biochem*, 2012. 23(9): pp. 1086–91.
120. Lippman, S.M. et al., Effect of selenium and vitamin E on risk of prostate cancer and other cancers: The Selenium and Vitamin E Cancer Prevention Trial (SELECT). *JAMA*, 2009. 301(1): pp. 39–51.
121. Klein, E.A. et al., Vitamin E and the risk of prostate cancer: The Selenium and Vitamin E Cancer Prevention Trial (SELECT). *JAMA*, 2011. 306(14): pp. 1549–56.
122. Larsen, E.H. et al., Speciation and bioavailability of selenium in yeast-based intervention agents used in cancer chemoprevention studies. *J AOAC Int*, 2004. 87(1): pp. 225–32.
123. Li, S. et al., Selenium sensitizes MCF-7 breast cancer cells to doxorubicin-induced apoptosis through modulation of phospho-Akt and its downstream substrates. *Mol Cancer Ther*, 2007. 6(3): pp. 1031–8.

124. Spallholz, J.E., B.J. Shriver, and T.W. Reid, Dimethyldiselenide and methylseleninic acid generate superoxide in an *in vitro* chemiluminescence assay in the presence of glutathione: Implications for the anticarcinogenic activity of L-selenomethionine and L-Se-methylselenocysteine. *Nutr Cancer*, 2001. 40(1): pp. 34–41.
125. Stewart, M.S. et al., Selenium compounds have disparate abilities to impose oxidative stress and induce apoptosis. *Free Radic Biol Med*, 1999. 26(1–2): pp. 42–8.
126. Wu, M. et al., Selenium compounds activate early barriers of tumorigenesis. *J Biol Chem*, 2010. 285(16): pp. 12055–62.
127. Wu, M. et al., Role for p53 in Selenium-Induced Senescence. *J Agric Food Chem*, 2011. 59(21): pp. 11882–7.
128. Rocourt, C.R. et al., The catalytic subunit of DNA-dependent protein kinase is downstream of ATM and feeds forward oxidative stress in the selenium-induced senescence response. *J Nutr Biochem*, 2013. 24(5): pp. 781–7.
129. Zeng, H., and G.F. Combs, Jr., Selenium as an anticancer nutrient: Roles in cell proliferation and tumor cell invasion. *J Nutr Biochem*, 2008. 19(1): pp. 1–7.
130. Bonasio, R., S. Tu, and D. Reinberg, Molecular signals of epigenetic states. *Science*, 2010. 330(6004): pp. 612–6.
131. Li, Y., and T.O. Tollefsbol, Impact on DNA methylation in cancer prevention and therapy by bioactive dietary components. *Curr Med Chem*, 2010. 17(20): pp. 2141–51.
132. Barrera, L.N. et al., Epigenetic and antioxidant effects of dietary isothiocyanates and selenium: Potential implications for cancer chemoprevention. *Proc Nutr Soc*, 2012. 71(2): pp. 237–45.
133. Lu, Q. et al., Epigenetics, disease, and therapeutic interventions. *Ageing Res Rev*, 2006. 5(4): pp. 449–67.
134. Davis, C.D., and E.O. Uthus, DNA methylation, cancer susceptibility, and nutrient interactions. *Exp Biol Med (Maywood)*, 2004. 229(10): pp. 988–95.
135. Choi, S.W., and S. Friso, Epigenetics: A new bridge between nutrition and health. *Adv Nutr*, 2010. 1(1): pp. 8–16.
136. Ho, E. et al., Dietary factors and epigenetic regulation for prostate cancer prevention. *Adv Nutr*, 2011. 2(6): pp. 497–510.
137. Fang, M.Z. et al., Tea polyphenol (-)-epigallocatechin-3-gallate inhibits DNA methyltransferase and reactivates methylation-silenced genes in cancer cell lines. *Cancer Res*, 2003. 63(22): pp. 7563–70.
138. Poirier, L.A., and T.I. Vlasova, The prospective role of abnormal methyl metabolism in cadmium toxicity. *Environ Health Perspect*, 2002. 110 Suppl 5: pp. 793–5.
139. Fiala, E.S. et al., Inhibition of DNA cytosine methyltransferase by chemopreventive selenium compounds, determined by an improved assay for DNA cytosine methyltransferase and DNA cytosine methylation. *Carcinogenesis*, 1998. 19(4): pp. 597–604.
140. Xiang, N. et al., Selenite reactivates silenced genes by modifying DNA methylation and histones in prostate cancer cells. *Carcinogenesis*, 2008. 29(11): pp. 2175–81.
141. Feng, Y. et al., Dietary selenium reduces the formation of aberrant crypts in rats administered 3,2′-dimethyl-4-aminobiphenyl. *Toxicol Appl Pharmacol*, 1999. 157(1): pp. 36–42.
142. Davis, C.D., and E.O. Uthus, Dietary selenite and azadeoxycytidine treatments affect dimethylhydrazine-induced aberrant crypt formation in rat colon and DNA methylation in HT-29 cells. *J Nutr*, 2002. 132(2): pp. 292–7.
143. Davis, C.D., E.O. Uthus, and J.W. Finley, Dietary selenium and arsenic affect DNA methylation *in vitro* in Caco-2 cells and *in vivo* in rat liver and colon. *J Nutr*, 2000. 130(12): pp. 2903–9.
144. Zeng, H. et al., Dietary selenomethionine increases exon-specific DNA methylation of the p53 gene in rat liver and colon mucosa. *J Nutr*, 2011. 141(8): pp. 1464–8.
145. Bi, X. et al., Selenium and sulindac are synergistic to inhibit intestinal tumorigenesis in Apc/p21 mice. *J Hematol Oncol*, 2013. 6: p. 8.

146. Ehrlich, M., DNA hypomethylation in cancer cells. *Epigenomics*, 2009. 1(2): pp. 239–59.
147. Davis, C.D., and E.O. Uthus, Dietary folate and selenium affect dimethylhydrazine-induced aberrant crypt formation, global DNA methylation and one-carbon metabolism in rats. *J Nutr*, 2003. 133(9): pp. 2907–14.
148. Uthus, E.O., and S.A. Ross, Dietary selenium affects homocysteine metabolism differently in Fisher-344 rats and CD-1 mice. *J Nutr*, 2007. 137(5): pp. 1132–6.
149. Tapp, H.S. et al., Nutritional factors and gender influence age-related DNA methylation in the human rectal mucosa. *Aging Cell*, 2013. 12(1): pp. 148–55.
150. Lee, O.J. et al., Hypermethylation and loss of expression of glutathione peroxidase-3 in Barrett's tumorigenesis. *Neoplasia*, 2005. 7(9): pp. 854–61.
151. Jee, C.D. et al., Identification of genes epigenetically silenced by CpG methylation in human gastric carcinoma. *Eur J Cancer*, 2009. 45(7): pp. 1282–93.
152. Zhang, X. et al., An 8-gene signature, including methylated and down-regulated glutathione peroxidase 3, of gastric cancer. *Int J Oncol*, 2010. 36(2): pp. 405–14.
153. Lodygin, D. et al., Functional epigenomics identifies genes frequently silenced in prostate cancer. *Cancer Res*, 2005. 65(10): pp. 4218–27.
154. Yu, Y.P. et al., Glutathione peroxidase 3, deleted or methylated in prostate cancer, suppresses prostate cancer growth and metastasis. *Cancer Res*, 2007. 67(17): pp. 8043–50.
155. Saga, Y. et al., Glutathione peroxidase 3 is a candidate mechanism of anticancer drug resistance of ovarian clear cell adenocarcinoma. *Oncol Rep*, 2008. 20(6): pp. 1299–303.
156. Falck, E. et al., Loss of glutathione peroxidase 3 expression is correlated with epigenetic mechanisms in endometrial adenocarcinoma. *Cancer Cell Int*, 2010. 10: p. 46.
157. Peng, D.F. et al., Silencing of glutathione peroxidase 3 through DNA hypermethylation is associated with lymph node metastasis in gastric carcinomas. *PLoS One*, 2012. 7(10): p. e46214.
158. Min, S.Y. et al., Prognostic significance of glutathione peroxidase 1 (GPX1) down-regulation and correlation with aberrant promoter methylation in human gastric cancer. *Anticancer Res*, 2012. 32(8): pp. 3169–75.
159. Kulak, M.V. et al., Transcriptional regulation of the GPX1 gene by TFAP2C and aberrant CpG methylation in human breast cancer. *Oncogene*, 2013. 32(34): pp. 4043–51.
160. Hu, Y. et al., Combination of selenium and green tea improves the efficacy of chemoprevention in a rat colorectal cancer model by modulating genetic and epigenetic biomarkers. *PLoS One*, 2013. 8(5): p. e64362.
161. Karelia, N. et al., Selenium-containing analogs of SAHA induce cytotoxicity in lung cancer cells. *Bioorg Med Chem Lett*, 2010. 20(22): pp. 6816–9.
162. Gowda, R. et al., Selenium-containing histone deacetylase inhibitors for melanoma management. *Cancer Biol Ther*, 2012. 13(9): pp. 756–65.
163. Kassam, S. et al., Methylseleninic acid antagonizes the cytotoxic effect of bortezomib in mantle cell lymphoma cell lines through modulation of Bcl-2 family proteins. *Br J Haematol*, 2011. 156(2): pp. 286–9.
164. Bermingham, E.N. et al., Post-weaning selenium and folate supplementation affects gene and protein expression and global DNA methylation in mice fed high-fat diets. *BMC Med Genomics*, 2013. 6: pp. 7.
165. Lee, J.I. et al., Alpha-keto acid metabolites of naturally occurring organoselenium compounds as inhibitors of histone deacetylase in human prostate cancer cells. *Cancer Prev Res (Phila)*, 2009. 2(7): pp. 683–93.
166. Wang, X.D. et al., Molecular mechanisms for hyperinsulinaemia induced by overproduction of selenium-dependent glutathione peroxidase-1 in mice. *Diabetologia*, 2008. 51(8): pp. 1515–24.
167. Paull, T.T., and J.H. Lee, The Mre11/Rad50/Nbs1 complex and its role as a DNA double-strand break sensor for ATM. *Cell Cycle*, 2005. 4(6): pp. 737–40.
168. Guo, Z. et al., ATM activation by oxidative stress. *Science*, 2010. 330(6003): p. 517–21.

169. Qi, Y. et al., Selenium compounds activate ATM-dependent DNA damage response via the mismatch repair protein hMLH1 in colorectal cancer cells. *J Biol Chem*, 2010. 285(43): pp. 33010–7.
170. Cheng, W.H. et al., Targeting Werner syndrome protein sensitizes U-2 OS osteosarcoma cells to selenium-induced DNA damage response and necrotic death. *Biochem Biophys Res Commun*, 2012. 420(1): pp. 24–8.
171. Pilsner, J.R. et al., Associations of plasma selenium with arsenic and genomic methylation of leukocyte DNA in Bangladesh. *Environ Health Perspect*, 2011. 119(1): pp. 113–8.
172. Bartkova, J. et al., Oncogene-induced senescence is part of the tumorigenesis barrier imposed by DNA damage checkpoints. *Nature*, 2006. 444(7119): pp. 633–7.

11 Proline
Metabolic Sensing and Parametabolic Regulation

James M. Phang, Wei Liu, Chad Hancock, and Meredith Harman

CONTENTS

11.1 INTRODUCTION

Dysregulated metabolism has emerged as a critical mechanism in carcinogenesis. Reprogramming of metabolism contributes to tumor requirements for energetics, redox homeostasis and substrates for cell mass. Previous emphasis has centered on glucose metabolism, but recent discoveries have focused on the metabolism of nonessential amino acids (NEAA), e.g., glutamine, serine, aspartic acid, proline. The contributions of NEAA to tumor metabolism may be conceptualized as "parametabolic regulation." In this context, proline, the only secondary proteinogenic amino (imino) acid has a special function. The proline metabolic system and the enzymes catalyzing synthesis and degradation are controlled by regulatory factors important in cancer, e.g., p53, PPAR-γ, AMPK, c-Myc, miRs, and which mediate redox signaling mechanisms for apoptosis, autophagy, redox homeostasis, and metabolic epigenetics.

11.1.1 Genetics and Cancer

The advances in human genetics during the last 30 years, and the exponential growth in genomics and informatics, have served to focus research in cancer.[1,2] The discovery of cancer genes based on Knudsen's "two-hit" hypothesis for Rb[3] as well as the identification of specific mutations for oncogenes and suppressor genes[4] formed the strategy for replacement or repair of the "driver genes" for cancer. It was soon evident, however, that the difficulties were overwhelming. Not only were there technical obstacles, but also the cancer genome, itself, presented formidable challenges.[2,5] Each cancer has its own complement of mutations with redundancies. Furthermore, the expression of these mutated genes may not only depend on the primary coding and noncoding sequences, but, in fact, may be regulated by epigenetic mechanisms that depend on the enzymatic modulation of both DNA and histone proteins in eukaryotic chromatin.[6,7]

The enzymatic mechanisms for epigenetic regulation are strongly linked to metabolism, and examples of this in cancer development and progression have been described.[8,9] It is increasingly recognized that these epigenetic mechanisms may be sensors for cellular metabolism such that normal gene regulation as well as the regulation or dysregulation of mutated genes in cancer may be sensitive to alterations in metabolism.[8,9] This concept has become the basis for new therapeutic strategies.

11.1.2 Signaling and Cancer

The attempt to dissect the signaling networks dysregulated in cancer has received considerable emphasis due to the early discovery that several oncogenes are mutated receptors, or autonomously activated tyrosine kinases. Direct inhibition of these kinases is an important target for successfully treating cancer.[10] The pursuit of the signaling network activated by these receptors showed not only redundancy of pathways but also cross-talk[11] that allow proliferation signals to bypass therapeutic inhibition. Although many of these signaling cascades terminate in transcriptional factors, providing escape from cell cycle check points,[2] many of them regulate metabolic pathways.[12,13] Some signaling cascades alter metabolic enzymes thereby reprogramming metabolism to mediate the substrates for increasing cell mass.[14] Of course, these signaling pathways also alter gene expression by altering metabolism-dependent epigenetics.

11.2 METABOLISM AND EPIGENETICS

Twenty years ago, before the enzymatic mechanisms of epigenetics were identified and their link with metabolic processes established, investigators showed that the malignant phenotype could be altered by nutritional manipulation.[15]

P53-knockout mice form rapidly growing tumors, comprised predominantly of lymphomas, such that all animals were euthanized by 37 weeks. However, if these animals were on a 40% calorie-restricted diet, their tumor formation was delayed. The median survival of mice was extended from 16 weeks for *ad libitum* fed mice to 25 weeks for calorie-restricted mice. This important finding showed that factors other than genotype are involved in the genetically determined course of carcinogenesis.[15]

11.2.1 DNA and Histone-Modifying Enzymes in Epigenetics

Over the last 15 years, the enzyme mechanisms altering the DNA/histone interaction in eukaryotic chromatin have been elucidated. Methylation of cytosines in DNA was defined, and gene expression occurred in hypomethylated regions, whereas hypermethylated regions were usually silent.[6,16] Histones were also shown to be methylated on specific lysines near the N-terminus on the outer surface of the chromosomes. Clearly, the expression of either normal or cancer mutated genes can be modulated by these two mechanisms. Methylation is mediated by DNA methyltransferases, which attach a methyl group onto a cytosine using *S*-adenosylmethionine. Demethylation is mediated by families of enzymes that hydroxylate the methyl group on cytosines prior to their removal. This activity is mediated by α-ketoglutarate (α-KG)-dependent dioxygenases. Another histone modification is acetylation mediated by lysine acetyltransferases with acetyl co-A as the cofactor.[17] Deacetylation by the NAD^+-dependent sirtuin family of enzymes transfers the acetyl group to NAD^+, forming 2′-*O*-acetyl-ADP-ribose and nicotinamide.

11.3 REPROGRAMMING OF TUMOR METABOLISM

11.3.1 Energetics

The initial interpretation by Warburg of tumor metabolism focused on altered energetics.[18] Glucose metabolism was dichotomized into oxidative and nonoxidative metabolism. Glycolysis, the conversion of glucose-6-phosphate to pyruvate, is the source of ATP during states of relative oxygen debt, e.g., during exercise. This inefficient but rapid pathway yields 2 ATP for each glucose-6-phosphate. Meanwhile, the conversion of pyruvate to acetyl-CoA and its complete oxidation to CO_2 in the tricarboxylic acid cycle, and the subsequent transfer of electrons into the electron transport chain generates 32 ATP per glucose-6-phosphate. This apparent paradox has been explained by various investigators as a sacrifice of metabolic efficiency in favor of production of cellular building blocks.[14,19]

11.3.2 Accumulation of Substrates for Cell Mass

Tumor cells are reprogrammed to function at a much lower level of energy efficiency via a pathway that recycles carbons rather than one that terminates in their excretion as CO_2. This conservation of carbons is critical for proliferating cells in need of carbon compounds as building blocks for cell mass.[14] Although this general principle has been widely accepted, the specific network of pathways mediating this rerouting of carbons differs among tumors and also may be differentially driven by various oncogenes.

11.3.3 Redox Homeostasis

In addition to rerouting of carbons into biosynthetic pathways, another important end point of tumor reprogramming is the maintenance of redox homeostasis.[20,21] The generation of NADPH by the oxidative arm of the pentose phosphate pathway (PPP) is a mechanism for redox protection through the glutathione-mediated mechanism, Some studies suggest that relatively little flux passes through the oxidative arm of the PPP to generate NADPH. It may be that high glucose concentrations (5× physiological) used in tissue culture inhibits glucose-6-phosphate dehydrogenase, yielding results biased by in vitro conditions.[22] However, alternative sources have been suggested by others who have shown that the malic enzyme (ME1) catalyzing the conversion of malate to pyruvate accompanied by the reduction of $NADP^+$ to NADPH may provide the necessary reducing potential for the glutathione system.[23] Nevertheless, nuclear factor erythroid 2–related factor 2 (Nrf-2), an important signaling factor for redox homeostasis, is a robust activator of the oxidative arm of the PPP, supporting its importance in redox balance.[22]

11.3.4 Upstream Modulators of Reprogramming

The expression of enzymes mediating metabolic reprogramming are regulated by diverse signaling pathways initiated by oncogenes or suppressor genes. For example, in non-small cell lung cancer the mutated receptor results in a constitutively activated tyrosine kinase.[24,25] This activation results not only in loss of cell cycle control but also markedly activated glucose and glutamine metabolism. Vousden's laboratory recently showed that p53 reprograms glucose metabolism by down-regulating oxidative metabolism and channeling glucose into the pentose phosphate pathway to generate ribose and ribonucleotides as substrates for DNA repair mechanisms.[26] Additionally, reprogramming of glutamine metabolism is well documented. The protooncogene MYC not only activates glucose metabolism but also the uptake and degradation of glutamine.[27,28] Recent reviews have focused on the reprogramming of hypoxia-inducible factor (HIF),[29–31] the principal responding mechanism to hypoxia, and "rat-sarcoma factor" (RAS),[23] the driving oncogene in a variety of tumors. Not only does metabolism respond to a variety of modulators,[9] but it acts as a sensor for epigenetics thereby closing the regulatory loop.[8] The link between metabolism and epigenetics has been recently reviewed. However, the specific combination of metabolism, cofactors, and reprogramming is only beginning to be understood.

11.3.5 Parametabolic Regulation

Epigenetic regulation may play a crucial role in the expression of cancer genes, and may represent sensing mechanisms for metabolism. Cofactors modulating epigenetics, i.e., SAM, acetyl-CoA, NAD^+, and α-KG, may be produced by glucose metabolizing pathways, but critical factors not directly derived from glucose may be contributed by ancillary pathways; these may be mechanisms acting as metabolic sensors and work through "parametabolic regulation." This concept emphasizes that the purpose of the conversion of precursor to product is not to generate the product, but to metabolize or recycle the cofactors or cosubstrates.[32] The focus is the metabolism, itself, rather than the canonical carbon product (Figure 11.1). Parametabolic regulation has been previously described,[33] but it may help us understand why some metabolic pathways have been reprogrammed in proliferating cancer cells.

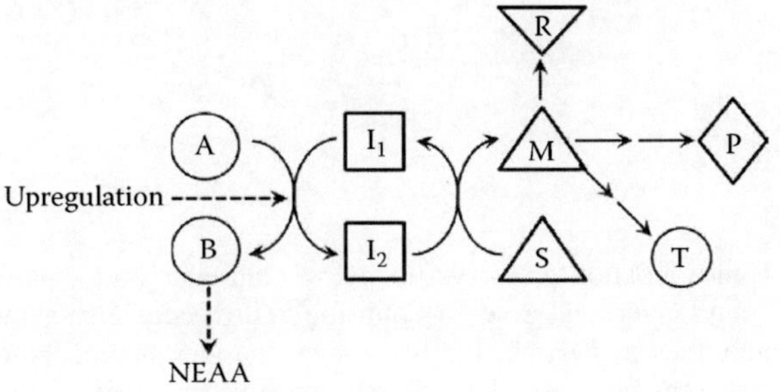

FIGURE 11.1 Schematic of parametabolic regulation. The NEAAs are synthesized by a variation of the proposed scheme. The biosynthetic pathway is upregulated under various conditions but pathway A to B is linked to the conversion of I_1 to I_2, which in turn increases the production of M. M can upregulate a number of processes: increased production of product R or upregulation of genetic mechanisms P or T.

11.4 NONESSENTIAL AMINO ACIDS

Considered nutritionally unnecessary, NEAA can be endogenously synthesized. In simple prokaryotes, all amino acids are nonessential, i.e., they can be synthesized de novo from carbon and nitrogen sources. However, as organisms evolved, all amino acids were available in ingested proteins. Thus, endogenous synthetic mechanisms were evolutionarily dispensable unless they provided some vital function other than for protein synthesis. For mammalian species, of the 20 proteinogenic amino acids, 11 have been found to be nonessential. These include alanine, arginine, asparagine, aspartate, cysteine, glutamate, glutamine, glycine, proline, serine, and tyrosine.[32–34] As was stated earlier, glutamine is a critical transporter of carbon and nitrogen and glutamate is both a precursor and product of glutamine. Furthermore, glutamate is an important participant as a nitrogen donor in transamination reactions where α-KG is the product.[27,35,36] Similarly, aspartate is the transamination partner with oxaloacetate. These two transamination reactions form the basis for redox transfers into and out of mitochondria because dicarboxylic acids can pass through mitochondrial membranes by a carrier mechanism.[23] By undergoing transamination and pyridine nucleotide reduction–oxidation, reducing equivalents can enter mitochondria to become electron donors to the electron transport chain (Figure 11.2). A transfer of glutamine-derived carbons into the cytosol with Ras-mediated regulation has been suggested as a component of Ras reprogramming for redox homeostasis.[23] Alanine and pyruvate also constitute a transamination pair and play a role in the transfer of carbons for the alanine cycle and gluconeogenesis.[37] Thus, transamination of amino acids can serve to affect cellular energetics and as effectors for Ras-mediated cancer reprogramming.[23]

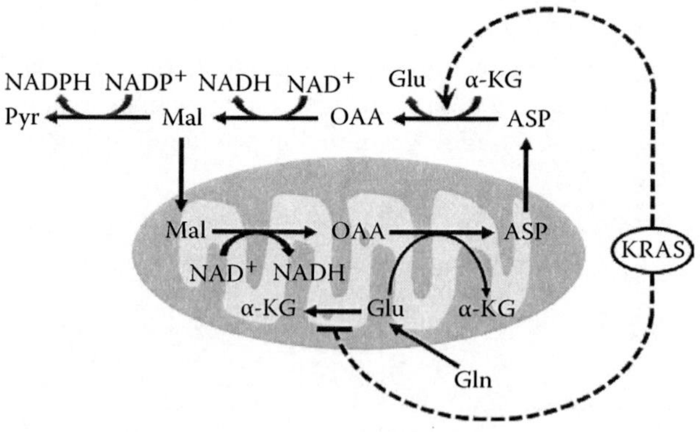

FIGURE 11.2 Transamination pathways involving glutamate and aspartate. The malate–aspartate shuttle plays a critical role in shuttling redox equivalents into mitochondria. Mitochondrial membranes are a barrier for pyridine nucleotides, but reducing equivalents generated by glycolysis in the cytosol are transferred into mitochondria by the transaminations involving glutamate and aspartate. A novel use of this system has been described in which kRAS inhibits mitochondrial glutamate dehydrogenase and upregulates cytosolic glutamate oxaloacetate transaminase (GOT1), Malate is converted to pyruvate by the malic enzyme to increase the production of NADPH for redox homeostasis. (Adapted and redrawn by permission from Macmillan Publishers Ltd., *Nature*, Son, J. et al., Glutamine supports pancreatic cancer growth through a KRAS-regulated metabolic pathway, 496, 101–5, copyright 2013.)

11.4.1 Nonessential Amino Acids: Glutamine

The reprogramming of metabolism has emphasized a number of substrates including glucose and fatty acids. Although amino acids are good candidates for parametabolic regulation, their importance in cancer is focused on their general availability as substrates for protein synthesis, i.e., by the mTOR pathway.[38] The importance of specific metabolic pathways remains largely unexplored. The one exception, of course, is glutamine, which has been recognized as an important player in the reprogramming of cancer metabolism.[27,35] Glutamine is known as the transporter of carbon and nitrogen from muscle, the principal site of proteolysis, to central organs, i.e., liver and kidneys to use proteins for energy and to process excess nitrogen.[35] Glutamine also is an important substrate for proliferating tissue, e.g., intestine, where it is the main source of energy, i.e., carbons recovered as CO_2. It is also converted to proline and citrulline.[36] Since glutamine is carried in the circulation, an inadequate blood supply demands glutamine from other sources. A model has been proposed for the synergistic metabolism of glutamine between tumor epithelial cells and stromal cells.[39]

11.4.1.1 Serine Biosynthesis

The biosynthetic pathways themselves may play a role in tumor cell proliferation. Recently, the synthesis of serine from glycolytic intermediates has been identified as critical for tumor cell proliferation.[40–42] Possemato and coworkers showed that knockdown of phosphoglycerate dehydrogenase (PHGDH), the enzyme that converts a glycolytic intermediate to serine precursors, markedly decreased the growth of xenograft tumors.[41] Additionally, others have shown that PHGDH is induced by oncogenic mechanisms and is involved in tumor metabolism reprogramming, which supports proliferation.[40] The interconversions of serine and glycine are involved in a variety of pathways, although the one most importantly linked to epigenetics is that involving the cycling of folate in the generation of methyl donors such as SAM, the substrate necessary for methylation of both DNA and histones (Figure 11.3).[42] Furthermore, serine was found to be an allosteric activator of pyruvate kinase M2,[43]

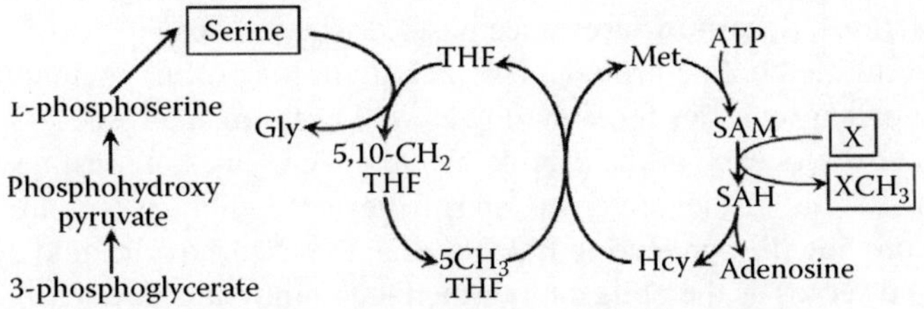

FIGURE 11.3 Serine biosynthesis. The first step in serine biosynthesis using a glycolytic intermediate is catalyzed by phosphoglycerate dehydrogenase, which converts 3-phosphoglycerate to phosphohydroxy pyruvate. The conversion of serine to glycine is linked to the generation of SAM, a critical substrate for methylation of DNA and histones. THF, tetrahydrofolate; 5,10-CH_2THF, 5,10 methylene tetrahydrofolate; 5 CH_3 THF, 5 methyl tetrahydrofolate; SAM, *S*-adenosylmethionine; SAH, *S*-adenosylhomocysteine; Hcy, homocysteine. Other abbreviations are standard.

a role for a NEAA directly regulating the enzymes of core glucose metabolism. This is a novel finding and gives rise to the possibility that metabolic intermediates of many pathways involving NEAA are involved in regulating the catalytic activity of strategic enzymes as well as parametabolically regulating the enzymes of epigenetics. Mechanisms are emerging for the understanding of NEAA and the evolutionary selection of their biosynthetic capabilities.

11.4.2 Proline as a Unique Nonessential Amino (Imino) Acid

Proline biosynthesis is present in all known species in both animal and plant kingdoms, suggesting that its metabolic pathways are important in parametabolic regulation.[33] The recognition of the importance of proline metabolism began in the 1980s.[44] First, we were attracted by the uniqueness of proline as the sole proteinogenic imino acid. With its peptide nitrogen sequestered in a pyrrolidine ring, proline contributes special features to protein structure as well as to intermediary metabolism.[45–47] The molecular stacking of multiple prolines forms the helical structure of collagen, and hydroxylation of proline by prolyl hydroxylase with subsequent crosslinking leads to the conformational stability of collagen found in higher organisms. Investigators have recognized that the cyclic structure of the proline side chain confers a conformational rigidity ideal for protein–protein recognition.[46] Proline in peptide bonds can exist in either *cis* or *trans* configuration. A prolyl peptidyl *cis–trans* isomerase catalyzes the transition with major alteration in protein structure. Investigators have proposed that this process plays a major role in phosphorylation-dependent protein modulation and function.[48]

11.5 METABOLISM OF PROLINE

11.5.1 Overview

Less well appreciated is the unique metabolism of proline among NEAA. Lacking an attached hydrogen on its α-nitrogen sequestered in the ring structure, proline is an unsuitable substrate for the generic amino acid metabolizing enzymes, e.g., the aminotransferases, decarboxylases, racemases. Instead, a special family of enzymes have evolved to metabolize proline. This paradigm for proline metabolism confers special metabolic properties for signaling as well as for anaplerosis.

Before we discuss each of the proline metabolic enzymes, it must be emphasized that GSA or P5C is at the center of intermediary metabolism (Figure 11.4). GSA and/or P5C provide the connecting link between the tricarboxylic acid cycle and the urea cycle and serves as the obligate intermediate in the transfer of carbons between the two major metabolic cycles.[47] This precursor/product relationship of P5C and proline is unique in amino acid metabolism, suggesting that their interconversions by two irreversible enzymes constitute a metabolic cycle.[44] The cycling of P5C and proline can serve as a mechanism for transferring redox equivalents across subcellular compartments. The existence of three P5C reductase isozymes with differential subcellular localizations and preferential use of reduced pyridine nucleotide provide channeling of cofactors for regulatory functions.[49,50] The structure of one of

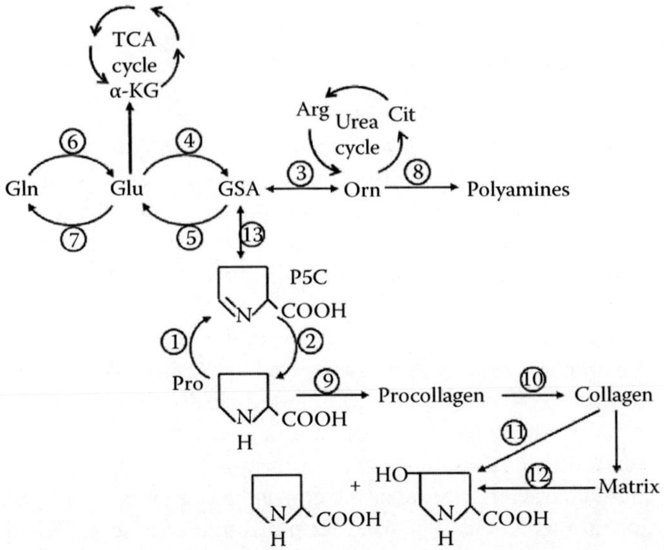

FIGURE 11.4 The proline metabolic system. A schematic showing the metabolism of proline and its connections to other metabolic pathways. Standard abbreviations are used for amino acids; GSA, glutamic-γ-semialdehyde; Cit, Citrulline. Numbers represent enzymes catalyzing the designated reaction. 1, Proline dehydrogenase or proline oxidase; 2, pyrroline-5-carboxylate reductase; 3, ornithine aminotransferase; 4, P5C synthase; 5, P5C dehydrogenase; 6, glutaminase; 7, glutamine synthetase; 8, ornithine decarboxylase and other enzymes; 9, collagen synthase, 10, collagen prolyl hydroxylase; 11, intracellular degradation; 12, matrix metalloproteinases, proteases, prolidase; 13, spontaneous.

these identified P5C reductases (human PYCR1) has been shown to be a decamer with a torus-like structure. Thus, it is likely to participate in transmembrane and chaperone functions.[51] The transfer of reducing–oxidizing potential may be compartmentalized by protein–protein interactions mediated by the family of P5C reductases. Ornithine, which can be derived from P5C in a reaction catalyzed by ornithine aminotransferase, is the direct precursor of polyamines, and investigators have suggested that the routing of P5C to polyamines is an alternative route for synthesis of this important family of compounds during activation of T cells.[52] Another important feature of the proline-metabolic system is the relationship to collagen.[46] Collagen is the most abundant protein by weight in the human body, and 25% of this protein's residues are either proline or hydroxyproline. Because collagen is the main constituent of the extracellular matrix, it can serve as a reservoir as well as a dump for proline. Thus, it is apparent that a very active metabolic system central to intermediary metabolism is plugged into a functional depot that is large but relatively inactive metabolically. This juxtaposition of a matrix to an active metabolizing system may account for some of the physiology and pathophysiology in various diseases that are only now beginning to be understood.

11.5.2 Specific Enzymes of Proline Metabolism

11.5.2.1 Biosynthetic Enzymes

11.5.2.1.1 Δ¹-Pyrroline-5-Carboxylate Synthase (P5CS)

Glutamine metabolism in cancer is well recognized. It provides the principal transport mechanism, synthesized in one tissue undergoing proteolysis for use in another tissue or for processing in central organs.[35] It is not surprising that glutamine serves as an important source of proline. Glutamine is converted to glutamate, which is then converted to glutamic-γ-semialdehyde (GSA) by P5C synthase (Figure 11.4).

The mammalian enzyme is a bifunctional protein with ATP and NADPH domains which catalyze the coupled phosphorylation and reduction, respectively.[53] There are two isoforms of the enzyme, long and short, formed by alternative splicing of RNA transcripts. The short form is found predominantly in the gut and is primarily involved in the conversion of glutamine to ornithine, a pathway requisite of ornithine aminotransferases. The short isoform is also sensitive to inhibition by ornithine.[53] These characteristics favor the conversion of glutamine to arginine. The intermediate, ornithine, can also be used for polyamine synthesis. The long form of P5CS is found in multiple tissues and used primarily for synthesis of proline.[53] These characteristics of P5CS have been recently reviewed.

11.5.2.1.2 Ornithine Aminotransferase (OAT)

Unlike generic aminotranferases, i.e., glutamic oxaloacetate transaminase; glutamic pyruvate transaminase; etc., which transfer nitrogen from the α position of the amino acid to an α-ketoacid acceptor, OAT transfers the delta nitrogen of ornithine to the α position of an acceptor α-ketoacid.[54] For ornithine, α-KG is the preferred acceptor. OAT converts ornithine and α-KG to products glutamate and glutamic semialdehyde (Figure 11.4). Additionally, the activity is reversible such that either GSA or ornithine can be precursor with glutamate and α-KG as their respective products. The enzyme is ubiquitous and located in the mitochondrial matrix. Kinetic studies show that preferred direction is to form GSA. However, in certain tissues such as the small intestine, ornithine and urea cycle intermediates are the major products of glutamine metabolism suggesting that there may be channeling of GSA.[36,50] The mechanism by which this occurs is not well understood. Deficiency of OAT in humans results in gyrate atrophy of the choroid and retina, an ocular disease in which cells in the choroid layer of the retina degenerate progressively resulting in blindness by the fourth decade of life. The production of urea cycle intermediates from glutamate is essential in neonates. Mice with OAT knockout succumb to hyperammonemia unless they are given arginine.[54] The production of ornithine from GSA appears to be a critical source of polyamines during the activation of T cells.[52]

11.5.2.1.3 Pyrroline-5-Carboxylate Reductase (PYCR)

Classic biochemical studies of pyrroline-5-carboxylate reductase activity showed that organically synthesized Δ¹-pyrroline-5-carboxylate is converted to proline in the presence of NAD(P)H.[55,56] The enzyme activity is ubiquitous and increased activity was found in tumor tissue (Figure 11.4).[57] Based on studies from the liver and kidney of cows and rats, the enzyme was thought to be primarily cytosolic. In physical fractionation studies, significant activity associated with the particulate fraction that solubilized into the supernatant with 0.25 M sodium phosphate, leading to the conclusion that the enzyme protein in situ had associated with membranes. The enzyme was purified from circulating human erythrocytes suggesting its cytosolic localization.[58,59] The presence of isoforms was suggested by functional differences based on feedback inhibition by proline and the preference for NADH or NADPH as cofactor. In lymphocytes undergoing activation, activity converted from proline-insensitive to one which was highly proline-sensitive.[60] A genetic mutant identified in lymphoblastoid leukemia cells (REH) as compared with a nonmalignant lymphoblastoid cell line showed that REH had a mutation in the isoform with a high Vmax, as defined by Michaelis-Menten kinetics, using NADH, whereas the NADPH-utilizing activity was preserved.[61] The latter activity was purified to homogeneity from human red cells and was shown to not only have several log units higher affinity (lower Km) for NADPH versus NADH, but also has higher affinity for P5C with NADPH. Merrill et al. suggested that the NADPH enzyme was functioning as a P5C-dependent NADPH dehydrogenase.[59] In metabolic studies, the ability of added P5C to markedly stimulate the activity of the oxidative arm of the PPP supports this hypothesis.[62,63] Furthermore, the production of phosphoribosyl pyrophosphate (PP-ribose-P), a critical intermediate in the synthesis and salvage of ribonucleotides, was increased by treatment with P5C.[64] The linkage with the oxidative arm of the PPP rather than the nonoxidative arm was convincingly shown by the absence of the P5C effect in erythrocytes from patients with glucose-6-phosphate dehydrogenase deficiency.[64]

The genes coding for PYCR1/2 were described by Daugherty et al.,[65] and a third gene designated PYCRL was subsequently cloned. The three isozymes have been characterized as to respective preference for pyridine nucleotide and sensitivity to inhibition by proline.[49] De Ingeniis and coworkers using cultured melanoma cells showed that PYCR1 and PYCR2 are associated with mitochondria whereas PYCRL is in the cytosol. Consistent with previous reports, PYCRL is insensitive to inhibition by proline whereas PYCR2 is most sensitive. Relative to the preferential use of pyridine nucleotides, proline formation is highest with PYCR2 using NADH whereas PYCR1 is somewhat lower, but NADH is the preferred cofactor. PYCRL, meanwhile, preferentially uses NADPH. These studies were performed with P5C 0.1 and 0.1 mM of the respective reduced pyridine nucleotide. Since the in situ concentrations of P5C and NADPH are much lower, the results do not take affinities into consideration. Their studies using ^{13}C-labeled ornithine or glutamate as precursor suggest that ornithine is the precursor of the proline formed by PYCRL in the cytosol. Whether these findings can be generalized to all proliferating cells or if they are a special feature of malignant melanoma cells need to be explored. Nevertheless, these findings correlate the gene product with functional characteristics, and are generally in keeping with earlier studies based on activity end points. The identification of a pedigree of patients with inherited deficiency of PYCR1 solidified some of the proposed regulatory functions of the P5C reductases.[66] These patients are characterized by cutis laxa, bony abnormalities, and intellectual deficiencies. Cultured fibroblasts from these patients show that they are especially susceptible to exposure to oxidants like hydrogen peroxide.[66]

11.5.2.2 Degradative Enzymes

11.5.2.2.1 Proline Dehydrogenase/Proline Oxidase (PRODH/POX)

PRODH/POX catalyzes the first step in the degradation of proline (Figure 11.4). The enzyme is tightly bound to mitochondrial inner membranes and transfers a pair of electrons from proline to protein-bound FAD.[46,67] Using the enzyme from *Thermus thermophilus*, Tanner's laboratory showed that the electrons from FAD can directly reduce solvent oxygen and produce superoxide.[68] Most of the electrons, however, enter the electron transport chain with CoQ as a carrier,[69] and eventually as a source of ATP. Electrons are transferred to CoQ from FAD. In addition, the oxidation of CoQ in complex III can be a source of ROS, which is produced in the intermembrane space and thereby accessible to the cytosol for ROS signaling.[70] These findings clearly establish that PRODH/POX can play an important role in redox signaling. Mutations and SNPs in PRODH/POX have been linked to psychiatric disorders including schizophrenia,[71] but the causative association in patients remains controversial.[72] Enzymes based on identified SNPs from patients showed decreased enzyme activity when translated in vitro, providing persuasive evidence of an association between catalytic activity and disease risk.[73,74]

11.5.2.2.2 P5C Dehydrogenase

The second step in the degradation of proline is catalyzed by P5C dehydrogenase, which converts P5C or GSA to glutamate (Figure 11.4). Thus, proline converted to α-KG can be anaplerotic for the TCA cycle as α-KG. The enzyme is localized to the mitochondrial matrix and uses NAD^+ as an electron acceptor.[74] Inherited deficiency of this enzyme results in markedly higher levels of P5C and proline, and both are excreted in the urine.[75] The activity of P5C dehydrogenase provides an outlet for the participating substrates of the proline cycle.[44,76] P5C dehydrogenase also links the parallel metabolism of proline and hydroxyproline. Proline dehydrogenase/oxidase and hydroxyproline dehydrogenase/oxidase as the first steps in proline and hydroxyproline metabolism, respectively, are catalyzed by distinct enzymes, which are the products of different genes. The second step in both pathways, the conversions of P5C and OH-P5C to glutamate and OH-GLU, respectively, are both catalyzed by P5C dehydrogenase.[67] In patients with deficient P5C dehydrogenase, OH-P5C is also elevated and excreted in the urine. Cultured fibroblasts and isolated lymphocytes from these patients have undetectable activity levels for dehydrogenation of both P5C and OH-P5C compared with cells from normal controls.[77]

11.5.2.3 Proline-P5C Cycle

Enzymes catalyzing the proline cycle include proline dehydrogenase/proline oxidase (PRODH/POX) and the PYCRs.[34,44,46,47] Early work included the use of partially purified erythrocyte extract, which only has PYCRL activity, and the use of purified organically or enzymatically synthesized P5C.[62] Originally, the cycle was conceived as a shuttle mechanism to transfer reducing potential into the mitochondria, but later findings added a number of ramifications.[78] Mitochondrial ROS and P5C are produced by PRODH/POX, and these may be transferred to the cytosol where the P5C is converted back to proline and is linked to the oxidation of NADPH.[79] In intact cells, the introduction of P5C markedly increased the activity of the oxidative arm of the PPP resulting in increased production of PP-ribose-P and ribonucleotides.[64,80] However, the recent work by De Ingeniis et al.[49] using melanoma cells showed that PYCR1/2 fractionate with mitochondria. PYCRL, meanwhile, is cytosolic and preferentially uses NADPH and P5C derived from ornithine.[49] These findings have stimulated revision of the functional cycle for redox regulation. Nevertheless, the identification of an inborn error with PYCR1 deficiency with increased vulnerability to oxidant challenge supports the hypothesis that one of the important parametabolic functions of the proline metabolic system involves redox regulation.[66]

11.5.3 PRODH/POX, Stress, and Cancer

The exceptional qualities of the proline metabolic system, especially those catalyzed by PRODH/POX, were recognized decades ago.[44] However, its niche in overall physiology and/or pathophysiology remained unknown until the last decade. The discovery that PRODH/POX was one of 14 genes responding rapidly and robustly to expression of p53 strongly suggested that the proline metabolic system was activated during stress (Figure 11.5a). Using serial analysis of gene expression (SAGE) and infection of DLD-1 cells by an adenoviral-p53, Polyak and coworkers[81] showed that PRODH/POX was induced with increases in both mRNA and protein. It was shown later that PRODH/POX catalytic activity paralleled the increase in protein and that this resulted in an increase in proline-dependent generation of ROS, a signaling mechanism associated with p53-induced apoptosis.[82,83] With the design of the DLD-tet off-POX cells by Yu and Hu,[82,84,85] a variety of studies were performed by several laboratories.[82,83,85] Signaling pathways, including blockade of the cell cycle, downregulation of MAPK signaling, COX-2 expression and activity and decreased β-catenin signaling were all mediated by proline-dependent ROS.[45] PRODH/POX is upregulated not only by p53 as a mediator of genotoxic stress, but also by PPAR-γ, a response to inflammatory stress.[86] Interestingly, overexpression of PRODH/POX activates a metabolic cascade resulting in decreased HIF signaling.[87] Measuring metabolic intermediates during the activation of PRODH/POX, we found that α-KG levels increased but fumarate and succinate levels were decreased. Increased α-KG may increase the hydroxylation of HIFs by prolyl hydroxylases in an α-KG-dependent dioxygenation and thereby increase the degradation of HIFs in proteasomes after VHL targeting. In cells undergoing hypoxia with or without glucose starvation, PRODH/POX is induced (Figure 11.5b).[88] But under these conditions, the mediator is not HIF-1 or HIF-2. Instead, AMPK signaling upregulates PRODH/POX to activate a prosurvival autophagic response. This suggests that the PRODH/POX paradigm is not a primary signal, but a two-edged link responding to the switching activated by microenvironmental stress.

11.5.3.1 DLD-POX Cells and Xenograft Tumors

The DLD-tet-off POX cells, under tissue culture conditions, were very useful to demonstrate downstream effectors from overexpression of PRODH/POX. Whether the suppressor-like activity can be translated into inhibition of xenograft tumor has yet to be demonstrated. We injected DLD-tet-off POX cells into the flanks of

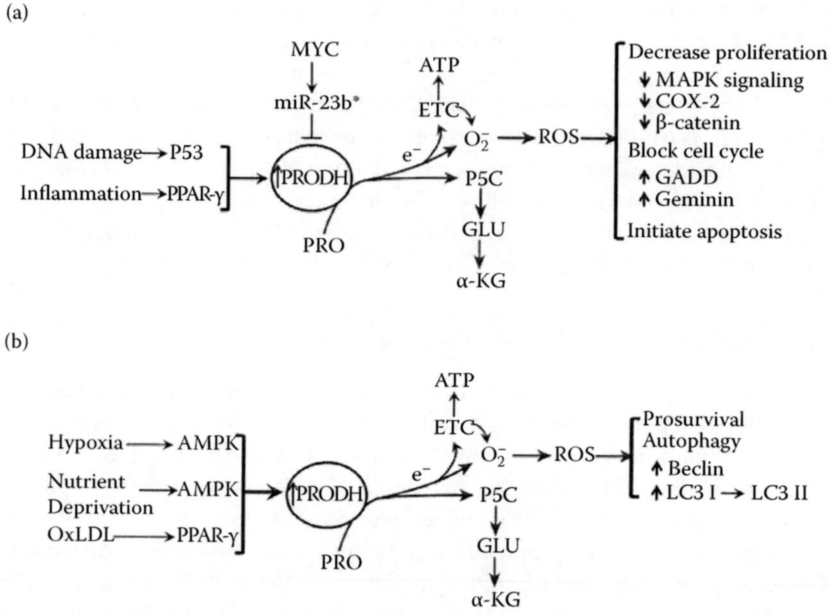

FIGURE 11.5 (a) PRODH/POX functions as a tumor suppressor. PRODH/POX is induced by p53 and PPAR-γ and downregulated by miR-23b*. Proline electrons can be donated to FAD and directly reduce oxygen to form superoxide. The bulk of the electrons are transferred to CoQ in the electron transport chain (ETC), which can be used to generate ATP. Alternatively superoxide can be formed at complex III. The proline-derived ROS can decrease proliferation, block the cell cycle, and initiate apoptosis. PPAR-γ, peroxisome proliferator–activated receptor γ; ETC, electron transport chain; MAPK, mitogen-activated protein kinase, COX-2, cyclooxygenase 2; GADD, growth arrest and DNA damage-inducible genes. (b) Under conditions of hypoxia or nutrient stress, PRODH/POX acts to promote survival. PRODH/POX can be upregulated by AMPK and PPAR-γ. The generated ROS in this context activates prosurvival autophagy. AMPK, AMP-activated kinase; OxLDL, oxidized low-density lipoprotein; LC3, microtubule associated protein 1 light chain 3.

immunodeficient mice and administered drinking water with and without doxycycline. The expression of the enzyme was documented immunohistochemically. DLD-1 colorectal cancer cells readily form tumors in immunodeficient mice. When the DLD-tet-off POX cells were injected, and the animals were given doxycycline in their drinking water to suppress PRODH/POX expression, tumors rapidly developed such that all the animals were euthanized by 4 weeks. In contrast, without doxycycline, PRODH/POX was robustly expressed and tumors were inhibited so that none of the mice had to be euthanized. In fact, using palpable tumors as an end point (100 mm^3 were detectable), only 25% of the animals not administered doxycycline to suppress PRODH/POX had palpable tumors by 5 weeks, whereas animals on doxycycline, without expression of PRODH, were all already euthanized by this time point. Clearly, the overexpression of PRODH/POX markedly inhibited xenograft tumor in immunodeficient mice.[87]

11.5.3.2 PRODH/POX in Human Tumors

Whether this suppressive effect of PRODH/POX could be demonstrated in human tumors was of great interest. We obtained 92 tumors (carcinomas, sarcomas, and lymphomas) paired with corresponding normal tissues from the same patient and monitored the presence of PRODH/POX protein immunohistochemically.[87] When all the tumors spanning a wide spectrum were compared with their normal counterparts, 56/92 (61%) showed decreased PRODH/POX expression. This was statistically significant at the $p < 0.05$ level. Moreover, when digestive tract and kidney tumors were examined, 28 (78%) of 36 showed markedly decreased PRODH/POX levels. This was significant at the $p < 0.001$ level. These findings suggest that the tumor suppressive effects of PRODH/POX must be downregulated for the tumor to be established and grow to clinical significance.

11.5.3.3 Mechanisms of Regulation: MicroRNA

The mechanism for downregulating PRODH/POX is an area of great interest. The PRODH coding region and promoter regions were checked for mutations in the tumor and for SNPs in adjacent normal tissue. None of the established SNPs correlating with changes in activity were found. Methylation sites in both the coding and promoter regions of PRODH/POX were compared and exhibited no difference between tumors and normal tissues. Noncoding RNA, specifically microRNAs became an attractive possibility.

Computer programs (miRBase, TargetScan, Microiinspector) were used to match the sequence in the 3′-UTR of PRODH/POX mRNA with known microRNA sequences. MiRNA microarrays compared normal kidney cells with renal cancer cells and mimic miRNAs were tested on control and cancer cells. We identified that miR-23b* decreased PRODH/POX expression in a sequence-dependent manner (Figure 11.5a). Transfection with miR-23b* not only decreased PRODH/POX levels but also decreased proline-dependent ROS levels and apoptosis.[89] To establish clinical relevance, we measured miR-23b* levels by qPCR and PRODH/POX by quantitative immunoblots in 16 clear cell renal carcinomas and corresponding normal tissues and showed that these two quantities were negatively correlated with a coefficient of −0.75 and a p value <0.001.[89] Thus, PRODH/POX downregulation in tumors is caused by increased levels of miR-23b*. We cannot say whether miR-23b* is the only miRNA regulating PRODH/POX expression; however, miR-23b* in renal tumors adequately explained our findings.

11.5.4 Reprogramming of Proline Biosynthesis

11.5.4.1 c-MYC and Enzymes of Biosynthesis

MiR-23 a/b are sibling miRNAs from the same transcript as miR-23b* and are regulated by c-MYC.[28] Because of this association, we asked whether MYC can regulate miR-23b* to downregulate PRODH/POX. Using P493-tet-off-MYC cells, we found that the expression of MYC not only increased the levels of glutaminase, but also increased the proline biosynthetic enzymes, P5C synthase, PYCR1, and OAT.[90] This effect was observed in protein levels as determined by immunoblot and also in their respective mRNAs. In contrast, the proline-degradative enzymes, PRODH/POX, and P5CDH were markedly decreased. Thus, MYC and p53 produced oppositely directed changes in PRODH/POX and P5C dehydrogenase.

11.5.4.2 Metabolomics: Glutamine to Proline

Metabolomic studies using $^{13}C_5$, $^{15}N_2$-labeled glutamine support that these changes are functionally significant. We traced the levels of isotopologues in P493 cells with and without MYC expression and found that proline synthesis from glutamine was markedly increased. For several of the isotopologues, proline synthesis was increased 8-fold.[90] The increased endogenous synthesis of proline from glutamine occurred in the face of medium (RPMI) containing physiological concentrations (0.1 mM) of proline. Proline is synthesized in excess of requirements for proteinogenesis and is excreted into the medium.

We hypothesized that proline functions as a stress substrate linked to redox regulation. We considered that this marked increase in proline biosynthesis was serving a parametabolic function and was not for producing proline as substrate for proteinogenesis. The biosynthesis of proline from glutamate converts 2 NAD(P)H molecules to $NAD(P)^+$, a NADPH by P5CS, and either NADH or NADPH by the respective isozyme of PYCR (Figure 11.6). Previous work has shown that the oxidative arm of the pentose phosphate pathway is tightly coupled to P5C reductase activity in certain cells. In patients with glucose-6-phosphate dehydrogenase deficiency, the production of PP-ribose-P is not only decreased but the response to exogenous P5C is abolished.[64] Additionally, the purified P5C reductase from human erythrocytes, presumably PYCRL, not only has higher affinities for NADPH over NADH, but also, the affinity for P5C with NADPH is an order of magnitude greater than that

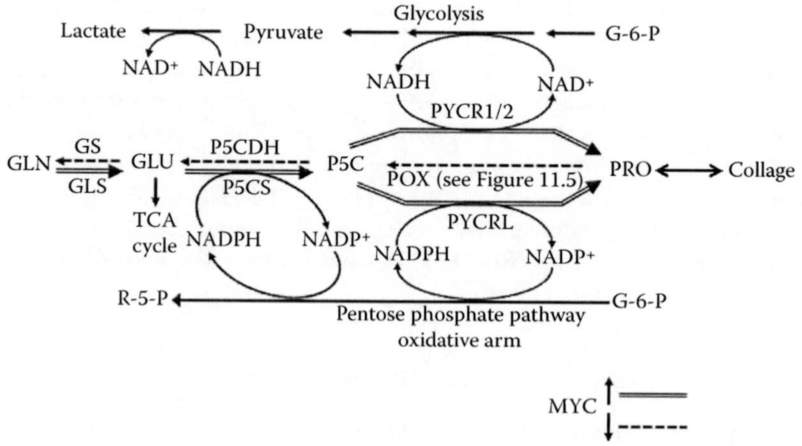

FIGURE 11.6 MYC regulation of proline regulatory axis promotes redox homeostasis. The enzymes upregulated by MYC are shown in double lines whereas those downregulated by MYC are shown by dashed lines. MYC channels glutamine to proline synthetic enzymes cycling pyridine nucleotides in a metabolic interlock with glucose metabolism. GLS, Glutaminase; P5CS, P5C synthase; PYCR, pyrroline-5-carboxylate reductase; POX, PRODH/POX; P5CDH, P5C dehydrogenase; GS, glutamine synthase.

with NADH.[59] The recent characterization of the three isozymes of P5C reductase in melanoma cells suggests that PYCR1/2 prefer NADH.[49] Structural studies[51] suggests that there may be channeling of substrates,[50] an idea that has been proposed by others. As was discussed under Section 11.3.5 on Parametabolic Regulation, the regulated endogenous biosynthesis of a NEAA is a parametabolic regulatory mechanism.

11.5.5 Linkage to Glucose as a Sensing Mechanism

There are several levels at which the proline metabolic axis can act as a sensing and response system. PRODH/POX is upregulated by p53, PPAR-γ, and AMPK.[32] Under conditions of nutrient availability and normoxia, proline can be used as a source of energy to generate ATP through the electron transport chain (ETC).[79,91] However, under stress conditions electrons from proline can reduce oxygen directly[68] or through the ETC generate ROS for signaling.[82] Coupled to an isozyme of PYCR, the cycling of proline and P5C can transfer reducing potential into mitochondria. Escape from this cycle can be provided by P5C dehydrogenase, which converts P5C to glutamate. When proline-derived electrons are transferred to ubiquinone (CoQ), they can serve as a source of ROS generated by Complex III for signaling. Complex III is located such that it accesses the intermembranous space and thus allows ROS to cross into the cytosolic compartment for signaling.[70]

With the activation of the biosynthetic pathways for proline, glutamine is converted sequentially to glutamate, P5C, and proline. Ornithine can be transaminated to P5C, with α-KG as the amino acceptor. However, the conversion of ornithine to P5C together with the recycling of glutamate to α-KG results in a reduced pyridine nucleotide in contrast to the pathway from glutamate to proline. For each glutamate converted to proline, 2 molecules of NAD(P)H are recycled. The $NAD(P)^+$ can be utilized in critical steps either in glycolysis or in the oxidative arm of the pentose phosphate pathway. However, if glucose is relatively depleted or inhibitors of these pathways de-link the utilization of glucose, then $NAD(P)^+$ will accumulate. NAD^+ is used as the acceptor of the acetyl group for deacetylating histones. The activation of the proline biosynthetic pathway participates in a metabolic interlock coupled to glucose metabolism as a sensor mechanism. Importantly, this sensing mechanism would be activated by signaling for cellular proliferation.[90]

11.5.6 The Proline Metabolic Paradigm and Collagen

The resurgence of research activity in cancer metabolism complements interest in the interaction between tumors and their microenvironment. Epigenetics is influenced by both metabolism[8,9] and the microenvironment,[92] and proline metabolism may act as the interface between these. A critical element in the microenvironment is the extracellular matrix that is predominantly collagen. The work of Matrisian[93,94] and others[95] defined the family of metalloproteinases involved in the turnover and processing of these matrix proteins. These enzymes and their linkage to tumor behavior make them useful as markers of malignant transformation.[96] Furthermore, the novel approaches of Bissell and others showing differential gene expression initiated by microenvironmental factors suggest these functions may be epigenetically important.[97] No one has made a plausible connection between these two important tumor functions, despite the elucidation of tumor cell metabolism and the processing of extracellular matrix proteins. The recent advances in the understanding of parametabolic regulation, epigenetics, and in particular, the role of proline metabolism in redox regulation and responses to stress, may make it possible to propose such a link.

11.5.6.1 Biosynthesis of Collagen and Proline: Maintenance of Redox Homeostasis

As was originally observed by Warburg in the 1920s, tumor tissue primarily uses aerobic glycolysis—rather than metabolizing glucose to CO_2 in mitochondria, they produce and excrete lactate. Presumably, lactate production is recycling NADH to NAD^+, a necessary cofactor for the early steps of glycolysis. The conversion of pyruvate to lactate regenerates NAD^+, paradoxically this mechanism is also diverting carbons necessary for creating cell mass for proliferation. Under in situ conditions, stromal cells accompany tumor cells, and lactate may be recycled to pyruvate, which can be used by the tumor cell for generating cell mass. Alternative pathways may be used for regenerating $NAD(P)^+$. The proline biosynthetic pathway with excretion of product proline, can also serve such a function. Since proline is a source of reducing potential, it can support glucose metabolism and may exit the metabolic pool by being incorporated into collagen. Under conditions of cellular proliferation, collagen can serve as a metabolic dump for excess reducing potential.

The critical question is whether collagen synthesis occurs concomitant to tumor cell proliferation. For certain tumors, this certainly is the case. In invasive breast cancer, the concomitant deposition of large amounts of collagen is the basis for their physical characteristic, i.e., a hard mass fixed to adjacent tissue.[98,99] A correlation between the aggressiveness of the tumor and increased collagen in breast tumors has been described.[100] Some investigators have suggested that collagen acts as a mechanoregulator for the tumor cells, providing a latticework or track, which the tumor cell uses for invasion.[101] This theory cannot be ruled out, however, more likely is the metabolic linkage of collagen synthesis to cancer metabolism. The deposition of collagen to facilitate invasion seems counterintuitive without mechanistic evidence.

The metabolic model has features advantageous to the tumor cell, however, are there examples of the metabolic model found in other diseases? Are there disease states in which proline synthesis can substitute for lactate formation and the deposition of proline into collagen functions as a substitute or complement for lactate production and excretion? In the case of parasitic infestations, the growth of parasites often precipitates a granulomatous response with increased collagen synthesis.[102–104] In alcohol- or CCl_4-induced cirrhosis, proline levels are increased concomitant to collagen synthesis.[105–107] Metabolically, alcohol ingestion results in an excess of reducing potential. The coupling of collagen synthesis with alcohol metabolism has been established and includes the downregulation of PRODH/POX.[108] Various disease states with fibrosis may be explored with this metabolic linkage in mind as a hypothetical model.

11.5.6.2 Collagen as Source of Stress Substrate

We propose that an inflammatory process such as PPAR-γ markedly induces PRODH/POX.[86] Collagen in this instance is a reservoir for proline, which can be used by PRODH/POX to generate ATP, redox signaling, and anaplerotic carbons in the form of α-KG.[79] During early wound healing, collagen levels in tissue rapidly decrease.[109] Since PRODH/POX responds to the stress of hypoxia, nutrient stress, and genotoxic damage through p53, it is not surprising that collagen degradation can serve as a source of the proline for this stress substrate.

11.5.6.3 Role of Hydroxyproline in Metabolism

Another feature of collagen is that it is the principle source of hydroxyproline in the body. Although there is no biosynthesis of free hydroxyproline, pyrroline-5-carboxylate reductase can reduce OH-P5C back to hydroxyproline.[56] Preformed hydroxyproline is not utilized for protein synthesis, therefore, this capability is puzzling. The activity is mediated by PYCR in animal cells, but not by the prokaryotic enzyme.[67] Whether this is true of all three PYCR isozymes or preferentially by one or the other remains unknown. It appears that a hydroxyproline cycle may complement the proline cycle. This may provide an explanation for the persistence of certain proline cycle-dependent functions in the face of mutations or knockdowns of PRODH/POX.

One of the important functions for collagen turnover may be the production of free hydroxyproline. Although the mechanism may be different than that for PRODH/POX, hydroxyproline oxidase is also upregulated by p53 and can be a source of ROS.[110] Increasing insight into the metabolic features of proline and hydroxyproline metabolism uncovers the inquiry that collagen synthesis and turnover may provide free hydroxyproline for some as yet undefined cellular function. With the importance of the proline metabolic system firmly established, the possibility that hydroxyproline metabolism also has special regulatory functions can be empirically tested.

11.6 CONCLUSION

The current emphasis on cancer metabolism promotes increased effort to identify mechanisms modulating tumor cell signaling, proliferation, or sensing for epigenetics. Amino acids, especially NEAAs have attracted research attention. In this context, we emphasize that the endogenous biosynthetic capacity for NEAAs was evolutionarily conserved for important reasons. These reasons supersede the production of necessary substrate for protein synthesis. A parametabolic function necessary for regulation is the end point driving selection for the NEAA. For each of the NEAAs, a possible function has been suggested. The imino acid, proline, has been discussed in depth as an archetype of this phenomenon. The proline metabolic pathway not only provides a mechanism for redox signaling and homeostasis, but also for optimizing the production of cell mass. These features are incorporated in the special metabolic enzymes evolved for this unique imino acid. Furthermore, the interconvertibility of proline, arginine, and glutamine contributes to its importance. Proline and its hydroxylated congener, hydroxyproline, constitute 25% of the residues in collagen. Thus, collagen can serve as a reservoir for a stress substrate as well as dump for transferring active equivalents out of the metabolic pool as a component of metabolic reprogramming.

ACKNOWLEDGMENTS

This work was supported by the Intramural Research Program of the NIH, National Cancer Institute, Center for Cancer Research. This project also has been funded in part with federal funds from the National Cancer Institute, NIH, under contract no. HHSN27612080001. The content of this review does not necessarily reflect the views or policies of the Department of Health and Human Services, nor does mention of trade names, commercial products, or organizations imply endorsements by the U.S. Government.

REFERENCES

1. Vogelstein B, Kinzler KW. The multistep nature of cancer. *Trends Genet* 1993; 9:138–41.
2. Vogelstein B, Kinzler KW. Cancer genes and the pathways they control. *Nat Med* 2004; 10:789–99.
3. Knudson AG, Jr. Mutation and cancer: Statistical study of retinoblastoma. *Proc Natl Acad Sci U S A* 1971; 68:820–3.
4. Friend SH, Dryja TP, Weinberg RA. Oncogenes and tumor-suppressing genes. *N Engl J Med* 1988; 318:618–22.
5. Marjanovic ND, Weinberg RA, Chaffer CL. Cell plasticity and heterogeneity in cancer. *Clin Chem* 2013; 59:168–79.
6. Baylin SB, Jones PA. A decade of exploring the cancer epigenome—biological and translational implications. *Nat Rev Cancer* 2011; 11:726–34.
7. Esteller M. Epigenetics in cancer. *N Engl J Med* 2008; 358:1148–59.
8. Yun J, Johnson JL, Hanigan CL, Locasale JW. Interactions between epigenetics and metabolism in cancers. *Front Oncol* 2012; 2:163.
9. Lu C, Thompson CB. Metabolic regulation of epigenetics. *Cell Metab* 2012; 16:9–17.
10. Jackman DM, Yeap BY, Lindeman NI et al. Phase II clinical trial of chemotherapy-naive patients > or = 70 years of age treated with erlotinib for advanced non-small-cell lung cancer. *J Clin Oncol* 2007; 25:760–6.
11. Yamaguchi H, Chang SS, Hsu JL, Hung MC. Signaling cross-talk in the resistance to HER family receptor targeted therapy. *Oncogene* 2014; 33:1073–81.
12. Dang CV, Kim JW, Gao P, Yustein J. The interplay between MYC and HIF in cancer. *Nat Rev Cancer* 2008; 8:51–6.
13. Ying H, Kimmelman AC, Lyssiotis CA et al. Oncogenic Kras maintains pancreatic tumors through regulation of anabolic glucose metabolism. *Cell* 2012; 149:656–70.
14. Vander Heiden MG, Cantley LC, Thompson CB. Understanding the Warburg effect: The metabolic requirements of cell proliferation. *Science* 2009; 324:1029–33.
15. Hursting SD, Perkins SN, Phang JM. Calorie restriction delays spontaneous tumorigenesis in p53-knockout transgenic mice. *Proc Natl Acad Sci U S A* 1994; 91:7036–40.
16. Baylin SB. DNA methylation and gene silencing in cancer. *Nat Clin Pract Oncol* 2005; 2 Suppl 1:S4–11.
17. Finkel T, Deng CX, Mostoslavsky R. Recent progress in the biology and physiology of sirtuins. *Nature* 2009; 460:587–91.
18. Warburg O. On the origin of cancer cells. *Science* 1956; 123:309–14.
19. Dang CV. Rethinking the Warburg effect with Myc micromanaging glutamine metabolism. *Cancer Res* 2010; 70:859–62.
20. Acharya A, Das I, Chandhok D, Saha T. Redox regulation in cancer: A double-edged sword with therapeutic potential. *Oxid Med Cell Longev* 2010; 3:23–34.

21. Mitsuishi Y, Taguchi K, Kawatani Y et al. Nrf2 redirects glucose and glutamine into anabolic pathways in metabolic reprogramming. *Cancer Cell* 2012; 22:66–79.
22. Zhang Z, Liew CW, Handy DE et al. High glucose inhibits glucose-6-phosphate dehydrogenase, leading to increased oxidative stress and beta-cell apoptosis. *FASEB J* 2010; 24:1497–505.
23. Son J, Lyssiotis CA, Ying H et al. Glutamine supports pancreatic cancer growth through a KRAS-regulated metabolic pathway. *Nature* 2013; 496:101–5.
24. Marchetti A, Martella C, Felicioni L et al. EGFR mutations in non-small-cell lung cancer: Analysis of a large series of cases and development of a rapid and sensitive method for diagnostic screening with potential implications on pharmacologic treatment. *J Clin Oncol* 2005; 23:857–65.
25. Yamamoto H, Toyooka S, Mitsudomi T. Impact of EGFR mutation analysis in non-small cell lung cancer. *Lung Cancer* 2009; 63:315–21.
26. Maddocks OD, Vousden KH. Metabolic regulation by p53. *J Mol Med (Berl)* 2011; 89:237–45.
27. Dang CV. MYC, microRNAs and glutamine addiction in cancers. *Cell Cycle* 2009; 8:3243–5.
28. Gao P, Tchernyshyov I, Chang TC et al. c-Myc suppression of miR-23a/b enhances mitochondrial glutaminase expression and glutamine metabolism. *Nature* 2009; 458:762–5.
29. Kaelin WG, Jr., Thompson CB. Q&A: Cancer: Clues from cell metabolism. *Nature* 2012; 465:562–4.
30. Semenza GL. Hypoxia-inducible factor 1 (HIF-1) pathway. *Sci STKE* 2007; 2007:cm8.
31. Gordan JD, Thompson CB, Simon MC. HIF and c-Myc: Sibling rivals for control of cancer cell metabolism and proliferation. *Cancer Cell* 2007; 12:108–13.
32. Payne SH, Loomis WF. Retention and loss of amino acid biosynthetic pathways based on analysis of whole-genome sequences. *Eukaryot Cell* 2006; 5:272–6.
33. Phang JM, Liu W, Hancock C, Christian KJ. The proline regulatory axis and cancer. *Front Oncol* 2012; 2:60.
34. Phang JM, Liu W, Hancock C. Bridging epigenetics and metabolism: Role of non-essential amino acids. *Epigenetics* 2013; 8:231–6.
35. Watford M. Glutamine metabolism and function in relation to proline synthesis and the safety of glutamine and proline supplementation. *J Nutr* 2008; 138:2003S–7S.
36. Windmueller HG, Spaeth AE. Uptake and metabolism of plasma glutamine by the small intestine. *J Biol Chem* 1974; 249:5070–9.
37. Schutz Y. Protein turnover, ureagenesis and gluconeogenesis. *Int J Vitam Nutr Res* 2011; 81:101–7.
38. Efeyan A, Zoncu R, Sabatini DM. Amino acids and mTORC1: From lysosomes to disease. *Trends Mol Med* 2012; 18:524–33.
39. Ko YH, Lin Z, Flomenberg N et al. Glutamine fuels a vicious cycle of autophagy in The tumor stroma and oxidative mitochondrial metabolism in epithelial cancer cells: Implications for preventing chemotherapy resistance. *Cancer Biol Ther* 2011; 12:1085–97.
40. Locasale JW, Grassian AR, Melman T et al. Phosphoglycerate dehydrogenase diverts glycolytic flux and contributes to oncogenesis. *Nat Genet* 2011; 43:869–74.
41. Possemato R, Marks KM, Shaul YD et al. Functional genomics reveal that the serine synthesis pathway is essential in breast cancer. *Nature* 2011; 476:346–50.
42. Kalhan SC, Hanson RW. Resurgence of serine: An often neglected but indispensable amino acid. *J Biol Chem* 2012; 287:19786–91.
43. Chaneton B, Hillmann P, Zheng L et al. Serine is a natural ligand and allosteric activator of pyruvate kinase M2. *Nature* 2012; 491:458–62.
44. Phang JM. The regulatory functions of proline and pyrroline-5-carboxylic acid. *Curr Top Cell Regul* 1985; 25:91–132.

45. Phang JM, Pandhare J, Liu Y. The metabolism of proline as microenvironmental stress substrate. *J Nutr* 2008; 138:2008S–15S.
46. Phang JM, Donald SP, Pandhare J, Liu Y. The metabolism of proline, a stress substrate, modulates carcinogenic pathways. *Amino Acids* 2008; 35:681–90.
47. Phang JM, Liu W, Zabirnyk O. Proline metabolism and microenvironmental stress. *Annu Rev Nutr* 2010; 30:441–63.
48. Lu KP, Finn G, Lee TH, Nicholson LK. Prolyl cis-trans isomerization as a molecular timer. *Nat Chem Biol* 2007; 3:619–29.
49. De Ingeniis J, Ratnikov B, Richardson AD et al. Functional specialization in proline biosynthesis of melanoma. *PLoS ONE* 2012; 7:e45190.
50. Arentson BW, Sanyal N, Becker DF. Substrate channeling in proline metabolism. *Front Biosci* 2012; 17:375–88.
51. Meng Z, Lou Z, Liu Z et al. Crystal structure of human pyrroline-5-carboxylate reductase. *J Mol Biol* 2006; 359:1364–77.
52. Wang R, Dillon CP, Shi LZ et al. The transcription factor Myc controls metabolic reprogramming upon T lymphocyte activation. *Immunity* 2011; 35:871–82.
53. Hu CA, Khalil S, Zhaorigetu S et al. Human Delta1-pyrroline-5-carboxylate synthase: Function and regulation. *Amino Acids* 2008; 35:665–72.
54. Valle D, Simell O. The hyperornithinemias. In: Scriver CR, Beaudet AL, Sly WS, Valle D, eds. *The Metabolic and Molecular Bases of Inherited Disease*. New York: McGraw-Hill, 2001.
55. Peisach J, Strecker HJ. The interconversion of glutamic acid and proline. V. The reduction of delta 1-pyrroline-5-carboxylic acid to proline. *J Biol Chem* 1962; 237:2255–60.
56. Adams E, Goldstone A. Hydroxyproline metabolism. III. Enzymatic synthesis of hydroxyproline from Delta 1-pyrroline-3-hydroxy-5-carboxylate. *J Biol Chem* 1960; 235:3499–503.
57. Herzfeld A, Mezl VA, Knox WE. Enzymes metabolizing delta1-pyrroline-5-carboxylate in rat tissues. *Biochem J* 1977; 166:95–103.
58. Yeh GC, Harris SC, Phang JM. Pyrroline-5-carboxylate reductase in human erythrocytes. *J Clin Invest* 1981; 67:1042–6.
59. Merrill MJ, Yeh GC, Phang JM. Purified human erythrocyte pyrroline-5-carboxylate reductase. Preferential oxidation of NADPH. *J Biol Chem* 1989; 264:9352–8.
60. Valle D, Blaese RM, Phang JM. Increased sensitivity of lymphocyte delta1-pyrroline-5-carboxylate reductase to inhibition by proline with transformation. *Nature* 1975; 253:214–6.
61. Lorans G, Phang JM. Proline synthesis and redox regulation: Differential functions of pyrroline-5-carboxylate reductase in human lymphoblastoid cell lines. *Biochem Biophys Res Commun* 1981; 101:1018–25.
62. Yeh GC, Phang JM. The function of pyrroline-5-carboxylate reductase in human erythrocytes. *Biochem Biophys Res Commun* 1980; 94:450–7.
63. Phang JM, Downing SJ, Yeh GC et al. Stimulation of the hexosemonophosphate-pentose pathway by pyrroline-5-carboxylate in cultured cells. *J Cell Physiol* 1982; 110:255–61.
64. Yeh GC, Roth EF, Jr., Phang JM et al. The effect of pyrroline-5-carboxylic acid on nucleotide metabolism in erythrocytes from normal and glucose-6-phosphate dehydrogenase-deficient subjects. *J Biol Chem* 1984; 259:5454–8.
65. Dougherty KM, Brandriss MC, Valle D. Cloning human pyrroline-5-carboxylate reductase cDNA by complementation in Saccharomyces cerevisiae. *J Biol Chem* 1992; 267:871–5.
66. Reversade B, Escande-Beillard N, Dimopoulou A et al. Mutations in PYCR1 cause cutis laxa with progeroid features. *Nat Genet* 2009; 41:1016–21.
67. Adams E. Metabolism of proline and of hydroxyproline. *Int Rev Connect Tissue Res* 1970; 5:1–91.
68. Tanner JJ. Structural biology of proline catabolism. *Amino Acids* 2008; 35:719–30.

69. Wanduragala S, Sanyal N, Liang X, Becker DF. Purification and characterization of Put1p from Saccharomyces cerevisiae. *Arch Biochem Biophys* 2010; 498:136–42.
70. Chandel NS. Mitochondrial complex III: An essential component of universal oxygen sensing machinery? *Respir Physiol Neurobiol* 2010; 174:175–81.
71. Liu H, Heath SC, Sobin C et al. Genetic variation at the 22q11 PRODH2/DGCR6 locus presents an unusual pattern and increases susceptibility to schizophrenia. *Proc Natl Acad Sci U S A* 2002; 99:3717–22.
72. Williams HJ, Williams N, Spurlock G et al. Detailed analysis of PRODH and PsPRODH reveals no association with schizophrenia. *Am J Med Genet B Neuropsychiatr Genet* 2003; 120B:42–6.
73. Bender HU, Almashanu S, Steel G et al. Functional consequences of PRODH missense mutations. *Am J Hum Genet* 2005; 76:409–20.
74. Hu CA, Bart Williams D, Zhaorigetu S et al. Functional genomics and SNP analysis of human genes encoding proline metabolic enzymes. *Amino Acids* 2008; 35:655–64.
75. Phang JM, Hu CA, Valle D. Disorders of proline and hydroxyproline metabolism. In: Scriver CR, Beaudet AL, Sly WS, Valle D, eds. *Metabolic and Molecular Basis of Inherited Disease*. New York: McGraw-Hill, pp. 1821–38, 2001.
76. Liang X, Zhang L, Natarajan SK, Becker DF. Proline mechanisms of stress survival. *Antioxid Redox Signal* 2013; 19:998–1011.
77. Valle D, Goodman SI, Harris SC, Phang JM. Genetic evidence for a common enzyme catalyzing the second step in the degradation of proline and hydroxyproline. *J Clin Invest* 1979; 64:1365–70.
78. Hagedorn CH, Phang JM. Transfer of reducing equivalents into mitochondria by the interconversions of proline and delta 1-pyrroline-5-carboxylate. *Arch Biochem Biophys* 1983; 225:95–101.
79. Pandhare J, Donald SP, Cooper SK, Phang JM. Regulation and function of proline oxidase under nutrient stress. *J Cell Biochem* 2009; 107:759–68.
80. Yeh GC, Phang JM. The stimulation of purine nucleotide production by pyrroline-5-carboxylic acid in human erythrocytes. *Biochem Biophys Res Commun* 1981; 103:118–24.
81. Polyak K, Xia Y, Zweier JL, Kinzler KW, Vogelstein B. A model for p53-induced apoptosis. *Nature* 1997; 389:300–5.
82. Donald SP, Sun XY, Hu CA et al. Proline oxidase, encoded by p53-induced gene-6, catalyzes the generation of proline-dependent reactive oxygen species. *Cancer Res* 2001; 61:1810–5.
83. Maxwell SA, Rivera A. Proline oxidase induces apoptosis in tumor cells, and its expression is frequently absent or reduced in renal carcinomas. *J Biol Chem* 2003; 278:9784–9.
84. Hu C-AA, Lin W-W, Donald SP, Sun X-Y, Almashanu S, Steel G, Phang J, Vogelstein B, Valle D. Overexpression of proline oxidase, a p53-induced gene (PIG6) induces reactive oxygen species generation and apoptosis in cancer cells. *Proc Am Assoc Cancer Res* 2001; 42:225.
85. Hu CA, Donald SP, Yu J et al. Overexpression of proline oxidase induces proline-dependent and mitochondria-mediated apoptosis. *Mol Cell Biochem* 2007; 295:85–92.
86. Pandhare J, Cooper SK, Phang JM. Proline oxidase, a proapoptotic gene, is induced by troglitazone: Evidence for both peroxisome proliferator-activated receptor gamma-dependent and -independent mechanisms. *J Biol Chem* 2006; 281:2044–52.
87. Liu Y, Borchert GL, Donald SP et al. Proline oxidase functions as a mitochondrial tumor suppressor in human cancers. *Cancer Res* 2009; 69:6414–22.
88. Liu W, Glunde K, Bhujwalla ZM et al. Proline oxidase promotes tumor cell survival in hypoxic tumor microenvironments. *Cancer Res* 2012; 72:3677–86.
89. Liu W, Zabirnyk O, Wang H et al. miR-23b* targets proline oxidase, a novel tumor suppressor protein in renal cancer. *Oncogene* 2010; 29:4914–24.

90. Liu W, Le A, Hancock C et al. Reprogramming of proline and glutamine metabolism contributes to the proliferative and metabolic responses regulated by oncogenic transcription factor c-MYC. *Proc Natl Acad Sci U S A* 2012; 109:8983–8.
91. Natarajan SK, Zhu W, Zhang L et al. Proline dehydrogenase is essential for proline protection against hydrogen peroxide-induced cell death. *Free Radic Biol Med* 2012; 53:1181–91.
92. Sotgia F, Martinez-Outschoorn UE, Howell A et al. Caveolin-1 and cancer metabolism in the tumor microenvironment: Markers, models, and mechanisms. *Annu Rev Pathol* 2012; 7:423–67.
93. McCawley LJ, Matrisian LM. Matrix metalloproteinases: Multifunctional contributors to tumor progression. *Mol Med Today* 2000; 6:149–56.
94. Lopez-Otin C, Matrisian LM. Emerging roles of proteases in tumour suppression. *Nat Rev Cancer* 2007; 7:800–8.
95. Tlsty TD, Coussens LM. Tumor stroma and regulation of cancer development. *Annu Rev Pathol* 2006; 1:119–50.
96. Kuivanen TT, Jeskanen L, Kyllonen L, Impola U, Saarialho-Kere UK. Transformation-specific matrix metalloproteinases, MMP-7 and MMP-13, are present in epithelial cells of keratoacanthomas. *Mod Pathol* 2006; 19:1203–12.
97. Ghajar CM, Bissell MJ. Extracellular matrix control of mammary gland morphogenesis and tumorigenesis: Insights from imaging. *Histochem Cell Biol* 2008; 130:1105–18.
98. Kao RT, Hall J, Stern R. Collagen and elastin synthesis in human stroma and breast carcinoma cell lines: Modulation by the extracellular matrix. *Connect Tissue Res* 1986; 14:245–55.
99. Schedin P, Keely PJ. Mammary gland ECM remodeling, stiffness, and mechanosignaling in normal development and tumor progression. *Cold Spring Harb Perspect Biol* 2010; 3:a003228.
100. Kakkad SM, Solaiyappan M, Argani P et al. Collagen I fiber density increases in lymph node positive breast cancers: Pilot study. *J Biomed Opt* 2012; 17:116017.
101. Provenzano PP, Inman DR, Eliceiri KW, Keely PJ. Matrix density-induced mechanoregulation of breast cell phenotype, signaling and gene expression through a FAK-ERK linkage. *Oncogene* 2009; 28:4326–43.
102. Isseroff H, Bock K, Owczarek A, Smith KR. Schistosomiasis: Proline production and release by ova. *J Parasitol* 1983; 69:285–9.
103. Wolf-Spengler ML, Isseroff H. Fascioliasis: Bile duct collagen induced by proline from the worm. *J Parasitol* 1983; 69:290–4.
104. Mark LG, Isseroff H. Levels of type I and type III collagen in the bile duct of rats infected with Fasciola hepatica. *Mol Biochem Parasitol* 1983; 8:253–62.
105. Kershenobich D, Fierro FJ, Rojkind M. The relationship between the free pool of proline and collagen content in human liver cirrhosis. *J Clin Invest* 1970; 49:2246–9.
106. Cerbon-Ambriz J, Cerbon-Solorzano J, Rojkind M. Regulation of collagen production in freshly isolated cell populations from normal and cirrhotic rat liver: Effect of lactate. *Hepatology* 1991; 13:551–6.
107. Savolainen ER, Leo MA, Timpl R, Lieber CS. Acetaldehyde and lactate stimulate collagen synthesis of cultured baboon liver myofibroblasts. *Gastroenterology* 1984; 87:777–87.
108. Ehrinpreis MN, Giambrone MA, Rojkind M. Liver proline oxidase activity and collagen synthesis in rats with cirrhosis induced by carbon tetrachloride. *Biochim Biophys Acta* 1980; 629:184–93.
109. Marian B, Mazzucco K. Dermal collagen metabolism during tumor promotion with 12-O-tetradecanoylphorbol-13-acetate in mouse skin. *Carcinogenesis* 1985; 6:501–4.
110. Cooper SK, Pandhare J, Donald SP, Phang JM. A novel function for hydroxyproline oxidase in apoptosis through generation of reactive oxygen species. *J Biol Chem* 2008; 283:10485–92.

12 Dietary Effects on Adipocyte Metabolism and Epigenetics

Kate J. Claycombe, Huawei Zeng, and Gerald F. Combs Jr.

CONTENTS

12.1 INTRODUCTION

Obesity risk appears to be perpetuated across generations by way of programmed DNA alterations that occur in utero and that affect gene expression throughout the life span.

Studies have demonstrated associations of maternal obesity and epigenetic changes, such as DNA methylation, histone modification, and chromatin remodeling, linked to adipose tissue growth and chronic disease risk in offspring. Diet has emerged as important in this regard, affecting epigenetic pathways in adipocytes. Understanding the metabolic bases by which diet affects the embryonic/fetal environment to regulate obesigenic epigenetic events is important for identifying maternal practices that can reduce obesity risk across generations. This overview summarizes current understanding of effects of diet on epigenetic control mechanisms in epigenetic regulation relevant to obesity.

12.2 EPIGENETIC REGULATION OF ADIPOGENESIS AND LIPOGENESIS

The onset of obesity is characterized by increased precursor cell differentiation into adipocytes, the numbers of which, thus, increase. This process can be mediated by increased adipogenic gene expression as a result of changes in gene promoter methylation pattern. The accumulation of intracellular fat is determined by the balance between fat synthesis (lipogenesis) and breakdown (lipolysis). De novo fatty acid synthesis is increased by activating fatty acid synthase, resulting in increased triglyceride accumulation in adipocytes [1]. In contrast, This process differs from adipogenesis, which refers to the differentiation of preadipocytes into mature fat cells [2]. Adipogenesis is initiated by activation of a well-known transcription factor, peroxisome proliferator–activated receptor γ (*PPAR*-γ). That step is followed by the binding of CAAT-enhancer binding protein β (C/EBPβ) to the *PPAR*-γ promoter [3], which leads to activation and binding of a subsequent set of adipogeneic transcription factors that activate C/EBPα, leading to the terminal differentiation of adipocytes [4]. Available data suggest that histone modification is associated with adipogeneic gene (e.g., *PPAR*-γ, *C/EBP*β, *C/EBP*α) activation. For example, PAX transactivation domain–interacting protein (PTIP), a protein known to associate with transcriptionally active chromatin by interacting with histone 3 lysine 4 (H3K4) methyltransferases, has been shown to upregulate *PPAR*-γ and *C/EBP*α expression during adipogenesis [5] (Figure 12.1).

Genomic imprinting is an epigenetic mechanism that, in diploid organisms, restricts the expression of a gene to one of the two parental chromosomes. Regardless of its parental origin, an imprinted gene affects both male and female offspring. Therefore, genomic imprinting is a consequence of parental inheritance, not of sex [6]. Imprinted genes with metabolic functions include mesoderm-specific transcript (*Mest*) [7,8], which is upregulated in white adipose tissue of obese mice of dietary or

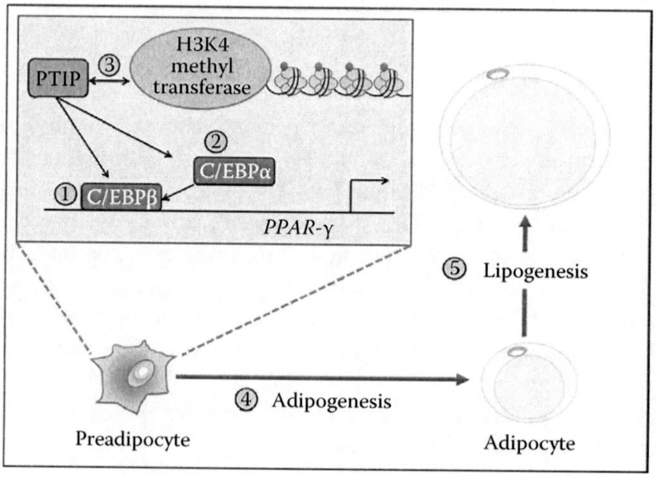

FIGURE 12.1 Proposed mechanism of epigenetic modulation of adipogenesis and lipogenesis. Activation of C/EBPβ leads to recruitment and subsequent activation of C/EBPα. Activation of C/EBPα and C/EBPβ leads to induction of *PPAR*-γ gene transcription and preadipocyte differentiation into adipocytes. By interacting with H3K4 methyl transferase, PTIP modulates the activation of *PPAR*-γ transcription through histone modification and increased accessibility to transcription factor binding to *PPAR*-γ regulatory DNA elements.

genetic origin [9], and preadipocyte factor 1 (*Pref*-1), which is an inhibitor of adipocyte differentiation [10,11]. Not all imprinted genes associated with obesity show epigenetic changes during enlargement of adipose tissue. For example, the expression of Mest was found to be positively associated with the level of energy intake, while its CpG methylation pattern was unaffected [12]. The extent of adipocyte differentiation is not correlated with methylation of adipogenic gene promoter DNA [13]; promoters of undifferentiated adipocyte stem cells are generally hypermethylated and remain so even when induced to differentiate ex vivo [14]. In contrast, precursor stem cells that can differentiate into adipogeneic lineage do not seem to have specific changes in DNA methylation pattern associated with adipocyte differentiation or alteration in gene expression in adipogenesis [14–17]. From these findings it is apparent that promoter DNA methylation of adipogeneic gene is poorly correlated with promoter activity and adipogenesis.

Recent studies have shown that microRNAs (miRNAs, small noncoding RNAs) play important roles in the development of obesity by regulating adipocyte differentiation [18,19]. Of a some hundred large number of miRNAs expressed in adipose tissue [18], only a few have been implicated in modulating such relevant metabolic functions as glucose tolerance regulation (miR-145m) [18] and lipogenesis (miR-181a) [19]. Although these miRNAs are of endogenous origin, nutrients, and other bioactive food components has been suggested as modulating their expression [20].

12.3 EPIGENETIC CHANGES CAN BE INDUCED BY HIGH-FAT MATERNAL DIETS

That prepregnancy maternal BMI is as a predictor of children's BMI [21] and that maternal consumption of excess energy in high-fat diets can increase risk for obesity in offspring [22–24] may be due to epigenetic modifications transmitted from parent to offspring. Aagaard-Tillery et al. [25] showed that feeding a calorically dense, high-fat diet to Japanese macaques could result in altered structure of fetal chromatin in such ways as to increase access of transcription factors to reach their respective DNA-binding sites. More recent studies have shown that maternal high-fat feeding can alter DNA methylation of other metabolically important organs. Dudley et al. [26] found that a high-fat diet (45% kcal from fat) reduced the expression of hepatocyte growth regulator gene *Cdkn1a*, which led to reduced liver cell size [26], suggesting detrimental, long-term consequences for offspring.

The major function of leptin is to regulate food intake and energy expenditure by working through hypothalamic signaling pathways [27–29]. Yokomori et al. [30] showed that expression of the leptin gene can be affected by high-fat feeding and by leptin gene promoter methylation in rat adipocytes; increases in adipose tissue mass were inversely associated with leptin promotor methylation. Further, CpG islands in the leptin gene promoter are hypermethylated in preadipocytes but become demethylated and, as a consequence, less active in differentiated 3T3–L1 adipocytes [30] and human adipocytes [31]. Treatment of preadipocytes with leptin reduces adipogenic transcription factor mRNA expression as well as adipocyte differentiation [32]. Taken together, these findings suggest that demethylation during adipocyte differentiation contributes to increased expression of the leptin gene [30], which then serves as a feedback inhibitor to reduce further adipogenesis.

Maternal diet can affect fetal growth. Chen et al. [33] found that maternal consumption of a diet high in both fat (30%) and sugar (36%) induced maternal obesity while reducing placental growth, nutrient transfer, and fetal growth. That diet also increased placental fatty acid transporter protein and placental growth regulatory imprinted genes including that for insulin-like growth factor 2 (*Igf2*) [33]. Dunn et al. [34] found that the progeny of high-fat fed dams had higher body weights and elevated fasting levels of glucose and leptin, with increased insulin resistance due to differential DNA methylation of growth hormone axis modulators including *Igf*-1 and growth hormone secretagogue receptor (*Ghsr*) genes [34].

Maternal high-fat feeding also affects, by epigenetic modification, proteins involved in embryonic osteoblast formation and fetal bone formation. Chen et al. [33] found that maternal high-fat feeding reduced fetal bone growth by increasing methylation of the homeodomain-containing factor/homeobox A-10 (*HoxA10*) gene promoter in the rat osteogenic cells while increasing adipogenic *PPAR*-γ gene expression. These results would suggest that maternal high-fat feeding can compromise low bone mass in offspring.

12.4 EFFECTS OF MATERNAL DIET ON EPIGENETIC CHANGES AND ENERGY METABOLISM

Clinical studies have shown that maternal protein-calorie malnutrition can contribute to the development of child obesity [35,36]. This involves intrauterine undernutrition, which results in intrauterine growth restriction (IUGR) as well as the epigenetic programming the offspring for survival in a nutrient-poor postnatal environment. Such effects typically also involve accelerated postnatal "catch-up" growth and consequential development of obesity [37–40] and type 2 diabetes (T2D) in later life [41]. Our findings in an animal model are consistent with these observations: the feeding of a low-protein (8%) diet to pregnant rats result in lower birth weight followed by accelerated adipose tissue catch-up growth accompanied by the development of obesity and insulin resistance by 3 months of age [42]. These effects were exacerbated by prenatal low-protein feeding of dams followed by postnatal high-fat feeding of pups, which showed increases in adipocyte number and size, and increased expression of *Igf2*, an imprinted gene associated with adiposity [43,44]. These effect would appear to involve increased methylation of CpG islands located within the differentially methylated region (DMR) of *Igf2/H19* locus [42], as hypermethylation [45] or mutation [46] of the ICR/H19 DMR is known to inhibit binding of the zinc-finger protein CCCTC-binding factor (CTCF) [47] to permit distal enhancer interaction with the *Igf2* promoter to induce *Igf2* gene transcription [48]. This system can also be affected by dietary folate [45,49,50], which is essential for the metabolic production of the methyl donor *S*-adenosylmethionine. Folate deprivation during pregnancy can, thus, reduce DNA methylation in the offspring [51–53], as in the case of *Igf2* DMR [54]. Such effects may be reversed by folate supplementation, as has been shown for the methylation of the *Igf2* gene in the very young child [52]. Thus, the methylated region (DMR) of *Igf2/H19* locus [42] can be affected by dietary protein, fat, and folate [51–53].

Igf2 gene expression can also be regulated by modifying acetylation of histone 4 (H4) [55] and polycomb repressive complex 2 (PRC2) [54]. The independent contributions of alterations in histone acetylation, PRC2 modulation in adipose tissue *Igf2* gene expression regulation have not yet been addressed.

The mechanism underlying the rapid adipose tissue "catch-up" weight gain may also involve increased expression of *Igf1*. Ong et al. [56] noted that children with relatively low birth weight had more rapid weight gains and greater concentrations of serum IGF1 by 5 years [56], suggesting adipose tissue-specific epigenetic effects. Like *Igf1*, the imprinted gene *Igf2* is positively associated with adipose tissue weight and obesity [43,44].

Other genes that control adipose tissue metabolism also appear to be regulated by maternal protein intake via epigenetic mechanisms. Lillycrop et al. [51] showed that offspring of protein-restricted rats had increased expression of *PPAR*-α, which was coupled with reduced methylation of the promoter region. As *PPAR*-α plays an important role in cellular fatty acid oxidation, its reduced expression may shift energy metabolism away from fat oxidation and toward lipid storage, resulting in obesity.

During pregnancy, maternal diet can influence mitochondrial development in a manner that produces heritable alterations in mitochondrial function [57,58]. Recent studies indicate that epigenetic modification of cytosine in mtDNA is much more common than previously believed, and promoter hypermethylation is related to mitochondrial DNA (mtDNA) alteration and nuclear DNA cross-talk [59]. Because mitochondria have a central role in regulating cellular energy metabolism, alterations in their development, particularly in tissues important in regulating energy balance, can be expected to predispose offspring to increased risk for obesity. Research has shown that both the quantity and quality of mtDNA can be influenced by the level of protein intake [60,61]. Park et al. [61] showed that the offspring of dams fed a low-protein diet during gestation and lactation had lower concentrations of mtDNA in liver and muscle, an effect not reversed by feeding adequate protein after weaning [61]. Similarly, Taylor et al. [62] found that offspring of rats fed a high-fat diet during gestation and lactation also had decreases in mtDNA copy number and expression of mitochondrial encoded subunits of cytochrome c oxidase mRNA in both liver and kidney. Recently, it has been reported that the methylation of genes involved in mitochondria function, such as *PPAR*-γ coactivator 1 α (*Pgc*-1α), were negatively correlated with Pgc-1α mRNA and mtDNA in type 2 diabetic patients [63]. Taken together, these effects were associated with subsequent development of insulin resistance and pancreatic β-cell dysfunction [62].

12.5 ROLE OF OXIDATIVE STRESS IN AFFECTING EPIGENETIC PATHWAYS AND ADIPOGENESIS

Maternal high-fat feeding may increase offspring adiposity by increasing oxidative stress. Sen and Simmons [64] found that offspring from dams fed a high-fat,

Western-type diet showed increased inflammation and adiposity and impaired glucose tolerance unless the maternal diet was supplemented with a mixture of antioxidant nutrients (vitamins A, E, C, and selenium). Those effects were reflected in the expression of adipogenic genes (acyl CoA synthase, fatty acid synthase) in adipose tissue, which increased were reduced by antioxidant supplementation. That these effects may have involved high-fat induced increases in reactive oxygen species (ROS) is suggested by the findings of Huh et al., who found that deletion of a mitochondrial peroxide scavenger peroxiredoxin isoform in differentiated 3T3–L1 adipocytes increased fat mass [65]. It has been shown that ROS regulate major epigenetic events including DNA methylation and histone acetylation via nucleophilic processes including DNA strand breakage and deoxycytosine halogenation [66]. As DNMT1 cannot distinguish methylated from halogenated cytosines in vitro [66], 5-halogenated cytosine can be a cause for inappropriate de novo DNA methylation. However, the mechanistic basis of ROS signaling in epigenetic processes linked to adipose tissue growth (e.g., acyl CoA synthase, fatty acid synthase) remains largely unclear.

Animals fed diets rich in highly polyunsaturated fatty acids (PUFA) such as eicosapentaenoic acid (EPA, 20:5 n3) and docosahexaenoic acid (DHA, 22:6 n3) have increased capacities to oxidized fat [67]. Thus, replacing saturated fats with n3 PUFAs in the maternal diets may be protective against obesigenic epigenetic programming caused by maternal obesity. It should be pointed out that, frequently, the high-fat diets used in these kinds of animal models include high levels of n-6 fatty acid-rich oils, e.g., corn oil. Grant et al. [68] showed that maternal feeding of a high-fat diet with an n-3:n6 ratio twice that of a low-fat, control diet reduced fasting plasma levels of DHA, EPA, and total n-3 fatty acids in both pregnant dams and fetuses, and increased fetal hepatic apoptosis in comparison to those controls. Thus, the question remains as to whether effects ascribed to "high-fat" feeding are those associated with excess caloric intake, fat level per se, or varying ratios of n-3:n-6 fatty acids. The dietary n-3:n-6 ratio has been shown to affect risk for metabolic syndrome [69–71]. A recent observational study found n-3 PUFA intake to be associated with increased CpG methylation in the regions of genes with the helicase-like transcription factor, α-actin2 smooth muscle/fas cell surface death receptor and serine proteinase. Methylation of CpG islands in those genes was significantly associated with cardiovascular disease risk factors (hyperlipidermia, hypercholesterolemia) [72]. Studies are needed to determine how much of the apparent maternal high-fat feeding effect on fetal development.

12.6 ROLES OF OTHER NUTRIENTS IN ADIPOCYTE METABOLISM

Recent studies have indicated that the micronutrient selenium (Se) may play a role in affecting adipogenesis. First, the expression of the Se transport protein, selenoprotein P (*Seppl*), is reduced in both adipose tissue of leptin-deficient obese (ob/ob) mice fed a high-fat diet, and primary adipose cells from obese, high-fat-fed Zucker rats [73]. In fact, adipose *Seppl* gene dose is inversely correlated with obesity and insulin resistance [73]. Second, dietary supplementation with Se has been shown to inhibit adipocyte differentiation by inducing transforming growth factor β1 (TGF-β1) signaling [74], which is suppressed in normal epithelial cells [75]. Many studies have shown that TGF-β is a potent adipocyte differentiation [76–78] via SMAD transcription factor–mediated pathways [79]. Because SMAD protein can inhibit C/EBP activity [80] and *PPAR*-γ [80], it is plausible that Se status may affect the TGF-β/smad signaling of adipocyte differentiation and body adiposity. That dietary Se may affect epigenetic mechanisms in adipose is suggested by findings in other tissues. Nutritional Se status, as indicated by the plasma Se level, has been inversely correlated with genomic DNA methylation in human leukocytes [81], and DNA hypomethylation in liver and colon has been reported in cases of both Se deprivation [82,83] and Se supplementation [84–86]. Se treatment of cultured tumor cells has been found to reduce methylated H3 while also reducing histone deacetylase activities, increasing levels of acetylated histone H3 lysine [87–89].

Because obesity is associated with increased inflammatory responses, a number of studies have addressed the roles of anti-inflammatory phytochemicals in modulating epigenetic mechanisms [90]. The soy isoflavone genistein and the flavonoid anthocyanin have each been shown to inhibit adipocyte differentiation [91–94], perhaps by affecting DNA methylation [90,95]. However, it remains unclear whether these or other phytochemicals may affect adipocyte differentiation via epigenetic mechanisms. Studies need to address this point, given the prevalence of such phytochemicals in many foods.

12.7 SUMMARY AND CONCLUSIONS

The effects of a maternal diet on offspring propensity to develop obesity are only now emerging. It is apparent that maternal diet can be a significant determinant of adipose epigenetic programming affecting offspring obesity risk obesity. Fetal growth can be increased by maternal high-fat feeding and decreased by maternal low-protein feeding, yet both conditions increase offspring obesity risk in later life. The effect of maternal low-protein feeding appears to involve the rapid postnatal catch-up growth of adipose in response to methylation of *Igf2* gene and stimulation by increased *Igf2*. Maternal high-fat feeding and/or exposures to other factors affecting adipocyte differentiation may further modulate postnatal adipose tissue epigenetic programming (Figure 12.1) to alter adipose tissue weight and, consequently, obesity. These bases of these effects of macronutrients may point to ways that micronutrients may have similar roles. The trace element selenium is one such possibility (Figure 12.2); other candidates are those involved in cellular antioxidant protection (vitamins E and C) and single-C metabolism (folate, vitamin B_{12}, choline, methionine). Although some studies have shown nutritional effects on epigenetic modifications that influence body adiposity by modulating adipogenesis and/or lipogenesis, that body of information is small. Additional, well designed studies are needed to determine whether nutrients and bioactive dietary factors can reduce the propensity toward adiposity and, particularly, that propensity across generations.

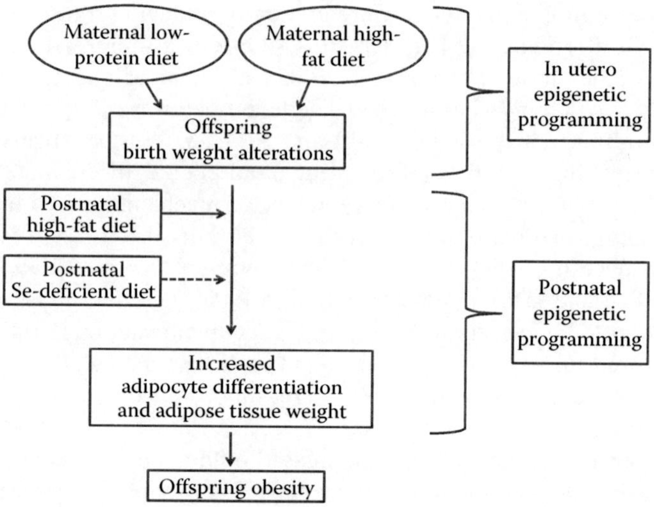

FIGURE 12.2 Potential mechanism of maternal and postnatal diets on adipose tissue epigenetic programming. Both maternal protein and fat intake levels alter adipose tissue weight, adipocyte differentiation, and obesity risk in the offspring. Maternal HF diet increases while maternal low-protein diet decreases fetal growth and birth weights. Maternal low-protein diet program offspring for rapid adipose tissue growth and obesity by increasing adipose tissue *Igf2* gene expression and DNA methylation [42]. Both of these result in obesity of offspring in later life.

REFERENCES

1. Moustaid N, Sul HS. Regulation of expression of the fatty acid synthase gene in 3T3-L1 cells by differentiation and triiodothyronine. *J Biol Chem.* 1991;266:18550–4.
2. Mandrup S, Lane MD. Regulating adipogenesis. *J Biol Chem.* 1997;272:5367–70.
3. Rosen ED, Spiegelman BM. PPARgamma: A nuclear regulator of metabolism, differentiation, and cell growth. *J Biol Chem.* 2001;276:37731–4.
4. Farmer SR. Transcriptional control of adipocyte formation. *Cell Metab.* 2006;4:263–73.
5. Cho YW, Hong S, Jin Q, Wang L, Lee JE, Gavrilova O et al. Histone methylation regulator PTIP is required for PPARgamma and C/EBPalpha expression and adipogenesis. *Cell Metab.* 2009;10:27–39.
6. Barlow DP, Bartolomei MS. Genomic imprinting in mammals. *Cold Spring Harb Perspect Biol.* 2014;6:a018382.
7. Lefebvre L, Viville S, Barton SC, Ishino F, Keverne EB, Surani MA. Abnormal maternal behaviour and growth retardation associated with loss of the imprinted gene Mest. *Nat Genet.* 1998;20:163–9.
8. Nishita Y, Sado T, Yoshida I, Takagi N. Effect of CpG methylation on expression of the mouse imprinted gene Mest. *Gene.* 1999;226:199–209.
9. Takahashi M, Kamei Y, Ezaki O. Mest/Peg1 imprinted gene enlarges adipocytes and is a marker of adipocyte size. *Am J Physiol Endocrinol Metab.* 2005;288:E117–24.
10. Takada S, Tevendale M, Baker J, Georgiades P, Campbell E, Freeman T et al. Delta-like and gtl2 are reciprocally expressed, differentially methylated linked imprinted genes on mouse chromosome 12. *Curr Biol.* 2000;10:1135–8.

11. Moon YS, Smas CM, Lee K, Villena JA, Kim KH, Yun EJ et al. Mice lacking paternally expressed Pref-1/Dlk1 display growth retardation and accelerated adiposity. *Mol Cell Biol*. 2002;22:5585–92.
12. Koza RA, Rogers P, Kozak LP. Inter-individual variation of dietary fat-induced mesoderm specific transcript in adipose tissue within inbred mice is not caused by altered promoter methylation. *Epigenetics*. 2009;4:512–8.
13. Boquest AC, Noer A, Collas P. Epigenetic programming of mesenchymal stem cells from human adipose tissue. *Stem Cell Rev*. 2006;2:319–29.
14. Sorensen AL, Jacobsen BM, Reiner AH, Andersen IS, Collas P. Promoter DNA methylation patterns of differentiated cells are largely programmed at the progenitor stage. *Mol Biol Cell*. 2010;21:2066–77.
15. Koch CM, Reck K, Shao K, Lin Q, Joussen S, Ziegler P et al. Pluripotent stem cells escape from senescence-associated DNA methylation changes. *Genome Res*. 2013;23:248–59.
16. Noer A, Sorensen AL, Boquest AC, Collas P. Stable CpG hypomethylation of adipogenic promoters in freshly isolated, cultured, and differentiated mesenchymal stem cells from adipose tissue. *Mol Biol Cell*. 2006;17:3543–56.
17. Noer A, Boquest AC, Collas P. Dynamics of adipogenic promoter DNA methylation during clonal culture of human adipose stem cells to senescence. *BMC Cell Biol*. 2007;8:18.
18. Kloting N, Berthold S, Kovacs P, Schon MR, Fasshauer M, Ruschke K et al. MicroRNA expression in human omental and subcutaneous adipose tissue. *PLoS One*. 2009;4:e4699.
19. Li H, Chen X, Guan L, Qi Q, Shu G, Jiang Q et al. MiRNA-181a regulates adipogenesis by targeting tumor necrosis factor-alpha (TNF-alpha) in the porcine model. *PLoS One*. 2013;8:e71568.
20. Ross SA, Davis CD. MicroRNA, nutrition, and cancer prevention. *Adv Nutr*. 2011; 2:472–85.
21. Schou Andersen C, Juhl M, Gamborg M, Sorensen TI, Nohr EA. Maternal recreational exercise during pregnancy in relation to children's BMI at 7 years of age. *Int J Pediatr*. 2012;2012:920583.
22. Franco JG, Fernandes TP, Rocha CP, Calvino C, Pazos-Moura CC, Lisboa PC et al. Maternal high-fat diet induces obesity and adrenal and thyroid dysfunction in male rat offspring at weaning. *J Physiol*. 2012;590:5503–18.
23. Caluwaerts S, Lambin S, van Bree R, Peeters H, Vergote I, Verhaeghe J. Diet-induced obesity in gravid rats engenders early hyperadiposity in the offspring. *Metabolism*. 2007;56:1431–8.
24. Benkalfat NB, Merzouk H, Bouanane S, Merzouk SA, Bellenger J, Gresti J et al. Altered adipose tissue metabolism in offspring of dietary obese rat dams. *Clin Sci (Lond)*. 2011;121:19–28.
25. Aagaard-Tillery KM, Grove K, Bishop J, Ke X, Fu Q, McKnight R et al. Developmental origins of disease and determinants of chromatin structure: Maternal diet modifies the primate fetal epigenome. *J Mol Endocrinol*. 2008;41:91–102.
26. Dudley KJ, Sloboda DM, Connor KL, Beltrand J, Vickers MH. Offspring of mothers fed a high fat diet display hepatic cell cycle inhibition and associated changes in gene expression and DNA methylation. *PLoS One*. 2011;6:e21662.
27. Campfield LA, Smith FJ, Guisez Y, Devos R, Burn P. Recombinant mouse OB protein: Evidence for a peripheral signal linking adiposity and central neural networks. *Science*. 1995;269:546–9.
28. Halaas JL, Gajiwala KS, Maffei M, Cohen SL, Chait BT, Rabinowitz D et al. Weight-reducing effects of the plasma protein encoded by the obese gene. *Science*. 1995;269:543–6.
29. Pelleymounter MA, Cullen MJ, Baker MB, Hecht R, Winters D, Boone T et al. Effects of the obese gene product on body weight regulation in ob/ob mice. *Science*. 1995;269:540–3.

30. Yokomori N, Tawata M, Onaya T. DNA demethylation modulates mouse leptin promoter activity during the differentiation of 3T3-L1 cells. *Diabetologia*. 2002;45:140–8.
31. Melzner I, Scott V, Dorsch K, Fischer P, Wabitsch M, Bruderlein S et al. Leptin gene expression in human preadipocytes is switched on by maturation-induced demethylation of distinct CpGs in its proximal promoter. *J Biol Chem*. 2002;277:45420–7.
32. Rhee SD, Sung YY, Jung WH, Cheon HG. Leptin inhibits rosiglitazone-induced adipogenesis in murine primary adipocytes. *Mol Cell Endocrinol*. 2008;294:61–9.
33. Chen JR, Zhang J, Lazarenko OP, Kang P, Blackburn ML, Ronis MJ et al. Inhibition of fetal bone development through epigenetic down-regulation of HoxA10 in obese rats fed high-fat diet. *FASEB J*. 2012;26:1131–41.
34. Dunn GA, Bale TL. Maternal high-fat diet promotes body length increases and insulin insensitivity in second-generation mice. *Endocrinology*. 2009;150:4999–5009.
35. Ravelli GP, Stein ZA, Susser MW. Obesity in young men after famine exposure in utero and early infancy. *N Engl J Med*. 1976;295:349–53.
36. Ozanne SE, Hales CN. The long-term consequences of intra-uterine protein malnutrition for glucose metabolism. *Proc Nutr Soc*. 1999;58:615–9.
37. Ernst KD, Radmacher PG, Rafail ST, Adamkin DH. Postnatal malnutrition of extremely low birth-weight infants with catch-up growth postdischarge. *J Perinatol*. 2003;23:477–82.
38. Muaku SM, Thissen JP, Gerard G, Ketelslegers JM, Maiter D. Postnatal catch-up growth induced by growth hormone and insulin-like growth factor-I in rats with intrauterine growth retardation caused by maternal protein malnutrition. *Pediatr Res*. 1997;42:370–7.
39. Bol VV, Reusens BM, Remacle CA. Postnatal catch-up growth after fetal protein restriction programs proliferation of rat preadipocytes. *Obesity (Silver Spring)*. 2008;16:2760–3.
40. Ozanne SE, Nicholas Hales C. Poor fetal growth followed by rapid postnatal catch-up growth leads to premature death. *Mech Ageing Dev*. 2005;126:852–4.
41. Berends LM, Fernandez-Twinn DS, Martin-Gronert MS, Cripps RL, Ozanne SE. Catch-up growth following intra-uterine growth-restriction programmes an insulin-resistant phenotype in adipose tissue. *Int J Obes (Lond)*. 2012;37:1051–1057.
42. Claycombe KJ, Uthus EO, Roemmich JN, Johnson LK, Johnson WT. Prenatal low-protein and postnatal high-fat diets induce rapid adipose tissue growth by inducing Igf2 expression in sprague dawley rat offspring. *J Nutr*. 2013;143:1533–1539.
43. Martin RM, Holly JM, Davey Smith G, Gunnell D. Associations of adiposity from childhood into adulthood with insulin resistance and the insulin-like growth factor system: 65-year follow-up of the Boyd Orr Cohort. *J Clin Endocrinol Metab*. 2006;91:3287–95.
44. Frystyk J, Skjaerbaek C, Vestbo E, Fisker S, Orskov H. Circulating levels of free insulin-like growth factors in obese subjects: The impact of type 2 diabetes. *Diabetes Metab Res Rev*. 1999;15:314–22.
45. Gao H, Sathishkumar KR, Yallampalli U, Balakrishnan M, Li X, Wu G et al. Maternal protein restriction regulates IGF2 system in placental labyrinth. *Front Biosci (Elite Ed)*. 2012;4:1434–50.
46. Pant V, Kurukuti S, Pugacheva E, Shamsuddin S, Mariano P, Renkawitz R et al. Mutation of a single CTCF target site within the H19 imprinting control region leads to loss of Igf2 imprinting and complex patterns of de novo methylation upon maternal inheritance. *Mol Cell Biol*. 2004;24:3497–504.
47. Pant V, Mariano P, Kanduri C, Mattsson A, Lobanenkov V, Heuchel R et al. The nucleotides responsible for the direct physical contact between the chromatin insulator protein CTCF and the H19 imprinting control region manifest parent of origin-specific long-distance insulation and methylation-free domains. *Genes Dev*. 2003;17:586–90.
48. Sasaki H, Ishihara K, Kato R. Mechanisms of Igf2/H19 imprinting: DNA methylation, chromatin and long-distance gene regulation. *J Biochem*. 2000;127:711–5.

49. Downing C, Johnson TE, Larson C, Leakey TI, Siegfried RN, Rafferty TM et al. Subtle decreases in DNA methylation and gene expression at the mouse Igf2 locus following prenatal alcohol exposure: Effects of a methyl-supplemented diet. *Alcohol.* 2011;45:65–71.
50. Gong L, Pan YX, Chen H. Gestational low protein diet in the rat mediates Igf2 gene expression in male offspring via altered hepatic DNA methylation. *Epigenetics.* 2010;5:619–26.
51. Lillycrop KA, Phillips ES, Jackson AA, Hanson MA, Burdge GC. Dietary protein restriction of pregnant rats induces and folic acid supplementation prevents epigenetic modification of hepatic gene expression in the offspring. *J Nutr.* 2005;135:1382–6.
52. Steegers-Theunissen RP, Obermann-Borst SA, Kremer D, Lindemans J, Siebel C, Steegers EA et al. Periconceptional maternal folic acid use of 400 microg per day is related to increased methylation of the IGF2 gene in the very young child. *PLoS One.* 2009;4:e7845.
53. McKay JA, Wong YK, Relton CL, Ford D, Mathers JC. Maternal folate supply and sex influence gene-specific DNA methylation in the fetal gut. *Mol Nutr Food Res.* 2011;55:1717–23.
54. Li T, Hu JF, Qiu X, Ling J, Chen H, Wang S et al. CTCF regulates allelic expression of Igf2 by orchestrating a promoter-polycomb repressive complex 2 intrachromosomal loop. *Mol Cell Biol.* 2008;28:6473–82.
55. Grandjean V, O'Neill L, Sado T, Turner B, Ferguson-Smith A. Relationship between DNA methylation, histone H4 acetylation and gene expression in the mouse imprinted Igf2-H19 domain. *FEBS Lett.* 2001;488:165–9.
56. Ong K, Kratzsch J, Kiess W, Dunger D. Circulating IGF-I levels in childhood are related to both current body composition and early postnatal growth rate. *J Clin Endocrinol Metab.* 2002;87:1041–4.
57. Reusens B, Theys N, Remacle C. Alteration of mitochondrial function in adult rat offspring of malnourished dams. *World J Diabetes.* 2011;2:149–57.
58. Ding Y, Lv J, Mao C, Zhang H, Wang A, Zhu L et al. High-salt diet during pregnancy and angiotensin-related cardiac changes. *J Hypertens.* 2010;28:1290–7.
59. Xie CH, Naito A, Mizumachi T, Evans TT, Douglas MG, Cooney CA et al. Mitochondrial regulation of cancer associated nuclear DNA methylation. *Biochem Biophys Res Commun.* 2007;364:656–61.
60. Jia Y, Li R, Cong R, Yang X, Sun Q, Parvizi N et al. Maternal low-protein diet affects epigenetic regulation of hepatic mitochondrial DNA transcription in a sex-specific manner in newborn piglets associated with GR binding to its promoter. *PLoS One.* 2013;8:e63855.
61. Park KS, Kim SK, Kim MS, Cho EY, Lee JH, Lee KU et al. Fetal and early postnatal protein malnutrition cause long-term changes in rat liver and muscle mitochondria. *J Nutr.* 2003;133:3085–90.
62. Taylor PD, McConnell J, Khan IY, Holemans K, Lawrence KM, Asare-Anane H et al. Impaired glucose homeostasis and mitochondrial abnormalities in offspring of rats fed a fat-rich diet in pregnancy. *Am J Physiol Regul Integr Comp Physiol.* 2005;288: R134–9.
63. Barres R, Osler ME, Yan J, Rune A, Fritz T, Caidahl K et al. Non-CpG methylation of the PGC-1alpha promoter through DNMT3B controls mitochondrial density. *Cell Metab.* 2009;10:189–98.
64. Sen S, Simmons RA. Maternal antioxidant supplementation prevents adiposity in the offspring of Western diet-fed rats. *Diabetes.* 2010;59:3058–65.
65. Huh JY, Kim Y, Jeong J, Park J, Kim I, Huh KH et al. Peroxiredoxin 3 is a key molecule regulating adipocyte oxidative stress, mitochondrial biogenesis, and adipokine expression. *Antioxid Redox Signal.* 2012;16:229–43.

66. Kawai Y, Morinaga H, Kondo H, Miyoshi N, Nakamura Y, Uchida K et al. Endogenous formation of novel halogenated 2′-deoxycytidine. Hypohalous acid-mediated DNA modification at the site of inflammation. *J Biol Chem*. 2004;279:51241–9.
67. Sadurskis A, Dicker A, Cannon B, Nedergaard J. Polyunsaturated fatty acids recruit brown adipose tissue: Increased UCP content and NST capacity. *Am J Physiol*. 1995;269:E351–60.
68. Grant WF, Gillingham MB, Batra AK, Fewkes NM, Comstock SM, Takahashi D et al. Maternal high fat diet is associated with decreased plasma n-3 fatty acids and fetal hepatic apoptosis in nonhuman primates. *PLoS One*. 2011;6:e17261.
69. Tishinsky JM. Modulation of adipokines by n-3 polyunsaturated fatty acids and ensuing changes in skeletal muscle metabolic response and inflammation. *Appl Physiol Nutr Metab*. 2013;38:361.
70. Clarke SD. Polyunsaturated fatty acid regulation of gene transcription: A mechanism to improve energy balance and insulin resistance. *Br J Nutr*. 2000;83 Suppl 1:S59–66.
71. Carpentier YA, Portois L, Malaisse WJ. n-3 fatty acids and the metabolic syndrome. *Am J Clin Nutr*. 2006;83:1499S–504S.
72. Aslibekyan S, Wiener HW, Havel PJ, Stanhope KL, O'Brien DM, Hopkins SE et al. DNA methylation patterns are associated with n-3 fatty acid intake in Yup'ik people. *J Nutr*. 2014;144:425–430.
73. Zhang Y, Chen X. Reducing selenoprotein P expression suppresses adipocyte differentiation as a result of increased preadipocyte inflammation. *Am J Physiol Endocrinol Metab*. 2011;300:E77–85.
74. Kim CY, Kim GN, Wiacek JL, Chen CY, Kim KH. Selenate inhibits adipogenesis through induction of transforming growth factor-beta1 (TGF-beta1) signaling. *Biochem Biophys Res Commun*. 2012;426:551–7.
75. Hannigan A, Smith P, Kalna G, Lo Nigro C, Orange C, O'Brien DI et al. Epigenetic downregulation of human disabled homolog 2 switches TGF-beta from a tumor suppressor to a tumor promoter. *J Clin Invest*. 2010;120:2842–57.
76. Sparks RL, Scott RE. Transforming growth factor type beta is a specific inhibitor of 3T3 T mesenchymal stem cell differentiation. *Exp Cell Res*. 1986;165:345–52.
77. Richardson RL, Hausman GJ, Gaskins HR. Effect of transforming growth factor-beta on insulin-like growth factor 1- and dexamethasone-induced proliferation and differentiation in primary cultures of pig preadipocytes. *Acta Anat (Basel)*. 1992;145:321–6.
78. Rahimi N, Tremblay E, McAdam L, Roberts A, Elliott B. Autocrine secretion of TGF-beta 1 and TGF-beta 2 by pre-adipocytes and adipocytes: A potent negative regulator of adipocyte differentiation and proliferation of mammary carcinoma cells. *In Vitro Cell Dev Biol Anim*. 1998;34:412–20.
79. Choy L, Skillington J, Derynck R. Roles of autocrine TGF-beta receptor and Smad signaling in adipocyte differentiation. *J Cell Biol*. 2000;149:667–82.
80. Zauberman A, Lapter S, Zipori D. Smad proteins suppress CCAAT/enhancer-binding protein (C/EBP) beta- and STAT3-mediated transcriptional activation of the haptoglobin promoter. *J Biol Chem*. 2001;276:24719–25.
81. Ferguson LR, Karunasinghe N, Zhu S, Wang AH. Selenium and its' role in the maintenance of genomic stability. *Mutat Res*. 2012;733:100–10.
82. Davis CD, Uthus EO. Dietary selenite and azadeoxycytidine treatments affect dimethylhydrazine-induced aberrant crypt formation in rat colon and DNA methylation in HT-29 cells. *J Nutr*. 2002;132:292–7.
83. Davis CD, Uthus EO, Finley JW. Dietary selenium and arsenic affect DNA methylation in vitro in Caco-2 cells and in vivo in rat liver and colon. *J Nutr*. 2000;130:2903–9.
84. Bermingham EN, Bassett SA, Young W, Roy NC, McNabb WC, Cooney JM et al. Post-weaning selenium and folate supplementation affects gene and protein expression and global DNA methylation in mice fed high-fat diets. *BMC Med Genomics*. 2013;6:7.

85. Armstrong KM, Bermingham EN, Bassett SA, Treloar BP, Roy NC, Barnett MP. Global DNA methylation measurement by HPLC using low amounts of DNA. *Biotechnol J*. 2011;6:113–7.
86. Zeng H, Yan L, Cheng WH, Uthus EO. Dietary selenomethionine increases exon-specific DNA methylation of the p53 gene in rat liver and colon mucosa. *J Nutr*. 2011;141:1464–8.
87. Kassam S, Goenaga-Infante H, Maharaj L, Hiley CT, Juliger S, Joel SP. Methylseleninic acid inhibits HDAC activity in diffuse large B-cell lymphoma cell lines. *Cancer Chemother Pharmacol*. 2011;68:815–21.
88. Xiang N, Zhao R, Song G, Zhong W. Selenite reactivates silenced genes by modifying DNA methylation and histones in prostate cancer cells. *Carcinogenesis*. 2008;29:2175–81.
89. Hu Y, McIntosh GH, Le Leu RK, Nyskohus LS, Woodman RJ, Young GP. Combination of selenium and green tea improves the efficacy of chemoprevention in a rat colorectal cancer model by modulating genetic and epigenetic biomarkers. *PLoS One*. 2013;8:e64362.
90. Vanden Berghe W. Epigenetic impact of dietary polyphenols in cancer chemoprevention: Lifelong remodeling of our epigenomes. *Pharmacol Res*. 2012;65:565–76.
91. Hwang JT, Park IJ, Shin JI, Lee YK, Lee SK, Baik HW et al. Genistein, EGCG, and capsaicin inhibit adipocyte differentiation process via activating AMP-activated protein kinase. *Biochem Biophys Res Commun*. 2005;338:694–9.
92. Pompei A, Toniato E, Innocenti P, D'Alimonte I, Cellini C, Mattoscio D et al. Cyanidin reduces preadipocyte differentiation and relative ChREBP expression. *J Biol Regul Homeost Agents*. 2012;26:253–64.
93. Kim HK, Kim JN, Han SN, Nam JH, Na HN, Ha TJ. Black soybean anthocyanins inhibit adipocyte differentiation in 3T3-L1 cells. *Nutr Res*. 2012;32:770–7.
94. Schreckinger ME, Wang J, Yousef G, Lila MA, Gonzalez de Mejia E. Antioxidant capacity and in vitro inhibition of adipogenesis and inflammation by phenolic extracts of Vaccinium floribundum and Aristotelia chilensis. *J Agric Food Chem*. 2010;58:8966–76.
95. Wang LS, Kuo CT, Cho SJ, Seguin C, Siddiqui J, Stoner K et al. Black raspberry-derived anthocyanins demethylate tumor suppressor genes through the inhibition of DNMT1 and DNMT3B in colon cancer cells. *Nutr Cancer*. 2013;65:118–25.

Section IV

Phytochemicals

13 Epigenetics of *BRCA-1*–Related Breast Tumorigenesis and Dietary Prevention

Donato F. Romagnolo and Ornella I. Selmin

CONTENTS

13.1 INTRODUCTION

Epigenetics refer to abnormalities in gene expression due to somatic heritable modifications in chromatin organization and alterations in profile and function of noncoding RNAs (Baylin and Jones, 2011). Research evidence produced during the past two decades clearly validated the concept that in addition to changes in genetic information, i.e., germline or somatic mutations and polymorphisms, an important modifier of cancer risk is the accumulation of epigenetic alterations, i.e., the cancer epigenome (Esteller, 2007). The significance of unraveling how epigenetic alterations contribute to the etiology of breast cancer is underscored by the evidence that most breast tumors are not linked to family history (Levy-Lahad and Friedman, 2007). Meanwhile, because epigenetic alterations are potentially reversible, they may offer a great many opportunities for therapeutic targeting (Kennedy et al., 2002).

Previous reviews have elegantly addressed the epigenetic events associated with breast cancer development and discussed opportunities for intervention using dietary approaches (Hardy and Tollefsbol, 2011; Khan et al., 2012; Stefanska et al., 2012). However, one of the major puzzles in breast and ovarian cancer biology is whether or not inactivation through epigenetic mechanisms of the tumor suppressor gene, *BRCA-1*, contributes to the development of sporadic tumors, or rather, if the silencing of *BRCA-1* expression is a late epigenetic event in multistage tumorigenesis (Butcher and Rodenhiser, 2007). Answering this question may provide important clues for the development of surveillance programs for cancer prevention as well as therapeutic strategies. In consideration of the role of *BRCA-1* in tumorigenesis, this chapter will focus on the epigenetic mechanisms of regulation of *BRCA-1* and their relevance to the development of breast and ovarian tumors. Second, we will discuss opportunities for cancer prevention with dietary compounds that target methylation and histone modifications at the *BRCA-1* gene, and noncoding RNA modifications that impinge on *BRCA-1* expression and its coregulators. Because of the limited space, we regret that not all relevant studies on *BRCA-1* could be cited and we refer the reader to excellent reviews on the topic of *BRCA-1* biology in normal breast development and tumorigenesis (Dworkin et al., 2009; Murphy and Moynahan, 2010; Parvin, 2004). We conclude that rich opportunities exist for epigenetic targeting of the *BRCA-1* gene and its regulatory network with food components. However, progress in this area is hampered by lack of knowledge of the mechanisms of epigenetic regulation of *BRCA-1* in normal cells and the mode of action of dietary compounds on the epigenetic machinery.

13.2 BRCAness

13.2.1 Hereditary Tumors

Perhaps, the *BRCA-1* gene offers one of the best examples of how epigenetic regulation may impact tumor development (Miki et al., 1994). The BRCA-1 protein participates in repair of DNA damage (Murphy and Moynahan, 2010) and transcription control (Parvin, 2004; Mullan et al., 2006). Germline mutations in the *BRCA-1* gene have a high penetrance and increase remarkably the risk of developing breast (~65%) (Ford et al., 1998) and ovarian (~40%) tumors (Easton et al., 1995). Women who carry a *BRCA-1* mutation have one functional *BRCA-1* allele, and its inactivation is referred to as loss of heterozygosity (LOH). The mechanisms responsible for LOH at the *BRCA-1* locus in *BRCA-1* mutation carriers remain largely unknown. *BRCA-1* tumors tend to be triple-negative basal-like characterized by lack of expression of estrogen receptor (ER), progesterone receptor (PR), and epidermal growth factor receptor 2 (HER2). They also have high levels of basal cell markers such as cytokeratin (CK) 5/6 and CK17, and reduced expression of luminal cytokeratins (i.e., CK8/18) (Melchor and Benitez, 2013). Remarkably, despite the high penetrance,

BRCA-1 mutations explain on average only 5–15% of all breast and ovarian cancer cases (Alsop et al., 2012; Berchuck et al., 1998; Genetics of Breast and Ovarian Cancer, 2013). Endocrine therapies with tamoxifen do not reduce breast cancer incidence among women with inherited *BRCA-1* mutations (King et al., 2001).

13.2.2 Sporadic Breast Tumors

Sporadic breast cancers, meanwhile, contribute 90–95% of all breast tumor cases. Although somatic mutations in *BRCA-1* are commonly not found in sporadic breast tumors, undisputed evidence suggests these cancers express low or undetectable levels of BRCA-1 (Magdinier et al., 1998; Rice et al., 1998; Seery et al., 1999; Taylor et al., 1998; Thompson et al., 1995; Wilson et al., 1999; Yoshikawa et al., 1999). Various degrees of *BRCA-1* promoter CpG methylation have been observed in sporadic breast cancers ranging from ~10% to 85% (Dobrovic and Simpfendorfer, 1997; Rice et al., 2000). Moreover, a high degree of correlation (~75%) has been documented between hypermethylation of the *BRCA-1* gene and reduced expression of ERα (Hosey et al., 2007) and hypermethylation of the *ER*α gene (Wei et al., 2008). In fact, sporadic breast tumors that show *BRCA-1* promoter hypermethylation cluster with the triple negative (ER, PR, HER2) subtype, and are resilient to endocrine therapies (Murphy and Moynahan, 2010). The fact that sporadic breast cancers with hypermethylated *BRCA-1* have overlapping tumor characteristics with *BRCA-1* mutated tumors, i.e., they possess BRCAness phenotype, raises the question whether or not epigenetic signatures among these tumors provide the basis for common management strategies (Lips et al., 2013) including those based on food components.

13.3 MECHANISMS OF EPIGENETIC REGULATION OF THE *BRCA-1* GENE

13.3.1 CpG Methylation

Table 13.1 summarizes data from representative studies that investigated the frequency of *BRCA-1* promoter CpG island hypermethylation in breast and ovarian breast tumors and its association with molecular and pathological signatures linked to tumor development. Figure 13.1 summarizes the various modes of epigenetic control over the *BRCA-1* gene that are summarized in detail below.

In hereditary breast cancers (Tapia et al., 2008), approximately 50% of *BRCA-1* mutation invasive ductal carcinomas have hypermethylated *BRCA-1*, and of these, ~40–50% are ERα- and PR-negative and 90% are Her2-negative. These observations have led to the generally accepted idea that hypermethylation of *BRCA-1* is a key feature of triple-negative hereditary breast cancers. Possibly, CpG methylation of the *BRCA-1* gene may function as a second hit-inactivating mechanism in *BRCA-1* mutation carriers.

The hypermethylation of the *BRCA-1* gene in sporadic breast tumors tends to correlate with both lower BRCA-1 and ERα protein expression. However, the frequency of CpG methylation varies considerably with tumor grade, with elevated frequency of hypermethylation (50–60%) in basal-like (Esteller et al., 2000; Matros et al., 2005;

TABLE 13.1
CpG Methylation of *BRCA-1* in Breast and Ovarian Tumors

Tumor Type	Frequency of CpG *BRCA-1* Methylation (%)	Molecular Markers and Biological End Points	Reference
		Breast Cancer	
Infiltrating adenocarcinoma	28	ERα-negative	Dobrovic and Simpfendorfer (1997)
Axillary node negative	14	NA	Rice et al (2000)
Medullary breast cancers	20	17/24 *ERα* hypermethylated	Wei et al (2008)
Metaplastic basal-like	63	Reduced BRCA-1 mRNA High ID4 expression	Turner et al (2007)
Invasive ductal carcinoma	38	*BRCA-1* hypermethylation Reduced BRCA-1 expression, Tumor Grade 3	Miyamoto et al (2002)
High grade	50	ERα-positive	Matros et al (2005)
Breast carcinomas	20	*BRCA-1* hypermethylation CpG methylation of CREB	Mancini et al (1998)
Rodent and human breast cancer cells	Reduced	BRCA-1 stimulates DNMT-1 expression	Shulka et al (2010)
Hereditary BRCA-1/ BRCA-2 invasive ductal carcinomas	50	*BRCA-1* hypermethylation correlates with HER-2 negative status	Tapia et al (2008)
Sporadic mucinous/ medullary	60	*BRCA-1* hypermethylated. More common with greater tumor grade	Esteller et al (2000)
Sporadic breast cancer UACC3199 cells	>60	*BRCA-1* methylated at 30 CpG sites	Rice et al (1998)
Sporadic	9	*BRCA-1* hypermethylated tumors with higher grade and infiltrating ductal type	Birgisdottir et al (2006)
Sporadic	45	Increased *BRCA-1* promoter methylation in PBCs for age >70 years	Bosviel et al (2012a, b)
Sporadic	5–30	3.5-fold higher risk with *BRCA-1* promoter methylation in PBCs	Wong et al (2011)
Sporadic	27–37	Triple negative	Lips et al (2013)
Sporadic	29	Triple negative	Jacot et al (2013)
Sporadic	25	Reduced BRCA-1 and ERα expression with hypermethylated *BRCA-1*	Bal et al (2012)

(*continued*)

TABLE 13.1 (Continued)
CpG Methylation of *BRCA-1* in Breast and Ovarian Tumors

Tumor Type	Frequency of CpG *BRCA-1* Methylation (%)	Molecular Markers and Biological End Points	Reference
		Ovarian, Cancer	
Sporadic	17	Reduced Survival with hypermethylated *BRCA-1*	Chiang et al. (2006)
Sporadic	16	Reduced *BRCA-1* expression with hypermethylated *BRCA-1*	Wilcox et al. (2005)
Sporadic	65	Reduced expression with hypermethylated *BRCA-1*	Chan et al. (2002)

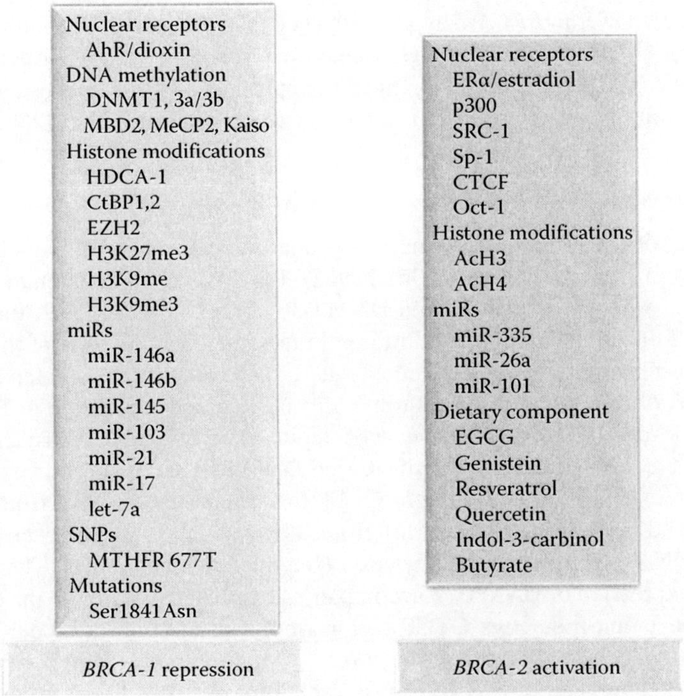

FIGURE 13.1 Summary of proposed epigenetic modifications and coregulators associated with repression or activation of BRCA-1 expression in breast tissue.

Turner et al., 2007) and higher histological grade (Miyamoto et al., 2002) tumors. A *BRCA-1* CpG site that is found to be hypermethylated in human sporadic tumors is a cAMP-responsive element (CRE=CGTCA)–binding (CREB) motif adjacent to the *BRCA-1* transcription start site (Mancini et al., 1998). The functional significance of the methylation event at the CpG–CRE site is unclear. Increased CREB expression and signaling have been observed in cancer tissues from patients with breast cancer (Xiao et al., 2010). Methylation of the CRE binding domain may prevent activation of *BRCA-1* expression by estrogen (Hockings et al., 2008). Meanwhile, CpG methylation of CRE-binding site(s) has been shown to create a binding site for c/EBPα, a transcription factor that induces differentiation (Rishi et al., 2010) and inhibits proliferation in breast cancer cells (Gery et al., 2005). These observations challenge the general assumption that CpG creates binding sites for proteins that repress transcription. In the case of the *BRCA-1* gene, CpG methylation at CRE sites may in fact generate core-binding elements for proteins that induce its transcription, as previously described for other genes (Chatterjee and Vinson, 2012). Detailed studies are needed to clarify the impact of site-specific methylation at the *BRCA-1* gene on BRCA-1 expression.

BRCA-1 promoter CpG methylation is also associated with poor ovarian cancer patient outcome (Chiang et al., 2006). In sporadic ovarian tumors, BRCA-1 expression is significantly lower in *BRCA-1* hypermethylated than unmethylated samples (Wilcox et al., 2005). Therefore, clarifying the mechanisms responsible for hypermethylation of the wild-type *BRCA-1* allele holds significant implications for the management of both breast and ovarian cancer in *BRCA-1* mutation carriers (Kennedy et al., 2002).

13.3.2 DNMTs and MBDs

The placing of CpG methylation marks is under the control of DNA methyltrasferases (DNMT) (Klose and Bird, 2006). DNMT-1 is involved in maintenance of DNA methylation, whereas DNMT-3a and DNMT-3b control de novo DNA methylation. Analyses of ductal and lobular infiltrating tumors revealed that 86% of the BRCA-1 low-expressing tumors were CpG methylated at the *BRCA-1* promoter and that a large subset of these tumors (19/24 tumors) had increased expression of DNMT-3b. Increased levels of DNMT3b have been reported in sporadic breast carcinomas (Girault et al., 2003). Tumors with increased DNMT3B show aberrant cytoplasmic localization of CTCF. The binding of CTCF to DNA insulates DNA from methylation (Bell and Felsenfeld, 2000; Graff et al., 1997). Conversely, the accessibility of CTCF to DNA is reduced by methylation (Baylin and Jones, 2011). Therefore, the increased expression of DNMT-3b accompanied by mislocalization to the cytoplasm of the methylation-protective CTCF factor are consistent with a model that links loss of methylation boundaries at the *BRCA-1* gene to increased hypermethylation and reduced expression of BRCA-1 in sporadic tumors. This model is corroborated by the observation that the displacement of CTCF from the nuclear matrix aids in the recruitment of histone deacetylases (HDACs) and histone methyl transferases (HMTs) (DiNardo et al., 2001).

Another transcription factor that protects CpG islands from methylation is Sp1. The *BRCA-1* promoters harbors several Sp1-binding sites (Butcher et al., 2004).

Recruitment of Sp-1 to the *BRCA-1* promoter is required for ERα-dependent activation of the *BRCA-1* transcription (Hockings et al., 2008). Consequently, conditions that displace CTCF and Sp-1 from the *BRCA-1* gene and elevate DNMT3b expression (Cui et al., 2009) may contribute to CpG methylation of the *BRCA-1* gene and tumorigenesis.

Not only is the *BRCA-1* gene a target for epigenetic regulation, but also the BRCA-1 protein itself affects global DNA methylation through activation of the *DNMT-1* gene. The BRCA-1 protein binds to the promoter region of *DNMT-1* through an Oct-1 site and stimulates DNMT-1 expression. Conversely, inactivation of *BRCA-1* expression lowers DNMT-1 levels and global hypomethylation, with significantly higher expression of oncogenes (i.e., c-Myc) (Shi et al., 2012; Shukla et al., 2010). Interestingly, BRCA-1 interacts also with Oct-1 at the promoter region of the *ERα* gene and stimulates its expression (Hosey et al., 2007). Meanwhile, the ERα protein is required for activation of *BRCA-1* transcription (Jeffy et al., 2005). As a result, loss of expression of ERα and BRCA-1 via hypermethylation at their respective genes may disrupt the regulatory circuitry between *BRCA-1* and *ERα* (Gorski et al., 2009), and induce global DNA hypomethylation through reduction of DNMT-1 expression. This may explain, at least in part, why tumors that have hypermethylated *BRCA-1* tend to be also ER-negative (Wei et al., 2008), and have reduced DNMT-1 levels (Shukla et al., 2010).

Nuclear factors that contribute to epigenetic regulation include methyl-CpG binding (MBD) and Kaiso proteins. The MBD family members MECP2, MBD1, MBD2, and MBD4 bind methylated DNA (Tate and Bird, 1993). Kaiso proteins recognize preferentially methylated CpGs and form complexes with the corepressor protein N-CoR (Yoon et al., 2003). The MBD2 protein was found to be associated with the methylated region of the *BRCA-1* promoter (Auriol et al., 2005). Also, increased association of MBD-2 with the hypermethylated *BRCA-1* promoter was observed in human breast cancer cells (Papoutsis et al., 2012) and rodent mammary tissue (Papoutsis et al., 2013) following exposure to ligands of the aromatic hydrocarbon receptor (AhR). The AhR is a nuclear receptor that binds environmental (i.e., dioxins, polycyclic aromatic hydrocarbons), endogenous (i.e., arachidonic acid), and a multitude of food compounds (i.e., resveratrol, indole-3-carbinol) (Denison and Nagy, 2003). Consequently, environmental, metabolic, and dietary conditions that increase the levels of AhR agonists may increase the risk of breast cancer through stimulation of CpG methylation-mediated repression of BRCA-1 (Romagnolo et al., 2010). Finally, studies documented that the BRCA-1 was a target for repression by the MeCP2 since overexpression of MeCP2 protein led to methylation-dependent inhibition of *BRCA-1* expression (Magdinier et al., 2000).

13.3.3 Histone Modifications

BRCA-1 transcription is induced by estradiol both in ERα-positive mammary and ovarian epithelial cells (Gudas et al., 1995; Romagnolo et al., 1998). Activation of *BRCA-1* transcription by estradiol triggers the recruitment of ERα and the histone acetylase p300 (Jeffy et al., 2005), and increases the association of acetylated H4 (AcH4) with the *BRCA-1* gene (Hockings et al., 2006). Conversely, the recruitment of the ligand-bound AhR to xenobiotic response elements (XRE) harbored in the *BRCA-1* gene hampers the association of the *BRCA-1* promoter with p300, SRC-1, and AcH4, while enhancing the occupancy by the histone deacetylase 1 (HDAC-1), and association with monomethylated (H3K9me) and trimethylated (H3K9me3) H3 at lysine 9 (Papoutsis et al., 2010). The latter are markers of repressed chromatin (Esteller, 2007).

The C-terminal binding proteins (CtBP) are NADH-dependent transcriptional repressors. Localization of CtBP proteins (CtBP-1 and CtBP-2) to the *BRCA-1* promoter assists in the recruitment of a corepressor complex containing HDAC-1 that reduces the association of AcH3 and AcH4 with the *BRCA-1* promoter. Conversely, the formation of the CtBP/HDAC-1 repressor complex at the *BRCA-1* promoter is disrupted by treatment with estrogen or in the presence of increased NAD+/NADH ratios (Di et al., 2010). These data provide important evidence that the *BRCA-1* gene is a target for epigenetic repression under metabolic conditions (i.e., high caloric intake) that would increase NADH levels (Di et al., 2013). The latter observation may have important implications for the development of dietary guidelines for *BRCA-1* mutation patients who are obese or on high carbohydrate/fat diet.

A component of the epigenetic machinery that influences *BRCA-1* expression is the enhancer of zeste homolog 2 (EZH2) protein, which is the catalytic subunit of the polycomb repressive complex 2 (PRC2). EZH2 functions as a histone methyltransferase responsible for trimethylation of histone H3 at lysine 27 (H3K27me3), a histone mark associated with transcription repression (Esteller, 2007). It is generally accepted that increased expression of EZH2 is a marker of aggressive breast cancer (Collett et al., 2006). The *BRCA-1* gene is a molecular target for EZH2, since EZH2 knockdown increases *BRCA-1* expression and reduces proliferation (Gonzalez et al., 2009). Conversely, EZH2 expression is higher in BRCA-1 deficient tumors (Puppe et al., 2009). The BRCA-1 protein binds to, and is a negative regulator of, EZH, and the reduced expression of BRCA-1 has been linked to uncontrolled activity of EZH2 and the development of basal-like breast tumors (Wang and Huang, 2013). Overall, nutritional targeting of EZH2 should be considered when developing prevention strategies against BRCA1-related tumors.

13.3.4 miRs

Alterations in microRNA (miR) expression have been linked to breast cancer development (Blenkiron et al., 2007; Bockmeyer et al., 2011; Murria et al., 2013). For example, levels of miR-146a and/or miR-146b-5p tend to be elevated in basal-like mammary tumor epithelial cell lines and in triple negative breast cancers, which mimic tumors arising in *BRCA-1* mutation carriers (Garcia et al., 2011). In particular, the binding of miR-146a and miR-146b-5p to the 3′-UTR of BRCA1 was reported to downregulate *BRCA-1* expression leading to increased proliferation and reduced homologous recombination rate, two processes controlled by BRCA1. Also, tumors of *BRCA-1* mutation carriers have been found to overexpress miR-17, let-7a, and miR-21 compared with sporadic breast cancers (Pinto et al., 2013). This differential pattern of miR expression may offer a biomarker of susceptibility to hereditary breast tumors and a plausible target in the therapeutic setting.

Recent studies reported that the BRCA-1 protein repressed epigenetically oncogenic miR-155 expression by decreasing the association of AcH2A and AcH3 and increasing the association of HDAC-2, with the miR-155 promoter (Chang et al., 2011). The upregulation of miR-155 has been documented in many cancers, including those of the breast (Iorio et al., 2005; Volinia et al., 2006) and has been linked to increased proliferation. Studies by Chang et al. (2011) revealed that miR-155 expression was upregulated in *BRCA-1* conditional knockout mice (Brodie et al., 2001) and two *BRCA-1*–deficient cell lines, HCC1937 and MDA-MB-436, compared with *BRCA-1* expressing MCF-7 cells, which have wild-type and hypomethylated *BRCA-1* (Rice et al., 1998).

MicroRNA-101 has also been implicated at least indirectly in regulation of *BRCA-1*, via stimulation of ATM, which contributes with BRCA-1 to regulation of DNA repair (Zheng et al., 2012). The genomic loss of miR-101 is accompanied by overexpression of EZH2 (Varambally et al., 2008) and the formation of repressive chromatin complexes at the *BRCA-1* promoter.

Through inhibition of EZH2, miR26a may prevent loss of BRCA-1. The expression of miR-26a was reported to be lower in breast tumor specimens compared with those of paired normal tissues (Zhang et al., 2011). The stable transfection of MCF-7 cells with miR-26a reduced EZH2 levels. Conversely, high miR-26a expression was associated with decreased proliferation through inhibition of EZH2 expression (Jansen et al., 2011; Lu et al., 2011). These cumulative data attribute to both miR-101 and miR-26a cancer protective effects through interference with EZH2 expression. Finally, accumulation of miR-335 has been correlated to increased levels of *BRCA-1* in MCF-7 cells (Heyn et al., 2011) suggesting miR-355 may exert cancer protective effects in ERα-positive cells through activation of *BRCA-1* expression.

Germline variants in the *BRCA-1* 3′-untranslated region (3′-UTR) are a potential source of epigenetic regulation for *BRCA-1* as they may create new binding sites for miRs. For example, a *BRCA-1* variant that introduces a functional miR-103 target site has been identified in breast cancer cases and the coexpression of miR-103 reduces *BRCA-1* transcription activity (Brewster et al., 2012). In human breast cancers, high levels of miR-103 have been linked to metastasis and poor outcome through induction of epithelial-to-mesenchymal transition (Martello et al., 2010). Overall, these cumulative observations suggest that noncoding RNAs play an important role in the epigenetic process and may contribute to the etiology of breast tumorigenesis related to *BRCA-1*. These findings also suggest that genetic variations in 3′-UTR may introduce an additional level of epigenetic regulation and influence the risk of breast cancer and success of therapies.

13.4 EPIGENETIC REGULATION OF *BRCA-1* BY DIETARY COMPOUNDS

Although prevention of epigenetic silencing and reactivation of silenced tumor suppressor genes are intensive areas of research (Fang et al., 2007; Li et al., 2013a), only a limited number of studies have focused on the epigenetic reactivation of *BRCA-1* with dietary compounds. This is somewhat surprising given the causative role of *BRCA-1* silencing in the etiology of familial, and possibly, sporadic breast cancers.

A study that examined the relationships between intake of fruits and vegetables and risk of breast cancer in a French-Canadian cohort of *BRCA-1* mutation carriers reported a considerable (80%) reduction in risk with the highest quintile of intake (Ghadirian et al., 2009). If confirmed, these findings suggest great potential for reducing the risk of breast cancer in *BRCA-1* mutation carriers with compounds present in fruits and vegetables.

13.4.1 Genistein

Genistein and (–)-epigallocatechin-3-gallate (EGCG) are food components that inhibit DNMT activity in vitro and reactivate via demethylation multiple tumor suppressor genes (i.e., p16 and retinoic acid receptor β) (Fang et al., 2003). Genistein was found to induce BRCA-1 protein expression (Fan et al., 2006), and its prepubertal protective effects against mammary tumorigenesis were dependent on expression of BRCA-1 (de Assis et al., 2011). To our knowledge, however, only one study documented that, at pharmaceutical doses, genistein (18.5 μM) and daidzein (78.7 μM) reversed DNA hypermethylation of the *BRCA-1* gene in breast cancer cells (Bosviel et al., 2012a, b). In fact, considerably lower human plasma levels of total isoflavones (2.0 μM), genistein (1.0–2.0 μM), and daidzein (0.5 μM) have been reported following supplementation with a single dose (20 g) of soy equivalent to 37 mg of isoflavones (Franke et al., 1998) or habitual soy intake (Fanti et al., 2003). Therefore, it remains unclear whether or not the average consumption of isoflavones may contribute genistein and daidzein at physiological concentrations sufficient to modulate *BRCA-1* promoter methylation in breast tissue.

Genistein is an inhibitor of DNMT and its consumption reduces global DNA methylation in rodent models (Day et al., 2002). Consequently, one may speculate that stimulation of *BRCA-1* gene expression by genistein and EGCG may be due to combined inhibition of DNMT activity (Fang et al., 2007) on the *BRCA-1* gene and reactivation of *ERα* expression (Belguise et al., 2007; Li et al., 2013b). In addition, the activation of *BRCA-1* expression may be achieved through inhibition of DNMT3b expression (Li et al., 2012).

The epigenetic effects of genistein may also be influenced by interactions between genotype and timing and dose of exposure (Dolinoy et al., 2006; Vanhees et al., 2011). For example, in rats carrying a mutated tuberous sclerosis complex 2 (TSC2) tumor suppressor gene, the postnatal exposure to genistein activated the pregenomic PI3/AKT signaling but repressed EZH2 through phosphorylation. These changes correlated with reduced levels of H3K27me3, increased expression of estrogen-responsive genes, and higher tumor incidence and multiplicity of uterine leiomyomas (Greathouse et al., 2012). Meanwhile, prepubertal genistein was found to upregulate *BRCA-1* mRNA and reduce mammary tumorigenesis (Cabanes et al., 2004; Murril et al., 1996). Furthermore, a study that evaluated the effects of supplementation in healthy premenopausal women with low (37.2 mg/day) and high dose (128.8 mg/day) of phytoestrogens including genistein (13.2 versus 45.3 mg/day) on mammary promoter methylation reported *RARβ2* and *CCND2* gene methylation were decreased with circulating levels of genistein <2.2 μM, ~600 ng/mL), and increased with higher concentrations (>600 ng/mL) (Qin et al., 2009). These examples clearly indicate that depending on context (i.e., genotype, timing of exposure, and dose) genistein may exert protective as well as harmful effects, which should be carefully evaluated before its use can be recommended for breast cancer prevention (Romagnolo and Selmin, 2012).

A mechanism that may contribute to the preventative effects of genistein on breast cancer is inhibition of protumorigenic miRs. For example, studies reported that genistein reduced miR-21 expression levels in culture and in tumors in nude mice (Zaman et al., 2012). In light of the fact miR-21 is overexpressed in *BRCA-1* tumors, this may be a useful epigenetic biomarker to assess the efficacy of genistein and other polyphenols-based therapies in subjects carrying mutated and/or hypermethylated *BRCA-1* gene.

13.4.2 Resveratrol

A dietary compound and phytoestrogen that modulates *BRCA-1* expression is resveratrol. In breast cancer MCF-7 cells, the repression of BRCA-1 transcription via histone deacetylation and CpG methylation induced by an AhR ligand (TCDD) was reversed by pretreatment with resveratrol (Papoutsis et al., 2010, 2012) at concentrations (1.0 μM) approaching those achieved (2.4 μM) in human pharmacokinetic studies (Boocock et al., 2007; Patel et al., 2010). The pretreatment with resveratrol during gestation antagonized the repressive effects of dioxin on *BRCA-1* expression and CpG promoter methylation in mammary tissue of rodent offspring (Papoutsis et al., 2013). These protective effects of resveratrol may be related to its antagonistic and partial agonist effects toward the AhR and ERα, respectively, on the *BRCA-1* gene, as well as activation of ERα expression via demethylation of the *ERα* gene (Lee et al., 2012). Additionally, resveratrol may induce *BRCA-1* expression via repression of miR-155 (Banerjee et al., 2012; Tili et al., 2010b) and miR-21 (Tili et al., 2010a, b).

13.4.3 Folate

Folate is a dietary compound that along with methionine impacts one-carbon metabolism and regulates epigenetic regulation. A protective role of folate (≥1272 dietary folate equivalents/day, 10-year average) against ERα-negative breast cancers was observed in postmenopausal women (Maruti et al., 2009). These results contrast with those of the Prostate Lung, Colorectal and Ovarian Screen Trial, which reported a greater risk of breast cancer among women consuming >853 dietary folate equivalents/day (Stolzenberg-Solomon et al., 2006). One possibility is that the effects of folate on breast cancer development may be influenced by *ERα* status and interactions with polymorphisms in the methyl-tetrahydrofolate reductase (MTHFR) gene. In fact, a study that examined the role of allele variants 677T and 1298C for the MTHFR gene reported that the presence of the 677T allele conferred an increased risk of breast cancer in *BRCA-1* mutation carriers (p = 0.007), which tend to be ERα-negative, whereas the presence of 1298C allele conferred an increased risk of breast cancer in sporadic cases (p = 0.015) (Pepe et al., 2007). Similarly, individuals with the variant TT genotype of the MTHFR gene were at increased risk for developing breast (1.5-fold) and ovarian (0.6-fold) tumors compared with subjects with the CC genotype (Jakubowska et al., 2007). Like in the case of genistein, variants in the *BRCA-1* gene may influence the response to dietary and supplemental folate.

Proteomic studies that examined the differential protein expression profiles in cells with the *BRCA-1* mutation Ser1841Asn, revealed that expression of this mutated form of *BRCA-1* was associated with increased expression of the folate receptor 1 (FOLR1) (Crugliano et al., 2007). The protein encoded by the FOLR1 gene is a member of the family of folate receptors that bind folic acid and its reduced derivatives, and transport 5-methyl-tetrafolate into cells. Overexpression of FOLR1 is associated with the development of triple-negative breast cancers and poor prognosis (Zhang et al., 2013). Overall, these cumulative observations support the hypothesis that interactions between genetics, i.e., mutations in *BRCA-1* and polymorphisms (i.e., MTHFR), and folate nutrition alter epigenetically the risk of breast cancer.

13.4.3.1 Vitamin C

The influence of vitamin C on breast cancer development remains a subject of debate and investigation (Misotti and Gnagnarella, 2013). Evidence from prospective studies did not support an association between vitamin C intake from diet or supplementation and breast cancer risk (Hutchinson et al., 2012). This evidence contrasted with that of a meta-analysis of retrospective case-control studies, which suggested a 12%–14% reduced risk associated with dietary vitamin C (WCRF/AICR 2007), or the results from a cohort of Swedish women (Michels et al., 2001), which indicated vitamin C had a protective effect against breast cancer in women who were overweight and with a large intake of linoleic acid. Conversely, results from the Women's Health Initiative (Cui et al., 2008) and a Danish nested-case control study (Nissen et al., 2003) suggested that total vitamin C intake was associated with an increased risk of breast cancer in postmenopausal and older women (on average 64 years). Clearly, more controlled studies are needed to ascertain the cancer effects of dietary and supplemental vitamin C on breast cancer risk.

To our knowledge, no studies have examined the impact of vitamin C on BRCA1 promoter methylation in *BRCA-1* mutation or sporadic breast tumors. One mechanism through which vitamin C may influence the epigenetic process and breast cancer risk is through enhancement of the catalytic activity of ten-eleven translocation (Tet) dioxygenases for the oxidation of 5-methylcytosine (Blaschke et al., 2013). This would be followed by DNA demethylation of gene promoters and upregulation of gene expression. Therefore, vitamin C may be protective against breast cancer by inducing Tet-dependent methylation reprogramming of tumor suppressors including *BRCA-1*. This notion is supported by evidence that blockade of Tet function by the miR-22 oncogene prevents demethylation and reactivation of expression of the tumor suppressor miR-20 leading to epithelial–mesenchymal transition (EMT) and enhancement of breast cancer stemness and metastasis (Song et al., 2013).

13.4.4 Other Food Components

Cancer protective effects have also been reported for quercetin (10 μM) through repression of miR-155 (Boesch-Saadatmandi et al., 2011) and stimulations of miR-26 (Lam et al., 2012); the short chain fatty acid butyrate (1–2 mM) by lowering of miR-17 levels (Hu et al., 2011); curcumin analogue (0.5 μM) through upregulation of miR-26a (Bao et al., 2012), and indol-3-carbinol (100–150 μM) via downregulation of miR-21 (Melkamu et al., 2010). Finally, preventative effects have been attributed to EGCG (1 mg/day, 3 times a week) through inhibition of miR-21 expression in tumor xenografts (Siddiqui et al., 2011), and compounds present in pomegranate juice (5%) via induction in cell culture of the metastasis suppressor miR-335 (Wang et al., 2011). Studies are needed to examine the impact of doses and bioavailability of these and other food components, alone or in association, on expression of miRs that target the *BRCA-1* network. For example, plasma bioavailability of indol-3-carbinol in animals following oral administration (250 mg/kg) approached ~30 μM, which is considerably lower than the concentrations (50–100 μM) commonly used in cell culture experiments (Howells et al., 2007). Meanwhile, the concentration of curcumin (0.5 μM) found to upregulate miR-26a levels (Bao et al., 2012) was similar to the bioavailability (1.7 μM) measured in patients with high-risk or pre-malignant lesions following oral administration of curcumin (Cheng et al., 2001).

13.5 SUMMARY AND CONCLUSIONS

Sporadic breast cancers with hypermethylated *BRCA-1* share tumor characteristics with *BRCA-1* mutated tumors, i.e., they have a BRCAness phenotype characterized by lack of ERα, PR, and HER2 expression. The frequency of CpG methylation at the *BRCA-1* gene is high in infiltrating tumors that are ERα-negative. Expression of *BRCA-1* is linked to coexpression of ERα, which, however, is frequently silenced in sporadic breast tumors. Therefore, maintenance or reactivation of *BRCA-1* expression may depend on the status of *ERα* gene methylation. Consequently, breast cancer preventative strategies based on food components should consider their impact on the functionality of the *ERα–BRCA-1* feedback loop. Food compounds that induce *BRCA-1* expression include the phytoestrogens genistein and resveratrol. Both compounds offer excellent proof of principle for regulation of *BRCA-1* through epigenetic mechanisms, and reactivation of ERα via CpG demethylation and repression of miRs that are associated with *BRCA-1* tumors. The risk of breast tumor development in *BRCA-1* mutation carriers is also influenced by dietary compounds that regulate one-carbon metabolism such as folate.

Although studies provide clear evidence that inactivation of *BRCA-1* is a "driver" event in breast tumorigenesis (Ford et al., 1998), dietary interventions would be expected to epigenetically affect a large number of genes, which together with BRCA-1 may contribute to modifying breast cancer risk. For example, studies with breast cancer cell lines reported that a mixture of isoflavones (genistein and daidzein) reduced CpG methylation at both *BRCA-1* and *BRCA-2* promoters (Bosviel et al., 2012a, b). Moreover, Dagdemir et al. (2013) found that the treatment of breast cancer cells with a mixture of isoflavones modulated histone methylation and acetylation at genes whose protein products impinge on epigenetic regulation of *BRCA-1* such as ERα, ERβ, p300, EZH2, and SRC3. Therefore, the full impact of intervention strategies based on dietary compounds on breast cancer risk can only be appreciated by integrating all the pathways involved in epigenetic regulation of genes that contribute to regulation of the *BRCA-1* and related network. Future studies designed to clarifying which epigenetic signatures confer a BRCAness phenotype to sporadic breast tumors may provide new targets for management strategies based on food components.

Finally, caution should be exercised when translating preclinical findings with dietary factors into recommendations for human consumption because the preventative or promoting effects of each food component against breast cancer may be highly influenced by synergies or antagonisms with other similar acting compounds for which the cumulative dose may be significantly higher than that of individual compounds with poor bioavailability (Howells et al., 2007).

ACKNOWLEDGMENTS

This research was supported by grants from the U.S. Army Medical Research and Material Command (DAMD 10-1-0215 to D.F.R), the American Institute for Cancer Research (10A058 to D.F.R.), and the Arizona Cancer Center Support Grant P30CA23074.

REFERENCES

Alsop K, Fereday S, Meldrum C et al. BRCA mutation frequency and patterns of treatment response in BRCA mutation-positive women with ovarian cancer: A report from the Australian Ovarian Cancer Study Group. *J Clin Oncol.* 2012 Jul 20;30(21):2654–63.

Auriol E, Billard LM, Magdinier F et al. Specific binding of the methyl binding domain protein 2 at the BRCA1-NBR2 locus. *Nucleic Acids Res.* 2005 Jul 28;33(13):4243–54.

Bal A, Verma S, Joshi K et al. BRCA1-methylated sporadic breast cancers are BRCA-like in showing a basal phenotype and absence of ER expression. *Virchows Arch.* 2012 Sep;461(3):305–12.

Banerjee N, Talcott S, Safe S et al. Cytotoxicity of pomegranate polyphenolics in breast cancer cells in vitro and vivo: Potential role of miRNA-27a and miRNA-155 in cell survival and inflammation. *Breast Cancer Res Treat.* 2012 Nov;136(1):21–34.

Bao B, Ali S, Banerjee S et al. Curcumin analogue CDF inhibits pancreatic tumor growth by switching on suppressor microRNAs and attenuating EZH2 expression. *Cancer Res.* 2012 Jan 1;72(1):335–45.

Baylin SB, Jones PA. A decade of exploring the cancer epigenome—Biological and translational implications. *Nat Rev Cancer.* 2011 Sep 23;11(10):726–34.

Belguise K, Guo S, Sonenshein GE. Activation of FOXO3a by the green tea polyphenol epigallocatechin-3-gallate induces estrogen receptor alpha expression reversing invasive phenotype of breast cancer cells. *Cancer Res.* 2007 Jun 15;67(12):5763–70.

Bell AC, Felsenfeld G. Methylation of a CTCF-dependent boundary controls imprinted expression of the Igf2 gene. *Nature.* 2000 May 25;405(6785):482–5.

Berchuck A, Heron KA, Carney ME et al. Frequency of germline and somatic BRCA1 mutations in ovarian cancer. *Clin Cancer Res.* 1998 Oct;4(10):2433–7.

Birgisdottir V, Stefansson OA, Bodvarsdottir SK et al. Epigenetic silencing and deletion of the BRCA1 gene in sporadic breast cancer. *Breast Cancer Res.* 2006;8(4):R38.

Blaschke K, Ebata KT, Karimi MM et al. Vitamin C induces Tet-dependent DNA demethylation and a blastocyst-like state in ES cells. *Nature.* 2013 Aug 8;500(7461):222–6.

Blenkiron C, Goldstein LD, Thorne NP et al. MicroRNA expression profiling of human breast cancer identifies new markers of tumor subtype. *Genome Biol.* 2007;8:R214.

Bockmeyer CL, Christgen M, Müller M et al. MicroRNA profiles of healthy basal and luminal mammary epithelial cells are distinct and reflected in different breast cancer subtypes. *Breast Cancer Res Treat.* 2011;130:735–45.

Boesch-Saadatmandi C, Loboda A, Wagner AE et al. Effect of quercetin and its metabolites isorhamnetin and quercetin-3-glucuronide on inflammatory gene expression: Role of miR-155. *J Nutr Biochem*. 2011 Mar;22(3):293–9.

Boocock DJ, Faust GE, Patel KR et al. Phase I dose escalation pharmacokinetic study in healthy volunteers of resveratrol, a potential cancer chemopreventive agent. *Cancer Epidemiol Biomarkers Prev*. 2007;16:1246–52.

Bosviel R, Dumollard E, Déchelotte P et al. Can soy phytoestrogens decrease DNA methylation in BRCA1 and BRCA2 oncosuppressor genes in breast cancer? *OMICS*. 2012a May;16(5):235–44.

Bosviel R, Durif J, Déchelotte P et al. Epigenetic modulation of BRCA1 and BRCA2 gene expression by equol in breast cancer cell lines. *Br J Nutr*. 2012b Oct;108(7):1187–93.

Brewster BL, Rossiello F, French JD et al. Identification of fifteen novel germline variants in the BRCA1 3′UTR reveals a variant in a breast cancer case that introduces a functional miR-103 target site. *Hum Mutat*. 2012 Dec;33(12):1665–75.

Brodie SG, Xu X, Qiao W et al. Multiple genetic changes are associated with mammary tumorigenesis in Brca1 conditional knockout mice. *Oncogene*. 2001;20:7514–23.

Butcher DT, Mancini-DiNardo DN, Archer TK et al. DNA binding sites for putative methylation boundaries in the unmethylated region of the BRCA1 promoter. *Int J Cancer*. 2004 Sep 20;111(5):669–78.

Butcher DT, Rodenhiser DI. Epigenetic inactivation of BRCA1 is associated with aberrant expression of CTCF and DNA methyltransferase (DNMT3B) in some sporadic breast tumours. *Eur J Cancer*. 2007 Jan;43(1):210–9.

Cabanes A, Wang M, Olivo S et al. Prepubertal estradiol and genistein exposures up-regulate BRCA1 mRNA and reduce mammary tumorigenesis. *Carcinogenesis*. 2004 May;25 (5):741–8.

Chan KY, Ozçelik H, Cheung AN et al. Epigenetic factors controlling the BRCA1 and BRCA2 genes in sporadic ovarian cancer. *Cancer Res*. 2002 Jul 15;62(14):4151–6.

Chang S, Wang RH, Akagi K et al. Tumor suppressor BRCA1 epigenetically controls oncogenic microRNA-155. *Nat Med*. 2011;17:1275–82.

Chatterjee R, Vinson C. CpG methylation recruits sequence specific transcription factors essential for tissue specific gene expression. *Biochim Biophys Acta*. 2012 Jul;1819(7):763–70.

Cheng AL, Hsu CH, Lin JK et al. Phase I clinical trial of curcumin, a chemopreventive agent, in patients with high-risk or pre-malignant lesions. *Anticancer Res*. 2001 Jul–Aug;21(4B):2895–900.

Chiang JW, Karlan BY, Cass L et al. BRCA1 promoter methylation predicts adverse ovarian cancer prognosis. *Gynecol Oncol*. 2006 Jun;101(3):403–10.

Collett K, Eide GE, Arnes J et al. Expression of enhancer of zeste homologue 2 is significantly associated with increased tumor cell proliferation and is a marker of aggressive breast cancer. *Clin Cancer Res*. 2006 Feb 15;12(4):1168–74.

Crugliano T, Quaresima B, Gaspari M et al. Specific changes in the proteomic pattern produced by the BRCA1-Ser1841Asn missense mutation. *Int J Biochem Cell Biol*. 2007;39(1):220–6.

Cui Y, Shikany JM, Liu S et al. Selected antioxidants and risk of hormone receptor-defined invasive breast cancers among postmenopausal women in the Women's Health Initiative Observational Study. *Am J Clin Nutr*. 2008;87:1009–18.

Cui M, Wen Z, Yang Z et al. Estrogen regulates DNA methyltransferase 3B expression in Ishikawa endometrial adenocarcinoma cells. *Mol Biol Rep*. 2009 Nov;36(8):2201–7.

Dagdemir A, Durif J, Ngollo M et al. Histone lysine trimethylation or acetylation can be modulated by phytoestrogen, estrogen or anti-HDAC in breast cancer cell lines. *Epigenomics*. 2013 Feb;5(1):51–63.

Day JK, Bauer AM, DesBordes C et al. Genistein alters methylation patterns in mice. *J Nutr*. 2002 Aug;132(8 Suppl):2419S–23S.

Denison MS, Nagy SR. Activation of the aryl hydrocarbon receptor by structurally diverse exogenous and endogenous chemicals. *Annu Rev Pharmacol Toxicol.* 2003;43:309–34.

de Assis S, Warri A, Benitez C et al. Protective effects of prepubertal genistein exposure on mammary tumorigenesis are dependent on BRCA1 expression. *Cancer Prev Res (Phila).* 2011 Sep;4(9):1436–48.

Di LJ, Fernandez AG, De Siervi A et al. Transcriptional regulation of BRCA1 expression by a metabolic switch. *Nat Struct Mol Biol.* 2010 Dec;17(12):1406–13.

Di LJ, Byun JS, Wong MM et al. Genome-wide profiles of CtBP link metabolism with genome stability and epithelial reprogramming in breast cancer. *Nat Commun.* 2013;4:1449.

DiNardo DN, Butcher DT, Robinson DP et al. Functional analysis of CpG methylation in the BRCA1 promoter region. *Oncogene.* 2001 Aug 30;20(38):5331–40.

Dobrovic A, Simpfendorfer D. Methylation of the BRCA1 gene in sporadic breast cancer. *Cancer Res.* 1997 Aug 15;57(16):3347–50.

Dolinoy DC, Weidman JR, Waterland RA et al. Maternal genistein alters coat color and protects Avy mouse offspring from obesity by modifying the fetal epigenome. *Environ Health Perspect.* 2006 Apr;114(4):567–72.

Dworkin AM, Huang TH, Toland AE et al. Epigenetic alterations in the breast: Implications for breast cancer detection, prognosis and treatment. *Semin Cancer Biol.* 2009 Jun;19(3): 165–71.

Easton DF, Ford D, Bishop DT et al. Breast and ovarian cancer incidence in BRCA1-mutation carriers. Breast Cancer Linkage Consortium. *Am J Hum Genet.* 1995 Jan;56(1):265–71.

Esteller M, Silva JM, Dominguez G et al. Promoter hypermethylation and BRCA1 inactivation in sporadic breast and ovarian tumors. *J Natl Cancer Inst.* 2000 Apr 5;92(7):564–9.

Esteller M. Cancer epigenomics: DNA methylomes and histone-modification maps. *Nat Rev Genet.* 2007 Apr;8(4):286–98.

Fan S, Meng Q, Auborn K et al. BRCA1 and BRCA2 as molecular targets for phytochemicals indole-3-carbinol and genistein in breast and prostate cancer cells. *Br J Cancer.* 2006 Feb 13;94(3):407–26.

Fang MZ, Wang Y, Ai N et al. Tea polyphenol (-)-epigallocatechin-3-gallate inhibits DNA methyltransferase and reactivates methylation-silenced genes in cancer cell lines. *Cancer Res.* 2003 Nov15;63(22):7563–70.

Fang M, Chen D, Yang CS et al. Dietary polyphenols may affect DNA methylation. *J Nutr.* 2007 Jan;137(1 Suppl):223S–8S.

Fanti P, Stephenson TJ, Kaariainen IM et al. Serum isoflavones and soya food intake in Japanese, Thai and American end-stage renal disease patients on chronic haemodialysis. *Nephrol Dial Transplant.* 2003 Sep;18(9):1862–8.

Ford D, Easton DF, Stratton M et al. Genetic heterogeneity and penetrance analysis of the BRCA1 and BRCA2 genes in breast cancer families. The Breast Cancer Linkage Consortium. *Am J Hum Genet.* 1998 Mar;62(3):676–89.

Franke AA, Custer LJ, Tanaka Y. Isoflavones in human breast milk and other biological fluids. *Am J Clin Nutr.* 1998 Dec;68(6 Suppl):1466S–73S.

Garcia AI, Buisson M, Bertrand P et al. Down-regulation of BRCA1 expression by miR-146a and miR-146b-5p in triple negative sporadic breast cancers. *EMBO Mol Med.* 2011 May;3(5):279–90.

Genetics of Breast and Ovarian Cancer. 2013. Available at http://www.cancer.gov.

Gery S, Xie D, Yin D et al. Ovarian carcinomas: CCN genes are aberrantly expressed and CCN1 promotes proliferation of these cells. *Clin Cancer Res.* 2005 Oct 15;11(20): 7243–54.

Ghadirian P, Narod S, Fafard E et al. Breast cancer risk in relation to the joint effect of BRCA mutations and diet diversity. *Breast Cancer Res Treat.* 2009 Sep;117(2):417–22.

Girault I, Tozlu S, Lidereau R et al. Expression analysis of DNA methyltransferases 1, 3A, and 3B in sporadic breast carcinomas. *Clin Cancer Res.* 2003 Oct 1;9(12):4415–22.

Gonzalez ME, Li X, Toy K et al. Downregulation of EZH2 decreases growth of estrogen receptor-negative invasive breast carcinoma and requires BRCA1. *Oncogene*. 2009 Feb 12;28(6):843–53.

Gorski JJ, Kennedy RD, Hosey AM et al. The complex relationship between BRCA1 and ERalpha in hereditary breast cancer. *Clin Cancer Res*. 2009 Mar 1;15(5):1514–8.

Graff JR, Herman JG, Myöhänen S et al. Mapping patterns of CpG island methylation in normal and neoplastic cells implicates both upstream and downstream regions in de novo methylation. *J Biol Chem*. 1997 Aug 29;272(35):22322–9.

Greathouse KL, Bredfeldt T, Everitt JI et al. Environmental estrogens differentially engage the histone methyltransferase EZH2 to increase risk of uterine tumorigenesis. *Mol Cancer Res*. 2012 Apr;10(4):546–57.

Gudas JM, Nguyen H, Li T et al. Hormone-dependent regulation of BRCA1 in human breast cancer cells. *Cancer Res*. 1995 Oct 15;55(20):4561–5.

Hardy TM, Tollefsbol TO. Epigenetic diet: Impact on the epigenome and cancer. *Epigenomics*. 2011 Aug;3(4):503–18.

Heyn H, Engelmann M, Schreek S et al. MicroRNA miR-335 is crucial for the BRCA1 regulatory cascade in breast cancer development. *Int J Cancer*. 2011;129:2797–806.

Hockings JK, Thorne PA, Kemp MQ et al. The ligand status of the aromatic hydrocarbon receptor modulates transcriptional activation of BRCA-1 promoter by estrogen. *Cancer Res*. 2006 Feb 15;66(4):2224–32.

Hockings JK, Degner SC, Morgan SS et al. Involvement of a specificity proteins-binding element in regulation of basal and estrogen-induced transcription activity of the BRCA1 gene. *Breast Cancer Res*. 2008;10(2):R29.

Hosey AM, Gorski JJ, Murray MM et al. Molecular basis for estrogen receptor alpha deficiency in BRCA1-linked breast cancer. *J Natl Cancer Inst*. 2007 Nov 21;99(22):1683–94.

Howells LM, Moiseeva EP, Neal CP et al. Predicting the physiological relevance of in vitro cancer preventive activities of phytochemicals. *Acta Pharmacol Sin*. 2007 Sep;28(9):1274–304.

Hu S, Dong TS, Dalal SR et al. The microbe-derived short chain fatty acid butyrate targets miRNA-dependent p21 gene expression in human colon cancer. *PLoS One*. 2011 Jan 20;6(1):e16221.

Hutchinson J, Lentjes MA, Greenwood DC et al. Vitamin C intake from diary recordings and risk of breast cancer in the UK Dietary Cohort Consortium. *Eur J Clin Nutr*. 2012 May;66(5):561–8.

Iorio MV, Ferracin M, Liu CG et al. MicroRNA gene expression deregulation in human breast cancer. *Cancer Res*. 2005;65:7065–70.

Jacot W, Thezenas S, Senal R et al. BRCA1 promoter hypermethylation, 53BP1 protein expression and PARP-1 activity as biomarkers of DNA repair deficit in breast cancer. *BMC Cancer*. 2013 Nov 5;13(1):523.

Jakubowska A, Gronwald J, Menkiszak J et al. Methylenetetrahydrofolate reductase polymorphisms modify BRCA1-associated breast and ovarian cancer risks. *Breast Cancer Res Treat*. 2007 Sep;104(3):299–308.

Jansen MP, Reijm EA, Sieuwerts AM et al. High miR-26a and low CDC2 levels associate with decreased EZH2 expression and with favorable outcome on tamoxifen in metastatic breast cancer. *Breast Cancer Res Treat*. 2011 Nov 18 [Epub ahead of print].

Jeffy BD, Hockings JK, Kemp MQ et al. An estrogen receptor-alpha/p300 complex activates the BRCA-1 promoter at an AP-1 site that binds Jun/Fos transcription factors: Repressive effects of p53 on BRCA-1 transcription. *Neoplasia*. 2005;7:873–82.

Kennedy RD, Quinn JE, Johnston PG et al. BRCA1: Mechanisms of inactivation and implications for management of patients. *Lancet*. 2002 Sep 28;360(9338):1007–14.

Khan SI, Aumsuwan P, Khan IA et al. Epigenetic events associated with breast cancer and their prevention by dietary components targeting the epigenome. *Chem Res Toxicol*. 2012 Jan 13;25(1):61–73.

King MC, Wieand S, Hale K et al. National Surgical Adjuvant Breast and Bowel Project. Tamoxifen and breast cancer incidence among women with inherited mutations in BRCA1 and BRCA2: National Surgical Adjuvant Breast and Bowel Project (NSABP-P1) Breast Cancer Prevention Trial. *JAMA*. 2001 Nov 14;286(18):2251–6.

Klose RJ, Bird AP. Genomic DNA methylation: The mark and its mediators. *Trends Biochem Sci*. 2006 Feb;31(2):89–97.

Lam TK, Shao S, Zhao Y et al. Influence of quercetin-rich food intake on microRNA expression in lung cancer tissues. *Cancer Epidemiol Biomarkers Prev*. 2012 Dec;21(12):2176–84.

Lee H, Zhang P, Herrmann A et al. Acetylated STAT3 is crucial for methylation of tumor-suppressor gene promoters and inhibition by resveratrol results in demethylation. *Proc Natl Acad Sci U S A*. 2012 May 15;109(20):7765–9.

Levy-Lahad E, Friedman E. Cancer risks among BRCA1 and BRCA2 mutation carriers. *Br J Cancer*. 2007 Jan 15;96(1):11–5.

Li H, Xu W, Huang Y et al. Genistein demethylates the promoter of CHD5 and inhibits neuroblastoma growth in vivo. *Int J Mol Med*. 2012 Nov;30(5):1081–6.

Li Y, Chen H, Hardy TM et al. Epigenetic regulation of multiple tumor-related genes leads to suppression of breast tumorigenesis by dietary genistein. *PLoS One*. 2013a;8(1):e54369.

Li Y, Meeran SM, Patel SN et al. Epigenetic reactivation of estrogen receptor-α (ERα) by genistein enhances hormonal therapy sensitivity in ERα-negative breast cancer. *Mol Cancer*. 2013b Feb 4;12:9.

Lips EH, Mulder L, Oonk A et al. Triple-negative breast cancer: BRCAness and concordance of clinical features with BRCA1-mutation carriers. *Br J Cancer*. 2013 May 28;108(10):2172–7.

Lu J, He ML, Wang L et al. MiR-26a inhibits cell growth and tumorigenesis of nasopharyngeal carcinoma through repression of EZH2. *Cancer Res*. 2011 Jan 1;71(1):225–33.

Magdinier F, Ribieras S, Lenoir GM et al. Down-regulation of BRCA1 in human sporadic breast cancer; analysis of DNA methylation patterns of the putative promoter region. *Oncogene*. 1998 Dec 17;17(24):3169–76.

Magdinier F, Billard LM, Wittmann G et al. Regional methylation of the 5′ end CpG island of BRCA1 is associated with reduced gene expression in human somatic cells. *FASEB J*. 2000 Aug;14(11):1585–94.

Mancini DN, Rodenhiser DI, Ainsworth PJ et al. CpG methylation within the 5′ regulatory region of the BRCA1 gene is tumor specific and includes a putative CREB binding site. *Oncogene*. 1998 Mar 5;16(9):1161–9.

Martello G, Rosato A, Ferrari F et al. MicroRNA targeting dicer for metastasis control. *Cell*. 2010 Jun 25;141(7):1195–207.

Maruti SS, Ulrich CM, White E. Folate and one-carbon metabolism nutrients from supplements and diet in relation to breast cancer risk. *Am J Clin Nutr*. 2009 Feb;89(2):624–33.

Matros E, Wang ZC, Lodeiro G et al. BRCA1 promoter methylation in sporadic breast tumors: Relationship to gene expression profiles. *Breast Cancer Res Treat*. 2005;91:179–86.

Melchor L, Benítez J. The complex genetic landscape of familial breast cancer. *Hum Genet*. 2013 Aug;132(8):845–63.

Melkamu T, Zhang X, Tan J et al. Alteration of microRNA expression in vinyl carbamate-induced mouse lung tumors and modulation by the chemopreventive agent indole-3-carbinol. *Carcinogenesis*. 2010 Feb;31(2):252–8.

Michels KB, Holmberg L, Bergkvist L et al. Dietary antioxidant vitamins, retinol, and breast cancer incidence in a cohort of Swedish women. *Int J Cancer*. 2001 Feb 15;91(4):563–7.

Miki Y, Swensen J, Shattuck-Eidens D et al. A strong candidate for the breast and ovarian cancer susceptibility gene BRCA1. *Science*. 1994 Oct 7;266(5182):66–71.

Misotti AM, Gnagnarella P. Vitamin supplement consumption and breast cancer risk: A review. *Ecancermedicalscience*. 2013 Oct 23;7:365. eCollection 2013.

Miyamoto K, Fukutomi T, Asada K et al. Promoter hypermethylation and post-transcriptional mechanisms for reduced BRCA1 immunoreactivity in sporadic human breast cancers. *Jpn J Clin Oncol*. 2002 Mar;32(3):79–84.

Mullan PB, Quinn JE, Harkin DP et al. The role of BRCA1 in transcriptional regulation and cell cycle control. *Oncogene*. 2006 Sep 25;25(43):5854–63.

Murphy CG, Moynahan ME. BRCA gene structure and function in tumor suppression: A repair-centric perspective. *Cancer J*. 2010 Jan–Feb;16(1):39–47.

Murria Estal R, Palanca Suela S, de Juan Jiménez I et al. MicroRNA signatures in hereditary breast cancer. *Breast Cancer Res Treat*. 2013 Nov;142(1):19–30.

Murrill WB, Brown NM, Zhang JX et al. Prepubertal genistein exposure suppresses mammary cancer and enhances gland differentiation in rats. *Carcinogenesis*. 1996 Jul;17(7):1451–7.

Nissen SB, Tjønneland A, Stripp C et al. Intake of vitamins A, C, and E from diet and supplements and breast cancer in postmenopausal women. *Cancer Causes Control*. 2003;14:695–704.

Papoutsis AJ, Lamore SD, Wondrak GT et al. Resveratrol prevents epigenetic silencing of BRCA-1 by the aromatic hydrocarbon receptor in human breast cancer cells. *J Nutr*. 2010 Sep;140(9):1607–14.

Papoutsis AJ, Borg JL, Selmin OI et al. BRCA-1 promoter hypermethylation and silencing induced by the aromatic hydrocarbon receptor-ligand TCDD are prevented by resveratrol in MCF-7 cells. *J Nutr Biochem*. 2012 Oct;23(10):1324–32.

Papoutsis AJ, Selmin OI, Borg JL et al. Gestational exposure to the AhR agonist 2,3,7,8-tetrachlorodibenzo-p-dioxin induces BRCA-1 promoter hypermethylation and reduces BRCA-1 expression in mammary tissue of rat offspring: Preventive effects of resveratrol. *Mol Carcinog*. 2013 Oct 17 doi:10.1002/mc.22095..

Parvin JD. Overview of history and progress in BRCA1 research: The first BRCA1 decade. *Cancer Biol Ther*. 2004 Jun;3(6):505–8.

Patel KR, Brown VA, Jones DJ et al. Clinical pharmacology of resveratrol and its metabolites in colorectal cancer patients. *Cancer Res*. 2010;70:7392–9.

Pepe C, Guidugli L, Sensi E et al. Methyl group metabolism gene polymorphisms as modifier of breast cancer risk in Italian BRCA1/2 carriers. *Breast Cancer Res Treat*. 2007 May;103(1):29–36.

Pinto R, Pilato B, Ottini L et al. Different methylation and microRNA expression pattern in male and female familial breast cancer. *J Cell Physiol*. 2013 Jun;228(6):1264–9.

Puppe J, Drost R, Liu X et al. BRCA1-deficient mammary tumor cells are dependent on EZH2 expression and sensitive to Polycomb Repressive Complex 2-inhibitor 3-deazaneplanocin A. *Breast Cancer Res*. 2009;11(4):R63.

Qin W, Zhu W, Shi H et al. Soy isoflavones have an antiestrogenic effect and alter mammary promoter hypermethylation in healthy premenopausal women. *Nutr Cancer*. 2009;61(2):238–44.

Rice JC, Massey-Brown KS, Futscher BW et al. Aberrant methylation of the BRCA1 CpG island promoter is associated with decreased BRCA1 mRNA in sporadic breast cancer cells. *Oncogene*. 1998;17:1807–12.

Rice JC, Ozcelik H, Maxeiner P et al. Methylation of the BRCA1 promoter is associated with decreased BRCA1 mRNA levels in clinical breast cancer specimens. *Carcinogenesis*. 2000 Sep;21(9):1761–5.

Rishi V, Bhattacharya P, Chatterjee R et al. CpG methylation of half-CRE sequences creates C/EBPalpha binding sites that activate some tissue-specific genes. *Proc Natl Acad Sci U S A*. 2010 Nov 23;107(47):20311–6.

Romagnolo D, Annab LA, Thompson TE et al. Estrogen upregulation of BRCA1 expression with no effect on localization. *Mol Carcinog*. 1998 Jun;22(2):102–9.

Romagnolo DF, Degner S, Selmin OI et al. Aryl hydrocarbon receptor-mediated carcinogenesis and modulation by dietary xenobiotic and natural ligands. In *Bioactive Compounds and Cancer*, Milner JH and Romagnolo DF (eds). Humana Press; New York, 2010; 761–82.

Romagnolo DF, Selmin OI. Flavonoids and cancer prevention: A review of the evidence. *J Nutr Gerontol Geriatr*. 2012;31(3):206–38.

Seery LT, Knowlden JM, Gee JM et al. BRCA1 expression levels predict distant metastasis of sporadic breast cancers. *Int J Cancer*. 1999 Jun 21;84(3):258–62.

Shi JF, Li XJ, Si XX et al. ERα positively regulated DNMT1 expression by binding to the gene promoter region in human breast cancer MCF-7 cells. *Biochem Biophys Res Commun*. 2012 Oct 12;427(1):47–53.

Shukla V, Coumoul X, Lahusen T et al. BRCA1 affects global DNA methylation through regulation of DNMT1. *Cell Res*. 2010 Nov;20(11):1201–15.

Siddiqui IA, Asim M, Hafeez BB et al. Green tea polyphenol EGCG blunts androgen receptor function in prostate cancer. *FASEB J*. 2011 Apr;25(4):1198–207.

Song SJ, Poliseno L, Song MS et al. MicroRNA-antagonism regulates breast cancer stemness and metastasis via TET-family-dependent chromatin remodeling. *Cell*. 2013 Jul 18;154(2):311–24.

Stefanska B, Karlic H, Varga F et al. Epigenetic mechanisms in anti-cancer actions of bioactive food components—The implications in cancer prevention. *Br J Pharmacol*. 2012 Sep;167(2):279–97.

Stolzenberg-Solomon RZ, Chang SC, Leitzmann MF et al. Folate intake, alcohol use, and postmenopausal breast cancer risk in the Prostate, Lung, Colorectal, and Ovarian Cancer Screening Trial. *Am J Clin Nutr*. 2006;83:895–904.

Tapia T, Smalley SV, Kohen P et al. Promoter hypermethylation of BRCA1 correlates with absence of expression in hereditary breast cancer tumors. *Epigenetics*. 2008 May–Jun;3(3):157–63.

Tate PH, Bird AP. Effects of DNA methylation on DNA-binding proteins and gene expression. *Curr Opin Genet Dev*. 1993 Apr;3(2):226–31.

Taylor J, Lymboura M, Pace PE et al. An important role for BRCA1 in breast cancer progression is indicated by its loss in a large proportion of non-familial breast cancers. *Int J Cancer*. 1998 Aug 21;79(4):334–42.

Thompson ME, Jensen RA, Obermiller PS et al. Decreased expression of BRCA1 accelerates growth and is often present during sporadic breast cancer progression. *Nat Genet*. 1995;9:444–50.

Tili E, Michaille JJ, Alder H et al. Resveratrol modulates the levels of microRNAs targeting genes encoding tumor-suppressors and effectors of TGFβ signaling pathway in SW480 cells. *Biochem Pharmacol*. 2010a Dec 15;80(12):2057–65.

Tili E, Michaille JJ, Adair B et al. Resveratrol decreases the levels of miR-155 by upregulating miR-663, a microRNA targeting JunB and JunD. *Carcinogenesis*. 2010b Sep;31(9):1561–6.

Turner NC, Reis-Filho JS, Russell AM et al. BRCA1 dysfunction in sporadic basal-like breast cancer. *Oncogene*. 2007;26:2126–32.

Vanhees K, Coort S, Ruijters EJ et al. Epigenetics: Prenatal exposure to genistein leaves a permanent signature on the hematopoietic lineage. *FASEB J*. 2011 Feb;25(2):797–807.

Varambally S, Cao Q, Mani RS et al. Genomic loss of microRNA-101 leads to overexpression of histone methyltransferase EZH2 in cancer. *Science*. 2008;322:1695–9.

Volinia S, Calin GA, Liu CG et al. A microRNA expression signature of human solid tumors defines cancer gene targets. *Proc Natl Acad Sci U S A*. 2006 Feb 14;103(7):2257–61.

Wang L, Alcon A, Yuan H et al. Cellular and molecular mechanisms of pomegranate juice-induced anti-metastatic effect on prostate cancer cells. *Integr Biol (Camb)*. 2011 Jul;3(7):742–54.

Wang L, Huang H. EZH2 takes the stage when BRCA1 loses. *Cell Cycle*. 2013 Dec 1;12(23):3575–6.

WCRF/AICR. *Food, Nutrition, Physical Activity, and the Prevention of Cancer: A Global Perspective*. AICR; Washington, DC, 2007.

Wei M, Xu J, Dignam J et al. Estrogen receptor alpha, BRCA1, and FANCF promoter methylation occur in distinct subsets of sporadic breast cancers. *Breast Cancer Res Treat.* 2008 Sep;111(1):113–20.

Wilcox CB, Baysal BE, Gallion HH et al. High-resolution methylation analysis of the BRCA1 promoter in ovarian tumors. *Cancer Genet Cytogenet.* 2005 Jun;159(2):114–22.

Wilson CA, Ramos L, Villaseñor MR et al. Localization of human BRCA1 and its loss in high-grade, non-inherited breast carcinomas. *Nat Genet.* 1999 Feb;21(2):236–40.

Wong EM, Southey MC, Fox SB et al. Constitutional methylation of the BRCA1 promoter is specifically associated with BRCA1 mutation-associated pathology in early-onset breast cancer. *Cancer Prev Res (Phila).* 2011 Jan;4(1):23–33.

Xiao X, Li BX, Mitton B et al. Targeting CREB for cancer therapy: Friend or foe. *Curr Cancer Drug Targets.* 2010 Jun;10(4):384–91.

Yoon HG, Chan DW, Reynolds AB et al. N-CoR mediates DNA methylation-dependent repression through a methyl CpG binding protein Kaiso. *Mol Cell.* 2003 Sep;12(3):723–34.

Yoshikawa K, Honda K, Inamoto T et al. Reduction of BRCA1 protein expression in Japanese sporadic breast carcinomas and its frequent loss in BRCA1-associated cases. *Clin Cancer Res.* 1999 Jun;5(6):1249–61.

Zaman MS, Shahryari V, Deng G et al. Up-regulation of microRNA-21 correlates with lower kidney cancer survival. *PLoS One.* 2012;7(2):e31060.

Zhang B, Liu XX, He JR et al. Pathologically decreased miR-26a antagonizes apoptosis and facilitates carcinogenesis by targeting MTDH and EZH2 in breast cancer. *Carcinogenesis.* 2011;32:2–9.

Zhang Z, Wang J, Tacha DE et al. Folate receptor α associated with triple-negative breast cancer and poor prognosis. *Arch Pathol Lab Med.* 2013 Sep 13 PMID:24028341.

Zheng Z, Ng WL, Zhang X et al. RNAi-mediated targeting of non-coding and coding sequences in DNA repair gene messages efficiently radiosensitizes human tumor cells. *Cancer Res.* 2012 Jan 11 72:1221–8.

14 Regulation of Histone Acyltransferases and Deacetylases by Bioactive Food Compounds for the Prevention of Chronic Diseases

Tho X. Pham and Ji-Young Lee

CONTENTS

14.1 INTRODUCTION TO EPIGENETIC MECHANISMS OF BIOACTIVE COMPOUNDS

Epigenetics is defined as the study of heritable changes in gene expression without underlying changes in DNA sequence [1–4]. DNA methylation and posttranslational modifications of histones are good examples for epigenetic mechanisms that can regulate gene expression. Chromatin modifiers that mediate the methylation, acetylation, and other modifications of histones represent targets to elicit changes in gene expression. The activities of chromatin-modifying enzymes are regulated, in part, by the concentration of intermediary metabolites that are utilized as substrates or cofactors for enzymatic activity [5,6]. It is conceivable that perturbation or changes in metabolism associated with obesity and aging could affect the activity of chromatin modifiers and thus gene expression. The opposite could also be true, where changes in the activity of chromatin modifiers could alter metabolism. Fortunately, many epigenetic modifications are reversible, leaving the door open for manipulation of chromatin-modifying enzymes through the consumption of bioactive food components as well as pharmacological agents.

Emerging evidence demonstrates that bioactive food components may elicit their health-promoting properties by epigenetic regulation. In this chapter, epigenetic mechanisms such as histone modifications and DNA methylation will be briefly reviewed, with a greater emphasis on histone acetylation. The family of enzymes that mediates the addition or removal of acetyl moieties, and their known roles in the pathogenesis of obesity-related metabolic diseases will be comprehensively discussed. Finally, bioactive compounds that have an epigenetic mode of action and their roles in the prevention of metabolic diseases will be reviewed.

14.2 EPIGENETIC REGULATION

In eukaryotic cells, 147 base pairs of DNA are wrapped around a histone octamer containing two copies of histones H2A, H2B, H3, and H4, which forms a nucleosome, the basic unit of chromatin [7]. The interactions between histones and DNA as well as between histones dictate chromatin compactness and ultimately regulate permissiveness or accessibility of DNA to transcription factors [8,9]. Each of the histone proteins contains a C-terminal globular domain that is buried in the interior of the protein and an unstructured N-terminal tail that protrudes from the globular C-terminal domain. The N-terminal tail is responsible for ~25%–30% of histone mass [10]. Both the globular and tail domains of histones undergo posttranslational modifications, which affect the interaction of histones and DNA, as well as histone-histone interactions. The posttranslation modifications of histones and resulting changes in chromatin structure are thought to be inherited in epigenetics [11,12]. DNA methylation and noncoding RNAs are other epigenetic mechanisms that can alter gene transcription. Methylated DNA at the promoters of genes are typically associated with heterochromatin, while noncoding RNAs can act as guiding molecules for the recruitment of chromatin modifying enzymes to specific genome locus [13,14].

Histones can undergo methylation, acetylation, phosphorylation, biotinylation, ubiqutination, sumyolation, ADP-ribosylation, glycosylation, succinylation, and malonylation that can affect their association with DNA and neighboring histones [15–17]. Of these modifications, histone methylation and acetylation are the best characterized. Much less is known about the roles of other modifications in regard to enzymes that mediate the reactions, as well as their physiological significance. Among histone modifications, histone acetylation was the first to be discovered [18]. Histone acetylation neutralizes positive charge of lysine residues on histone tails, which weakens charge-dependent interactions between histones and DNA or adjacent histones [15]. Consequently, chromatin is relaxed and transcriptional machinery can have easier access to DNA to increase gene expression. Histones can also be monomethylated, dimethylated, or trimethylated on their tails. Methylation on the lysine residues of histones does not alter its positive charge and, therefore, is thought to have less of a direct effect on the electrostatic interactions of DNA and histones compared with acetylation [15]. Furthermore, the role of histone methylation in the regulation of gene transcription remains unclear because inhibition of trimethylation of histone H3 lysine 4 has a minimal effect on the transcription of certain genes [19]. Instead, it has been suggested that histone methylation may function to recruit specific repressor complexes, such as polycomb repressive complex 2 and other repressor complexes, to the promoters of genes for silencing [15]. Like DNA methylation, histone methylation was originally thought to be irreversible, but only in 2004 was the first histone demethylase, lysine-specific demethylase 1 (LSD1), discovered to remove methyl marks from histone H3 lysine 4 [20]. Since then, others histone demethylases belonging to the jumonji domain containing 2 (JMJD2) family of histone demethylases have been discovered. Some of histone demethylases could remove methyl groups from trimethylated lysine, a mark that was considered irreversible and that LSD1 could not catalyze [21,22]. Of particular interest to human health and nutrition is that the enzymatic activities of LSD1 and the JMJD2 family are dependent on intermediary metabolites. LSD1 is dependent on availability of flavin adenine dinucleotide (FAD) as a cofactor, while the JMJD2 family of enzymes are dependent on α-ketoglutrate and Fe(II), tying the activity of these enzymes to cellular energy status [23,24].

DNA methylation is the best characterized epigenetic modification, whereby a cytosine at a CpG dinucleotide gets methylated [25]. This modification was first suggested to have a role on epigenetics in 1975 when DNA methylation was shown to be responsible for the inactivation of chromosome X by Riggs et al. [26]. The silencing effect of DNA methylation on gene expression was also independently reported by Holliday et al. [27]. It is estimated that approximately 70% of gene promoters reside within CpG islands, which are stretches of ~1000 base pairs of DNA that are rich in CpGs [28]. The majority of CpG islands are unmethylated in cells [29] and unmethylated CpG islands are thought to act as promoters, contain transcriptional start sites, and serve as binding sites for certain transcription factors [28]. It is suggested that gene repression due to DNA methylation is attributed to direct inhibition of transcription factor binding or mediated by methyl-binding proteins that recruit chromatin-modifying complexes [30,31]. In agreement with these concepts, it has been demonstrated that DNA methylation plays an important role in the regulation cell type-specific gene expression, which is dependent, in part, on hypomethylated or hypermethylated status of a gene promoter [32].

It is now known that DNA methylation is mediated by the enzyme family of DNA methyltransferase (DNMT), which includes DNMT1, DNMT3a, and DNMT3b [33]. It is believed that DNMT3a and DNMT3b are responsible for the initial CpG methylation, while DNMT1 plays a role in maintaining methylation patterns throughout replication [34]. Of the three types of epigenetic mechanism, i.e., DNA methylation, histone modification, microRNAs, the heritability of DNA methylation is the best understood. It is now clear that during the semiconservative process of DNA replication, the parental DNA strand containing 5-methylated cytosine serves as the template for the synthesis of the daughter strand. This yields double-stranded DNA, containing a parental strand that is still methylated and a daughter strand that is unmethylated, called hemi-methylated DNA. DNMT1 is able to recognize hemi-methylated DNA and then methylate the cytosine on the daughter strand and thus maintaining methylation state of DNA during replication [35]. Meanwhile, currently, there are no known enzymes that can mediate the direct removal of the methyl group from cytosine. Rather, DNA demethylation is suggested to occur in multiple steps. It has been demonstrated that 5-methylcytosine is hydroxylated to 5-hydroxymethylcytosine (5-hmC) by the ten-eleven translocation (TET) family of enzymes [36]. 5-hmC is then recognized by the base excision repair or the nucleotide excision repair system, replacing it with an unmethylated cytosine [37,38]. Similarly, to the JMJD2 family of enzymes, the TET family is also dependent on α-ketoglutrate and Fe(II) for catalytic activity, which suggests that this family of enzymes may also be regulated by cellular energy states [24,36].

14.3 HISTONE ACETYLTRANSFERASES AND DEACETYLASES

14.3.1 Histone Acetyltransferases

The acetylation of histones is mediated by histone acetyltransferases (HATs), which is a diverse family of proteins that utilizes acetyl-CoA as the acetyl group donor [39]. In humans, HATs can be divided into four families. All HATs contain an acetyltransferase domain, and sequence comparison and additional motifs are used to categorize them into subfamilies [40]. Family members share significant protein sequence homology, while members between families have very little sequence similarities [41]. Members of the general control non-derepressible 5 (GCN5)–related *N*-acetyltransferase (GNAT) family consist of GCN5 and its ortholog p300/cAMP response element binding protein (CREB)–binding protein (CBP)–associated factor (PCAF) [42]. The MYST family, named after its founding members, such as MOZ, Ybf2/Sas3, Sas2, and Tip60, includes monocytic leukemia zinc finger protein (MOZ), TAT-interactive protein with mass of 60kda (TIP60), MOZ-related factor (MORF), and HAT bound to ORC1 (HBO1) [42]. The p300/CBP family consists of p300 and CBP. The fourth family of HAT is the steroid receptor coactivators (SRC) family that consists of SRC-1, nuclear receptor coactivator (ACTR), SRC-3, and TATA box-binding (TBP)–associated factor (TAF) [41–43].

The regulation of HATs is as diverse as the members of this class of enzymes [42]. The GNAT family functions primarily as coactivators or adaptors in transcriptional complexes to facilitate gene transcription. They also contain bromodomains, which have been demonstrated to physically interact with acetyl-lysines [44,45]. Members of the MYST family have diverse functions. Tip60 has been shown to be a coregulator for several nuclear receptors, such as androgen receptor and Rev-erbβ, and to interact with cytoplasmic receptors, e.g., interleukin 9 receptor [46–48]. The MYST member, MOZ is linked to recurrent acute myeloid leukemia, where gene rearrangement produces fusion proteins of MOZ and other HATs such as CBP and p300 [49]. Among the HAT families, the CBP/p300 family has less substrate specificity than the other HAT enzymes because recombinant CBP and p300 have been shown to acetylate all four soluble histones as well as histones bound to DNA [50]. In contrast to CBP/p300, GCN5 preferentially acetylated histone H3 and H4 [51]. Furthermore, CBP and p300 have been found to acetylate a variety of different transcription factors, such as hepatocyte nuclear factor 4 and nuclear factor κB (NF-κB), to regulate their transcriptional activity [52,53]. The SRC family consists primarily of coactivators that possess intrinsic HAT activity and associates with nuclear hormone receptors, such as estrogen receptors and androgen receptors [43,54].

14.3.2 Histone Deacetylases

Histone deacetylases (HDACs) are the enzymes that facilitate the removal of acetyl groups from acetylated histones, which is an opposing effect of HATs [55]. HDACs are categorized into two families and four classes. The classical zinc-dependent HDACs of the Rpd3/Hda1 family are made up of class I, II and IV, while the NAD^+-dependent sirtuin (SIRT) family is categorized as class III HDACs [56]. Class I HDACs include HDAC1, HDAC2, HDAC3, and HDAC8. Class II HDACs are further subcategorized into class IIa and IIb: members of class IIa are HDAC4, HDAC5, HDAC7, and HDAC9; and class IIb members are HDAC6 and HDAC10. Class III HDACs are SIRT1 through SIRT7 and the only member of class IV is HDAC11 [56,57].

Class I HDACs are ubiquitously expressed in various cell types, while the expression of class II HDACs is varied among cell types and at different stages of the cell cycle. HDAC isoforms display distinct subcellular localization, which is important for their function and activity. The subcellular localization of each isoform is signal-dependent as well as cell type-dependent. In general, with the exception of HDAC3, class I HDACs are found predominately in the nucleus. Both HDAC1 and HDAC2 contain nuclear localization sequences, but lack a nuclear export signal [56,58]. In contrast, HDAC3 has both nuclear import and export signals, and has been shown to translocate back and forth between the cytoplasm and the nucleus in response to cellular signals that have not been fully identified [59,60]. Interestingly, it has been demonstrated that HDAC3 is recruited to corepressor complexes by HDAC4, HDAC5, and HDAC7 in vitro [61]. Originally, overexpression of HDAC8 suggested that it was localized to the nucleus, but more recently, HDAC8 was found to reside within the cytoplasm where it interacts with cytoskeleton proteins and plays an important role in cell motility [62,63].

Compared with class I, class II HDACs shuttle in and out of the nucleus in response to cellular signaling. In particular, HDAC4 and HDAC5 are known to translocate from the nucleus to the cytoplasm upon phosphorylation by a calcium/

calmodulin-dependent protein kinase [64]. Phosphorylation of HDAC4 and HDAC5 facilitates the binding of 14-3-3 protein to the HDACs for export out of the nucleus [65]. Among class II HDAC isoforms, HDAC7, HDAC9, and HDAC10 possess splice variants [66–68]. In particular, one splice variant of HDAC9 that does not contain a catalytic domain, called histone deacetylase–related protein (HDRP)– or myocyte enhancer factor 2 (MEF-2)–interacting transcription repressor (MITR), has been shown to function as an adaptor in corepressor complexes and facilitate the recruitment of other HDACs to the complexes [69,70]. Interestingly, it has been suggested that the primary role of class IIa HDACs is the crucial recruitment of HDAC3 to corepressor complexes, which can provide the deacetylase activity in a signal-dependent manner [71,72]. This is supported by findings that in the absence of class I HDACs, the class IIa HDAC4 displays a 100-fold less of its enzymatic activity [73], suggesting that class I HDACs is required for optimal deacetylating activity of class IIa HDACs.

14.3.3 Dynamic Histone Acetylation for the Control of Gene Transcription

Acetylation of histones plays an important role in controlling the transcription of genes [74]. The steady state of acetylated histones is dependent on the balance between the enzymatic activities of HATs and HDACs [75]. The recruitment of HATs or HDACs to the promoters of genes by DNA-binding proteins generates a localized domain of modified histones that can influence transcriptional activity. HATs are typically found associated with DNA-binding coactivators, while HDACs are associated with corepressors. Therefore, typically, histone acetylation induces gene expression, while histone deacetylation represses the transcription of genes [74,76].

The turnover of acetylation on histones is relatively quick, with half-lives estimated to range from 4 to 87 min [77]. The variability in the half-lives can be explained by differences in cell type and whether specific histones or bulk histones were monitored [77–79]. Independent of cell type and methods, the turnovers of acetylation on histones are generally slower than phosphorylation, but faster than methylation [77]. In *Saccharomyces cerevisiae*, the acetylated group of histone H3 is removed within 1.5 min after acetylation of histone H3 and it takes 5–8 min for the deacetylated histones to be acetylated again [80]. Despite the rapid rate of acetylation and deacetylation, histone acetylation has been suggested to be heritable [81]. Furthermore, the acetylation state of histones can recruit histone methyltransferase and demethylase to induce more stable epigenetic modifications, such as histone methylation [82,83].

Nuclear receptors are ligand-dependent transcription factors that mediate upregulation and downregulation of gene transcription based on their association with coactivators or corepressors, respectively [84,85]. In the unliganded state, some nuclear receptors can repress gene transcription through their association with corepressors, such as nuclear receptor corepressor 1 (NCOR1) and silencing mediator of retinoic acid and thyroid hormone receptor (SMRT) [86,87]. These corepressors have been shown to associate with HDAC1, HDAC2, HDAC3, HDAC4, HDAC5, HDAC7, and HDAC9 in multiprotein complexes, where HDAC activity was found to be integral to the repression of gene transcription by nuclear receptors [68,88,89].

In contrast, binding of ligands to nuclear receptors alters their conformation to prevent corepressor binding and to promote their association with coactivators, such as ACTR and CREB [43,90]. These coactivators have been demonstrated to either possess intrinsic histone acetylation activity or recruit HATs to a coactivator complex [43]. Furthermore, the association of coactivators and corepressors are not limited to nuclear receptors, rather, the switching of coregulators as a mechanism to regulate gene transcription also extends to general transcription factors as well [91,92]. Coactivator and corepressor switching is regulated, in part, by the acetylation status of the transcription factor themselves. Both HATs and HDACs were originally discovered to acetylate histones, but their nonhistone protein targets have since been identified [93]. Nonhistone targets of HATs and HDACs includes, but are not limited to, general transcription factors, nuclear receptors, chaperones, signal transducers, and cytoskeletal proteins [94,95]. The acetylation of transcription factors can increase their association with coactivators by allowing them to be recognized by the bromodomain of coactivators, such as those of CBP [96]. In most cases, the acetylation of transcription factors and nuclear receptors increases their DNA-binding ability and association with coactivators, but reduction of DNA binding of acetylated transcription factors has also been found [52,97–100]. Thus, the interplay of HATs and HDACs regulates gene expression on multiple levels: from the electrostatic interaction of DNA and histones to the acetylation of transcription factors and their association with coregulators. In addition to the roles of HATs and HDACs in gene expression, the list of their nonhistone targets are ever expanding, implicating them as regulators of multiple cellular processes.

14.4 ROLES OF HDACs AND METABOLIC DISEASES

14.4.1 Ties of HATs and HDACs to Metabolism

In recent years, there has been an explosion of interest in epigenetics and its contribution to the development of obesity-related metabolic diseases, such as type 2 diabetes, nonalcoholic fatty liver disease (NAFLD), and cardiovascular disease (CVD). Pathogenesis of the metabolic diseases, arising from metabolic dysregulation, may involve epigenetic mechanisms [3,101]. In particular, acetylation of histones by HATs is closely tied to energy status because of its dependence on the availability of acetyl-CoA as the acetyl group donor [39]. In cells, distinct pools of acetyl-CoA exist, where cytoplasmic and nuclear acetyl-CoA is primarily used for acetylation by HATs, while mitochondrial acetyl-CoA is used for energy production [102]. Conversely, deacetylation of histones by HDACs releases free acetate, which can be used to regenerate acetyl-CoA by the action of acetyl-CoA synthetase1 and 2 [55]. In addition, HDAC activity can also be modulated by intermediates of energy metabolism, such as acetyl-CoA, malonyl-CoA, 3-hydroxy-3-methylglutaryl-CoA, and free CoA, which further link histone acetylation/deacetylation to energy status [103]. The reliance of HATs and other chromatin modifier such as the TET and the jumonji family on intermediary metabolites as cofactors for their enzymatic activity represents an avenue whereby metabolic perturbation, can affect the epigenome.

Epigenetic modifications have been a focus of cancer research, where minute accumulation of epigenetic changes over time has been thought to lead to the repression of tumor suppressor genes, consequently leading to the development of tumors [104]. In this regard, HDAC inhibitors have been successfully exploited and approved by the U.S. Food and Drug Administration (FDA) for their use in the treatment of cutaneous T-cell lymphoma, and there are ongoing clinical trials to test the use of HDAC inhibitors in the treatment of other type of cancers [3,105,106]. The epigenetic basis for the development of type 2 diabetes, NAFLD, and CVD has also been proposed to follow such a model, where epigenetic modifications lead to suppression of genes important for glucose utilization, maintenance of insulin signaling, and proper fat storage [3,107–109]. Currently, the use of HDAC inhibitors for the treatment of metabolic diseases is gaining support based on emerging evidence that HDACs play an important role in the progression of these diseases.

14.4.2 Dysregulation of HATs and HDACs in Metabolic Diseases

In healthy individuals, insulin is released by pancreatic β cells to facilitate the removal of glucose from the systemic circulation and to signal for the storage of excess energy in the form of triglycerides in the fed state [110,111]. In contrast, in the fasting state, glucagon is secreted by pancreatic α cells into the bloodstream, where its main physiological role is to stimulate hepatic glucose output by increasing glycogenolysis and gluconeogenesis [112]. Furthermore, a decrease in insulin-mediated suppression of lipolysis and increased catecholamine enhance triglyceride hydrolysis and consequently free fatty acid release from adipose tissue for energy utilization [113]. However, in obesity, excess fat accumulation leads to insulin resistance and metabolic dysfunction in adipose tissue, which ultimately triggers the onset of systemic insulin resistance [114]. Impaired systemic insulin signaling can lead to uncontrolled hepatic glucose output regardless of feeding state, and contribute to the development of type 2 diabetes and NAFLD [115]. One of primary factors that cause hepatic insulin resistance and the development of NAFLD is the accumulation of adipocyte-derived lipids, which have been estimated to account for ~10%–50% of free fatty acids in the liver [116]. Thus, the ability for adipose tissue to properly store fat is important for the prevention of obesity-related metabolic diseases.

It has been shown that HDAC3 regulates hepatic gluconeogenesis, a process that is dysregulated in type 2 diabetes [117]. It appears that HDAC3 is recruited to the promoter of lipogenic genes, such as fatty acid synthase and steroyl-CoA desaturase, to silence their expression and thus metabolic precursors are rerouted for gluconeogenesis instead of lipogenesis in the fasted state. In the fed state, HDAC3 is absent from the promoter of lipogenic genes, allowing their transcription and consequently lipogenesis. In contrast, when mice with a liver-specific deletion of *Hdac3* were fed a high fat diet for 4 weeks, they were more insulin-sensitive and had lower blood glucose levels than wild-type mice [117]. Therefore, liver HDAC3 appears to play an important role in high-fat diet–induced insulin resistance. Class IIa HDACs have also been implicated in the control of hepatic glucose production. It was demonstrated that during fasting, HDAC4, HDAC5, and HDAC7 translocate into the nucleus to the promoters of gluconeogenic enzymes, such as glucose-6-phosphotase and phosphoenolpyruvate carboxykinase [72]. This, in turn, induces transcriptional activation of gluconeogenic genes by recruiting HDAC3 to the promoter region, consequently increasing hepatic glucose production. The induction of gluconeogenic genes by the HDACs was not attributed to histone deacetylation, rather to the deacetylation of the transcription factor, forkhead box protein O1, which increases the activity of the transcription factor. Utilizing adenoviral small hairpin RNA to knockdown HDAC4, HDAC5, and HDAC7 in *db/db* mice, a strain of mice that eats excessively due to mutation in leptin receptor, it was demonstrated that suppression of all three HDACs could ameliorate hyperglycemia and improve insulin sensitivity [72].

Impaired glucose utilization by skeletal muscle largely contributes to elevated plasma glucose levels in type 2 diabetes [118]. In response to insulin, skeletal muscle is responsible for 75% of glucose disposal in humans and could contribute up to 95% in hyperglycemic conditions [119]. It has been demonstrated that MEF2, a transcription factor that controls muscle differentiation [120], associates with HDACs to regulate muscle differentiation as well as glucose transport in muscle. MEF2 has been shown to form a complex with HDAC4 and HDAC5 in the nucleus of undifferentiated muscle cells, and the association induces the repression of MEF2-dependent gene transcription. In response to calcium/calmodulin-dependent protein kinase phosphorylation, HDAC4 and HDAC5 are shuttled out of the nucleus, allowing the activation of MEF-2 for the induction of muscle differentiation program [64]. In resting muscle, MEF2 interacts with HDAC5, which represses the transcription of glucose transporter 4 (GLUT4). During exercise, HDAC5 is phosphorylated by adenosine monophosphate-activated kinase and shuttled out of the nucleus into the cytoplasm, allowing MEF2 to associate with coactivators to transcribe GLUT4 in skeletal muscle [120]. Knockdown of HDAC5 in human primary muscle cells increased glucose uptake, at least in part, by increasing the expression of GLUT4 [121]. Furthermore, Scriptaid, a novel pan-HDAC inhibitor, was able to increased histone H3 acetylation in muscle myotubes and increased glucose uptake through the induction of GLUT4 expression. Thus, it appears that certain HDACs play an essential role in the differentiation of muscle and the inhibition of HDAC activity may be utilized to increase glucose uptake.

Following myocardial infarction, cardiac remodeling helps maintain cardiac output, but ultimately leads to a decline in function due to cardiomyocyte hypertrophy and fibrosis [122]. Class IIa HDACs have been shown to play a role in the suppression of cardiomyocte hypertrophy through repression of MEF2 [123]. Initial heart weight and size of *Hdac9* knockout mice were similar to those of wild-type control, but after 8 months, the knockout mice had a 46% increase in the heart to body weight ratio compared with the wild type [123]. In this study, transgenic mice over expressing calcineurin, a kinase that phosphorylates class IIa HDACs for their export out of the nucleus, also displayed cardiac hypertrophy. Furthermore, mice with *Hdac9* deletion and calcineurin transgene had an increased cardiac mass by ~220% compared with controls. Interestingly, several studies in rats and mice utilizing aortic banding to induce myocardial infraction demonstrated that administration of pan-HDAC inhibitors can block cardiac hypertrophy following myocardial infraction, suggesting that following myocardial infarctions, administration of HDAC inhibitors can be used to avert the development of cardiac hypertrophy [124–126]. The aforementioned studies appear to be contradictory as to the roles of class IIa HDAC in the regulation of myocyte hypertrophy. Potentially, it can be explained that class IIa HDACs may suppress cardiac hypertrophy through a mechanism independent of HDAC activity or different classes of HDACs, i.e., class I, IIa, and IIb, may have differential effects on myocyte development in the heart.

In addition to the roles of HDACs in insulin signaling and energy utilization of muscle and the liver and cardiac remodeling, HDACs have also been implicated in the development of nonalcoholic steatohepatitis (NASH). Liver fibrosis, a key feature of NASH, involves the transdifferentiation of hepatic stellate cells (HSCs) to myofibroblasts [127]. Transforming growth factor β1 (TGFβ1) is an important cytokine that mediates HSC transdifferentiation and induces the expression of α-smooth muscle actin (α-SMA), a marker for myofibroblasts [128]. Studies in skin and lung fibroblasts demonstrated that HDAC4 is necessary for the induction of α-SMA by TGFβ1 [129,130]. In both studies, knockdown of HDAC4 abolished the TGFβ1-induced expression of α-SMA. As liver fibrosis is the result of excessive accumulation of extracellular matrix (ECM), the balance between ECM synthesis and breakdown is of importance [131]. Primary quiescent rat HSCs highly express matrix metalloproteinase 9 (MMP9) and MMP13 that break down ECM to maintain ECM balance [132]. In activated HSCs, the expression of MMP9 and MMP13 was repressed, whereas HDAC4 protein levels were increased. Similarly, overexpression of *HDAC4* in quiescent HSCs also reduced the expression of the MMPs compared with the empty vector control [132]. Taken together, these studies suggest that HDAC4 plays a role in the activation of HSCs and that specific inhibition of HDAC4 may be a promising therapeutic target for the treatment of NASH by inhibiting transdifferentiation of HSCs into myofibroblasts.

The role of HATs in the pathogenesis of metabolic diseases has not been well investigated compared with HDACs. Nonetheless, there are several studies that implicate specific HATs in the regulation of energy metabolism. It was demonstrated that chronic ingestion of a high-fat diet led to an increase in the expression of SRC-1 in white adipose tissue, while the high-fat diet reduced the mRNA abundance of SRC-1 in brown adipose tissue of C57BL/6J mice [133]. *Src-1* knockout mice on a high-fat diet displayed higher body weight, massive lipid accumulation in brown adipose tissue, lower rectal temperature after cold exposure, and lower respiratory quotient compared with wild-type mice. Furthermore, the knockout mice developed insulin resistance. This study suggests that SRC-1 may play an important role in energy expenditure through the regulation of thermogenesis in brown adipose tissue, which is disrupted by consumption of a high-fat diet [133]. In contrast to SRC-1, *Src-3* knockout mice on a high-fat diet had lower plasma levels of total cholesterol, triglycerides, and free fatty acids and displayed better insulin sensitivity than their wild-type counterparts [134]. It was demonstrated that SRC-3 induced the expression of GCN5, which inactivates peroxisome proliferator–activated receptor (PPAR-γ) coactivator 1α (PGC-1α) through acetylation. As PGC-1α is an important coactivator that increases fatty acid β-oxidation and mitochondria biogenesis [135], the decreased PGC-1α activity by GCN5 may alter energy expenditure. This notion was corroborated by the finding that obese C57BL/6J mice had higher expression of both

SRC-3 and GCN5 in muscle, suggesting that obesity was, at least in part, the result of lower energy expenditure induced by SRC-3 [134].

It is evident that the manifestation of obesity-related metabolic diseases involves histone acetylation and deacetylation. The aforementioned studies support that HAT and HDAC inhibitors may be effectively used to prevent the pathogenesis of metabolic diseases. Pharmaceutical HAT and HDAC inhibitors for the treatment of metabolic diseases are still being developed and have yet met FDA approval. Furthermore, currently approved HDAC inhibitors for the treatment of cancer are not without adverse side effects [136]. Bioactive compounds found in foods are safe and some of which are known to have HDAC- and HAT-inhibiting properties. Several of these compounds have already been demonstrated to prevent metabolic diseases via epigenetic modes of action in animal models. Interestingly, these bioactive compounds encompass a diverse group of structures, including fatty acids, polyphenols, and organosulfur compounds.

14.5 BIOACTIVE FOOD COMPONENTS

14.5.1 Short-Chain Fatty Acids

14.5.1.1 Butyrate

Butyrate is a short-chain fatty acid (SCFA) that can be found in butter and cheese. It is abundantly produced by gut microbiota through fermentation of dietary fibers and rapidly absorbed by passive diffusion and active transport by ion exchange transporters in the large intestine [137–139]. The incubation of butyrate with HeLa cells, a cervical cancer cell line, increased histone H3 and H4 acetylation by inhibiting HDACs [140,141]. Subsequent investigation revealed that butyrate is a noncompetitive inhibitor of HDACs, which means that it does not directly bind to the catalytic site of HDACs [142]. Despite decades after discovery of the inhibitory effect of butyrate on HDACs, the exact molecular mechanism by which it inhibits HDACs remains to be discovered.

Butyrate has been extensively studied for its potential use as a cancer chemopreventive, but only in the last decade has research focused on its effect on obesity-related metabolic diseases [143]. It is well-known that high-fat diets impair glucose tolerance and insulin sensitivity in mice [144]. When C57BL/6J mice were fed a high-fat diet supplemented with 5% butyrate by weight for 16 weeks, they showed lower body weight despite consuming more calories than controls [137]. The butyrate-fed mice exhibited increased energy expenditure, which corresponded to increases in mRNA and protein levels of PGC-1α and uncoupling protein 1 (UCP-1) in brown adipose tissue, suggesting that butyrate may increase thermogenesis to dissipate energy as heat. Butyrate supplementation also improved glucose tolerance with lower plasma levels of fasting glucose and insulin when compared with high fat controls. Furthermore, HDAC activity was decreased by 50% in the muscle tissue of the butyrate-fed mice, which suggests its inhibitory effect on HDACs could be responsible for the anti-obesogenic effects in mice [137]. In another study, when butyrate was injected at a dosage of 500 mg/kg body weight into high fat-fed C57BL/6J mice, hepatic expression of fibroblast growth factor 21 (FGF21) was upregulated, resulting in increased fatty acid oxidation and ketone production [145]. FGF21 is a hormone that is secreted by the liver and is known to stimulate gluconeogenesis, fatty acid oxidation, and ketogenesis [146]. In human embryonic kidney 293 (HEK293) cells, HDAC3 knockdown increased FGF21 promoter activity by 5-fold, while knockdown of HDAC1 and HDAC2 had minimal effect on the promoter activity. Chromatin immunoprecipitation confirmed that HDAC3 interacts with PPAR-α at the promoter of FGF21 [145]. Therefore, the results suggest that the induction of FGF21 in response to the injection of butyrate may be due to the inhibition of HDAC3, which is a part of a corepressor complex that inhibits PPAR-α–dependent transcription of FGF21 [145].

In obese individuals, hypertrophied adipocytes secrete large amount of monocyte chemoattractant protein 1 that recruits monocytes to the adipose tissue, where they differentiate into macrophages [147]. The macrophages secrete tumor necrosis factor α (TNF-α) and interleukin 6 (IL-6) that impair the ability of adipocytes to store triglycerides and increase lipolysis [114]. The effects of these cytokines are not limited to the adipose tissue and they trigger systemic insulin resistance and metabolic dysfunction in other tissues [148]. As such, the lipid-laden adipose tissue becomes a source of circulating proinflammatory cytokines, ultimately inducing chronic inflammation. Repressive effect of butyrate on obesity-induced inflammation has been reported in vivo. C57BL/6J mice on a high-fat diet for 10 weeks were given 2 g/kg of teributyrin, a prodrug of butyrate, 3 times a week [149]. Teributyrin administration decreased fasting plasma glucose and insulin levels compared with high fat controls, which was attributed to its anti-inflammatory effects. Teributyrin supplementation also attenuated basal as well as lipopolysaccharide (LPS)-induced TNF-α and IL-1β secretion from isolated peritoneal macrophages. Furthermore, the expression of TNF-α and IL-1β was significantly reduced in the adipose tissue of teributryin-treated mice, suggesting decreased macrophage infiltration into the adipose tissue [149]. This result is in agreement with reports that HDAC inhibitors have potent anti-inflammatory effects [150,151].

Supporting the notion that HDAC inhibitors can suppress maladaptive cardiac remodeling following myocardial infarction, butyrate has been shown to ameliorate myocardial damage following ischemia–reperfusion [152]. In this study, Sprague-Dawley rats that were injected with 10 mg/kg of sodium butyrate prior to the induction of ischemia and reperfusion displayed a reduction in infarct size and cardiomyocyte apoptosis.

Taken together, data suggest that butyrate is a multifaceted bioactive that can improve insulin resistance, and reduce inflammation and reperfusion damage. Therefore, consumption of foods rich in fibers that can be fermented to butyrate may prevent the development of obesity-related metabolic diseases through an HDAC-dependent manner.

14.5.2 Polyphenols

14.5.2.1 Curcumin

Curcumin is a polyphenolic compound abundantly found in the spice, turmeric. Turmeric is derived from the rhizomes of *Curcuma longa*, which is a member of the ginger family [153]. Curcumin has been suggested to possess antioxidant, anti-inflammatory, and anticancer properties [153]. It is a potent scavenger of reactive oxygen species as well as reactive nitrogen species in vitro, but its antioxidant capacity in vivo is poorly understood [154]. Several clinical trials in humans have indicated that curcumin has low bioavailability [155]. When healthy adults were given a capsule containing 0.5 to 12 g of curcumin and blood samples were taken at 1, 2, and 4 h after oral ingestion of the capsule, consumption of 0.5 g up to 8 g of curcumin resulted in undetectable serum curcumin levels at all of the time points, while a dose of 10 or 12 g increased serum curcumin levels to 50.5, and 51.2 ng/mL at 4 h, respectively [156]. Despite the concern on its low bioavailability, curcumin is still one of the most studied bioactive food components.

It was discovered that curcumin is a HAT inhibitor in 2004 [157]. Curcumin was demonstrated to inhibit p300/CBP-mediated transfer of the acetyl group of acetyl-CoA to purified core histones, at a concentration of 20–100 μM. This inhibition of acyltransferase activity by curcumin did not appear to affect PCAF. Curcumin was also shown to decrease acetylation of histones in vitro [157]. Furthermore, treatment of neuronal cancer cells with 5 μg/mL of curcumin induced apoptosis in the cells through the induction of caspases or p53 signaling. This effect was attributed to curcumin's ability to inhibited p300/CBP [158].

Curcumin possesses a potent anti-inflammatory effect that can ameliorate chronic low-grade inflammation in metabolic diseases [159]. Elevated blood glucose levels in diabetic conditions can activate NF-κB in monocytes, which is a major transcription factor involved in inflammation by inducing the transcription of various proinflammatory cytokines and chemokines [160]. As NF-κB is known to associate with p300/CBP to induce proinflammatory gene expression, the inhibition of p300/CBP by curcumin may inhibit NF-κB transcriptional activity [161]. When human Mantel cell lymphoma, an aggressive B cell non-Hodgkin lymphoma, was treated with 10 μM curcumin, p300/CBP HAT activity was inhibited, concomitantly decreasing NF-κB activity [162]. Furthermore, THP-1 monocytes exposed to hyperglycemic conditions displayed increased HAT activity, while HDAC activity was decreased compared with monocytes exposed to normoglycemic conditions [163]. The high glucose-induced alteration in HAT and HDAC activities increased NF-κB transcriptional activity and consequently the production of TNF-α and IL-6. Curcumin treatment at a concentration of 1.5–12.5 μM, however, abrogated the activation of NF-κB [164].

The anti-inflammatory effects of curcumin, possibly through the modulation of HAT and HDAC activities, were also found in vivo. Male Sprague-Dawley rats on a high-fat diet for 60 days developed hyperglycemic, hyperinsulinemia, and insulin resistance [164]. Curcumin supplementation at a daily dose of 80 mg/kg in the rats, however, maintained fasting blood glucose levels similar to those of rats fed a normal chow diet. Furthermore, curcumin was able to attenuate high fat-induced increases in plasma concentrations of free fatty acids and TNF-α. In streptozotocin-induced diabetic rats, intraperitoneal injection of 150 mg/kg/day of curcumin attenuated the expression of endothelial nitric oxide synthase (eNOS) and TGFβ1 in the kidneys. The decrease in the expression of eNOS and TGFβ1 was associated with a decrease in the expression of p300/CBP and NF-κB nuclear translocation [165]. In a human intervention study of prediabetic individuals, 9 months of curcumin supplementation at a dose of 750 mg/day was able to prevent the development of diabetes, while

16.4% of prediabetic subjects in the placebo group developed type 2 diabetes [166]. Subjects who consumed curcumin supplement exhibited lower plasma C peptide levels, higher homeostasis model assessment (HOMA) β, lower HOMA-IR, and higher plasma adiponectin levels, which suggests that curcumin improved pancreatic β-cell function and insulin sensitivity when compared with the placebo group.

As more research points to chronic low-grade inflammation as an integral step in the development of metabolic diseases, the anti-inflammatory effect of curcumin could be exploited as a treatment to alleviate inflammatory conditions. Although the exact molecular mechanism responsible for the anti-inflammatory effect of curcumin needs further investigation, its epigenetic mode of action through the inhibition of p300/CBP may play an important role in exerting the anti-inflammatory effect. One major concern on the usage of curcumin as an effective dietary supplement for gaining health benefits is its poor intestinal absorption. Efforts to increase its bioavailability through special formulation and encapsulation are currently well underway [153,167,168].

14.5.2.2 Epigallocatechin-3-Gallate

Green tea is one of the most widely consumed beverages around the world, and its several health benefits are well recognized. Green tea and other teas derived from the plant *Camellia sinensis* are rich in flavonoid polyphenols. Among these green tea polyphenols, catechin, and epigallocatechin-3-gallate (EGCG) have been attributed to many of green tea's biological effects. EGCG represents ~50%–70% of the total catechin content of green tea, which makes up 30%–40% of the dry weight of a cup of green tea [169,170].

Evidence that supports the effect of EGCG on the inhibition of HATs was first suggested by the finding that EGCG repressed the acetylation of histone H3 lysine 9 at the promoter of human telomerase reverse transcriptase in MCF-7 breast cancer cells [171]. Subsequent screening for natural compounds that can inhibit HAT activity identified EGCG as a HAT inhibitor. EGCG can inhibit CBP/p300 as well as Tip60 and PCAF in a dose dependent manner with no effect on HDAC activity in vitro [172]. Other related compounds, such as catechin, epicatechin, and epigallocatechin, also possess HAT inhibitory activity but with much lower potency than EGCG. Although EGCG was shown not to inhibit HDAC activity [172], green tea polyphenols containing ~65% EGCG reduced HDAC activity in human prostate cancer LNCap cells and PC-3 cells when the cells were treated with 10–80 μg/mL of green tea polyphenols for 24 h [173,174]. The reduction of HDAC activity by green tea polyphenols was attributed to its repressive effect on class I HDAC protein expression. The discrepancy between studies on the inhibitory effect of EGCG on HDAC activity could be due to the presence of polyphenols other than EGCG in green tea, which could be responsible for the HDAC inhibition. Another possibility for the discrepancy is differences on how the HDAC activity assay was done in the studies. The assay that was used to show that green tea polyphenols inhibited HDAC activity utilized lysates from cells treated with EGCG prior to the assay, which would reflect the downregulation of HDAC proteins levels [173]. This is in contrast to the assay that showed EGCG did not inhibit HDAC activity, where lysates from cells without being pretreated with EGCG were incubated in the absence or presence of EGCG [172]. Another possibility for the discrepancy is that the effects of EGCG on HDAC activity is cell-type specific.

The uses of green tea polyphenols and EGCG for the prevention of metabolic diseases have been demonstrated in vivo. Seventeen weeks of EGCG supplementation in C57BL/6 mice on a high-fat diet with or without 3.2 g EGCG/kg body weight resulted in reduced body weight gain, and lower plasma levels of fasting glucose and circulating cytokines, such as IL-6 and granulocyte colony–stimulating factor, when compared with control [175]. The reduction in obesity-associated inflammation by EGCG supplementation may be due to its inhibitory effects on HAT activity. Acetylation of NF-κB by p300 increases its transcriptional activity [98]. In contrast, inhibition of p300 by EGCG has been shown to decrease NF-κB acetylation and therefore its transcriptional activity [172]. The protective effects of EGCG against high fat-induced metabolic syndrome was also observed in Sprague-Dawley rats fed a high-fat diet with supplementation of green tea polyphenols in their drinking water (0.5%, wt/vol) [176]. After 8 weeks, the rats consumed water containing green tea polyphenols had a significant reduction in body weight, and serum levels of IL-1β and IL-6 compared with control rats.

Green tea polyphenols, EGCG in particular, appear to be an inhibitor of HAT. This provides a molecular mechanism by which EGCG can directly attenuate chronic inflammation independent of its antioxidant properties. It is apparent from the aforementioned studies that EGCG supplementation has protective effects against high fat-induced metabolic diseases. One hindrance in the use of EGCG supplementation for the prevention of metabolic diseases is hepatoxicity, which could be due to its prooxidant effects [177]. While it appears to happen, hepatotoxicity has only been reported in response to use of green tea supplementation as a weight loss product in humans and is still extremely rare [178]. Therefore, the consumption of EGCG through beverages such as green tea remains an attractive means to prevent metabolic diseases.

14.5.3 Organosulfur Compounds

14.5.3.1 Sulforaphane

Sulforaphane is an isothiocynate derived from the hydrolysis of glucoraphanin, a glucosinolate precursor. It can be found in cruciferous vegetable such as broccoli, cabbage, and kale [179]. Sulforaphane is a strong inducer of endogenous antioxidant defense through its ability to activate nuclear factor E2–related factor 2 (NRF2), which is a transcription factor that plays a central role in the regulation of basal and inducible expression of phase II detoxification enzymes as well as endogenous antioxidant enzymes [180–182]. NRF2 is normally localized in the cytoplasm where it is bound to Kelch-like ECH-associated protein 1 (Keapl), which sequesters NRF2 from translocating to the nucleus, and promotes proteasomal degradation of NRF2. Under oxidative stress, the dissociation of Keapl from NRF2 allows NRF2 to translocate to the nucleus and binds to the antioxidant response element (ARE) in the promoters of antioxidant enzymes [183]. Consistent with the roles of NRF2 in the regulation of antioxidant enzymes, *Nrf2* knockout mice exhibit impaired responses to oxidative stress and increased inflammation that led to increased incidence of cancer

[184,185]. However, in athymic mice, oral administration of 12 mg/kg body weight of sulforaphane twice a day for 5 weeks after xenografts significantly reduced tumor sizes and increased apoptosis in the tumors [186]. Furthermore, in another xenograft study, dietary supplementation of sulforaphane at an average daily dose of 7.5 μM per animal for 3 weeks slowed tumor growth compared with control mice. Thus, these studies suggest that sulforaphane possesses anticancer properties, but whether the anticancer properties of sulforaphane is related to its ability to prevent oxidative stress through the activation of NRF2 pathway requires further investigation.

The anticancer activities of sulforaphane are corroborated by the finding that it has an inhibitory effect on HDAC activity [187]. In HCT116 colon cells and HEK293 cells, treatment of 3–35 μM of sulforaphane showed a dose-dependent decrease in HDAC activity [187,188]. The inhibition of HDAC activity in HCT116 cells was suggested to be due to decreases in HDAC2 and HDAC3 protein levels [188]. Furthermore, a single dose of sulforaphane at 10 μM reduced HDAC activity in the colonic mucosa within 6 h of treatment, which was followed by increases in histone H3 and H4 acetylation [189]. In C57BL/6J mice fed a diet supplemented with sulforaphane at a level of 443-mg/kg diet for 10 weeks, the acetylation of histone H3 and H4 was increased in the ileum, colon, prostate, and peripheral blood mononuclear cells [189]. The exact mechanisms by which sulforaphane inhibits HDAC activity remain unknown, but it has been suggested that sulforaphane-*N*-acetylcysteine and sulforaphane-cysteine, metabolites of sulforaphane, are responsible for the inhibition of HDACs [187,189].

Recent studies have suggested that inhibition of HDACs plays a role in the activation of NRF2, raising the question whether sulforaphane could activate NRF2 by inhibiting HDACs [190,191]. When microglial cells isolated from Sprague-Dawley rats were treated with 1 mM of valporic acid or 10 nM of trichostatin A (TSA), pan-HDAC inhibitors, NRF2 transcription activity was increased [190]. In RAW 264.7 macrophages, treatment of 30 nM of TSA decreased Keap1 protein levels and consequently increased the NRF2-dependent expression of heme oxygenase 1 [191]. Although these findings suggest that sulforaphane may induce NRF2 activation through an HDAC-dependent pathway, this aspect remains to be directly tested.

14.5.3.2 Diallyl Disulfide

Organosulfur compounds have emerged as HDAC inhibitors and sulforaphane described above belongs to this group of compounds. Other organosulfer compounds, including diallyl disulfide (DADS) and allyl mercaptan (AM), have also been shown to possess inhibitory activities toward HDACs [192]. DADS is a major organosulfur found in garlic and makes up ~40%–60% of garlic oil [193]. Alliins in garlic are compounds that can be converted to diallyl thiosulfate (DADSO) by the enzyme alliinase when garlic cloves are cut or crushed. Ultimately, DADSO is transformed to DADS, which is metabolized in the liver to AM and other metabolites [193]. It was observed that DADS treatment on DS19 mouse erythroleukemic cells, as well as rat McA-RH7777 hepatoma cells induced histone hyperacetylation [194,195]. Using a cell-free HDAC activity assay, it was demonstrated that DADS significantly inhibited HDAC activity by 29%, while its metabolite, AM, inhibited 92% of HDAC activity in both Caco-2 and HeLa cell lysates [196]. Furthermore, treatment with

DADS has been shown to increase histone acetylation in vitro and in vivo [196,197]. Male Wistar rats were administered 200 mg/kg body weight of DADS or vehicle by intracecal perfusion that delivered DADS directly to colon epithelial cells or through gavage [198]. Both routes of DADS administration resulted in significant increases in histone H3 and H4 acetylation compared with vehicle control, with intracecal perfusion resulting in higher levels of acetylation compared with gavage.

Both DADS and AM have been extensively studied for their chemopreventive effects. However, their roles in the prevention of metabolic diseases have not been well studied. In relation to the chemopreventive effects of DADS and AM, it was demonstrated that they do not possess peroxyl radical scavenging ability and therefore cannot be considered antioxidants [199]. Even so, DADS has been shown to upregulate the activity of antioxidant enzymes such as glutathione peroxidase, glutathione reductase, and superoxide dismutase in the rats fed a high-fat diet supplemented with DADS [200]. The increased activity of antioxidant enzymes could be due to activation of NRF2, which is known to be activated by HDAC inhibitors [191,201]. Male Sprague-Dawley rats fed a 65% fructose diet showed that supplementation of 250 mg/kg/day of raw garlic homogenate for 8 weeks significantly lowered plasma fasting glucose and triglyceride levels compared with control rats [202]. Recently, a comparative study of diabetic humans showed that garlic extract had anti-hyperglycemic and lipid-lowering effects similar to those found in rats [203]. In this study, obese patients diagnosed with type 2 diabetes were given 500 mg metformin tablets twice a day or the same dose of metformin together with 250 mg garlic capsule per day. After 12 weeks, both groups showed lower fasting and postprandial blood glucose levels compared with baseline, but the garlic-supplemented group had significantly lower blood glucose levels than those of the control group. Plasma total cholesterol, triglycerides, and low-density lipoprotein cholesterol (LDL-C) levels were all significantly lower, while high-density lipoprotein cholesterol (HDL-C) was significantly higher in the garlic-supplemented group compared with the control. In another human study, patients with type 2 diabetes were prescribed only 500 mg of metformin twice daily or together with 300 mg tablets of garlic extracts 3 times a day [204]. At the end of 24 weeks, subjects who consumed garlic extracts displayed lower plasma levels of fasting blood glucose, total cholesterol, triglycerides, and LDL-C, while HDL-C levels were significantly increased by 6.72% compared with those of the control group [204]. Although it appears that garlic supplementation could be effective in the treatment of type 2 diabetes, currently it is not possible to conclude whether DADS or AM contribute to the beneficial effects of garlic supplementation because garlic contains many other compounds. Furthermore, whether garlic mediates its effects through HDAC inhibition remains to be demonstrated.

14.6 CONCLUSION

HATs and HDACs are dynamic effectors of epigenetics that play an essential role in the maintenance of cellular transcription programs. The enzymes have also been implicated in the pathogenesis of metabolic diseases, namely type 2 diabetes, CVD, and NAFLD. Owing to their direct ties with energy metabolism and the reversibility of acetylation, modulation of HATs and HDACs are considered an attractive means for nutritional intervention to prevent or treat obesity-associated metabolic diseases. Bioactive components, including short-chain fatty acids, polyphenols, and organosulfur compounds, have been shown to ameliorate inflammation, and high fat-induced obesity and insulin resistance, at least in part, by regulating the activities of HATs and HDACs as summarized in Figure 14.1a. Bioactive compounds not only alter the acetylation status of histones but that of transcription factors, which can influence DNA binding of transcriptional factors as well as their interaction with coactivators or corepressors (Figure 14.1b). However, our current understanding of the roles of nutrients and bioactive foods in the regulation of HATs and HDACs, as well as the consequences of their regulation to the epigenome, is very limited.

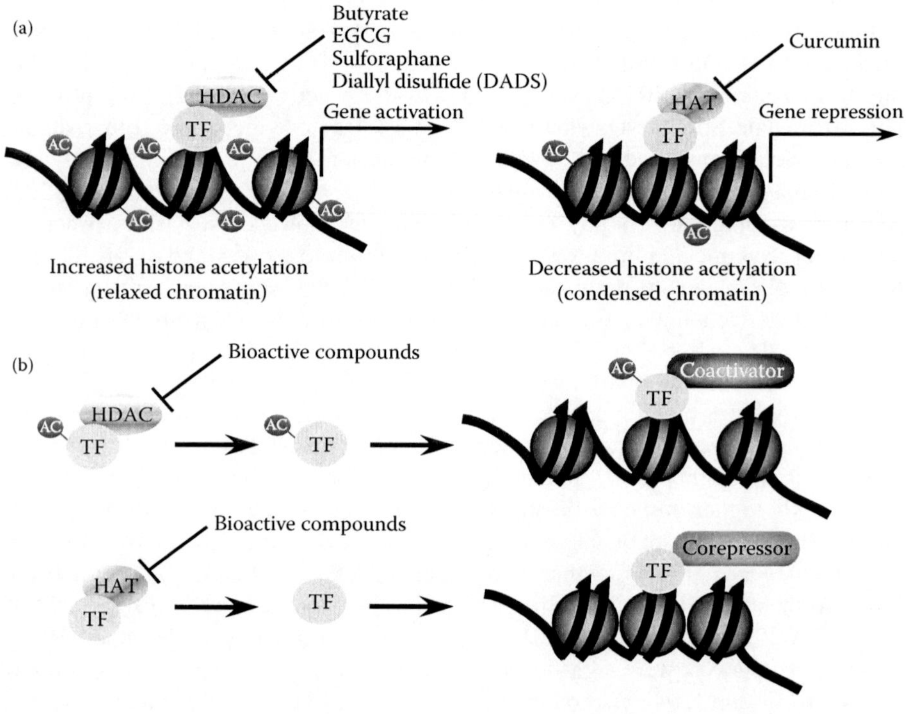

FIGURE 14.1 Summary for the regulation of HATs and HDACs by bioactive compounds. (a) The steady-state level of HAT and HDAC activities controls the expression of genes. (Left) Inhibition of HDACs by bioactive components, such as butyrate, EGCG, sulforaphane, and DADS, can disrupt the balance between HATs and HDACs and increase histone acetylation, leading to transcriptional activation of genes that are otherwise repressed. (Right) In contrast, inhibition of HATs by bioactives, e.g., curcumin, can inhibit HATs, decreasing histone acetylation for gene repression. (b) Bioactive compounds can alter the acetylation status of transcription factors, which affects their association with coactivators or corepressors. (Upper) The inhibition of HDACs by bioactive compounds allows for the recognition of the acetyl moiety on transcription factors by the bromodomain of coactivators, facilitating the association of a transcription factor and a coactivator. (Lower) In contrast, the inhibition of HATs by bioactive compounds can lead to unacetylated transcription factors and enhance their binding to corepressors.

Further understanding of how HATs and HDACs mediate epigenetic changes in lipid metabolism, inflammatory pathways, and endogenous antioxidant defense mechanism will aid in the exploitation of nutritional components for maintaining optimal health and preventing the development of metabolic diseases.

REFERENCES

1. Fu, Y. and M.W. Nachtigal, Analysis of epigenetic alterations to proprotein convertase genes in disease. *Methods Mol Biol*, 2011. **768**: pp. 231–45.
2. Reynolds, R.M., Glucocorticoid excess and the developmental origins of disease: Two decades of testing the hypothesis—2012 Curt Richter Award Winner. *Psychoneuroendocrinology*, 2013. **38**(1): pp. 1–11.
3. Lorenzen, J.M., F. Martino and T. Thum, Epigenetic modifications in cardiovascular disease. *Basic Res Cardiol*, 2012. **107**(2): p. 245.
4. Vaissiere, T., C. Sawan and Z. Herceg, Epigenetic interplay between histone modifications and DNA methylation in gene silencing. *Mutat Res*, 2008. **659**(1–2): pp. 40–8.
5. Lu, C. and C.B. Thompson, Metabolic regulation of epigenetics. *Cell Metab*, 2012. **16**(1): pp. 9–17.
6. Meier, J.L., Metabolic mechanisms of epigenetic regulation. *ACS Chem Biol*, 2013.
7. Kumar, A. et al., Structural delineation of histone post-translation modifications in histone-nucleosome assembly protein complex. *J Struct Biol*, 2012. **180**(1): pp. 1–9.
8. Yan, C. and D.D. Boyd, Histone H3 acetylation and H3 K4 methylation define distinct chromatin regions permissive for transgene expression. *Mol Cell Biol*, 2006. **26**(17): pp. 6357–71.
9. Shrader, T.E. and D.M. Crothers, Effects of DNA sequence and histone-histone interactions on nucleosome placement. *J Mol Biol*, 1990. **216**(1): pp. 69–84.
10. Zheng, C. and J.J. Hayes, Structures and interactions of the core histone tail domains. *Biopolymers*, 2003. **68**(4): pp. 539–46.
11. Mersfelder, E.L. and M.R. Parthun, The tale beyond the tail: Histone core domain modifications and the regulation of chromatin structure. *Nucleic Acids Res*, 2006. **34**(9): pp. 2653–62.
12. Lee, K.K. and J.L. Workman, Histone acetyltransferase complexes: One size doesn't fit all. *Nat Rev Mol Cell Biol*, 2007. **8**(4): pp. 284–95.
13. Nakao, M., Epigenetics: Interaction of DNA methylation and chromatin. *Gene*, 2001. **278**(1–2): pp. 25–31.
14. Buhler, M. and D. Moazed, Transcription and RNAi in heterochromatic gene silencing. *Nat Struct Mol Biol*, 2007. **14**(11): pp. 1041–8.
15. Zentner, G.E. and S. Henikoff, Regulation of nucleosome dynamics by histone modifications. *Nat Struct Mol Biol*, 2013. **20**(3): pp. 259–66.
16. Choi, S.W. and S. Friso, Epigenetics: A new bridge between nutrition and health. *Adv Nutr*, 2010. **1**(1): pp. 8–16.
17. Xie, Z. et al., Lysine succinylation and lysine malonylation in histones. *Mol Cell Proteomics*, 2012. **11**(5): pp. 100–7.
18. Phillips, D.M., The presence of acetyl groups of histones. *Biochem J*, 1963. **87**: pp. 258–63.
19. Lenstra, T.L. et al., The specificity and topology of chromatin interaction pathways in yeast. *Mol Cell*, 2011. **42**(4): pp. 536–49.
20. Shi, Y. et al., Histone demethylation mediated by the nuclear amine oxidase homolog LSD1. *Cell*, 2004. **119**(7): pp. 941–53.
21. Cloos, P.A. et al., Erasing the methyl mark: Histone demethylases at the center of cellular differentiation and disease. *Genes Dev*, 2008. **22**(9): pp. 1115–40.

22. Cloos, P.A. et al., The putative oncogene GASC1 demethylates tri- and dimethylated lysine 9 on histone H3. *Nature*, 2006. **442**(7100): pp. 307–11.
23. Hino, S. et al., FAD-dependent lysine-specific demethylase-1 regulates cellular energy expenditure. *Nat Commun*, 2012. **3**: p. 758.
24. Xiao, M. et al., Inhibition of alpha-KG-dependent histone and DNA demethylases by fumarate and succinate that are accumulated in mutations of FH and SDH tumor suppressors. *Genes Dev*, 2012. **26**(12): pp. 1326–38.
25. Choudhuri, S., From Waddington's epigenetic landscape to small noncoding RNA: Some important milestones in the history of epigenetics research. *Toxicol Mech Methods*, 2011. **21**(4): pp. 252–74.
26. Riggs, A.D., X inactivation, differentiation, and DNA methylation. *Cytogenet Cell Genet*, 1975. **14**(1): pp. 9–25.
27. Holliday, R. and J.E. Pugh, DNA modification mechanisms and gene activity during development. *Science*, 1975. **187**(4173): pp. 226–32.
28. Deaton, A.M. and A. Bird, CpG islands and the regulation of transcription. *Genes Dev*, 2011. **25**(10): pp. 1010–22.
29. Saxonov, S., P. Berg and D.L. Brutlag, A genome-wide analysis of CpG dinucleotides in the human genome distinguishes two distinct classes of promoters. *Proc Natl Acad Sci U S A*, 2006. **103**(5): pp. 1412–7.
30. Klose, R.J. and A.P. Bird, Genomic DNA methylation: The mark and its mediators. *Trends Biochem Sci*, 2006. **31**(2): pp. 89–97.
31. Bogdanovic, R. et al., Transient type 1 pseudo-hypoaldosteronism: Report on an eight-patient series and literature review. *Pediatr Nephrol*, 2009. **24**(11): pp. 2167–75.
32. Futscher, B.W. et al., Role for DNA methylation in the control of cell type specific maspin expression. *Nat Genet*, 2002. **31**(2): pp. 175–9.
33. Moore, L.D., T. Le and G. Fan, DNA methylation and its basic function. *Neuropsychopharmacology*, 2013. **38**(1): pp. 23–38.
34. Chen, T. and E. Li, Establishment and maintenance of DNA methylation patterns in mammals. *Curr Top Microbiol Immunol*, 2006. **301**: pp. 179–201.
35. Bestor, T.H., Activation of mammalian DNA methyltransferase by cleavage of a Zn binding regulatory domain. *EMBO J*, 1992. **11**(7): pp. 2611–7.
36. Tahiliani, M. et al., Conversion of 5-methylcytosine to 5-hydroxymethylcytosine in mammalian DNA by MLL partner TET1. *Science*, 2009. **324**(5929): pp. 930–5.
37. Zhu, J.K., Active DNA demethylation mediated by DNA glycosylases. *Annu Rev Genet*, 2009. **43**: pp. 143–66.
38. Guo, J.U. et al., Hydroxylation of 5-methylcytosine by TET1 promotes active DNA demethylation in the adult brain. *Cell*, 2011. **145**(3): pp. 423–34.
39. Wellen, K.E. and C.B. Thompson, A two-way street: Reciprocal regulation of metabolism and signalling. *Nat Rev Mol Cell Biol*, 2012. **13**(4): pp. 270–6.
40. Khan, S.N. and A.U. Khan, Role of histone acetylation in cell physiology and diseases: An update. *Clin Chim Acta*, 2010. **411**(19–20): pp. 1401–11.
41. Marmorstein, R., Structure of histone acetyltransferases. *J Mol Biol*, 2001. **311**(3): pp. 433–44.
42. Roth, S.Y., J.M. Denu and C.D. Allis, Histone acetyltransferases. *Annu Rev Biochem*, 2001. **70**: pp. 81–120.
43. Chen, H. et al., Nuclear receptor coactivator ACTR is a novel histone acetyltransferase and forms a multimeric activation complex with P/CAF and CBP/p300. *Cell*, 1997. **90**(3): pp. 569–80.
44. Dhalluin, C. et al., Structure and ligand of a histone acetyltransferase bromodomain. *Nature*, 1999. **399**(6735): pp. 491–6.
45. Ornaghi, P. et al., The bromodomain of Gcn5p interacts *in vitro* with specific residues in the N terminus of histone H4. *J Mol Biol*, 1999. **287**(1): pp. 1–7.

46. Sliva, D. et al., Tip60 interacts with human interleukin-9 receptor alpha-chain. *Biochem Biophys Res Commun*, 1999. **263**(1): pp. 149–55.
47. Brady, M.E. et al., Tip60 is a nuclear hormone receptor coactivator. *J Biol Chem*, 1999. **274**(25): pp. 17599–604.
48. Wang, J. et al., The orphan nuclear receptor Rev-erbbeta recruits Tip60 and HDAC1 to regulate apolipoprotein CIII promoter. *Biochim Biophys Acta*, 2008. **1783**(2): pp. 224–36.
49. Champagne, N., N. Pelletier and X.J. Yang, The monocytic leukemia zinc finger protein MOZ is a histone acetyltransferase. *Oncogene*, 2001. **20**(3): pp. 404–9.
50. Timmermann, S. et al., Histone acetylation and disease. *Cell Mol Life Sci*, 2001. **58**(5–6): pp. 728–36.
51. Kuo, M.H. et al., Transcription-linked acetylation by Gcn5p of histones H3 and H4 at specific lysines. *Nature*, 1996. **383**(6597): pp. 269–72.
52. Soutoglou, E., N. Katrakili and I. Talianidis, Acetylation regulates transcription factor activity at multiple levels. *Mol Cell*, 2000. **5**(4): pp. 745–51.
53. Chen, L. et al., Duration of nuclear NF-kappaB action regulated by reversible acetylation. *Science*, 2001. **293**(5535): pp. 1653–7.
54. Spencer, T.E. et al., Steroid receptor coactivator-1 is a histone acetyltransferase. *Nature*, 1997. **389**(6647): pp. 194–8.
55. Sakakibara, I. et al., Fasting-induced hypothermia and reduced energy production in mice lacking acetyl-CoA synthetase 2. *Cell Metab*, 2009. **9**(2): pp. 191–202.
56. de Ruijter, A.J. et al., Histone deacetylases (HDACs): Characterization of the classical HDAC family. *Biochem J*, 2003. **370**(Pt 3): pp. 737–49.
57. Gregoretti, I.V., Y.M. Lee and H.V. Goodson, Molecular evolution of the histone deacetylase family: Functional implications of phylogenetic analysis. *J Mol Biol*, 2004. **338**(1): pp. 17–31.
58. Hong, S. et al., A novel domain in histone deacetylase 1 and 2 mediates repression of cartilage-specific genes in human chondrocytes. *FASEB J*, 2009. **23**(10): pp. 3539–52.
59. Gao, Z. et al., Regulation of nuclear translocation of HDAC3 by IkappaBalpha is required for tumor necrosis factor inhibition of peroxisome proliferator–activated receptor gamma function. *J Biol Chem*, 2006. **281**(7): pp. 4540–7.
60. Gupta, P. et al., HDAC3 as a molecular chaperone for shuttling phosphorylated TR2 to PML: A novel deacetylase activity-independent function of HDAC3. *PLoS One*, 2009. **4**(2): p. e4363.
61. Fischle, W. et al., Enzymatic activity associated with class II HDACs is dependent on a multiprotein complex containing HDAC3 and SMRT/N-CoR. *Mol Cell*, 2002. **9**(1): pp. 45–57.
62. Waltregny, D. et al., Expression of histone deacetylase 8, a class I histone deacetylase, is restricted to cells showing smooth muscle differentiation in normal human tissues. *Am J Pathol*, 2004. **165**(2): pp. 553–64.
63. Waltregny, D. et al., Histone deacetylase HDAC8 associates with smooth muscle alpha-actin and is essential for smooth muscle cell contractility. *FASEB J*, 2005. **19**(8): pp. 966–8.
64. McKinsey, T.A. et al., Signal-dependent nuclear export of a histone deacetylase regulates muscle differentiation. *Nature*, 2000. **408**(6808): pp. 106–11.
65. Grozinger, C.M. and S.L. Schreiber, Regulation of histone deacetylase 4 and 5 and transcriptional activity by 14-3-3-dependent cellular localization. *Proc Natl Acad Sci U S A*, 2000. **97**(14): pp. 7835–40.
66. Tong, J.J. et al., Identification of HDAC10, a novel class II human histone deacetylase containing a leucine-rich domain. *Nucleic Acids Res*, 2002. **30**(5): pp. 1114–23.
67. Kao, H.Y. et al., Isolation and characterization of mammalian HDAC10, a novel histone deacetylase. *J Biol Chem*, 2002. **277**(1): pp. 187–93.

68. Petrie, K. et al., The histone deacetylase 9 gene encodes multiple protein isoforms. *J Biol Chem*, 2003. **278**(18): pp. 16059–72.
69. Chatterjee, T.K. et al., Histone deacetylase 9 is a negative regulator of adipogenic differentiation. *J Biol Chem*, 2011. **286**(31): pp. 27836–47.
70. Zhang, C.L. et al., Association of COOH-terminal-binding protein (CtBP) and MEF2-interacting transcription repressor (MITR) contributes to transcriptional repression of the MEF2 transcription factor. *J Biol Chem*, 2001. **276**(1): pp. 35–9.
71. Lobera, M. et al., Selective class IIa histone deacetylase inhibition via a nonchelating zinc-binding group. *Nat Chem Biol*, 2013. **9**(5): pp. 319–25.
72. Mihaylova, M.M. et al., Class IIa histone deacetylases are hormone-activated regulators of FOXO and mammalian glucose homeostasis. *Cell*, 2011. **145**(4): pp. 607–21.
73. Lahm, A. et al., Unraveling the hidden catalytic activity of vertebrate class IIa histone deacetylases. *Proc Natl Acad Sci U S A*, 2007. **104**(44): pp. 17335–40.
74. Wang, Z. et al., Combinatorial patterns of histone acetylations and methylations in the human genome. *Nat Genet*, 2008. **40**(7): pp. 897–903.
75. Peserico, A. and C. Simone, Physical and functional HAT/HDAC interplay regulates protein acetylation balance. *J Biomed Biotechnol*, 2011. **2011**: p. 371832.
76. Struhl, K., Histone acetylation and transcriptional regulatory mechanisms. *Genes Dev*, 1998. **12**(5): pp. 599–606.
77. Evertts, A.G. et al., Quantitative dynamics of the link between cellular metabolism and histone acetylation. *J Biol Chem*, 2013. **288**(17): pp. 12142–51.
78. Waterborg, J.H., Dynamics of histone acetylation in Saccharomyces cerevisiae. *Biochemistry*, 2001. **40**(8): pp. 2599–605.
79. Zhang, D.E. and D.A. Nelson, Histone acetylation in chicken erythrocytes. Rates of acetylation and evidence that histones in both active and potentially active chromatin are rapidly modified. *Biochem J*, 1988. **250**(1): pp. 233–40.
80. Katan-Khaykovich, Y. and K. Struhl, Dynamics of global histone acetylation and deacetylation *in vivo*: Rapid restoration of normal histone acetylation status upon removal of activators and repressors. *Genes Dev*, 2002. **16**(6): pp. 743–52.
81. Smith, C.M. et al., Heritable chromatin structure: Mapping "memory" in histones H3 and H4. *Proc Natl Acad Sci U S A*, 2002. **99 Suppl 4**: pp. 16454–61.
82. Cervoni, N. and M. Szyf, Demethylase activity is directed by histone acetylation. *J Biol Chem*, 2001. **276**(44): pp. 40778–87.
83. Zhang, Y. and D. Reinberg, Transcription regulation by histone methylation: Interplay between different covalent modifications of the core histone tails. *Genes Dev*, 2001. **15**(18): pp. 2343–60.
84. Xu, L., C.K. Glass and M.G. Rosenfeld, Coactivator and corepressor complexes in nuclear receptor function. *Curr Opin Genet Dev*, 1999. **9**(2): pp. 140–7.
85. Rachez, C. et al., Ligand-dependent transcription activation by nuclear receptors requires the DRIP complex. *Nature*, 1999. **398**(6730): pp. 824–8.
86. Chen, J.D. and R.M. Evans, A transcriptional co-repressor that interacts with nuclear hormone receptors. *Nature*, 1995. **377**(6548): pp. 454–7.
87. Horlein, A.J. et al., Ligand-independent repression by the thyroid hormone receptor mediated by a nuclear receptor co-repressor. *Nature*, 1995. **377**(6548): pp. 397–404.
88. Guenther, M.G., O. Barak and M.A. Lazar, The SMRT and N-CoR corepressors are activating cofactors for histone deacetylase 3. *Mol Cell Biol*, 2001. **21**(18): pp. 6091–101.
89. Ishizuka, T. and M.A. Lazar, The N-CoR/histone deacetylase 3 complex is required for repression by thyroid hormone receptor. *Mol Cell Biol*, 2003. **23**(15): pp. 5122–31.
90. Lee, Y.H. et al., Synergy among nuclear receptor coactivators: Selective requirement for protein methyltransferase and acetyltransferase activities. *Mol Cell Biol*, 2002. **22**(11): pp. 3621–32.

91. Jepsen, K. and M.G. Rosenfeld, Biological roles and mechanistic actions of co-repressor complexes. *J Cell Sci*, 2002. **115**(Pt 4): pp. 689–98.
92. Li, B. et al., FOXP3 interactions with histone acetyltransferase and class II histone deacetylases are required for repression. *Proc Natl Acad Sci U S A*, 2007. **104**(11): pp. 4571–6.
93. Brandl, A., T. Heinzel and O.H. Kramer, Histone deacetylases: Salesmen and customers in the post-translational modification market. *Biol Cell*, 2009. **101**(4): pp. 193–205.
94. Singh, B.N. et al., Nonhistone protein acetylation as cancer therapy targets. *Expert Rev Anticancer Ther*, 2010. **10**(6): pp. 935–54.
95. Das, C. and T.K. Kundu, Transcriptional regulation by the acetylation of nonhistone proteins in humans—a new target for therapeutics. *IUBMB Life*, 2005. **57**(3): pp. 137–49.
96. Mujtaba, S. et al., Structural mechanism of the bromodomain of the coactivator CBP in p53 transcriptional activation. *Mol Cell*, 2004. **13**(2): pp. 251–63.
97. Fu, M. et al., Acetylation of androgen receptor enhances coactivator binding and promotes prostate cancer cell growth. *Mol Cell Biol*, 2003. **23**(23): pp. 8563–75.
98. Chen, L.F., Y. Mu and W.C. Greene, Acetylation of RelA at discrete sites regulates distinct nuclear functions of NF-kappaB. *EMBO J*, 2002. **21**(23): pp. 6539–48.
99. Ito, K. et al., Histone deacetylase 2-mediated deacetylation of the glucocorticoid receptor enables NF-kappaB suppression. *J Exp Med*, 2006. **203**(1): pp. 7–13.
100. Yao, Y.L., W.M. Yang and E. Seto, Regulation of transcription factor YY1 by acetylation and deacetylation. *Mol Cell Biol*, 2001. **21**(17): pp. 5979–91.
101. Drong, A.W., C.M. Lindgren and M.I. McCarthy, The genetic and epigenetic basis of type 2 diabetes and obesity. *Clin Pharmacol Ther*, 2012. **92**(6): pp. 707–15.
102. Wellen, K.E. et al., ATP-citrate lyase links cellular metabolism to histone acetylation. *Science*, 2009. **324**(5930): pp. 1076–80.
103. Vogelauer, M. et al., Stimulation of histone deacetylase activity by metabolites of intermediary metabolism. *J Biol Chem*, 2012. **287**(38): pp. 32006–16.
104. Johnstone, R.W., Histone-deacetylase inhibitors: Novel drugs for the treatment of cancer. *Nat Rev Drug Discov*, 2002. **1**(4): pp. 287–99.
105. Sato, A., T. Asano and K. Ito, Vorinostat and bortezomib synergistically cause ubiquitinated protein accumulation in prostate cancer cells. *J Urol*, 2012. **188**(6): pp. 2410–8.
106. Fiskus, W. et al., Co-treatment with vorinostat synergistically enhances activity of Aurora kinase inhibitor against human breast cancer cells. *Breast Cancer Res Treat*, 2012. **135**(2): pp. 433–44.
107. Caro, J.F. et al., Liver glucokinase: Decreased activity in patients with type II diabetes. *Horm Metab Res*, 1995. **27**(1): pp. 19–22.
108. Ling, C. and L. Groop, Epigenetics: A molecular link between environmental factors and type 2 diabetes. *Diabetes*, 2009. **58**(12): pp. 2718–25.
109. Sookoian, S. et al., Epigenetic regulation of insulin resistance in nonalcoholic fatty liver disease: Impact of liver methylation of the peroxisome proliferator–activated receptor gamma coactivator 1alpha promoter. *Hepatology*, 2010. **52**(6): pp. 1992–2000.
110. Seino, S., T. Shibasaki and K. Minami, Dynamics of insulin secretion and the clinical implications for obesity and diabetes. *J Clin Invest*, 2011. **121**(6): pp. 2118–25.
111. Ducharme, N.A. and P.E. Bickel, Lipid droplets in lipogenesis and lipolysis. *Endocrinology*, 2008. **149**(3): pp. 942–9.
112. Jiang, G. and B.B. Zhang, Glucagon and regulation of glucose metabolism. *Am J Physiol Endocrinol Metab*, 2003. **284**(4): pp. E671–8.
113. Jaworski, K. et al., Regulation of triglyceride metabolism. IV. Hormonal regulation of lipolysis in adipose tissue. *Am J Physiol Gastrointest Liver Physiol*, 2007. **293**(1): pp. G1–4.
114. Guilherme, A. et al., Adipocyte dysfunctions linking obesity to insulin resistance and type 2 diabetes. *Nat Rev Mol Cell Biol*, 2008. **9**(5): pp. 367–77.

115. Utzschneider, K.M. and S.E. Kahn, Review: The role of insulin resistance in nonalcoholic fatty liver disease. *J Clin Endocrinol Metab*, 2006. **91**(12): pp. 4753–61.
116. Kjaergard, H. et al., Coronary artery bypass grafting within 30 days after treatment of acute myocardial infarctions with angioplasty or fibrinolysis—a surgical substudy of DANAMI-2. *Scand Cardiovasc J*, 2004. **38**(3): pp. 143–6.
117. Sun, Z. et al., Hepatic Hdac3 promotes gluconeogenesis by repressing lipid synthesis and sequestration. *Nat Med*, 2012. **18**(6): pp. 934–42.
118. DeFronzo, R.A. and D. Tripathy, Skeletal muscle insulin resistance is the primary defect in type 2 diabetes. *Diabetes Care*, 2009. **32 Suppl 2**: pp. S157–63.
119. Klip, A. and M.R. Paquet, Glucose transport and glucose transporters in muscle and their metabolic regulation. *Diabetes Care*, 1990. **13**(3): pp. 228–43.
120. McGee, S.L. and M. Hargreaves, Exercise and skeletal muscle glucose transporter 4 expression: Molecular mechanisms. *Clin Exp Pharmacol Physiol*, 2006. **33**(4): pp. 395–9.
121. Raichur, S. et al., Histone deacetylase 5 regulates glucose uptake and insulin action in muscle cells. *J Mol Endocrinol*, 2012. **49**(3): pp. 203–11.
122. Cohn, J.N., R. Ferrari and N. Sharpe, Cardiac remodeling—concepts and clinical implications: A consensus paper from an international forum on cardiac remodeling. Behalf of an International Forum on Cardiac Remodeling. *J Am Coll Cardiol*, 2000. **35**(3): pp. 569–82.
123. Zhang, C.L. et al., Class II histone deacetylases act as signal-responsive repressors of cardiac hypertrophy. *Cell*, 2002. **110**(4): pp. 479–88.
124. Kee, H.J. et al., Inhibition of histone deacetylation blocks cardiac hypertrophy induced by angiotensin II infusion and aortic banding. *Circulation*, 2006. **113**(1): pp. 51–9.
125. Kong, Y. et al., Suppression of class I and II histone deacetylases blunts pressure-overload cardiac hypertrophy. *Circulation*, 2006. **113**(22): pp. 2579–88.
126. Cao, D.J. et al., Histone deacetylase (HDAC) inhibitors attenuate cardiac hypertrophy by suppressing autophagy. *Proc Natl Acad Sci U S A*, 2011. **108**(10): pp. 4123–8.
127. Brenner, D.A., Molecular pathogenesis of liver fibrosis. *Trans Am Clin Climatol Assoc*, 2009. **120**: pp. 361–8.
128. Mann, D.A. and D.E. Smart, Transcriptional regulation of hepatic stellate cell activation. *Gut*, 2002. **50**(6): pp. 891–6.
129. Glenisson, W., V. Castronovo and D. Waltregny, Histone deacetylase 4 is required for TGFbeta1-induced myofibroblastic differentiation. *Biochim Biophys Acta*, 2007. **1773**(10): pp. 1572–82.
130. Guo, W. et al., Abrogation of TGF-beta1-induced fibroblast-myofibroblast differentiation by histone deacetylase inhibition. *Am J Physiol Lung Cell Mol Physiol*, 2009. **297**(5): pp. L864–70.
131. Bataller, R. and D.A. Brenner, Liver fibrosis. *J Clin Invest*, 2005. **115**(2): pp. 209–18.
132. Qin, L. and Y.P. Han, Epigenetic repression of matrix metalloproteinases in myofibroblastic hepatic stellate cells through histone deacetylases 4: Implication in tissue fibrosis. *Am J Pathol*, 2010. **177**(4): pp. 1915–28.
133. Picard, F. et al., SRC-1 and TIF2 control energy balance between white and brown adipose tissues. *Cell*, 2002. **111**(7): pp. 931–41.
134. Coste, A. et al., The genetic ablation of SRC-3 protects against obesity and improves insulin sensitivity by reducing the acetylation of PGC-1{alpha}. *Proc Natl Acad Sci U S A*, 2008. **105**(44): pp. 17187–92.
135. Feige, J.N. and J. Auwerx, Transcriptional coregulators in the control of energy homeostasis. *Trends Cell Biol*, 2007. **17**(6): pp. 292–301.
136. Bruserud, O. et al., Histone deacetylase inhibitors in cancer treatment: A review of the clinical toxicity and the modulation of gene expression in cancer cell. *Curr Pharm Biotechnol*, 2007. **8**(6): pp. 388–400.

137. Gao, Z. et al., Butyrate improves insulin sensitivity and increases energy expenditure in mice. *Diabetes*, 2009. **58**(7): pp. 1509–17.
138. Arora, T., R. Sharma and G. Frost, Propionate. Anti-obesity and satiety enhancing factor? *Appetite*, 2011. **56**(2): pp. 511–5.
139. Velazquez, O.C., H.M. Lederer and J.L. Rombeau, Butyrate and the colonocyte. Production, absorption, metabolism, and therapeutic implications. *Adv Exp Med Biol*, 1997. **427**: pp. 123–34.
140. Boffa, L.C. et al., Suppression of histone deacetylation *in vivo* and *in vitro* by sodium butyrate. *J Biol Chem*, 1978. **253**(10): pp. 3364–6.
141. Riggs, M.G. et al., n-Butyrate causes histone modification in HeLa and Friend erythroleukaemia cells. *Nature*, 1977. **268**(5619): pp. 462–4.
142. Cousens, L.S., D. Gallwitz and B.M. Alberts, Different accessibilities in chromatin to histone acetylase. *J Biol Chem*, 1979. **254**(5): pp. 1716–23.
143. Sengupta, S., J.G. Muir and P.R. Gibson, Does butyrate protect from colorectal cancer? *J Gastroenterol Hepatol*, 2006. **21**(1 Pt 2): pp. 209–18.
144. Winzell, M.S. and B. Ahren, The high-fat diet-fed mouse: A model for studying mechanisms and treatment of impaired glucose tolerance and type 2 diabetes. *Diabetes*, 2004. **53 Suppl 3**: pp. S215–9.
145. Li, H. et al., Sodium butyrate stimulates expression of fibroblast growth factor 21 in liver by inhibition of histone deacetylase 3. *Diabetes*, 2012. **61**(4): pp. 797–806.
146. Woo, Y.C. et al., Fibroblast growth factor 21 as an emerging metabolic regulator: Clinical perspectives. *Clin Endocrinol (Oxf)*, 2013. **78**(4): pp. 489–96.
147. Sartipy, P. and D.J. Loskutoff, Monocyte chemoattractant protein 1 in obesity and insulin resistance. *Proc Natl Acad Sci U S A*, 2003. **100**(12): pp. 7265–70.
148. Wisse, B.E., The inflammatory syndrome: The role of adipose tissue cytokines in metabolic disorders linked to obesity. *J Am Soc Nephrol*, 2004. **15**(11): pp. 2792–800.
149. Vinolo, M.A. et al., Tributyrin attenuates obesity-associated inflammation and insulin resistance in high-fat-fed mice. *Am J Physiol Endocrinol Metab*, 2012. **303**(2): pp. E272–82.
150. Cantley, M.D. et al., Histone deacetylase inhibitors as suppressors of bone destruction in inflammatory diseases. *J Pharm Pharmacol*, 2012. **64**(6): pp. 763–74.
151. Ku, C.S. et al., Edible blue-green algae reduce the production of pro-inflammatory cytokines by inhibiting NF-kappaB pathway in macrophages and splenocytes. *Biochim Biophys Acta*, 2013. **1830**(4): pp. 2981–8.
152. Lim, S.H., K.S. Song and J. Lee, Butyrate and propionate, short chain fatty acids, attenuate myocardial damages by inhibition of apoptosis in a rat model of ischemia-reperfusion. *J Korean Soc Appl Biol Chem*, 2010. **53**(5): pp. 570–7.
153. Sharma, R.A., A.J. Gescher and W.P. Steward, Curcumin: The story so far. *Eur J Cancer*, 2005. **41**(13): pp. 1955–68.
154. Sreejayan and M.N. Rao, Nitric oxide scavenging by curcuminoids. *J Pharm Pharmacol*, 1997. **49**(1): pp. 105–7.
155. Anand, P. et al., Bioavailability of curcumin: Problems and promises. *Mol Pharm*, 2007. **4**(6): pp. 807–18.
156. Lao, C.D. et al., Dose escalation of a curcuminoid formulation. *BMC Complement Altern Med*, 2006. **6**: p. 10.
157. Balasubramanyam, K. et al., Curcumin, a novel p300/CREB-binding protein-specific inhibitor of acetyltransferase, represses the acetylation of histone/nonhistone proteins and histone acetyltransferase-dependent chromatin transcription. *J Biol Chem*, 2004. **279**(49): pp. 51163–71.
158. Kang, S.K., S.H. Cha and H.G. Jeon, Curcumin-induced histone hypoacetylation enhances caspase-3-dependent glioma cell death and neurogenesis of neural progenitor cells. *Stem Cells Dev*, 2006. **15**(2): pp. 165–74.

159. Shehzad, A. et al., New mechanisms and the anti-inflammatory role of curcumin in obesity and obesity-related metabolic diseases. *Eur J Nutr*, 2011. **50**(3): pp. 151–61.
160. Shanmugam, N. et al., High glucose-induced expression of proinflammatory cytokine and chemokine genes in monocytic cells. *Diabetes*, 2003. **52**(5): pp. 1256–64.
161. Gerritsen, M.E. et al., CREB-binding protein/p300 are transcriptional coactivators of p65. *Proc Natl Acad Sci U S A*, 1997. **94**(7): pp. 2927–32.
162. Shishodia, S. et al., Curcumin (diferuloylmethane) inhibits constitutive NF-kappaB activation, induces G1/S arrest, suppresses proliferation, and induces apoptosis in mantle cell lymphoma. *Biochem Pharmacol*, 2005. **70**(5): pp. 700–13.
163. Yun, J.M., I. Jialal and S. Devaraj, Epigenetic regulation of high glucose-induced proinflammatory cytokine production in monocytes by curcumin. *J Nutr Biochem*, 2011. **22**(5): pp. 450–8.
164. El-Moselhy, M.A. et al., The antihyperglycemic effect of curcumin in high fat diet fed rats. Role of TNF-alpha and free fatty acids. *Food Chem Toxicol*, 2011. **49**(5): pp. 1129–40.
165. Chiu, J. et al., Curcumin prevents diabetes-associated abnormalities in the kidneys by inhibiting p300 and nuclear factor-kappaB. *Nutrition*, 2009. **25**(9): pp. 964–72.
166. Chuengsamarn, S. et al., Curcumin extract for prevention of type 2 diabetes. *Diabetes Care*, 2012. **35**(11): pp. 2121–7.
167. Belcaro, G. et al., Product-evaluation registry of Meriva(R), a curcumin-phosphatidylcholine complex, for the complementary management of osteoarthritis. *Panminerva Med*, 2010. **52**(2 Suppl 1): pp. 55–62.
168. Belcaro, G. et al., Efficacy and safety of Meriva(R), a curcumin-phosphatidylcholine complex, during extended administration in osteoarthritis patients. *Altern Med Rev*, 2010. **15**(4): pp. 337–44.
169. Kanwar, J. et al., Recent advances on tea polyphenols. *Front Biosci (Elite Ed)*, 2012. **4**: pp. 111–31.
170. Lambert, J.D. and C.S. Yang, Mechanisms of cancer prevention by tea constituents. *J Nutr*, 2003. **133**(10): pp. 3262S–7S.
171. Berletch, J.B. et al., Epigenetic and genetic mechanisms contribute to telomerase inhibition by EGCG. *J Cell Biochem*, 2008. **103**(2): pp. 509–19.
172. Choi, K.C. et al., Epigallocatechin-3-gallate, a histone acetyltransferase inhibitor, inhibits EBV-induced B lymphocyte transformation via suppression of RelA acetylation. *Cancer Res*, 2009. **69**(2): pp. 583–92.
173. Thakur, V.S., K. Gupta and S. Gupta, Green tea polyphenols causes cell cycle arrest and apoptosis in prostate cancer cells by suppressing class I histone deacetylases. *Carcinogenesis*, 2012. **33**(2): pp. 377–84.
174. Pandey, M., S. Shukla and S. Gupta, Promoter demethylation and chromatin remodeling by green tea polyphenols leads to re-expression of GSTP1 in human prostate cancer cells. *Int J Cancer*, 2010. **126**(11): pp. 2520–33.
175. Chen, Y.K. et al., Effects of green tea polyphenol (-)-epigallocatechin-3-gallate on newly developed high-fat/Western-style diet-induced obesity and metabolic syndrome in mice. *J Agric Food Chem*, 2011. **59**(21): pp. 11862–71.
176. Lu, C. et al., Green tea polyphenols reduce body weight in rats by modulating obesity-related genes. *PLoS One*, 2012. **7**(6): p. e38332.
177. Lambert, J.D. et al., Hepatotoxicity of high oral dose (-)-epigallocatechin-3-gallate in mice. *Food Chem Toxicol*, 2010. **48**(1): pp. 409–16.
178. Mazzanti, G. et al., Hepatotoxicity from green tea: A review of the literature and two unpublished cases. *Eur J Clin Pharmacol*, 2009. **65**(4): pp. 331–41.
179. Sasaki, K. et al., Quantitative profiling of glucosinolates by LC-MS analysis reveals several cultivars of cabbage and kale as promising sources of sulforaphane. *J Chromatogr B Analyt Technol Biomed Life Sci*, 2012. **903**: pp. 171–6.

180. Zhang, Y. et al., A major inducer of anticarcinogenic protective enzymes from broccoli: Isolation and elucidation of structure. *Proc Natl Acad Sci U S A*, 1992. **89**(6): pp. 2399–403.
181. McMahon, M. et al., The Cap'n'Collar basic leucine zipper transcription factor Nrf2 (NF-E2 p45-related factor 2) controls both constitutive and inducible expression of intestinal detoxification and glutathione biosynthetic enzymes. *Cancer Res*, 2001. **61**(8): pp. 3299–307.
182. Posner, G.H. et al., Design and synthesis of bifunctional isothiocyanate analogs of sulforaphane: Correlation between structure and potency as inducers of anticarcinogenic detoxication enzymes. *J Med Chem*, 1994. **37**(1): pp. 170–6.
183. Copple, I.M., The Keap1-Nrf2 cell defense pathway—a promising therapeutic target? *Adv Pharmacol*, 2012. **63**: pp. 43–79.
184. Cheung, K.L. et al., Nrf2 knockout enhances intestinal tumorigenesis in Apc(min/+) mice due to attenuation of anti-oxidative stress pathway while potentiates inflammation. *Mol Carcinog*, 2012.
185. Osburn, W.O. et al., Increased colonic inflammatory injury and formation of aberrant crypt foci in Nrf2-deficient mice upon dextran sulfate treatment. *Int J Cancer*, 2007. **121**(9): pp. 1883–91.
186. Wang, F. and Y. Shan, Sulforaphane retards the growth of UM-UC-3 xenographs, induces apoptosis, and reduces survivin in athymic mice. *Nutr Res*, 2012. **32**(5): pp. 374–80.
187. Myzak, M.C. et al., A novel mechanism of chemoprotection by sulforaphane: Inhibition of histone deacetylase. *Cancer Res*, 2004. **64**(16): pp. 5767–74.
188. Rajendran, P. et al., Histone deacetylase turnover and recovery in sulforaphane-treated colon cancer cells: Competing actions of 14-3-3 and Pin1 in HDAC3/SMRT corepressor complex dissociation/reassembly. *Mol Cancer*, 2011. **10**: p. 68.
189. Myzak, M.C. et al., Sulforaphane inhibits histone deacetylase *in vivo* and suppresses tumorigenesis in Apc-minus mice. *FASEB J*, 2006. **20**(3): pp. 506–8.
190. Correa, F. et al., Activated microglia decrease histone acetylation and Nrf2-inducible anti-oxidant defence in astrocytes: Restoring effects of inhibitors of HDACs, p38 MAPK and GSK3beta. *Neurobiol Dis*, 2011. **44**(1): pp. 142–51.
191. Wang, B. et al., Histone deacetylase inhibition activates transcription factor Nrf2 and protects against cerebral ischemic damage. *Free Radic Biol Med*, 2012. **52**(5): pp. 928–36.
192. Myzak, M.C., E. Ho and R.H. Dashwood, Dietary agents as histone deacetylase inhibitors. *Mol Carcinog*, 2006. **45**(6): pp. 443–6.
193. Germain, E. et al., Hepatic metabolism of diallyl disulphide in rat and man. *Xenobiotica*, 2003. **33**(12): pp. 1185–99.
194. Lea, M.A., V.M. Randolph and M. Patel, Increased acetylation of histones induced by diallyl disulfide and structurally related molecules. *Int J Oncol*, 1999. **15**(2): pp. 347–52.
195. Lea, M.A. and V.M. Randolph, Induction of histone acetylation in rat liver and hepatoma by organosulfur compounds including diallyl disulfide. *Anticancer Res*, 2001. **21**(4A): pp. 2841–5.
196. Druesne, N. et al., Diallyl disulfide (DADS) increases histone acetylation and p21(waf1/cip1) expression in human colon tumor cell lines. *Carcinogenesis*, 2004. **25**(7): pp. 1227–36.
197. Druesne-Pecollo, N., C. Chaumontet and P. Latino-Martel, Diallyl disulfide increases histone acetylation in colon cells *in vitro* and *in vivo*. *Nutr Rev*, 2008. **66 Suppl 1**: pp. S39–41.
198. Druesne-Pecollo, N. et al., *In vivo* treatment by diallyl disulfide increases histone acetylation in rat colonocytes. *Biochem Biophys Res Commun*, 2007. **354**(1): pp. 140–7.
199. Amorati, R. and G.F. Pedulli, Do garlic-derived allyl sulfides scavenge peroxyl radicals? *Org Biomol Chem*, 2008. **6**(6): pp. 1103–7.

200. Sheen, L.Y. et al., Effects of garlic oil and its organosulfur compounds on the activities of hepatic drug-metabolizing and antioxidant enzymes in rats fed high- and low-fat diets. *Nutr Cancer*, 1999. **35**(2): pp. 160–6.
201. Chen, C. et al., Induction of detoxifying enzymes by garlic organosulfur compounds through transcription factor Nrf2: Effect of chemical structure and stress signals. *Free Radic Biol Med*, 2004. **37**(10): pp. 1578–90.
202. Padiya, R. et al., Garlic improves insulin sensitivity and associated metabolic syndromes in fructose fed rats. *Nutr Metab (Lond)*, 2011. **8**: p. 53.
203. Kumar, R. et al., Antihyperglycemic, antihyperlipidemic, anti-inflammatory and adenosine deaminase- lowering effects of garlic in patients with type 2 diabetes mellitus with obesity. *Diabetes Metab Syndr Obes*, 2013. **6**: pp. 49–56.
204. Ashraf, R., R.A. Khan and I. Ashraf, Garlic (Allium sativum) supplementation with standard antidiabetic agent provides better diabetic control in type 2 diabetes patients. *Pak J Pharm Sci*, 2011. **24**(4): pp. 565–70.